GW01184590

JOHN F. PATES
15 CARTMEL ROAD
BEXLEYHEATH
KENT DA7 5EA
TEL:- ERITH 42903

HOROLOVAR

400-DAY CLOCK REPAIR GUIDE

HOROLOVAR

400-DAY CLOCK REPAIR GUIDE

by

Charles Terwilliger

edited by

H. W. Ellison

TENTH EDITION
Over 66,000 Copies Sold

The Horolovar Company, Publisher

Author: *Charles Terwilliger*
Editors: *H. W. Ellison*
Escapement Section: *Henry B. Fried*
Photography: *Charles Terwilliger, Peter Ireland*
Design: *John Harrington*
Typography and Mechanicals: *Different By Design, Inc.*
Lithography: *Braun-Brumfield, Inc.*
Publisher: *The Horolovar Company*

ISBN: 0-916316-04-1

Library of Congress
Catalog Card Number 83-81592

Printed in U. S. A.

First Edition	July 1953
Second Edition	July 1964
Third Edition	September 1956
Fourth Edition	January 1959
Fifth Edition	September 1965
Sixth Edition	March 1972
Seventh Edition	April 1974
Eighth Edition	April 1978
Ninth Edition	January 1984
Tenth Edition	June 1991

CONTENTS

Foreword 7

SECTION 1 History of the 400-Day Clock in the United States 8

SECTION 2 400-Day Clock Events by Years 17

SECTION 3 History of Striking 400-Day Clocks 22

SECTION 4 400-Day Clock Patents 24

SECTION 5 Problems of Temperature Change and Timekeeping 31

SECTION 6 Some Outstanding Pre-World War I 400-Day Clocks ... 33

SECTION 7 Adjustment of the 400-Day Clock Escapement 42

- Introduction 42
- Graham (Dead Beat) Escapement 43
- Pin Pallet Escapement 44
- **Explanation of the Graham (Dead Beat) Escapement** 44
- Examining and Testing the Graham Escapement 45
- Making Adjustments to the Graham Escapement 45
- About Adjusting the Pallets 46
- Summary of Adjustments 47
- Trouble-Shooting Chart 47
- Table of Adjustments 47
- Repairing Damaged Graham Escape Wheel Teeth 47
- Repairing Pallets in the Graham Escapement 48
- Repairing an Escape Wheel in the Graham Escapement 48
- **Explanation of the Pin Pallet Escapement** 49
- Repairing the Pin Pallet Escapement 50
- Summary of Adjustments 51

SECTION 8 Answers to the Most Frequently Asked Questions About 400-Day Clock Repair 53

1—How do you adjust a 400-Day Clock Escapement? 53

2—How can you turn a tight eccentric nut without chewing up the sides of the slot 53

3—How can you stop the anchor from fluttering? 53

4—How much clearance should there be between the anchor pin and the fork tines? 54

5—How should a broken anchor pin be replaced? 54

6—How do you put the pendulum in beat? 54

7—How much should the pendulum rotate? 56

8—Is there any trick to regulating the pendulum so that the clock will keep good time? 57

9—Do all 400-Day Clock pendulums turn 8 per minute? 57

10—How long should the suspension spring be? 57

11—What is the best way to determine whether a replaced suspension spring is of the correct strength? 58

12—How do you replace a suspension spring in a clock when the suspension blocks have no screws? 58

13—How tight should the hands be? 59

14—How important is the tension washer behind the cannon pinion? 60

15—How is it possible for the hour hand to lose time even though it is tight on the hour wheel pipe? 60

16—When you clean a clock, is it always necessary to remove the mainspring? 60

17—What is the purpose of the extra barrel and mainspring attached to the third wheel in some clocks? 61

18—Where should you oil the clock? 61

19—Can 400-Day Clock parts of different manufacture be interchanged? 62

20—What do you do when, in a brand new, correctly set up clock, (1) the pendulum touches the base, (2) you've shortened the suspension spring to correct for this, and (3) you still can't get the pendulum to swing slowly enough? 62

21—How can you diagnose the fault when an apparently well-ordered clock still fails to run? 62

22—I've been in the repair business over 20 years and can make a nice profit on clocks that sell for $75 or more, but I've given up taking in 400-Day Clocks. How can I make a profit when I have to cut my repair charges so much? 63

23—Does the eccentric nut always carry the *anchor* pivot hole? 63

24—I've always turned the regulating nut on a 400-Day Clock pendulum *clockwise to slow the clock down.* However, I've noticed that with some of the newer clocks, I have to turn counter-clockwise to slow down. This is very confusing. Can you list the plate numbers of the clocks in the Guide with pendulums having regulating nuts that must be turned *counter-clockwise* to make them go *slower*? 64

25—Are there any special instructions for the repair of the Schatz *1000-Day* Clock? 64

26—What strength Horolovar suspension spring should be used for the 8-Day Staiger torsion pendulum clock? 64

27—I have a clock in the shop (Plate 1087) which I have cleaned and checked over carefully. It runs perfectly when the hands are off, but with the hands on, the clock always stops when the minute and hour hands are being raised. Can you suggest where the loss of power might be? 64

28—My biggest problem is not with the clock. It's with their owners! They take the clock home without knowing how to set it up, they bend the spring which throws the pendulum out of beat, then they bring it back complaining that it wasn't repaired right! How can I convince them that it was their fault? 64

29—How can you identify a genuine Horolovar suspension spring? 65

30—Is it possible to tighten the two movement-holding screws too much? 65
31—How can I tell where to put the fork on a suspension spring unit that is not illustrated in the Guide? 65
32—I have a Kundo Midget clock and used Horolovar Unit 5F as recommended. However, with the regulating nut turned to maximum slow, the clock runs very fast. Shouldn't I use a lighter spring? 67
33—I need to thin a suspension spring slightly. What is the best way to do this? 68
34—A customer brought in a pre-World War I 400-Day Clock with 4-ball pendulum to repair. Can this pendulum be original with the clock? About what year did 4-ball pendulums replace the old disc pendulums? 68
35—I am at the end of the pendulum adjustment and the clock is running too fast. How can I slow the clock? 68
36—I find it impossible to adjust the escapement on some clocks so that the anchor pin will go smoothly from one side to the other. How can a jerk be eliminated? 68
37—What is the best way to lacquer the brass parts of the clock after they have been cleaned and polished? 68
38—When viewed from the end of the anchor, the anchor pin is angled to one side. Is this proper? 68
39—Why doesn't the old suspension spring in my clock match the dimensions of the recommended suspension spring shown in Section 9 of the Guide? 69
40—Do you have any suggestions for installing a Horolovar Mainspring in a mainspring barrel? 69

SECTION 9 400-Day Clock Movement Back Plate Illustrations 70

SECTION 10 400-Day Clock Suspension Unit Identification 178

SECTION 11 Myths and Facts About 400-Day Clock Repair 190

SECTION 12 Names and Trade Marks on 400-Day Clock Dials 192

SECTION 13 400-Day Clock Pendulums 194

SECTION 14 400-Day Clock Ratchet Wheels 201

SECTION 15 400-Day Clock Suspension Brackets and Saddles 202

SECTION 16 400-Day Clock Spare Parts and Repairs 205

SECTION 17 Horolovar 400-Day Clock Products 206

SECTION 18 Appendix 211

FOREWORD

In late 1988, I purchased the Horolovar Company from the estate of the founder, Mr. Charles Terwilliger who had died earlier in that year. Among the items that were included with the company were the notes that Mr. Terwilliger had gathered for the next edition of the **400-Day Clock Repair Guide**. Based on these notes and my experiences in operating The Horolovar Company, I decided to carry out Mr. Terwilliger's plans and publish a Tenth Edition of the **Repair Guide**.

In preparing the Tenth Edition, it was necessary to investigate the background of the **Repair Guide**. The First Edition of what was to become the **Horolovar 400-Day Clock Repair Guide** was published in 1953, a 48 page paperback book called **The Horolovar 400-Day Clock Suspension Spring Guide**. Since then, there have been seventeen printings and nine editions of the **Repair Guide** and a change of name to indicate its wider scope. In the First Edition of the **Guide**, the back plate illustrations were arranged in a very generalized order and were then given a sequential number. In the Second Edition, the plates were alphabetized based on the manufacturer's identification and were assigned every fourth number in order to permit additions. This numbering system has been continued since that time and it is interesting to note how the back plate illustrations have increased in number, particularly as evidenced by the need to add suffixes to several illustration numbers.

The Ninth Edition of the **Repair Guide**, with its Sections concerning the history and identification of 400-Day Clocks had proven quite popular and it was decided to follow the pattern of the Ninth Edition in this next edition. While additions were made where new information was available, changes were minimized when ever possible so that the **Repair Guide** will remain familiar to frequent users.

In addition to inclusion of approximately 50 "new" back plate illustrations, this edition of the **Repair Guide** contains information concerning the number of beats per minute which the pendulum swings. This information permits the easier and quicker regulation of 400-Day Clocks, and when combined with a new approach to setting the pendulum beat, should save the repairer much time. Throughout this edition, all references to clock "repairman" have been removed. A large number of women participate and contribute actively to the repair and collecting of 400-Day Clocks and clock repair and collecting is no longer (if it ever was) an exclusively male domain.

Mr. Terwilliger's suggestions on how to get the most out of this book in the shortest period of time are still valid: First, read quickly through the "Appendix", Section 18. Second, do the same for "Answers to the Most Frequently Asked Questions," Section 8. (Please take particular note of Question 21 as it may save you considerable time and frustration.) Third, review Section 7 since Mr. Fried's explanation of the Dead Beat Escapement remains the clearest description of this mechanism available. Finally, thumb through the entire book, reading parts as you go in order to become familiar with the contents of each section and to become able to locate specific references quickly. With this orientation, you will find the book a valuable aid.

There is still much to learn about 400-Day Clocks. There are significant gaps in the history of these clocks and several of the patents referred to on the back plates of clocks remain a mystery. Still, information continues to be discovered. Much of this discovery is due to help of collectors and repairers around the world. Their help in preparing this edition is acknowledged and future help is solicited. In like manner, I welcome any questions pertaining to 400-Day Clocks for it is only through an exchange of information, ideas, and questions that we can advance the knowledge of these fascinating clocks.

Bill Ellison
June 1991

SECTION 1 History of the 400-Day Clock in the United States

The story of the 400-Day Clock is not easy to document, as records are sparse and confusing. Even the origin of the name is uncertain. The earliest use of the words "400-Day" to describe the clock is on the dial of a torsion pendulum striking year clock made by Jahresuhrenfabrik,[1] who modified the patent granted to F.A.L. de Gruyter in 1884 for such a clock. A later striking clock, made by Richard Schneckenburger,[2] has "400 days clock" on its dial. The clocks have also been referred to as Year Clocks, and "Anniversary" was a trade mark,[3] registered in 1901 by a promotion-minded U.S. importer who saw the clocks as birthday or wedding gifts that could be wound each year on the anniversary of the event.

More important than the name, however, is the history and development of the timepiece to which the name "400-Day Clock" was later given. There were several "inventors" of this form of clock movement using a torsion spring.

Unquestionably the first was Aaron D. Crane of Newark, New Jersey, who was given U.S. Patent No. 1973[4] on 10 February 1841 for a clock using a torsion pendulum. His idea was used successfully to make 8-day, 30-day and year clocks. One model of the latter is an ornate astronomical clock of which only four are known to have been made; one of these is in the Smithsonian Institution in Washington. Crane torsion pendulum clocks were manufactured by The Year Clock Company, Newark, New Jersey, but at least one astronomical clock was made in Boston after Crane moved there in the 1850's.

The second American to patent a torsion pendulum clock was Silas B. Terry of Plymouth, Connecticut, who was given U.S. Patent No. 9310[5] on 5 October 1852. His patent was for a "Marine Clock," so called because the design of his pendulum made it unnecessary for the clock to be in a level position and thus it could be used on a boat. These clocks were made by the Terry Clock Company, Terryville, Connecticut, and would run for only thirty hours.

A third American was John W. Hile of Waterville, Kansas, about whom nothing is known except that he was assigned U.S. Patent No. 176309[6] for a torsion pendulum clock on 29 March 1876. The illustration in Hile's patent looks more like a 400-Day Clock than either Crane's or Terry's and part of the text reads, "In a clock, the combination, with dead beat escapement, of a weight (the drawing shows a flat pendulum) suspended by a torsion spring, rotating alternately from right to left..." If Hile had also filed his patent in Germany, he might have been hailed as the inventor of the 400-Day Clock. However, it would take a great deal of imagination to see any connection between the Hile patent and German patent No. 2437,[7] issued 21 March 1879, three years later, to Lorenz Jehlen of Sackingen, Germany, for a torsion *pendulum,* not a clock.

In Germany, the origin of the 400-Day Clock is usually credited to Anton Harder[8] of Ransen, but the information about Harder and his invention is confused. *Praktisches Handbuch fur Uhrmacher,* published by Grosch, Dietzschold & Huttig in 1917 states that Harder, a simple farmer, was watching a steam boiler hanging from a chain. As the boiler was given a turn by a workman, it rotated; Harder timed the movements as closely as he could and found that the *period* of rotation was the same even though the *amount* of rotation diminished.

An article, "Year Clocks," which appeared in *Deutsches Uhrmacher Zwitung* in 1926, states that "a nobleman named Harder provided the stimulus to the production" of the 400-Day Clock. Harder purportedly watched the long recoiling period of a multi-candled chandelier after the lamplighter had twisted it around to reach the last candle and then released it. It is said that he visualized the principle of a rotation mass as making it possible to construct a clock "with long-going time."

Both versions convey the idea of a rotating pendulum, with strong overtones of Galileo's observations almost three centuries earlier.

Finally, the 1949 catalog of a leading manufacturer of 400-Day Clocks says that "Richard Schneckenburger, born at Muhlheim a.D., near Tuttlingen, Wurtemberg, was the inventor of the 400-Day Clock." While it is known that Schneckenburger manufactured striking pendulum year clocks[9] in the late 1890's, there appear to be no valid records to support the contention that he was the 400-Day Clock's inventor.

To return to Anton Harder, it seems that he worked with Lorenz Jehlen, the holder of the German torsion pendulum

[1]Section 3, Clock 1
[2]Section 3, Clock 5
[3]Section 4, Patent 16
[4]Section 4, Patent 1
[5]Section 4, Patent 2
[6]Section 4, Patent 3
[7]Section 4, Patent 4
[8]Section 18, No. 112
[9]Section 9, Plate 1504

patent, but it is not known whether there was any formal association between the two. However, when Lorenz Jehlen died in November, 1879, the rights to his patent were taken over by Anton Harder.

Harder apparently made a model of a torsion pendulum clock, as he received a patent for such a clock in Austria-Hungary in 1880. No Harder German patent has been found, possibly because the authorities saw some conflict with the Jehlen patent. But Harder later covered himself in the United States with U.S. Patent No. 269052[10] issued on 12 December 1882 for the clock.

Again, according to "Year Clocks," Harder turned his model over to the firm of A. Willmann Company in Freiburg (Silesia) in 1879 "to initiate regular production." During the next two years two other factories were involved in the first experiments: Gustav Becker, also of Freiburg, and Fortuna Company, Triberg, Black Forest.

It appears that the three factories made several hundred clocks, with verge, cylinder or Graham escapements. Most of these clocks were sold to the trade for ultimate purchase by consumers, but a large number of the clocks proved to be mechanically defective. Since none of these early production clocks was identifed by its maker, it is not possible to tell which factory made each of the several models that can be seen today, any of which have been put into good condition and running order.

The earliest of these clocks, with glass dome, is eighteen inches high and is obviously experimental.[11] The movement, with trapezoid plates, has a verge escapement, four-inch diameter disc pendulum and twelve-inch suspension spring. The movement is on a handmade steel platform attached to a round, wood base that had been used previously. The thin brass of the pendulum covers a split cast iron ring which serves to equalize differences in temperature. (Temperature changes were apparently seen as a problem from the first.) The five-inch diameter enamel dial is marked "Harder/ Ransen bei Steinau a/O."

Two smaller clocks, obviously later, have movements with oblong plates and disc pendulum. One has a verge escapement and is marked "Harder Ransen bei Steinau a/O/D.R. Patent No. 2437" (Jehlen's patent) on the dial;[12] the other has a cylinder escapement with a very large (⅝" diameter) escape wheel.[13] The dial of this clock is a replacement and has no markings.

The most interesting of these early clocks is one about fourteen inches high, having a movement with round plates and verge escapement, in a very fine, French, four-glass oval case with champleve decoration.[14] The case is of the quality that was used for the finest clocks in the early 1900's and this movement, made in the 1880's, may have indeed been fitted into the case after 1900.

Finally, in 1881, Harder and a group of workers in Triberg (some of whom had probably worked for Fortuna) formed a new company known as Jahresuhrenfabrik (Year Clock Factory). The group was headed by August Schatz, grandfather of Charles and Kurt Schatz, principals of Jahresuhrenfabrik (Aug. Schatz & Sohne), a major manufacturer of 400-Day Clocks in the post World War II period.

The Jahresuhrenfabrik clocks used the movement with Graham escapement illustrated in the Harder U.S. patent. At last, it appeared that production would be successful and it is known that regular deliveries of the clock started in March 1882.

The United States has always been the largest market for 400-Day Clocks and many of these first Jahresuhrenfabrik clocks were soon in the hands of a U.S. importer. In July 1883, an advertisement of The New Haven Clock Company appeared in *The Jewelers' Circular and Horological Review* featuring "Harder's Torsion-Pendulum Clocks...Run Accurately 400 Days." In it, the advertising copywriter wrote a glowing description of the clock:

> However astonishing the improvements recorded in the history of clock making since the days of Huygens, i.e., 250 years ago, no one until now ever succeeded in producing clocks of plain construction and general usefulness that would run longer than a week, or a fortnight at the utmost. This Clock has astounded everybody wherever it has been introduced, for it is of very simple construction, and runs with great accuracy throughout the whole 400 days *(sic)*, and it has an enormous sale in other countries. There have been former inventions of clock movements to run as long a time, but never on any such principle, or with any success, as they were never found to be at all accurate, and could only be bought at a very high price. Harder's patent does away with both of these objections, and we guarantee it to perform all that we claim for it.
>
> The Clock with rotary torsion-pendulum owes its origin to the ingenious idea of replacing the swinging pendulum by a time measure of equal exactitude, viz., the rotary disc, which, in its rotation, meets with hardly any impediment from atmospheric resistance. The clock makes a very handsome showing in itself, and is a very great novelty; for this reason, every jeweler should have at least one. It attracts an unusual amount of attention in show rooms and windows.
>
> The price is such as will enable the jeweler to sell it with heavy profit to himself. Each clock is securely packed by itself in a strong wooden box, with full instructions for setting up.

The advertisement points out that The New Haven Clock Company was the sole agent for the United States. The New

---

[10]Section 4, Patent 5
[11]Section 6, Clock 1
[12]Section 6, Clock 2
[13]Section 6, Clock 3
[14]Section 6, Clock 4

Haven Company must have received many complaints, as the accuracy of these first clocks, with their steel suspension springs, was deplorable. Nevertheless, they were constructed with much more precision than mass-produced clocks which were to follow and the brass parts took a superior polish. They are distinguished by dials with patent numbers in a circle around the center reading: D.R. Patent 2437 R.L. Patent 2182 (so far unidentified) U.S. Patent 269052.[15] The pendulum is a flat disc with two round regulating weights, without the decorative gallery that was added later, and the top surface of the base is covered with fabric. A surprising number of these early Jahresuhrenfabrik clocks are still in existence today.

There is no information on the number of clocks that Jahresuhrenfabrik produced, nor on how many were exported to the U.S., England or other countries. Whatever the figures, perhaps because he was discouraged by sales and prospects for the future after his long struggle to get his clock on the market, Harder sold the patent rights to the invention in 1884 to F.A.L. deGruyter of Amsterdam, who became the sole agent and distributor of the Jahresuhrenfabrik clocks. (In that same year, deGruyter was issued German patent No. 29348[16] for a torsion pendulum *striking* year clock. This clock was later made by Jahresuhrenfabrik.)

In 1885, the first change was made in the appearance of the clock with a three-pillar gallery added to the pendulum.[17] This made the rotation of the pendulum more attractive, but the movement had to be raised to accommodate the gallery. This was the model that Aug. Schatz & Sohne reproduced in 1981 to commemorate their one hundred years of clockmaking.[18]

In 1887, a major event took place that changed the manufacture and marketing of 400-Day Clocks. For reasons unknown, unless he felt that the clocks had no future, deGruyter failed to pay the patent renewal fee, and the patent was cancelled. This, of course, opened the door for any manufacturer to compete with the Jahresuhrenfabrik product. And compete they did! In 1888, Lenskirch, an established Black Forest clock manufacturer, is reputed to have been the first to make a competitve clock, but no indentifiable example of this production exists today.

In 1889, Badische Uhrenfabrik clocks appeared. They were the first 400-Day Clocks to be made with pin pallet escapement and the wheel train used lantern pinions rather than the solid type used by Jahresuhrenfabrik. Also, they initiated the use of a ball pendulum, rather than disc, but the Badische pendulum had three balls[19] rather than the four that became popular about ten years later.

Prior to 1900, more than a dozen manufacturers had made 400-Day Clocks, but only rarely can their clocks be identified. They include:

Badische Uhrenfabrik	Jahresuhrenfabrik
Gustav Becker	Wilhelm Kohler
Fortuna Clock Co.	Lenskircher Uhrenfabrik
Thomas Haller	Schatz & Wintermantel
Phil. Hauck	Richard Schneckenburger
Andreas Huber	Wendes, Metzger & Co.
Ideal Clock Co.	A. Willmann Co.

Some of the round plate movements in early 400-Day Clocks appear to be French made, and the French clock *case* industry played an important part in early 400-Day Clock sales. Many of these square[20] and oval[21] four-glass cases were of equal quality to those made for mercury pendulum clocks and, in fact, were probably made by the same manufacturers. Most of these cases were shipped from France to the German clock manufacturers where the 400-Day Clock movements were installed, but in the case of one U.S. importer, Bowler & Burdick, the cases were shipped to the United States where Jahresuhrenfabrik movements and pendulums were put into them. The French-cased clocks, because of their considerably higher cost, were sold almost exclusively by leading jewelers such as Tiffany & Co.; Caldwell; Bailey, Banks & Biddle; Bigelow, Kennard & Co.; and Shreve, Crump & Low. There seemed to be no consistency in case markings. Some say "France," other "Made in France," yet others, obviously French, have no markings at all.

While The New Haven Clock Company may have been the first U.S. importer of the 400-Day Clock, the most successful importer was Bowler & Burdick, an alert company whose base of operation was not New York, as one might expect, but Cleveland, Ohio. Bowler & Burdick was quick to see the sales opportunity which the clock offered and, for several years, remained ahead of the competition in 400-Day Clock sales. They did three innovative things: (1) They made an exclusive contract with Jahresuhrenfabrik to import movement and pendulum units as well as complete clocks. (2) As noted previously, they made arrangements with a still unidentified Paris case maker to ship a large selection of very fine four-glass cases[22] to Cleveland where the Jahresuhrenfabrik movement and pendulum units were installed. By doing this, they developed a product of high quality and attractive appearance that soon found its way into the best jewelry stores in the country. And (3) in a stroke of genius, they thought of, and copyrighted, a trade mark for their imported clocks—just one word: ANNIVERSARY[3]—that automatically suggested the clock as a gift for all kinds of anniversaries.

[15]Section 6, Clock 6
[16]Section 4, Patent 7
[17]Section 6, Clock 9
[18]Section 6, Clock 86
[19]Section 13, Pendulum 28
[20]Section 6, Clock 37
[21]Section 6, Clock 59
[22]Section 6, Clock 68

The "Anniversary" trade mark was registered December 1901 and Bowler & Burdick's first full page advertisement for the clocks appeared in November 1903 issue of *The Keystone,* leading jewelry trade paper of the day. Part of the copy reads:

> THE ANNIVERSARY CLOCK is now so well known to both the Trade and the public that further introduction seems unnecessary. A timepiece of merit. Noiseless. Sells at sight for wedding and birthday gifts. Wind it only on each Anniversary day. No up-to-date Jewelry store can afford to be without them in stock. Buy only the genuine, and beware of imitations. All real Anniversary Clocks are plainly stamped "Anniversary Trade Mark Registered."

The best clock illustrated in the advertisement was the "Colonial"

> of Mexican Onyx and Gilded Bronze with Anniversary Movement. This style is also made in White Marble and Mahogany in place of Onyx. We have Eighteen Styles of the Best and Newest Mantel Clock Cases, made expressly for us in Paris, in which we fit the Anniversary Clock. Send for illustrations and prices.

While Bowler & Burdick emphasized the more expensive clocks in the advertisement, they illustrated, in a lower corner, the less expensive 400-Day Clock under a glass dome that they called the "Globe Cased Clock."

Up to this time, clocks were intended to be retailed by jewelers and 400-Day Clocks were no exception. The customer expected the selling jeweler to set-up the clock in the customer's home. Since the jeweler could be expected to have some experience with the clock, instructions were brief and to the point. For example, Figure 1 illustrates the instructions provided with a 1905 Gustav Becker 400-Day Clock.

There were at least eleven manufacturers making 400-Day Clocks between 1900 and World War I. They included:

Badische Uhrenfabrik	Gebr. Junghans
Gustav Becker	Kieninger & Obergfell
C. Grivolas (France)	Kienzle Clock Factories
Phillipp Haas	Franz Vosseler
Andreas Huber	Wintermantel Uhrenfabrik
Jahresuhrenfabrik	

With each manufacturer making many different 400-Day Clock models (the Jahresuhrenfabrik 1905 catalog illustrated 83; the C. Grivolas 1910 catalog illustrated 51), the varieties of case styles seem endless. In addition, the bewildering variety of cases contain almost equally diverse movements, dials and pendulums.

World War I put a stop to the manufacture of 400-Day Clocks. Most of the factories re-tooled to make war materiel or civilian necessities. Jahresuhrenfabrik, for instance, made hundreds of thousands of alarm clocks, among other things.

Between World Wars I and II, sales of 400-Day Clocks were relatively slow and, for the first time, the United States was not a major market. At first, for obvious reasons, products made in Germany were slow to gain reacceptance. A few thousand clocks made by Kieninger & Obergfell were imported by George Borgfeldt,[23] an importer from New York City.

From 1929 through the depression years, the United States was not a large consumer of luxury products, and clocks were considered to be in this category. During this period, Kienzle Clock Factories sold its 400-Day Clock tools and dies to its plant foreman, Karl Kern, who went into partnership with J. Link to form Kern & Link, but the partnership was dissolved in 1932. Karl Kern continued to manufacture 400-Day Clocks under his name.

Again in 1939, clock production was superseded by the demands of the German war effort. After World War II, only three of the original factories resumed the manufacture of 400-Day Clocks, but there were changes in names. Jahresuhrenfabrik became better known by its firm name, Aug. Schatz & Sohne; Kieninger & Obergfell by its better known trade name, Kundo (K und O) and Karl Kern had been joined by his sons to form Kern & Sohne.

A new and big wave of 400-Day Clocks began to enter the United States early in 1949. The first importer was Forestville Clock Co., of New York, who brought in clocks made by Kundo.[24] Just a few months later, Schatz clocks[25] appeared, marketed through the nation-wide outlets of the Electric Shaver Division of Remington Rand under a reciprocal agreement by which Schatz made parts for the Remington razor. By the end of 1949, Kern clocks[26] were imported by W. & J. Overocean Traders, also of New York.

Shortly after this year of new importations, 400-Day Clocks enjoyed a phenomenal sale and ultimately became one of the most popular gift items ever produced. The United States was again the largest market, but second was that created by U.S. Military Post Exchanges all over the world. Servicemen bought 400-Day Clocks by the tens of thousands at the Exchanges, at very low prices, and sent them home to their families and friends in the United States.

As the 400-Day Clocks began to enter the U.S. in large numbers, there was a fundamental change in the retailing of these clocks. As noted previously, prior to World War II clocks were generally sold by jewelers and by a few of the larger department stores. After the war, almost everyone got into the selling act. Any Remington Rand electric shaver outlet, mass marketers, and the military PX stores all began selling 400-Day Clocks. Chances were very slim that the customer would have anyone to turn to with questions

[23]Section 9, Plate 1151
[24]Section 9, Plate 1107
[25]Section 9, Plate 1278
[26]Section 9, Plate 1343

Instructions how to set going and to regulate The 400 Day Clocks

The Advantage of the G. B. 400 Day Clock is

the Safety Bridge A

which protects the Suspension Spring during transit, the springs in this class of clock, if not protected in this way, very often getting twisted or bent in packing, when sent out from the Warehouses.

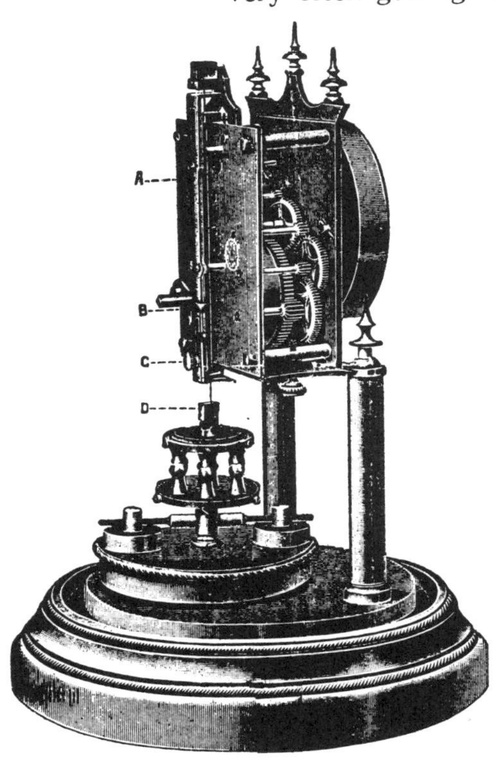

The Bridge **A** entirely covers the Spring, and protects it from any possible injury.

At the bottom of the Bridge there is an extension slide **B,** fitted with a screw **C.** This slide is let down and rests on the pin of the Suspension Block **D,** thus keeping the Suspension taut and rigid.

<u>To set the Clock going:</u>

All that is necessary is to slightly loosen the screw **C,** push the slide **B** up and then tighten the screw **C** again, thus freeing the Suspension Spring.

Then hang the Pendulum on the block **D.** After winding the Clock, turn the Pendulum from left to right and let it swing, when it will go for 400 Days.

If the Clock, goes too fast, turn the screw on Pendulum with the winding key to „**S**" and it will go slower.

If it goes slow, turn the screw towards „**F**" to make it go faster.

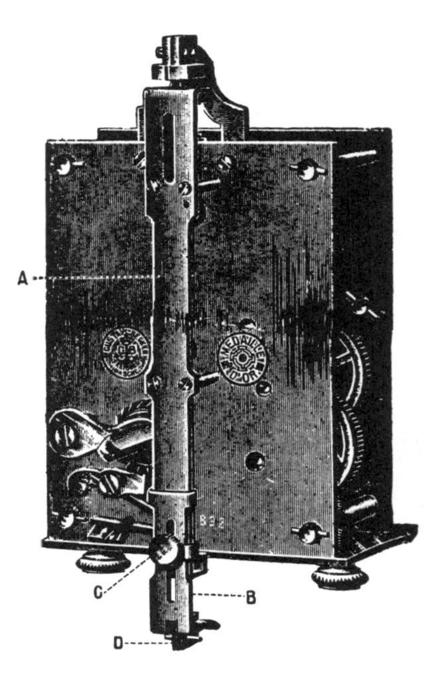

The Clock must stand quite level so that the Suspension Spring can circulate freely, without rubbing against the sides of the Bridge.

This free circulation of the Suspension Spring affords at the same time the Control that the Clock is standing on the level.

P. SCHARSIG
SOLE AGENT AND DISTRIBUTOR FOR
GUSTAV BECKER CLOCKS
JEWELERS BLDG., CHICAGO

Figure 1 1905 Gustav Becker Instructions

Instructions
400 Day Clock
with case

Unpacking

1. Take out the clock with its pendulum locked and case.
2. Do not throw away any paper packing before removing key.
3. Do not throw away packing material, you might need it for a possible transportation of the clock at a later date.
4. **Fig. 1**
   Carefully lift top of the case. To do this, turn both levers (1) located on both sides under the bottom plate to the open position. When lifting or replacing top of case, be careful not to touch the hands of the clock.
   Fig. 2
   Unlock the decoration screws (2) on the bottom plate, then carefully lift top of the case.
   Fig. 3
   Remove carefully the back panel.
   To do this the spring (1) on the back panel must be pressed backwards. Be careful that the back panel does not fall out. Then gently pull the back panel down and simultaneously outward (2)
5. Remove cardboard layers under the side glasses.
6. Remove the paper protection No. 1, then pull out the paper protection No. 2 in the direction of the arrow. (Fig. 4)

Starting the clock (Fig. 5)

1. Place your clock where it is the least exposed to vibrations.
2. Check whether clock is fully wound by turning the key to the left.
3. Pull out the cotter pin (No. 3) Fig. 4
4. Hold the pendulum (a) and push locking lever (b) as far to the left as possible.
5. Slowly lower pendulum into cup (e) until it is suspended freely.
6. Level the clock by turning the adjustable feet (c) so that the pendulum center post (d) is exactly in the center of the cup (e) or hole.
7. Set the clock in motion by turning the pendulum slowly 1 turn in either direction, and let it swing back by itself.
   CAREFULLY OBSERVE THE INSTRUCTIONS TO PROTECT THE DELICATE SUSPENSION SPRING. DO NOT SPIN PENDULUM!
8. Setting the clock is done by turning the large (minute) hand clockwise or counterclockwise to the desired time. Do not touch the small (hour) hand. Pay attention that the hands do not touch each other or the dial.
9. Carefully place the case over the clock. Do not forget to lock the levers under the bottom plate, to tighten the decoration screws or to insert the back panel.

Regulating the clock (Fig. 5)

1. Watch the timekeeping of your clock for several days before you start regulating it.
2. If your clock gains time, turn the knurled regulating disk (f) towards the minus (–) sign to make clock run slower.
   To do this, hold the pendulum when it changes its rotation. After the adjustment release the pendulum.
3. If your clock looses time, turn the regulating disk (f) towards the plus (+) sign to make clock run faster.

1/4 turn of the regulating disk results in a time variation of $\pm$ 4 minutes within 24 hours.

Winding of clock

The 400-day clock, as the name says, will run approx. 400 days. However it is advisable to wind it up once a year, perhaps on some special occasion such as birthday or an anniversary. To wind the clock, set the key firmly on the square winding post extending from the back plate of the movement. Turn the key in counterclockwise direction until you feel you cannot wind it any further.

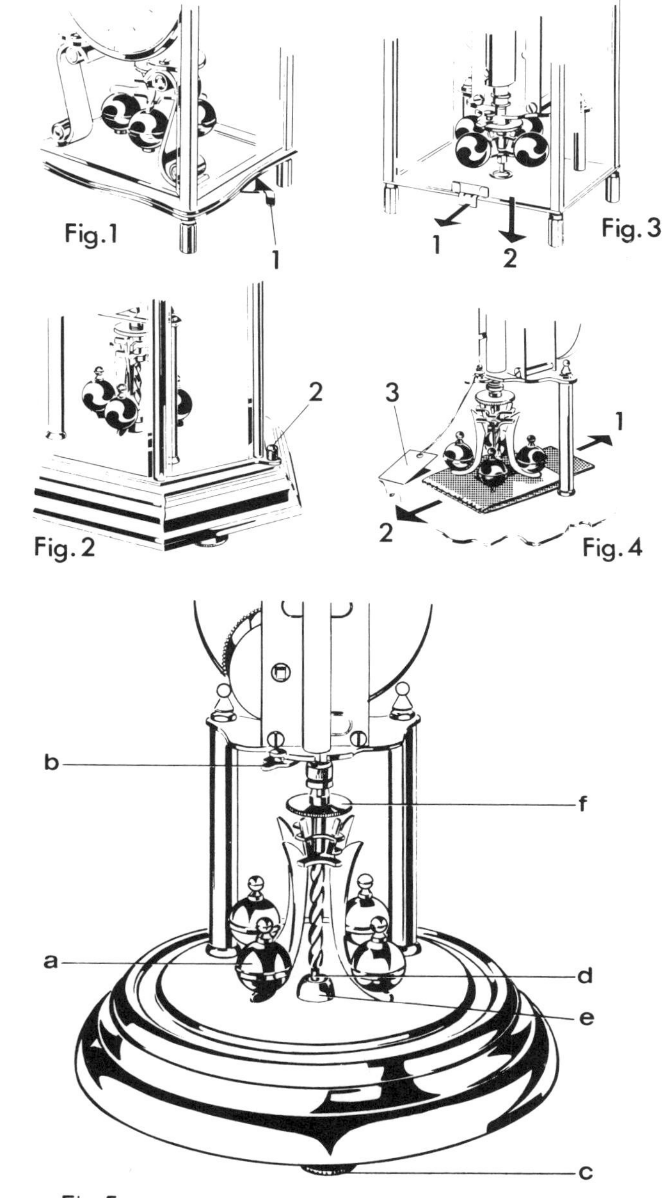

Transportation of clock

The following points should be observed if clock has to be shipped:

1. It is important that the pendulum is locked in place by means of the locking lever (b). To do this, lift pendulum (a) until the locking lever (b) can be pushed to the right into the "shut" position.
2. Fasten the adjustable feets.
3. Carefully pack up the clock.
   If possible use the original packing.

All our 400-day clocks are equipped with a special Nivarox suspension spring. If this spring is damaged, be sure to have it replaced by your clockmaker.

Figure 2 1955 Kundo Instructions. (These instructions were repeated in French and German. Only English instructions are shown here.)

about setting up their new clock. As a consequence, the manufacturers instructions became much more detailed as is demonstrated by a copy of the Kundo instructions shown as Figure 2.

Major improvements and innovations in the production of 400-Day Clocks came in a flood in the early 1950's. Among them were:

Pendulum guide cups. Kundo pioneered the use of a small brass cup attached to the top of the clock's base, positioned directly below the pendulum so that the tip of the pendulum was within, but not touching, the cup. When the pendulum tip was in the proper position, the pendulum had to be hanging straight and the clock had to be level. It was hoped that the use of the pendulum guide cup would improve the clock's operation. In any event, pendulum cups made it easier to avoid damage to the clock's suspension spring when the clock had to be moved. The idea was either not patented or was not patentable, because all clock manufacturers soon used pendulum guide cups. That the guide cups were not entirely successful, however, is shown by the fact that many clocks of this period will run only when their bases are raised slightly at points around their perimeter by inserting cardboard shims.

Base leveling screws. A brisk, if short-lived, mail order business was conducted by at least two quick-witted entrepreneurs aware of the leveling problem, who sold "400-Day Clock Leveling Bases." One was a plastic ring with three feet, adjustable in height, on which the clock was set. Another, with a similar leveling ring, was attached to the under side of the base and could not be seen. All manufacturers soon adopted some means of leveling the clock. Most popular was a type that consisted of three knurled thumb discs around the base used to raise or lower the feet in order to level the clock and center the pendulum over the guide cup. A similar, but more complicated form of base leveling, with invisible feet, was used by Schatz. This design added considerable factory cost to the clock as the leveling device has eighteen separate parts. Even more costly levelers were used in clocks made by Uhrenfabrik J. Kaiser in which the discs in the base actuate the leveling feet through gears.

Pendulum locking devices. Suspension spring breakage or bending was a major problem to both buyers and sellers of 400-Day Clocks. Many clocks were returned to the dealer after purchase because "the clock won't run." To preserve customer good will, the dealer often exchanged the "defective" clock for a new one even though it was obvious that the condition of the spring was brought about by the user's careless handling. This was costly to the dealer and discouraged his future purchases from importers. Many people, in fact, had (and still have today) difficulty in attaching the relatively heavy pendulum to the suspension spring, which is only slightly thicker than a human hair.

Kern & Sohne was the first manufacturer to make a clock with a permanently attached pendulum.[27] The pendulum guide cup was redesigned with threads and a thumb nut so that it could be raised or lowered. When the clock was to be moved, the thumb nut was turned so that the pendulum was forced against a bracket just above it and held immobile, regardless of the position of the clock. When the cup was lowered, the pendulum was again free to be rotated.

Soon, almost like lightning, all clocks appeared with some form of pendulum locking device. Schatz used one on their miniature and 1000-day clocks that was operated by a lever extending from under the base. When moved to the far right, the lever caused the cup to raise the pendulum against a bracket. When the clock was to be shipped, the pendulum was further locked in this position with a thumb screw. There were many other forms of pendulum locking devices, of which a most unusual one was on a model made by Georg Wurthner exclusively for the John Wanamaker stores. Here the pendulum was automatically locked whenever the clock was picked up. This method, rather interestingly, was based on a patent of Aug. Schatz & Sohne.[28]

Miniature models. The first miniature model to appear on the market was one with pin pallet escapement made by Kundo. (The escapement was soon changed to Graham.) Other miniature models, some with Graham escapements and some with pin pallet, soon followed, made by Schatz, Kern, Koma, Herr, Reiner, Hermle, Henn, Haller, Wurthner and Link. Most were of excellent quality, but some were inexpensively made, having 100-day and 30-day movements. A few manufacturers even made midget models, most of which had pin pallet escapements.

Case styles. Most popular, of course, were the models under glass domes. Each manufacturer had a standard clock with his largest movement, that was usually his most popular model. The most widely used standard dome was 5½" x 8½". Other standard models used other, usually larger, domes. Kundo's Model 5 used a 7" x 11" dome. A tall pillared model[29] that was made originally by several manufacturers, and was revived in the 1950's by Kern, Koma, Link and possibly others, used the largest 400-Day Clock dome, 7½" x 15".

---

[27]Section 4, Patent 36
[28]Section 4, Patent 42
[29]Section 6, Clock 73

Many other case styles were also available. Several were of the four-glass variety, others six-sided and some were "coach" styles. One model by Link was in an enameled Neuchatel case with wall bracket.

Pendulums. The majority of both standard and miniature clocks had four-ball pendulums, but three-ball pendulums were used by Schatz,[30] Koma and others. There were other varieties of pendulums, including a few discs, some having two or four small plastic figures of elves, ballet dancers or men and women in regional dress on them,[31] made by Kern, Koma and Petersen. A popular model by Henn with disc pendulum and four truncated regulating weights[32] could be regulated only when the pendulum was locked. This proved to be a poor design as the pendulum was awkward to manipulate and the regulating linkage broke frequently.

Perhaps the one thing that caused more confusion to dealers than to clock buyers was that on most pendulums, one turned the regulating nut clockwise to make the clock run slower, while on other pendulums, the nut had to be turned counter-clockwise.

Hour strike. Uhrenfabrik Herr made a model of their 1952 standard 400-Day Clock available with a brass bell at the top to be struck by a hammer once an hour.[33] Again, it was based on a patent of Aug. Schatz & Sohne, who had made the clock, discovered its fault and stopped production before any clocks were sold. The clock would strike when it was fully wound, but after only a week or ten days, the strike would stop as there was insufficient power in the movement to lift the hammer. This model was soon discontinued.

Calendar/moon phase dials. One clock with calendar dial was made by Link,[34] but it suffered the same fate as the Herr striker because the movement had insufficient power. Another calendar 400-Day Clock with two wheel trains was made about 1954 by W. Petersen.[35] The set-up tool cost of this clock must have been tremendous, as it had to have a much larger movement to accommodate the two trains and the mechanism controlling the calendars for day-of-week, day-of-month and month. It is believed that none of these clocks ever ran properly.

However, there were two clocks with adaptations of moon phase dials that were successful. One was a model with moon phase dial similar to the type used on some tall case clocks, again made by Georg Wurthner[36] exclusively for John Wanamaker. The other was a well-made clock by Uhrenfabrik J. Kaiser[37] with a three-dimensional "moon" about ¾" in diameter that was situated at the top of the dial frame and made a complete turn every month.

In 1951, two things happened that made it possible for 400-Day Clocks to be sold in the United States at lower prices. First, the U.S. Government reduced the import duty on clocks from 65% to 32½% as a result of an international agreement known as the Torquay Protocol. Second, the 400-Day Clock manufacturers reduced the width of the movement plates to under 1.77", allowing the clocks to be classified as watches, on which the duty was lower.

Kundo was the first to have its clocks with narrow plate movement[38] pass through U.S. Customs (on this model, thin brass "wings" were attached to make the plate appear wider), but the sales advantage Kundo enjoyed lasted only until the other manufacturers re-tooled to follow suit. Seemingly overnight, the United States was almost literally flooded with clocks. German manufacturers, wishing to garner their share of the lucrative U.S. market, sprang up like mushrooms and by 1953, just four years after the new wave of importations began, there were fourteen manufacturers in full production. They included, to use their trade names, Kundo, Schatz, Kern, Koma, Herr, Reiner, Petersen, Kaiser, Wurthner, Hermle, Linco, Haller and Crescent, a Japanese manufacturer.

The boom sales of 400-Day Clocks in the early 1950's were of short duration. In 1949, clocks were being wholesaled at $35 and retailed from $60 to $80, depending on the store. But these prices did not last. The importers wanted lower factory prices and, if they could not get them from one manufacturer, they would get them from another. To lower prices, the manufacturers found ways to lower production costs. They used less brass and had their original heavy glass domes made thinner to save weight and shipping cost. Most important of all, they streamlined production. But with buyers from the large retail stores playing one manufacturer against another with firm orders for 25,000 or more clocks at a time, some factories made the mistake of accepting orders at prices below cost.

John Wanamaker & Co., with their big department stores in Philadelphia and New York, was the largest buyer and price-cutter of 400-Day Clocks. Their Philadelphia clock buyer told Uhrenfabrik Reiner that John Wanamaker would buy 75,000 clocks if Reiner would deliver a clock that could be sold at retail for $13.75. Wanamaker expected to make only a small profit on the sale, but would receive considerable bènefit from publicity. However, it meant that the factory price of the clock had to be down to an incredible $8.00!

Reiner accepted the order and, quite surprisingly, delivered a clock of good appearance[39] that operated as well as other 400-Day Clocks being imported at the time. In fact, at a distance, it looked very much like the clock that Kundo was

[30]Section 13, Pendulum 55
[31]Section 13, Pendulum 91
[32]Section 13, Pendulum 59
[33]Section 9, Plate 1071
[34]Section 9, Plate 1077
[35]Section 9, Plate 1674
[36]Section 9, Plate 1591
[37]Section 9, Plate 1309
[38]Section 9, Plate 1377
[39]Section 18, No. 114

wholesaling for $35. How was it possible to make the clock so inexpensively? Reiner had in their line a miniature clock with miniature movement. They put the miniature movement[40] on their standard clock platform and added two "U" shaped brass "wings" which covered both top and bottom in addition to the sides of the back plate. These wings gave the clock a more substantial appearance. They also used their standard pendulum, but reduced the weight of the balls[41].

The year of the $13.75 Wanamaker clock was 1953[42]. There may be no connection, but within two years John Wanamaker had closed its New York store and Uhrenfabrik was bankrupt. In January 1955, 6,800 of the clocks were sold at auction by Parke-Bernet when the stock of Wanamaker's New York store was liquidated. Almost all were bought in case lots. A cut-rate retailer, S. Klein-on-the-Square in New York, promoted the sale of these clocks in *The New York Post* later that year for the all-time low price of $6.99.[43]

As a result of the publicity that spread across the country, sales of 400-Day Clocks slowed to a trickle. Who would pay $60 for a Kundo or Schatz clock when 400-Day Clocks were selling in New York for $7.00? The effect on retailers is easy to visualize.

By 1956, many manufacturers had either become bankrupt or removed the 400-Day Clock from their lines. Henn and Link had closed their doors. A new factory, Uhrenfabrik Neueck, which had superseded Herr and Reiner, found itself in financial trouble in 1958 and went into receivership from which it never recovered.

There were fourteen manufacturers of key-wind 400-Day Clocks in 1953, nine by 1960, and only seven in 1970. By the mid-1980's the producers had shrunk to three. Today, there is very limited production of key-wind 400-Day Clocks although several manufacturers are producing quartz controlled 400-Day Clocks. The key-wind 400-Day Clocks are being recognized as collectible by antique clock enthusiasts and many of the approximately 15 million 400-Day clocks which were produced during the slightly more than 100 years that the clocks were manufactured are being repaired and restored and are still functioning as timekeepers.

Epilogue

In 1981 Aug. Schatz & Sohne made available a reproduction of their 1885 Jahresuhrenfabrik clock with disc pendulum and three-pillar gallery[18] to commemorate their firm's one hundred years of clockmaking. This clock might be thought of as a monument to mark a fascinating century of horological history.

---

[40]Section 9, Plate 1325, 1326
[41]Section 18, No. 124
[42]Section 18, No. 114
[43]Section 18, No. 115

SECTION 2 400-Day Clock Events by Years

1841 —10 February. U.S. Patent No. 1973 issued to Aaron D. Crane of Newark, New Jersey, for a torsion pendulum clock. Subsequently, clocks were manufactured in 8-day, 30-day and year duration models, many of which have survived in good running order. (Section 4, Patent 1)

1852 —5 October. U.S. Patent No. 9310 issued to Silas B. Terry of Plymouth, Connecticut, for a Marine (torsion pendulum) clock. The movement was used in a variety of shelf and wall cases and would operate perfectly even through not set in a level position. Many of these clocks have survived in good running order. (Section 4, Patent 2)

1876 —29 March. U.S. Patent No. 176309 filed by John W. Hile of Waterville, Kansas, for a clock, "the combination, with dead beat escapement, of a weight (pendulum) suspended by a torsion spring, rotating alternately from right to left..." There is no indication that any clocks were made. (Section 4, Patent 3)

1879 —21 March. German Patent No. 2437 issued to Lorenz Jehlin of Sackingen, Germany, for a torsion *pendulum.* While the patent is in Jehlin's name, it was generally known that Anton Harder, usually credited with the invention of the 400-Day Clock, had a major interest in it. (Section 4, Patent 4)

—30 November. Lorenz Jehlin died. Rights to the patent went to Anton Harder.

—Harder made arrangements with A. Willmann & Co. of Freiburg (Silesia), to make torsion pendulum clocks. About 300 to 400 clocks were made in several different sizes and models using verge, cylinder and Graham escapements. Pendulums were disc-type with two round, flat regulating weights controlled by a left- and right-threaded regulating rod. From a production standpoint, the clocks were considered unsatisfactory. One experimental example with verge escapement, which *may* have been made by Wilmann, exists today in running order. (Section 6, Clock 1; Section 9, Plate 1727)

1880 —17 August. Austro-Hungarian Patents Nos. 2591 and 16111 issued to Anton Harder of Ransen, Germany, for a torsion pendulum clock. (See 12 December 1882)

—Harder made arrangements with Gustav Becker of Freiburg (Silesia) and also with Fortuna Clock Co. of Triberg (Black Forest), to make torsion pendulum clocks, but again without production success. None of these Becker or Fortuna clocks is identified. (The name "Fortuna" appears on dials of clocks of later manufacture.) Two clocks that *may* have been made

1841 Crane torsion pendulum year clock based upon his original 1841 patent.

1852 Terry torsion pendulum "Marine" clock based upon his original 1852 patent.

Erloschen.

ROTATIONSPENDEL

1887 Lorenz Jehlin's 1879 patent with "Erloschen" (Cancelled) stamped on the cover.

by Becker or Fortuna (one with verge escapement, the other cylinder), exist today in running order. (Section 6, Clocks 2 and 3; Section 9, Plates 1633 and 1731)

1881 —The firm Jahresuhrenfabrik (Year clock factory), Triberg (Black Forest), was organized by a group of workers, some possibly from Fortuna. Harder, the obvious instigator, was assured of this group's "know how." The group was headed by Aug. Schatz, grandfather of Kurt and Charles Schatz, principals in the later-day firm Aug. Schatz & Sohne, a major 400-Day Clock manufacturer.

1882 —17 March. Jahresuhrenfabrik delivered the first twelve clocks with Graham escapement movements—the first successful production of 400-Day Clocks. (Section 6, Clock 6)

—12 December. U.S. Patent No. 269052 issued to Anton Harder for a torsion pendulum clock. (See 17 August 1880) The patent drawings illustrate, with minor variations, the keywind 400-Day Clock as it is known today. (Section 4, Patent 5; Section 9, Plate 1475)

—Jahresuhrenfabrik made full striking 400-Day Clocks based upon deGruyter's 1884 patent. The dials were marked 400 DAYS STRIKING DE GRUYTER'S PATENT. This is the first recorded reference to "400 Day." (Section 3, Clock 1; Section 9, Plate 1303)

1887 —Major turning point in 400-Day Clock manufacture. Original patent was cancelled because deGruyter failed to pay renewal fee. With patent cancelled, the door was open for anyone to make 400-Day Clocks in competition with Jahresuhrenfabrik.

1888 —Lenzkirche Uhren Fabrik 400-Day Clocks appeared. If any exist today, they are not identified.

1889 —Badische Uhrenfabrik 400-Day Clocks appeared, the first to use pin-pallet escapement and lantern pinions. (Section 9, Plate 1016A)

1890 —Wendes & Metzger 400-Day Clocks appeared. If any exist today, they are not identified.

1891 —Huber Uhren 400-Day Clocks appeared. Huber's early clocks are not identified, but the Huber name appeared later on clocks made by Badische Uhrenfabrik and others. (Section 9, Plates 1251 and 1251A)

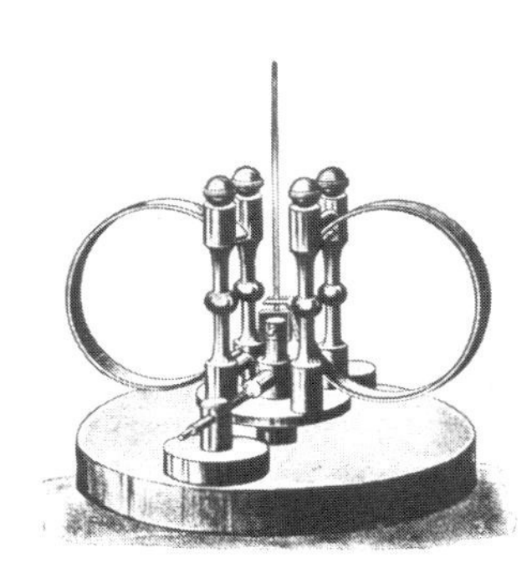

1902 Promotion for Andreas Huber's "twin loop" temperature compensating pendulum in August 1903 issue of *The Keystone*.

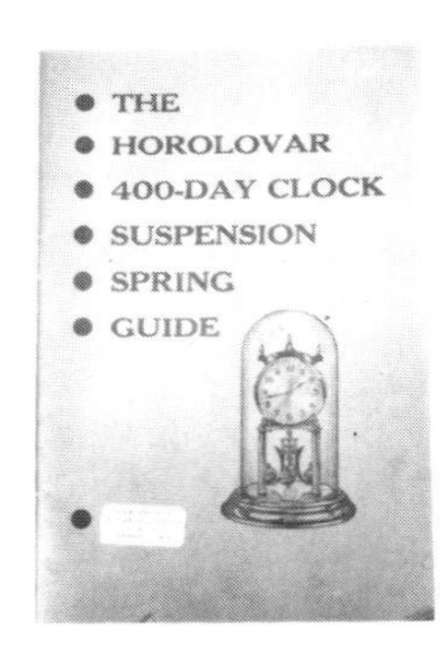

1953 First *The Horolovar 400-Day Clock Repair Guide*.

1954 Uhrenfabrik J. Kaiser's "Universe" model 400-Day Clock.

1884 —Harder sold patent rights to clock to F.A.L. deGruyter of Amsterdam, who became exclusive sales agent and distributor of Jahresuhrenfabrik clocks.

—18 May. DeGruyter was issued German Patent No. 29348 for a full *striking* torsion pendulum year clock. (Section 4, Patent 7)

1885 —A three-pillared gallery was added to the plain disc pendulum for added attraction. (Section 6, Clock 9)

1892 —22 June. Swiss Patent No. 5172 issued to J.J. Meister of Paris, France, for a chronometer-type temperature compensating pendulum. This appears to be the earliest attempt to correct the poor timekeeping of the clock. (Section 4, Patent 13)

1894 —R. Schneckenburger, successor to Muller & Co. of Mulheim, a. D., made full striking 400-Day Clock movements. (Section 9, Plate 1504)

1895 —11 May. Wilhelm Kohler of Furth, Germany, issued

German Patent No. 10519 for a detent escapement used with a full striking 400-Day Clock movement. (Section 4, Patent 14; Section 9, Plate 1504)

1897 —8 November. Christian Bauer of Furth, Germany, issued Swiss Patent No. 15752 for a duplex escapement used with a full striking 400-Day Clock movement. (Section 4, Patent 15; Section 9, Plate 1303)

1898 —First disc pendulum designed with a hook to attach to pinned bottom suspension block.

1900 —First pendulum with 6-pillar gallery.

—First 4-ball pendulum.

—First click with steady pin.

1901 —24 December. Bowler & Burdick of Cleveland, Ohio, issued U.S. Patent Registration No. 37484 for trade mark "Anniversary." (Section 4, Patent 16; Section 9, Plate 1009)

1902 —2 April. Louis Wille of Leipzig, Germany, issued German patent D.R.P. 144687 (D.R.F.G. 208297) for a temperature compensating pendulum. (Section 4, Patent 17)

—5 April. Andreas Huber of Munich, Germany, issued German patent 144688 for "twin loop" temperature compensating pendulum. (Section 4, Patent 18; Section 13, Pendulum 10)

—First disc pendulum with decorative, ball-topped gallery with curved arms. (Section 13, Pendulum 19)

1903 —1 December. Gustav Becker of Freiburg (Silesia), issued German patent No. 157384 for a temperature compensating pendulum. (Section 4, Patent 19) There is no record of its ever having been produced.

1904 —June. Louis Wille's temperature compensating pendulum is promoted with advertisement in *The Keystone,* U.S. jewelry trade paper.

—8 December. Edouard Guillaume of Sevres, France, issued Swiss patent No. 32895 for "Elenvar," a temperature compensating alloy. (Section 4, Patent 21) Apparently no through was given to its use for 400-Day Clock suspension springs.

1905 —Jahresuhrenfabrik made full striking year clocks having movements with Graham escapements, much less complicated than the deGruyter movement. (Section 9, Plate 1181)

—First tubular suspension spring guards. Original type had large, flared end to hold friction-tight cup for protection of bottom suspension block during shipping. (Section 6, Clock 18)

—Wintermantel Uhrenfabrik 400-Day Clocks appeared. (Section 9, Plate 1579AA)

—Several manufacturers made brass cased clocks. (Section 6, Clocks 11, 14, 26, 27, 31, 32)

1906 —Gustav Becker applied a beat-adjusting overhead suspension to some of its 400-Day Clocks.

—Gustav Becker introduced a flat suspension guard with bottom suspension locking feature. (Section 18, No. 11)

1907 —Gustav Becker made a skeleton model 400-Day Clock. (Section 6, Clock 15; Section 9, Plate 1189)

—First self-leveling suspension saddle with gimbals. (Section 15, Bracket 14)

—Phillipp Haas made first 400-Day Clock with miniature movement. (Section 6, Clock 13; Section 9, Plate 1425)

1908 —3 January. Claude Grivolas of St. Cloud, France, issued Swiss patent No. 43059 for a "long running" torsion pendulum clock. (Section 4, Patent 24; Section 6, Clock 62; Section 9, Plate 1471A)

1910 —Gebr. Junghans made pendulum with "governor" -type regulation, erroneously promoted as temperature compensating. (Section 6, Clock 50; Section 9 Plate 1171)

—Jahresuhrenfabrik made small, 30-Day torsion pendulum clock in pillared case. (Section 6, Clock 78; Section 9, Plate 1257)

—Franz Vosseler made small, 30-Day torsion pendulum clock in pillard case. (Section 6, Clock 77; Section 9, Plate 1145)

1911 —Kienzle Clock Factories made 30-Day miniature clock only five inches high. (Section 6, Clock 82; Section 9, Plate 1175)

1912 —Kienzle Clock Factories made calendar model with day-of-week and day-of-month dials. (Section 6, Clock 21; Section 9, Plate 1602)

1913 —Kienzle Clock Factories made a "very cheap and higher quality" model. (Section 9, Plate 1383)

—October. U.S. Customs regulations require that the country of origin be marked on all imported products. Some clock manufacturers had stamped "Germany" or "Made in Germany" before this time.

1914 —Production of all clocks discontinued.

1929 —Kienzle Clock Factories sold its 400-Day Clock tools and dies to Karl Kern, shop foreman. Kern, with partner J. Link, formed Kern & Link to produce 400-Day Clocks. (Section 9, Plate 1342)

1932 —Kern and Link dissolve partnership. Kern continued to manufacture 400-Day Clocks under his name alone. (These clocks have no identification marks.)

1937 —Karl Kern joined with his sons to form Kern & Sohne, continuing to make 400-Day Clocks.

1949 —New wave of U.S. 400-Day Clock imports:

—Kieninger & Obergfell clocks imported by Forestville Clock Co., New York. (Section 9, Plate 1107)

—Aug. Schatz & Sohne clocks imported by the Electric Shaver Division of Remington Rand Co., Bridgeport, Connecticut. (Section 9, Plate 1278)

—Kern & Sohne clocks imported by W. & J. Overocean Traders, New York. (Section 9, Plate 1343)

—First 400-Day Clock bases with pendulum guide cups.

1950 —Beginning of 400-Day Clock sales through military post exchanges throughout the world. Thousands of clocks, purchased by "GI's" for about $8.00 were sent to the U.S. as gifts to families and friends.

—Konrad Mauch began production of 400-Day Clocks. (Section 9, Plate 1394)

1951 —Horolovar Co. made its *400-Day Clock Temperature Compensating Suspension Springs* available.

—First bases appeared with three leveling screws.

—Movement plates reduced to 1.77" width to lower import duty. Forrestville Clock Co. was first to take advantage of this loop hole.

—Georg Wurthner added 400-Day Clocks to its line. (Section 9, Plate 1587)

—Franz Hermle added 400-Day Clocks to its line. (Section 9, Plate 1103)

—Edgar Henn began production of 400-Day Clocks; promoted them in the U.S. as equipped with *Horolovar Temperature Compensating Suspension Springs.* (Section 9, Plate 1059)

—Uhrenfabrik M. Reiner began production of 400-Day Clocks. (Section 9, Plate 1573)

—Uhrenfabrik Herr began production of 400-Day Clocks. (Section 9, Plate 1241)

—U.S. reduced import duty on clocks from 65% to 32½%.

1952 —Horolovar sold 10,000 of its *Temperature Compensating Suspension Springs* to Aug. Schatz & Sohne.

—W. Petersen began production of 400-Day Clocks. (Section 9, Plate 1583)

—J. Link (formerly with Kern & Link) began production of 400-Day Clocks; promoted its clocks in the U.S. as equipped with *Horolovar Temperature Compensating Suspension Springs.* (Section 9, Plate 1111)

1953 —First Edition of the "The Horolovar 400-Day Clock Repair Guide", titled "The Horolovar 400-Day Clock Suspension Spring Guide" published. The "Repair Guide" made possible the practical repair of 400-Day Clocks by professional repairers.

—John Wanamaker purchased 75,000 400-Day Clocks from M. Reiner Uhrenfabrik to retail for $13.75. (Section 9, Plate 1325 and 1326)

1954 —John Wanamaker's New York store went out of business.

—W. Petersen marketed, unsuccessfully, an automatic-change calendar 400-Day Clock. Movement had insufficient power to operate the calendar mechanism. (Section 9, Plate 1674)

—Uhrenfabrik J. Kaiser added 400-Day Clocks to its line. (Section 4, Patent 46; Section 9, Plate 1310)

—Uhrenfabrik Herr went into bankruptcy.

—Edgar Henn went out of business.

1955 —Uhrenfabrik M. Reiner went into bankruptcy.

—At the auction of John Wanamaker's inventory, 6800 Reiner clocks were sold.

—S. Klein-on-the-Square, New York, purchased most of the Reiner clocks and later sold them for $6.99.

1956 —Uhrenfabrik Neueck was organized to take over the bankrupt Herr and Reiner factories; sold several thousand clocks through the "Around-the-World Shopper's Club." (Section 9, Plate 1469C)

1957 —Uhrenfabrik Neueck went out of business.

--W. Petersen went out of business.

—J. Link went out of business.

—Georg Wurthner discontinued production of 400-Day and 30-Day torsion pendulum clocks.

1958 —Only seven post-1949 manufacturers of 400-Day Clocks remained in business: Kundo, Schatz, Kern, Koma, Hermle, Kaiser and Haller.

1962 —J. Kaiser eliminated 400-Day Clocks from its line.

1965 —Nisshindo Watch Co. of Tokyo, Japan, made "Master" 400-Day Clocks (Section 9, Plate 1461D) and "New Master" 100-Day Clocks (Section 9, Plate 1469E). At first, sales were chiefly through U.S. Navy stores in Viet Nam, Bangkok and other Near East cities. Later, some clocks were sold direct to U.S. importers.

1966 —Attempts made, unsuccessfully, by Schatz, Kundo, Kern and others to use remontoir-type battery movements in 400-Day Clocks. The electrical contacts were not dependable.

1971 —Franz Hermle and Aug. Schatz & Sohne gave up production of key-wind 400-Day Clocks and made electronic 400-Day Clocks exclusively.

1972 —Konrad Mauch made electronic 400-Day Clocks using Hermle electronic movements.

1973 —Nisshindo gave up production of key-wind 400-Day and 100-Day clocks and made electronic clocks exclusively.

1975 —By now, Kundo, Schatz, Kern, Koma, Hermle, Haller and Nisshindo were all manufacturing their own quartz 400-Day Clocks. Key-wind clocks continued to be made only by Kundo, Kern, Koma and Haller.

1981 —Aug. Schatz & Sohne made 3000 reproduction key-wind clocks, based upon their 1885 model, to commemorate 100 years of torsion pendulum year-clockmaking (1881-1981). (Section 6, Clock 86; Section 9, Plate 1614)

1985 —Aug. Schatz & Sohne went out of business.

1986 —Kern stops production of Key-wind 400-Day Clocks. All production will be quartz movements.

SECTION 3 History of Striking 400-Day Clocks

Before 1900, two manufacturers made striking torsion pendulum year clocks. One was by Jahresuhrenfabrik, based on the deGruyter patent, with either bell or gong strike.[1] These clocks have 400 DAY STRIKING DE GRUYTER'S PATENT in a circle around the center of the dial, although their movements were considerably simplified by Jahresuhrenfabrik from the specifications and drawings in the deGruyter 1884 patent.[2] The pendulum had a three-pillar gallery and the clock was covered by an oval glass dome (Clock 1).

The other manufacturer was R. Schneckenburger (successor to Muller & Co.) who made clocks[3] with both detent[4] escapements, patented in 1895 by Wilhelm Kohler (Clocks 2, 3, 5, 7), and duplex[5] escapements, patented in 1897 by Johann Christian Bauer (Clock 6). These clocks were excellently made and were in wood wall cases of various designs. The pendulum is regulated by turning a small knurled nut in the hook at the top of the pendulum, shortening or lengthening the suspension spring. A wall clock with the Schneckenburger movement is in the collection of the Smithsonian Institution in Washington, D.C. Its wooden case contains an enamel plaque (4) which is also seen in Clock 5. The wall cases have largely disappeared but occasional movements with dial and pendulum mounted on wood bases under glass domes can be found. One of these with duplex escapement (Clock 6) has on its dial: "400 days clock."

Year striking 400-Day Clocks were short lived. Their lack of acceptance was probably due to the fact that the strike trains were controlled by count wheel rather than rack and snail. A mere adjusting of the hands would often trip the train to strike the next hour or half-hour. Since the clocks, with their steel suspension springs, were poor time-keepers, the hands needed frequent adjusting, with the result that the clock often had to be stopped until the time caught up, or the hands turned ahead through 23 half-hour periods while

[1]Section 9, Plate 1473
[2]Section 4, Patent 7
[3]Section 9, Plate 1504
[4]Section 4, Patent 14; Appendix, No 32
[5]Section 4, Patent 15; Appendix, No 32

waiting for each hour and half-hour to strike. Of course, another reason for their lack of popularity could have been because they were relatively expensive. These clocks have become desirable collector's items, but everyone who owns one will confirm the fact that it is difficult to get the clock to strike properly.

The only other full striking torsion pendulum year clock was made in 1905 by Jahresuhrenfabrik,[6] in another futile attempt to market these more expensive clocks (Clock 8-Pendulum not original). Other than the fact that the clock is illustrated in the Jahresuhrenfabrik catalog for 1905, very little is known about it. The clock had no better success with the public than its predecessors, although it had a less complicated movement with Graham escapement and a glass dome case.

5 6 7 8

[6]Section 9, Plate 1181

SECTION 4 Early 400-Day Clock Patents

The illustrations from patents, with their brief descriptions, in this Section represent only a portion of the patents relating specifically to torsion pendulum clocks issued to individuals and companies from 1841 to 1955.

For one to make a complete file of patents in this category, he would need to comb the patent records not only of the U.S. and Germany, but also of Austro-Hungary, France, Switzerland and many other countries. The enamel plaque in the wooden cases of some c. 1896 Schneckenburger year striking clocks (for an example, see Section 3, No. 4), reads in part, "...Patented in nearly all civilised States," and illustrates the coats of arms of nine different countries. Just what other countries were thought not to be "civilised" in 1896 is not made clear.

Many of the objects of these patents never saw the light of day, chiefly because the ideas they documented were impractical or cost too much to produce. In the important area of temperature compensation and later in pendulum locking devices, there was considerable competition among those who felt that their method was the way to solve the problems of poor timekeeping and suspension spring breakage.

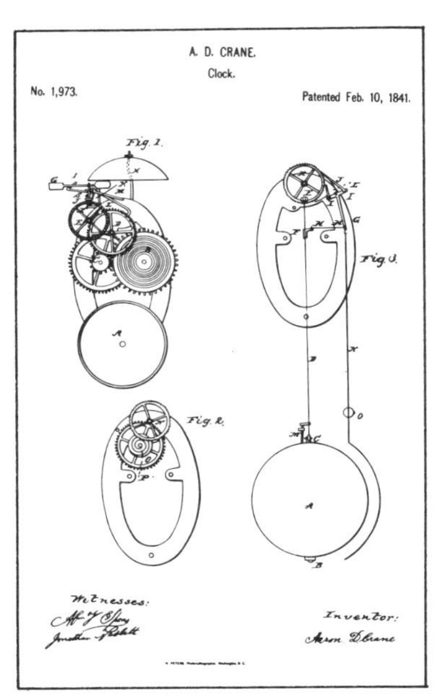

1 U.S. Patent No. 1973 issued to A.D. Crane of Newark, New Jersey, on 10 February 1841 for a (torsion pendulum) clock.

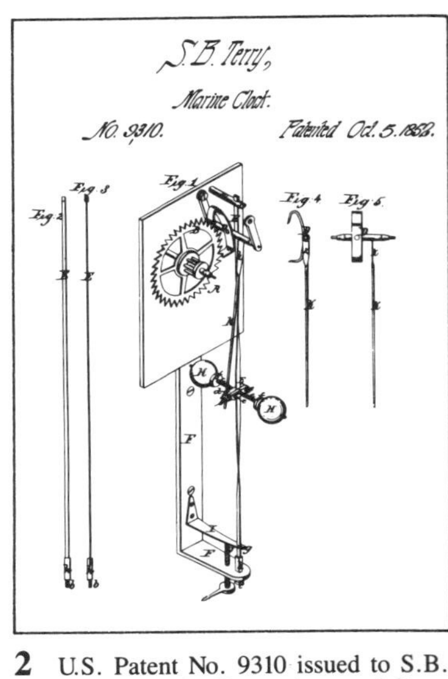

2 U.S. Patent No. 9310 issued to S.B. Terry of Plymouth, Connecticut, on 5 October 1852 for a (torsion pendulum) marine clock.

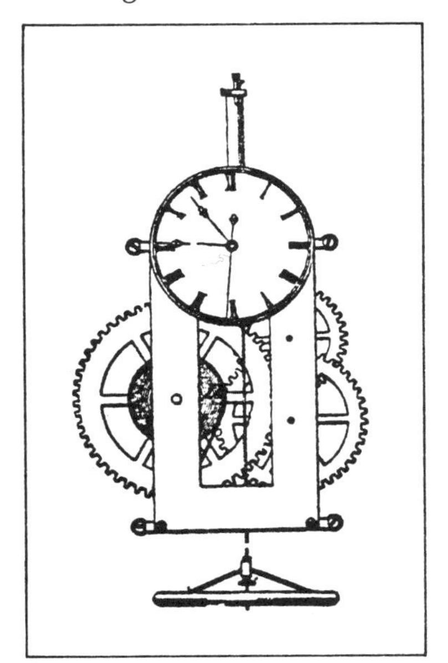

3 U.S. Patent No. 176309 issued to John W. Hile of Waterville, Kansas, on 29 March 1876 for a clock (with torsion pendulum).

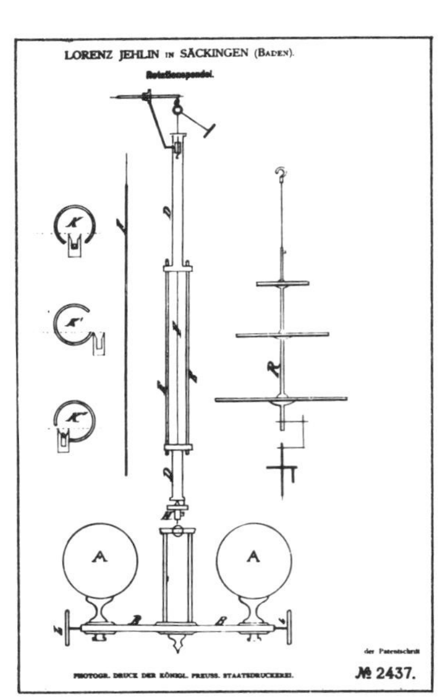

4 German patent No. 2437 issued to Lorenz Jehlin of Sackingen (Baden), Germany, on 21 March 1879 for a "Rotationspendel" (torsion pendulum).

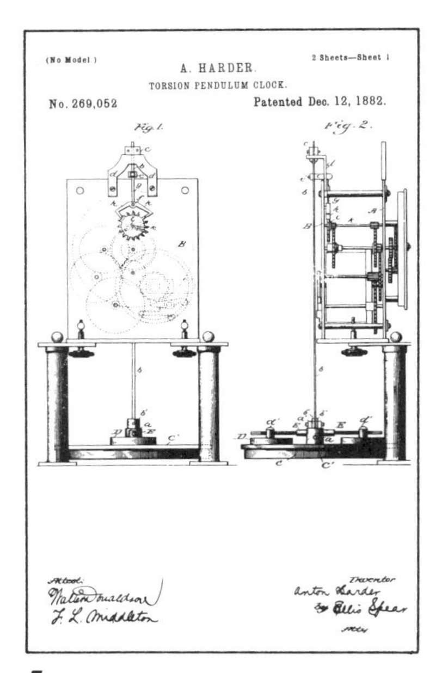

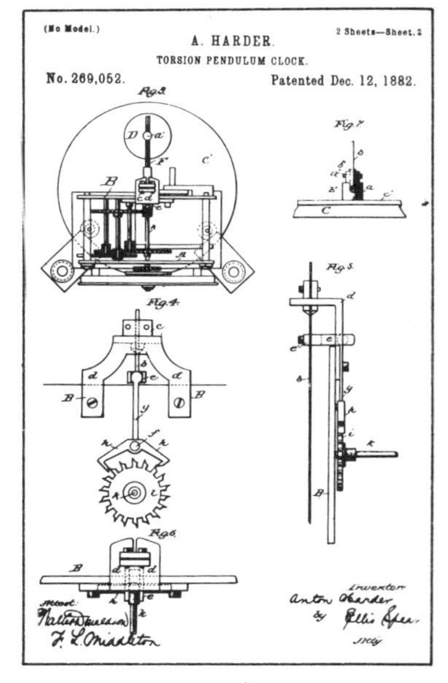

5 U.S. Patent No. 269052 issued to A. Harder of Ransen, Germany, on 12 December 1882 for a torsion pendulum clock.

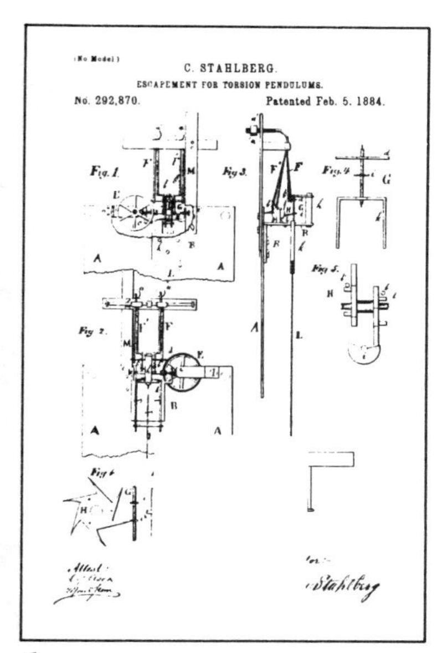

6 U.S. Patent No. 292870 issued to C. Stahlberg of Corsicana, Texas, on 5 February 1884 for a torsion pendulum clock escapement.

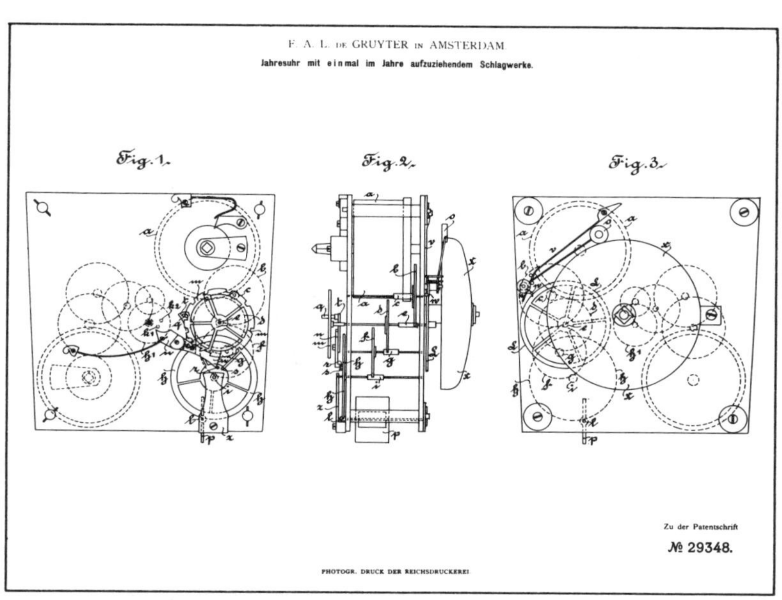

7 German patent No. 29348 issued to F.A.L. deGruyter of Amsterdam, Holland, on 18 May 1884 for a full striking (torsion pendulum) year clock.

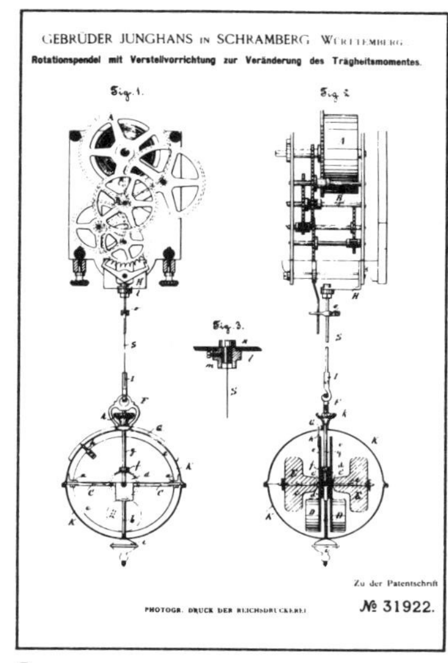

8 German patent No. 31922 issued to Gebr. Junghans of Schramberg, Wurttemberg, Germany, on 9 September 1884 for a torsion pendulum regulating device.

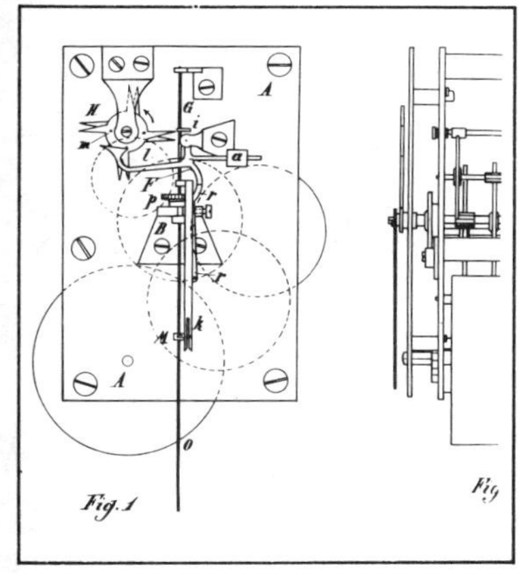

9 German patent No. 30189 issued to Charles Stahlberg of Corsicana, Texas, on 28 January 1885 for a torsion pendulum clock escapement.

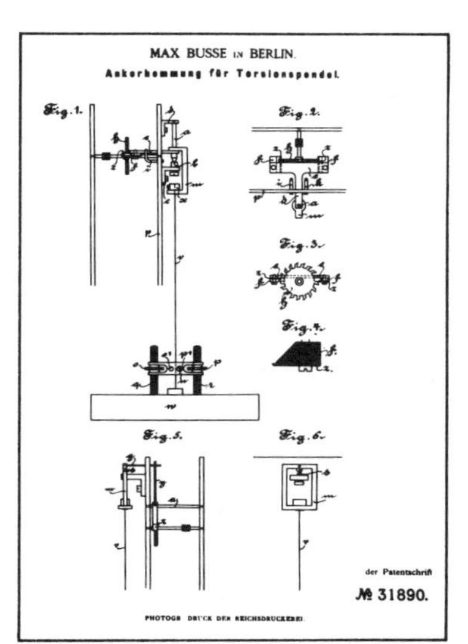

10 German patent No. 31890 issued to Max Busse of Berlin, Germany, on 12 June 1885 for a torsion pendulum clock escapement.

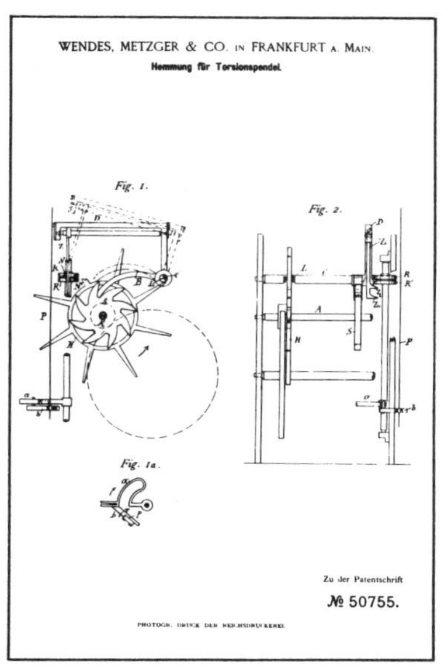

11 German patent No. 50755 issued to Wendes, Metzger & Co. of Frankfort, Germany, on 12 February 1890 for a torsion pendulum clock escapement.

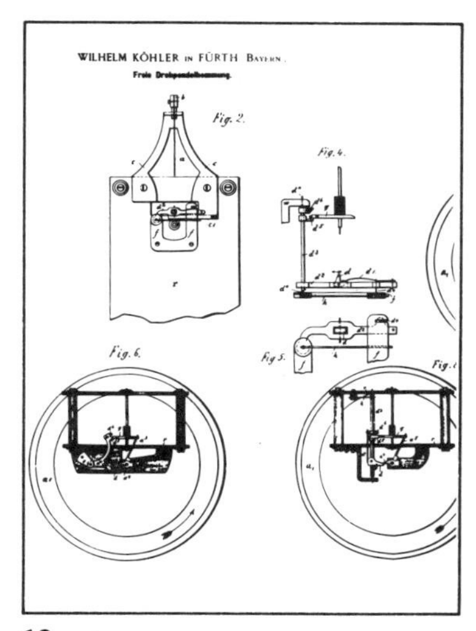

12 German patent No. 57492 issued to Wilhelm Kohler of Furth (Bavaria), Germany, on 8 July 1891 for an escapement (for torsion pendulum clock).

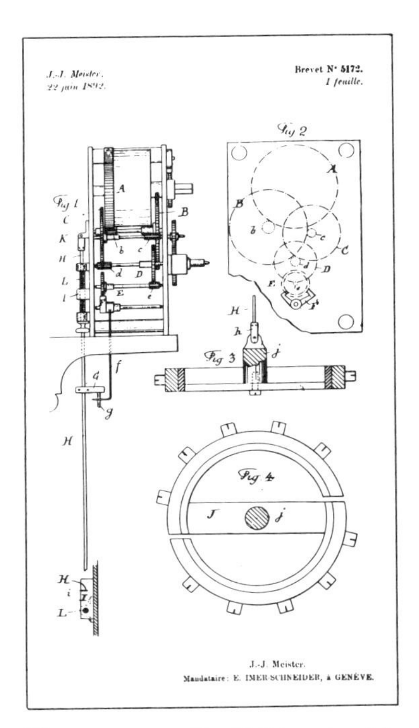

13 Swiss Patent No. 5172 issued to J.J. Meister of Paris, France, on 22 June 1892 for a temperature compensating pendulum.

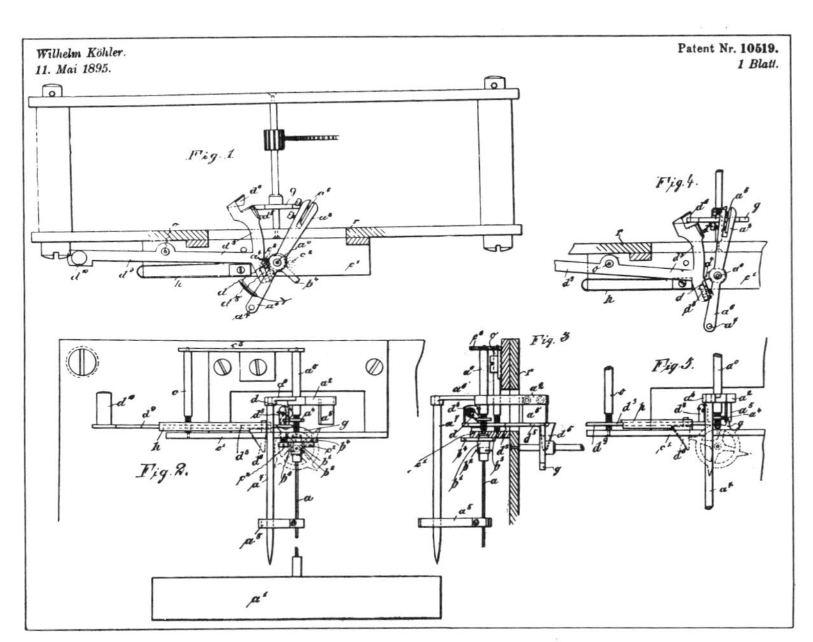

14 German patent No. 10519 issued to Wilhelm Kohler of Furth (Bavaria), Germany, on 11 May 1895 for a detent escapement for full-striking (torsion pendulum) year clock.

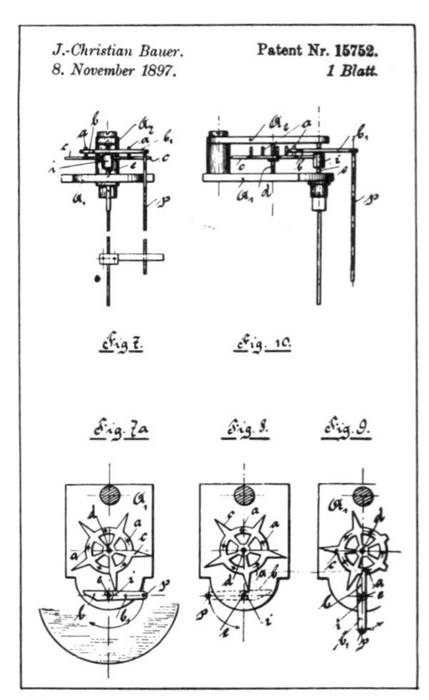

15 Swiss patent No. 15752 issued to J. Christian Bauer of Furth (Bavaria), Germany, on 8 November 1897 for a duplex escapement for (torsion pendulum) clock.

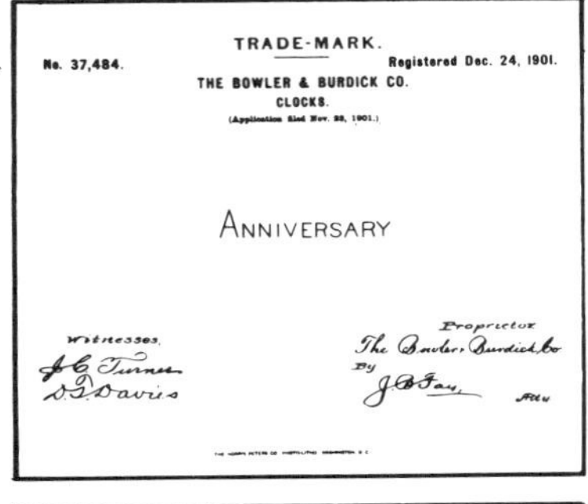

TRADE-MARK.

No. 37,484. Registered Dec. 24, 1901.

THE BOWLER & BURDICK CO.

CLOCKS.

ANNIVERSARY

Witnesses: Proprietor:

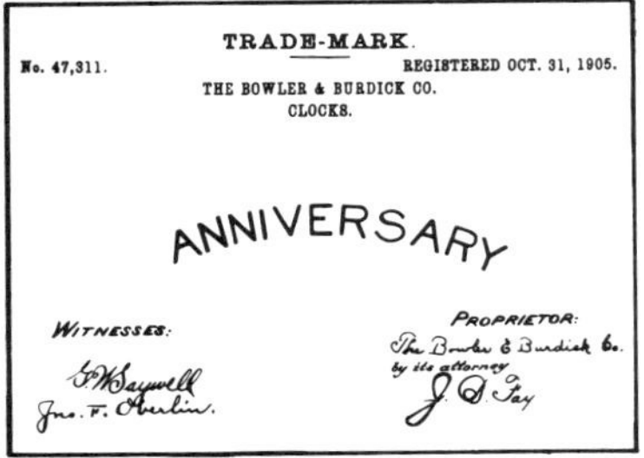

TRADE-MARK.

No. 47,311. REGISTERED OCT. 31, 1905.

THE BOWLER & BURDICK CO.

CLOCKS.

ANNIVERSARY

WITNESSES: PROPRIETOR:

16 U.S. Patent No. 37484 issued to Bowler & Burdick of Cleveland, Ohio, on 24 December 1901 for the trade mark "Anniversary."

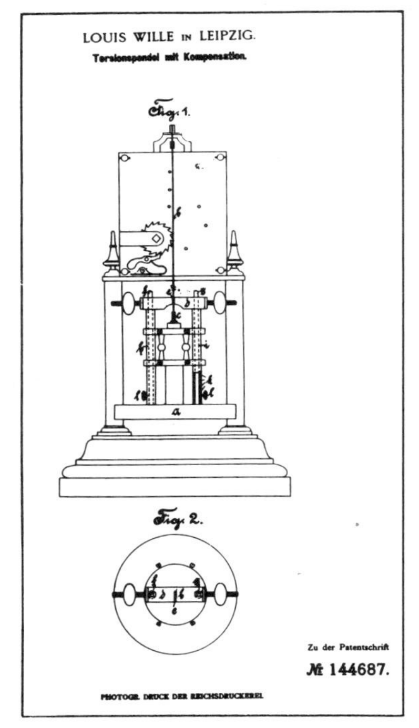

17 German patent No. 144687 issued to Louis Wille of Leipzig, Germany, on 2 April 1902 for a temperature compensating pendulum.

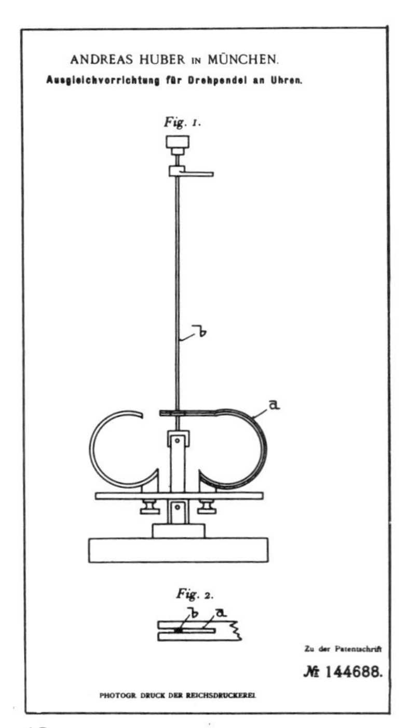

18 German patent No. 144688 issued to Andreas Huber of Munich, Germany, on 5 April 1902 for a temperature compensating pendulum. (U.S. patent No. 751686 was issued to H. Sattler of Munich, Germany, on 9 February 1904 for the same pendulum.)

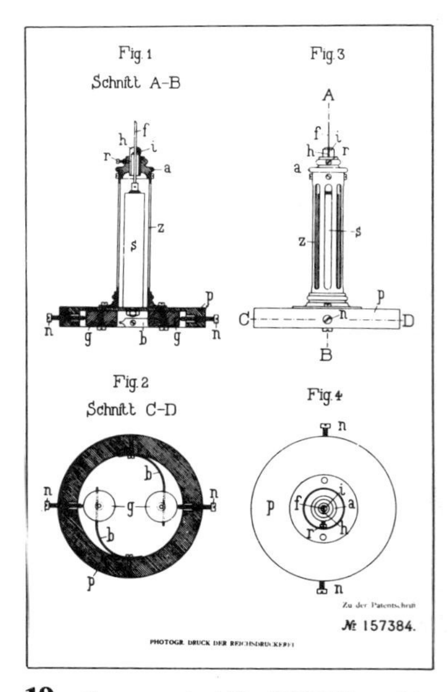

19 German patent No. 157384 issued to Gustav Becker of Freiburg (Silesia), Germany, on 1 December 1903 for a temperature compensating pendulum.

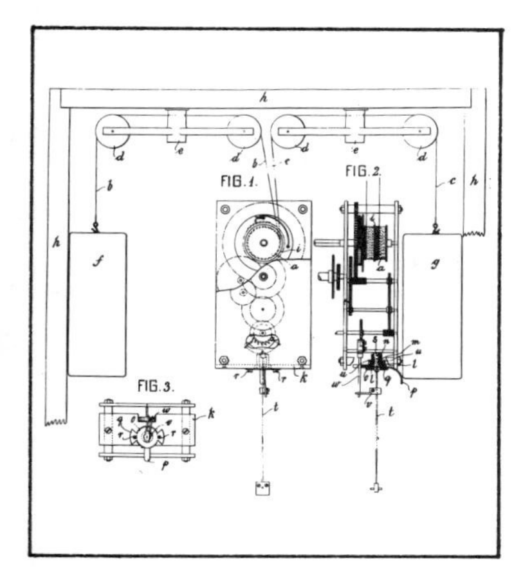

20 English patent No. 7059 issued to John Baptiste Rombach of London, England, on 28 April 1904 for a weight-driven torsion pendulum clock.

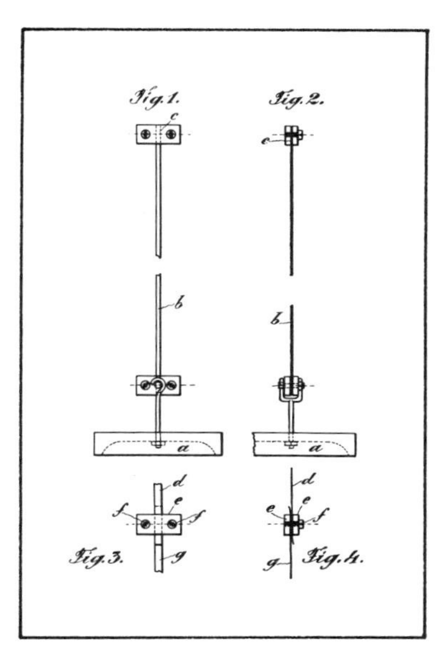

21 Swiss patent No. 32895 issued to Charles Edouard Guillaume of Sevres, France, on 8 December 1904 for a temperature compensating alloy (later given the trade name *Elenvar*).

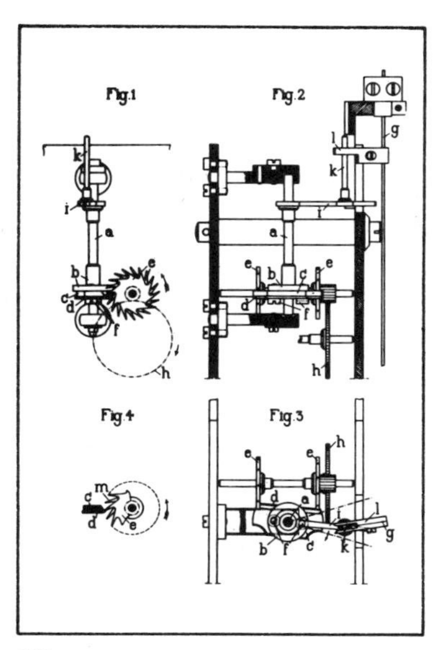

22 German patent No. 183392 issued to Gustav Becker of Freiburg (Silesia), Germany, on 4 April 1907 for a torsion pendulum clock escapement.

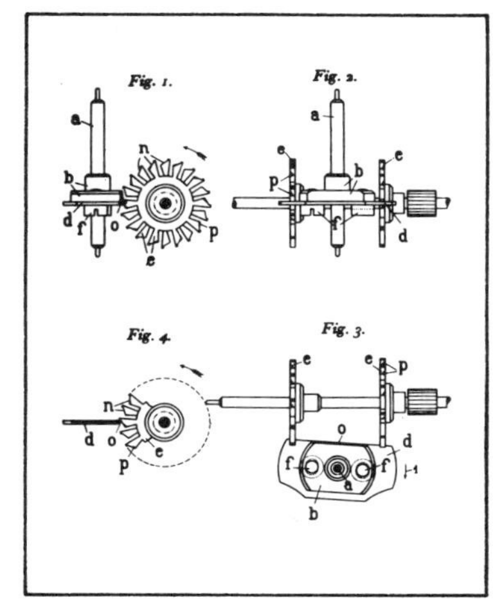

23 German patent No. 188952 issued to Gustav Becker of Freiburg (Silesia), Germany, on 16 September 1907 for a torsion pendulum clock escapement.

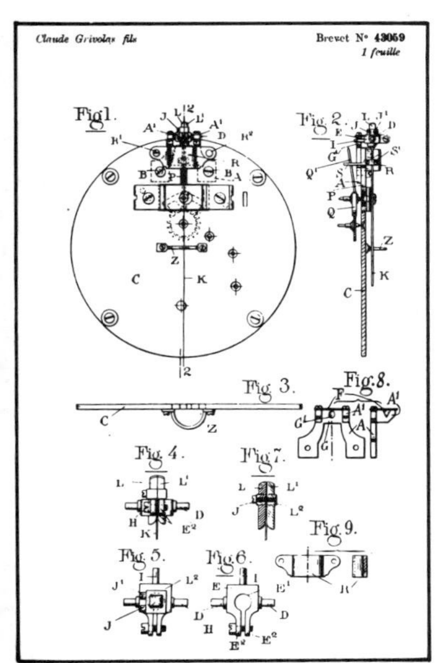

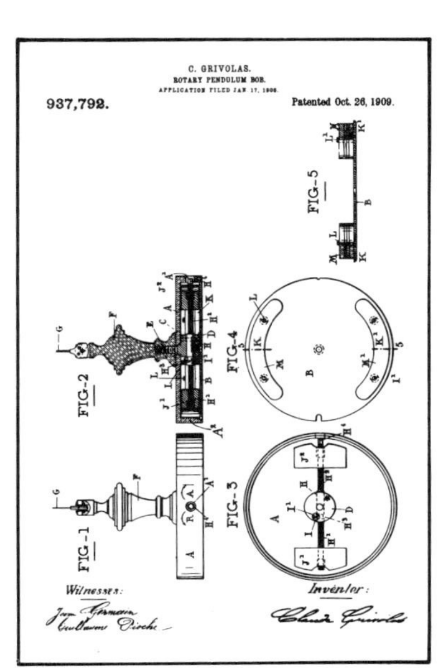

24 Swiss patent No. 43059 (3 January 1908) and U.S. patent No. 937792 (26 October 1909) issued to Claude Grivolas Fils of St. Cloud, France, for a torsion pendulum clock.

25 German patent No. 244910 issued to Louis Wille of Schaffhausen, Switzerland, on 19 March 1912 for a shock absorbing device to minimize the effect of a heavy pendulum on regulation.

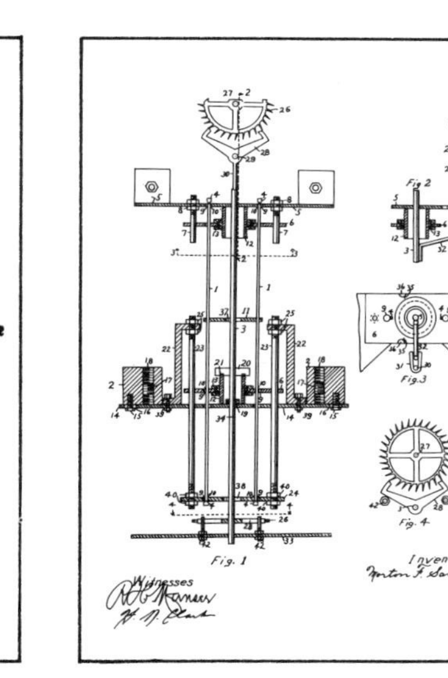

26 U.S. patent No. 1409242 issued to Morton F. Sanborn of Pleasantville, New York, on 14 March 1922 for an oscillating, rotary (torsion) pendulum with improvements over other types.

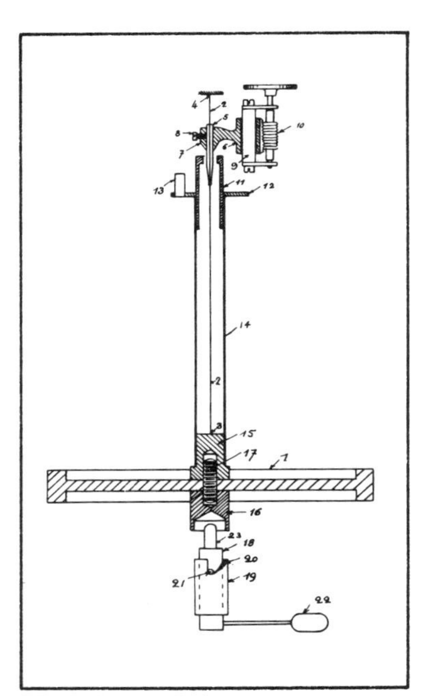

27 English patent No. 356216 issued to Jean Leon Reutter of Paris, France, on 4 September 1931 for a regulating, stabilising and locking device for a torsion pendulum.

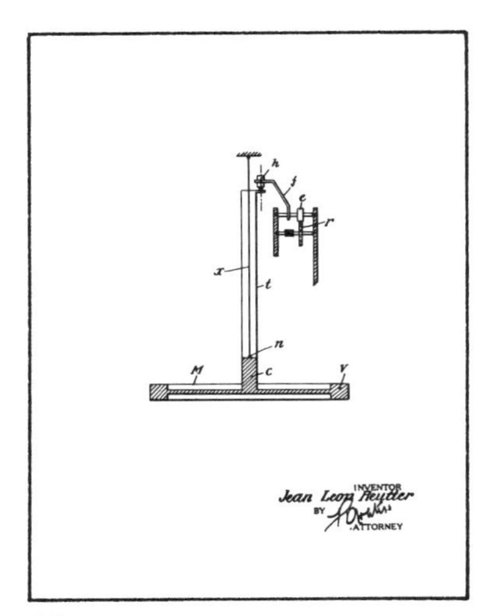

28 U.S. patent No. 1904169 issued to Jean Leon Reutter of Paris, France, on 18 April 1933 for a torsion pendulum that receives its impulses direct from the escapement (rather than through the suspension spring).

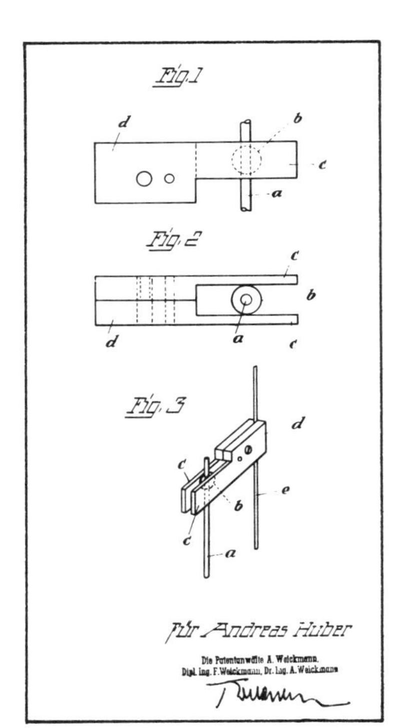

29 German patent No. 1611236 issued to Andreas Huber of Munich, Germany, on 15 June 1950 for a frictionless contact between anchor pin and suspension fork of a torsion pendulum clock.

30 German patent No. 1642179 issued to Walter Petersen Uhrenfabrik of Schwenningen/Neckar, Germany, on 10 June 1952 for a torsion pendulum with miniature figures of men and women on it.

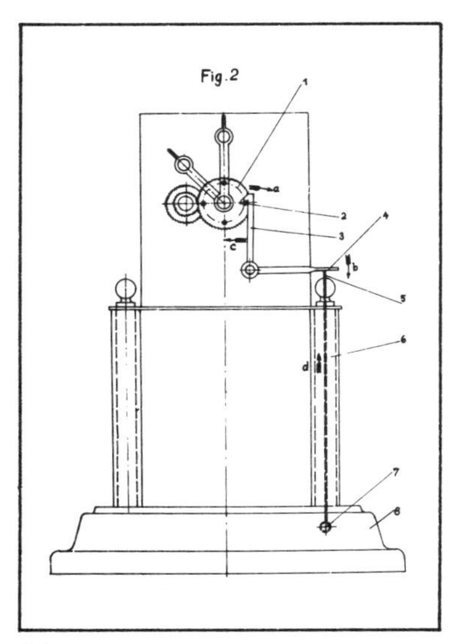

31 German patents No. 1649500 issued to Konrad Mauch of Schwenningen/Neckar, Germany, on 5 January 1953 for a musical box in the base of a torsion pendulum clock to play every quarter hour.

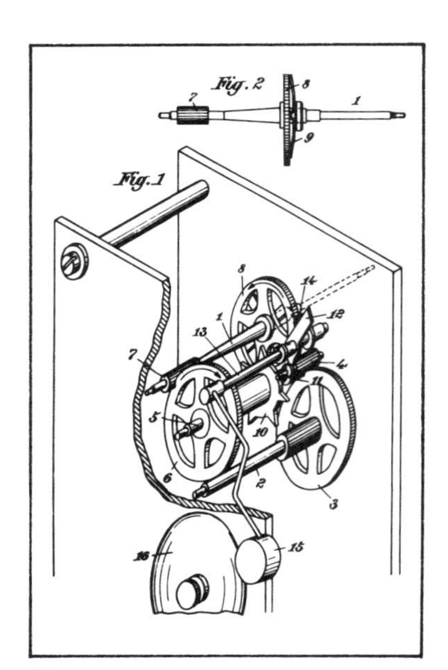

32 German patent No. 1656460 issued to Aug. Schatz & Sohne of Triberg (Black Forest), Germany, on 28 May 1953 for an hour striking attachment for a torsion pendulum clock.

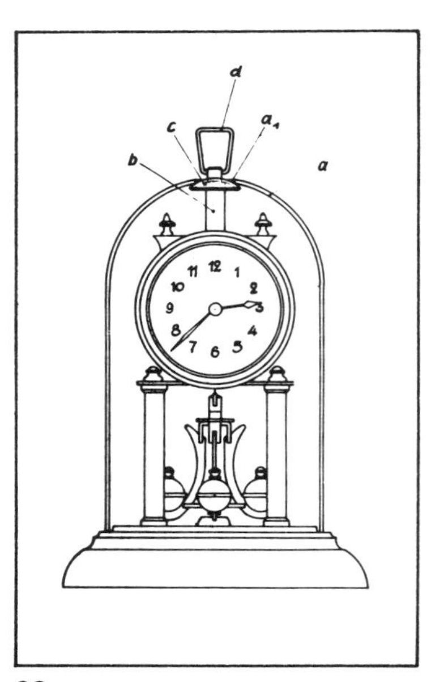

33 German patent No. 1660112 issued to Aug. Schatz & Sohne of Triberg (Black Forest), Germany, on 30 July 1953 for a torsion pendulum clock case design.

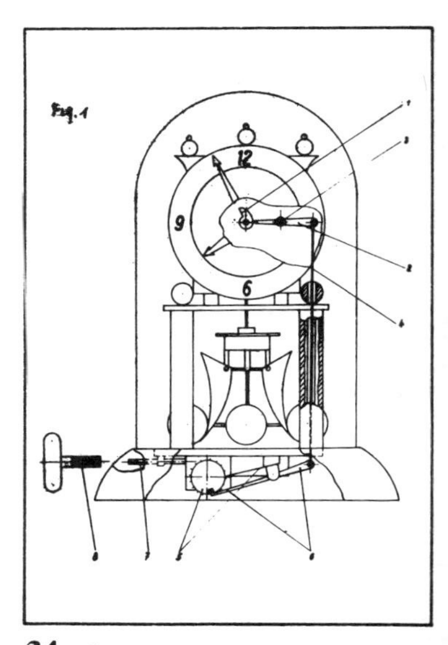

34 German patent No. 1661680 issued to Badische Uhrenfabrik of Furtwangen (Black Forest), Germany, on 20 August 1953 for a musical box in the base of a torsion pendulum clock.

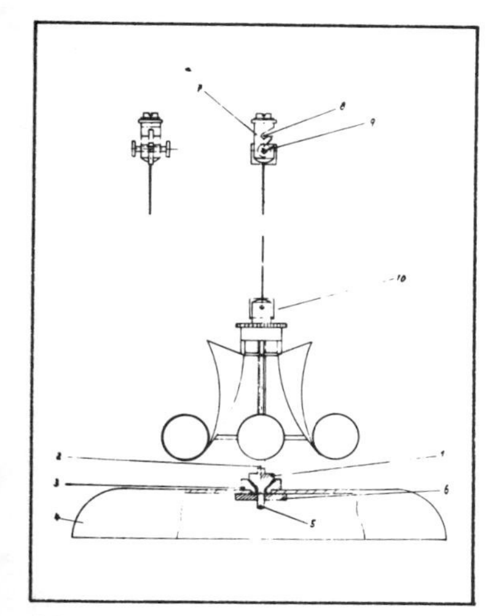

35 German patent No. 1668079 issued to Badische Uhrenfabrik of Furtwangen (Black Forest), Germany, on 3 December 1953 for a torsion pendulum locking device.

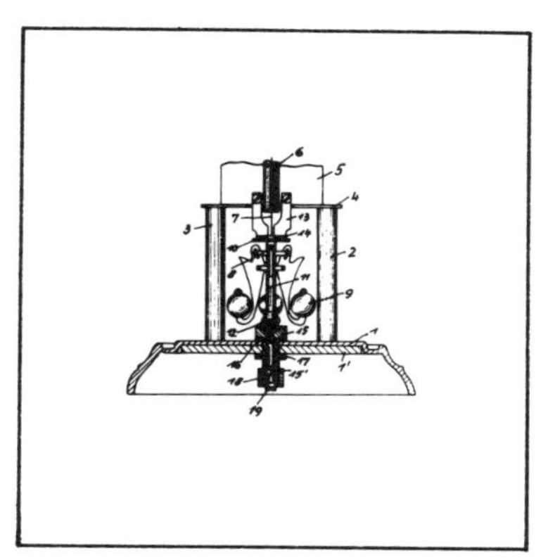

36 German patent No. 1667839 issued to Kern & Sohne Jahresuhrenfabrik of VS-Marbach (Baden), Germany, on 3 December 1953 for a (torsion) pendulum locking device.

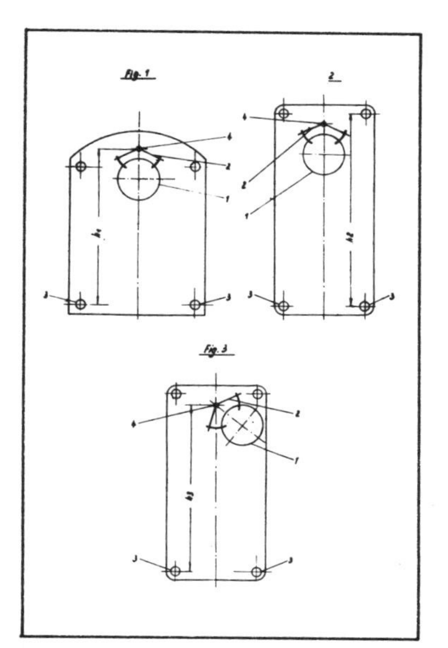

37 German patent No. 1669760 issued to Badische Uhrenfabrik of Furtwangen (Black Forest), Germany, on 7 January 1954 for a rearrangement of a torsion pendulum clock gear train to make narrower movement plates possible.

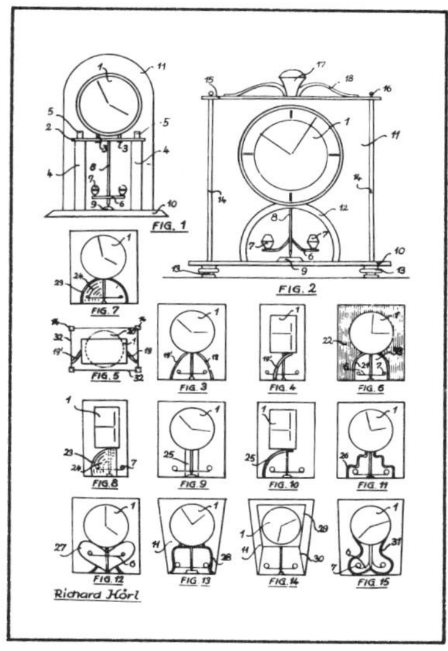

38 German patent No. 1669855 issued to Richard Horl of Augsburg (Goggingen), Germany, on 7 January 1954 for (torsion pendulum) year clock case designs.

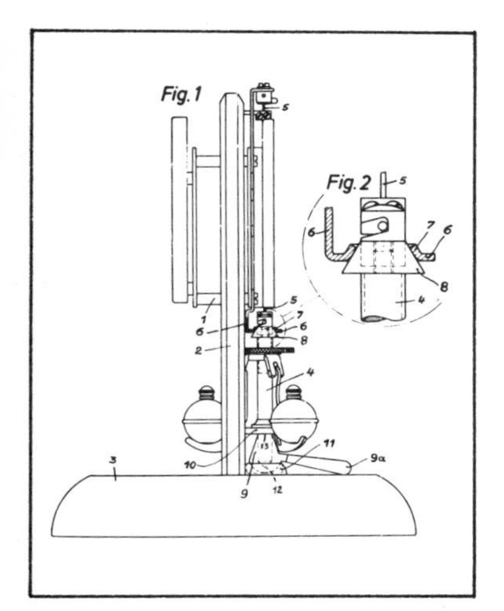

39 German patent No. 1672860 issued to Aug. Schatz & Sohne of Triberg (Black Forest), Germany, on 4 March 1954 for a torsion pendulum locking device.

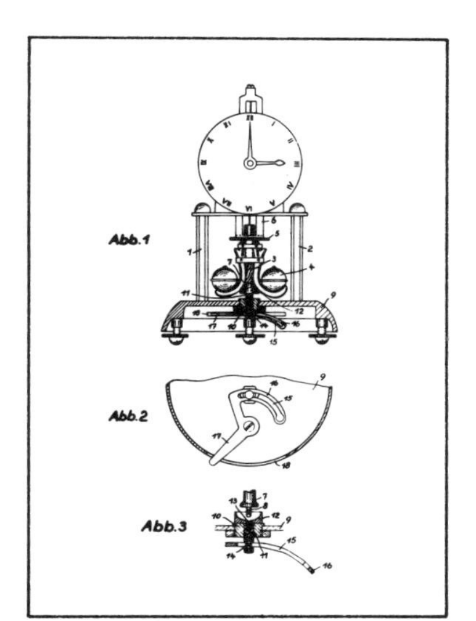

40 German patent No. 1674697 issued to Franz Hermle & Shone Uhrenfabrik of Gosheim (Wurttemberg), Germany, on 8 April 1954 for a torsion pendulum locking device.

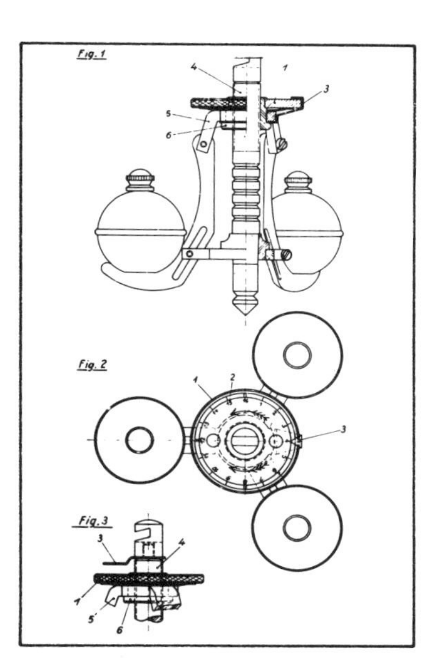

41 German patent No. 1678536 issued to Aug. Schatz & Sohne of Triberg (Black Forest), Germany, on 24 June 1954 for a device on a torsion pendulum that will gauge the amount of regulation.

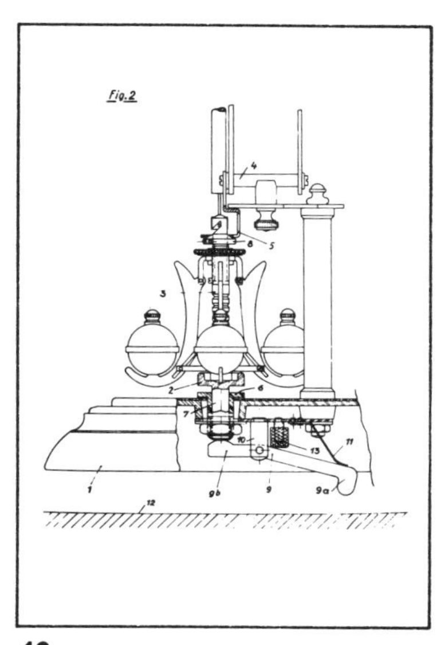

42 German patent No. 1678537 issued to Aug. Schatz & Sohne of Triberg (Black Forest), Germany, on 24 June 1954 for a device that locked the torsion pendulum when the clock was picked up.

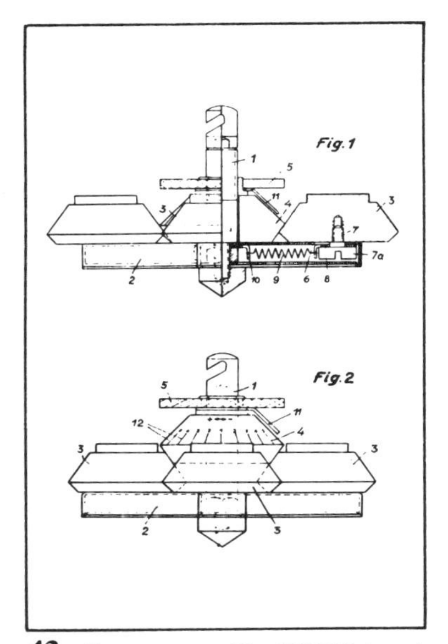

43 German patent No. 1685000 issued to Aug. Schatz & Sohne of Triberg (Black Forest), Germany, on 14 October 1954 for a torsion pendulum design.

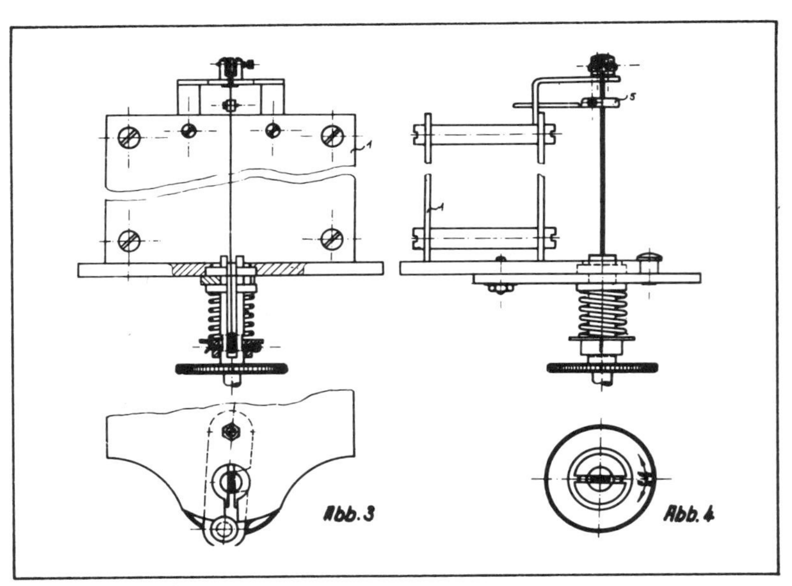

44 German patent No. 1692297 issued to Kieninger & Obergfell of St. Georgen (Black Forest), Germany, on 27 January 1955 for a torsion pendulum locking device.

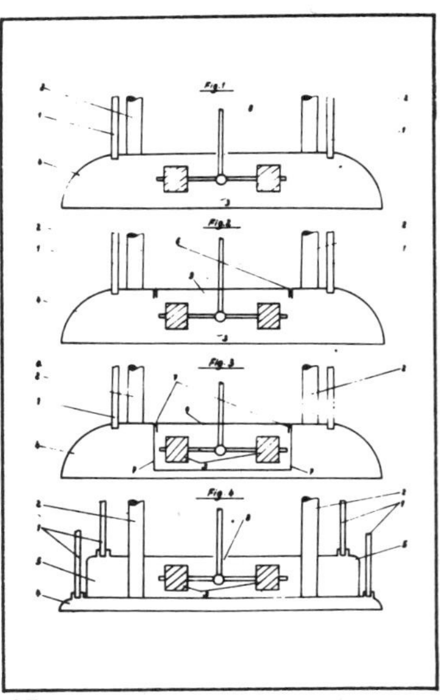

45 German patent No. 1695279 issued to Badische Uhrenfabrik of Furtwangen (Black Forest), Germany, on 24 March 1955 for a torsion pendulum located in the base of a clock where it cannot be seen.

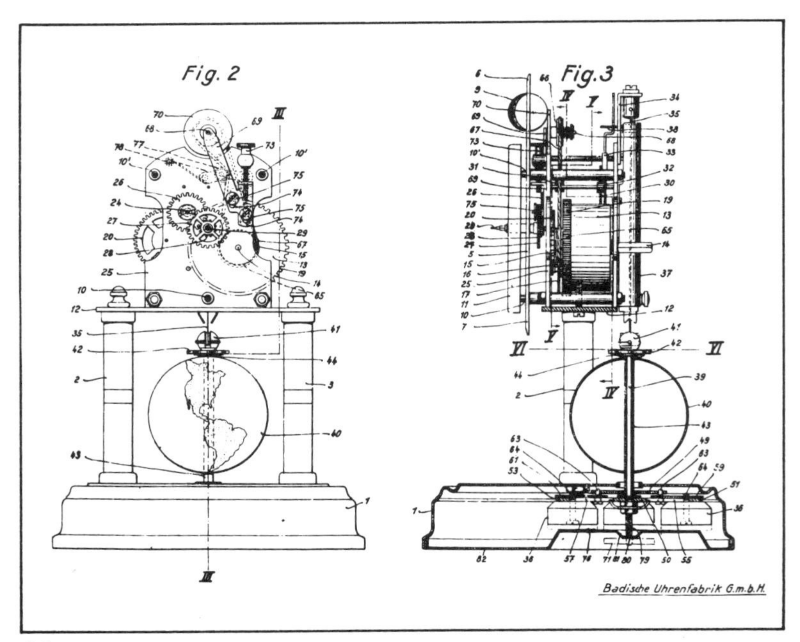

46 German patent No. 1698596 issued to Badische Uhrenfabrik of Furtwangen (Black Forest), Germany, on 18 May 1955 for a torsion pendulum year clock.

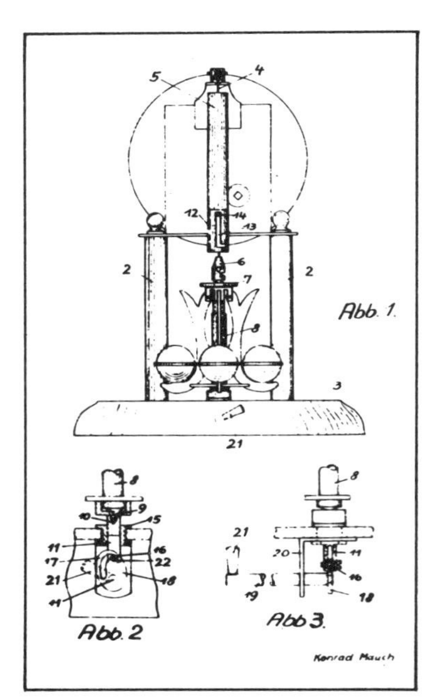

47 German patent No. 1697699 issued to Konrad Mauch Uhrenfabrik of Schwenningen/Neckar, Germany, on 5 May 1955 for a torsion pendulum locking device.

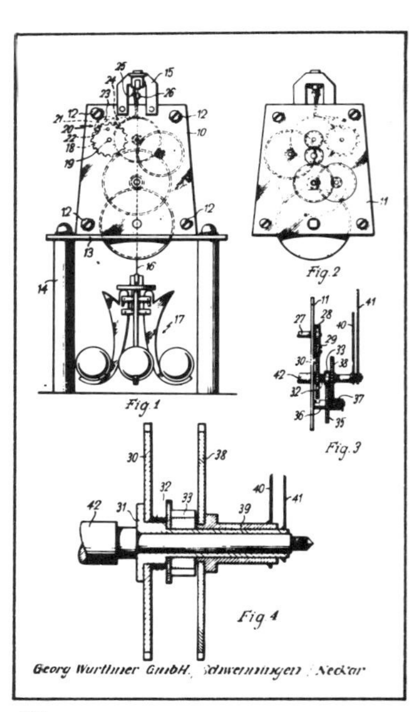

48 German patent No. 1705531 issued to Georg Wurthner of Schwenningen/Neckar, Germany, on 25 August 1955 for a motion train in a torsion pendulum clock that eliminates the need for a clutch (hand tension).

SECTION 5 The Problems of Temperature Change and Timekeeping with 400-Day Clocks

Despite repeated efforts and minor changes over the years, the clocks could not be regulated to keep accurate time. When the temperature rose a few degrees, they lost time; when the temperature went down, they gained.

The first person who attempted to do something about this problem was J.J. Meister[1] of Paris, who was issued Swiss patent No. 5172 in June 1892 for a temperature compensating pendulum, which had a bi-metal split ring that expanded and contracted with temperature changes and looked like a big watch balance wheel with regulating screws. Another pendulum with bi-metal split ring and two large outer weights has been attributed to Phil. Hauck of Munich. As this pendulum can be found on the clocks of different manufacturers[2], it was apparently sold to any manufacturer who would purchase it.

In 1902, Louis Wille[3] of Lepzig, Germany, entered the competition. He was issued German patent No. 144687 for a temperature compensating pendulum that was promoted in the June 1904 issue of *The Keystone* in an advertisement for clocks imported by Geo. Kuehl & Co., Chicago. This pendulum was built with a slotted arm through which the suspension spring was threaded before it was attached to the pendulum. Any change in temperature would cause the arm to bend, making the suspension spring shorter (and the pendulum turn faster), or longer (slower), thus compensating for the effect of temperature change on the pendulum. This pendulum has been seen in the clocks of only one manufacturer.

Another type, with crossed mercury tubes[4], is known to have been made by Gustav Becker, but the temperature compensating pendulum most frequently seen is one patented in 1902 by Andreas Huber of Munich[5]. It is often referred to as the "twin loop" pendulum. The loops are bi-metal and, as with the Louis Wille arm, the suspension spring is threaded through one loop before the pendulum is attached. With temperature change, the loop will expand or contract, shortening or lengthening the spring. The second loop is a dummy, on the clock only for appearance and balance.

Andreas Huber was careful to protect his interest in the patent and insisted on having the manufacturers who used it stamp the numbers of his German and U.S. patents on the movement back plates. This is the reason that the stamping "D.R.P. No. 144688/U.S.P. No. 751686" appears on the back plates of some of the clocks made from about 1905 to 1908 by Jahresuhrenfabrik[6], Badische Uhrenfabrik[7] and Phillipp Haas[8]. A number of clocks carrying these patent numbers do not have the temperature compensating pendulum; possibly some importers may have elected not to pay its additional cost.

None of the temperature compensating pendulums remained on the market long, probably because of their added cost and the fact that they did little to improve the clock's poor timekeeping.

From the first, the clock's torsion spring, from which the pendulum was suspended, has been made of steel and this

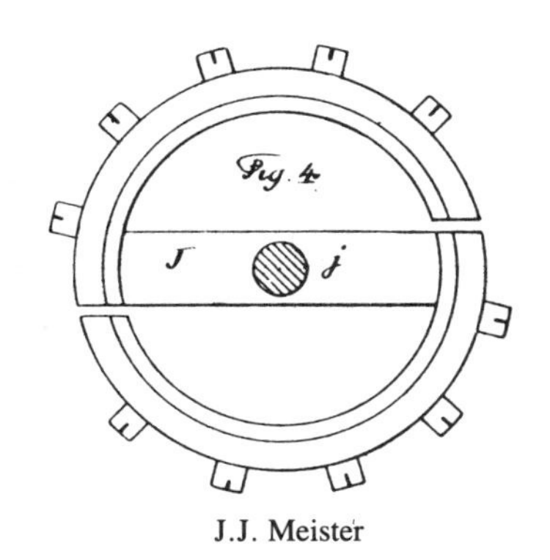

J.J. Meister

Phil. Hauck

Gustav Becker

Gustav Becker

TEMPERATURE COMPENSATING PENDULUMS

[1]Section 4, Patent 13
[2]Section 6, Clocks 18, 66
[3]Section 4, Patent 17
[4]Section 6, Clock 17
[5]Section 4, Patent 18
[6]Section 9, Plate 1047
[7]Section 9, Plate 1041
[8]Section 9, Plate 1043

continued to be the material from which springs were made by all manufacturers prior to World War I. When minor production was resumed between the two wars, at least one manufacturer, Kieninger & Obergfell, used suspension springs made of bronze. While bronze springs are also affected by changes in temperature, they are much easier to fabricate and therefore less costly to make than steel. In 1949, the year the clocks began their big wave of importation into the United States, bronze was the material used for suspension springs by all manufacturers.

Mr. Charles O. Terwilliger became interested in improving the timekeeping of 400-Day Clocks about 1947. Although temperature compensating alloys were known, no German manufacturers were using such alloys in their suspension springs. The 1910 catalog of C. Grivolas[9], the only French manufacturer of "pendules 400 jours," promoted the fact that their clocks used a suspension spring made of a new metal "compensator" called *Invar* that made "marvelous precision regulation" possible. (An illustration in the catalog shows a strip of *Invar* clamped to the lower third of the spring.) Obviously , the use of *Invar* was not successful in solving the temperature problem.

The problem was presented to the U.S. National Bureau of Standards in Washington, D.C. and they pointed out that the timekeeping problem was not a matter of expansion and contraction of the suspension spring caused by temperature change, but rather had to do with the spring's change in its *modulus of elasticity*, a stiffening or weakening of the spring caused by temperature changes. In fact, variations in the elasticity of the suspension spring have, for each degree of temperature change, twenty times the effect of the expansion and contraction of the spring on a 400-Day Clock's timekeeping. The Bureau suggested the use of patented nickel-steel-chromium alloys *Elenvar or Ni Span "C"*, which have a nearly constant modulous of elasticity with respect to change in temperature. Either of these alloys, if they could be fabricated into 400-Day Clock suspension springs, would make compensation for temperature changes possible. (In fact, if heat treated before fabrication, these alloys tend to actually become stiffer as the temperature increases making it possible to compensate for the expansion of the pendulum.)

Interestingly, the recommendations of the Bureau of Standards had returned the search for improved timekeeping to consideration of an alloy patented by Charles Eduard Guillaume[10] of Sevres, France in 1904. It remains an unanswered question as to why this alloy, trade marked "Elenvar" was not used by 400-Day Clock manufacturers prior to World War I instead of steel, or later, instead of bronze.

Based on both availability and ease of fabrication, Ni Span "C" became the alloy that Mr. Terwilliger used during a year's experimentation with springs and pendulums in clocks being imported at the time from Kundo, Schatz, Kern and others. As a result of these experiments, *Horolovar Temperature Compensating 400-Day Clock Suspension Springs*[11] were introduced in 1951. By 1953, there were twelve different thicknesses of Horolovar springs available, and as the factories made new clocks or new models available, new springs, if necessary, were drawn and rolled to fit. Today the number of Horolovar springs totals twenty-four, making a spring available for every 400-Day Clock—and clocks with a Horolovar spring can be regulated to keep time.

It did not take long for the news of the springs to reach Germany. Schatz purchased 10,000 for their standard clock and later, when clocks made by Edgar Henn and J. Link first appeared in the U.S., they were both promoted by the importer, Forestville Clock Company, as being excellent timekeepers because they were equipped with *Horolovar Temperature Compensating Suspension Springs*. However, German import duty on the springs made their cost too high, so the clock manufacturers found a less expensive domestic alloy that was adequate.

Horolovar Temperature Compensating Suspension Springs have played an important part in the history of the 400-Day Clock by making possible the first improvement in the timekeeping of these clocks since the torsion pendulum was patented.

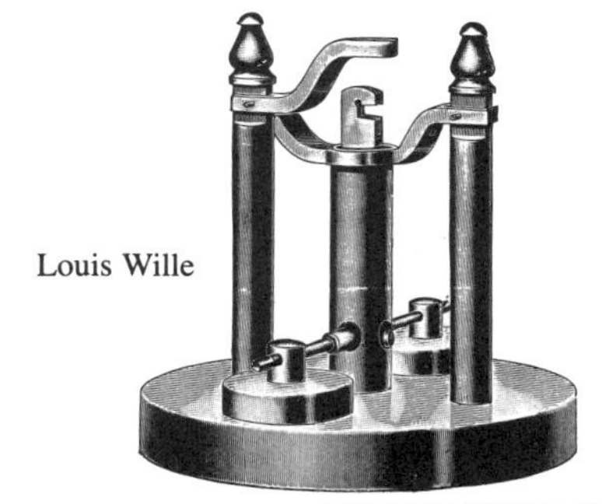

Louis Wille

Andreas Huber

TEMPERATURE COMPENSATING PENDULUMS

[9]Section 9, Plate 1471A
[10]Section 4, Patent 21
[11]Section 17

SECTION 6 Some Outstanding Pre-World War I 400-Day Clocks

The number of different models of key-wind torsion pendulum 400-Day Clocks that have been made during the past one hundred years totals thousands. Even if limited to the Pre-World War I years in this section (1897-1914), the number of different examples seems endless. Just when you think that you have seen them all, another model turns up.

In this section are several of the very early models made by factories whose identity has not yet been verified, some early wall models, twenty-four models illustrated in Jahresuhrenfabrik 1905 catalog, probably the most diversified collection of the time, and a miscellaneous group of other 400-Day Clocks in no special order. (Their approximate dates are with their movement back plates in Section 9.)

The most popular type of case is the 5½" x 10½" glass dome that covered what became known as the "standard" size clock. First used with clocks as early as 1881, the same size dome is still used today. Although similar in size, these glass domes covered models that differ in appearance with changes in the size and shape of dials, arches, finials and pendulums, and sometimes the style of hands. Certain distinguishing features became recognized as being allied with certain manufacturers.

Popular "fashions" in clock cases—wood cases, brass cases, square and oval four-glass cases, Art Nouveau and Art Deco cases—were followed at various times as a variation from glass domes.

The most attractive 400-Day Clock cases were those made in France for German movements (Nos. 38-45; 68). These were the clocks sold in better jewelry stores.

1—See Section 9, Plate 1727
A. Willmann Co.?

2—See Section 9, Plate 1731
Gustav Becker?

3—See Section 9, Plate 1633
A. Willmann Co.?

4—See Section 9, Plate 1732
A. Willmann Co.?

5—See Section 9, Plate 1150
Unknown

6—See Section 9, Plate 1475
Jahresuhrenfabrik

7—See Section 9, Plate 1475
Jahresuhrenfabrik

8—See Section 9, Plate 1680
Thomas Haller

9—See Section 9, Plate 1475
Jahresuhrenfabrik

10—See Section 9, Plate 1471
Jahresuhrenfabrik

11—See Section 9, Plate 1439
Jahresuhrenfabrik

12—See Section 9, Plate 1055
Jahresuhrenfabrik

13—See Section 9, Plate 1425
Phillipp Haas

14—See Section 9, Plate 1619
Unknown

15—See Section 9, Plate 1189
Gustav Becker

16—See Section 9, Plate 1207A
Gustav Becker

17—See Section 9, Plate 1207
Gustav Becker

18—See Section 9, Plate 1579
Kienzle Clock Factories

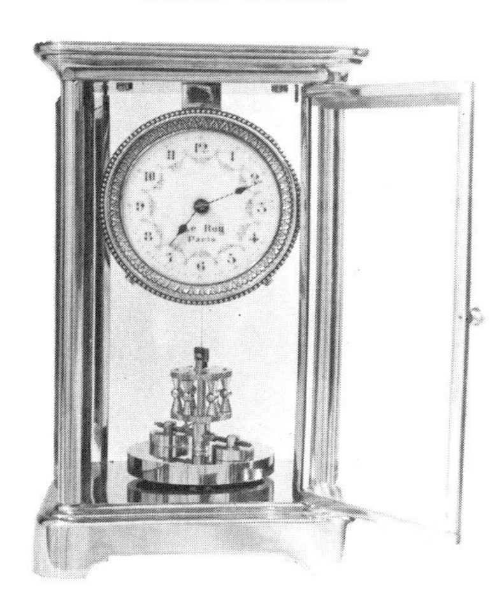

19—See Section 9, Plate 1631
Jahresuhrenfabrik

20—See Section 9, Plate 1617
Jahresuhrenfabrik

21—See Section 9, Plate 1602
Kienzle Clock Factories

22—Clock No. 112 in 1905
Jahresuhrenfabrik Catalog

23—Clock No. 109 in 1905
Jahresuhrenfabrik Catalog

24—Clock No. 111 in 1905
Jahresuhrenfabrik Catalog

25—Clock No. 247 in 1905
Jahresuhrenfabrik Catalog

26—Clock No. 235 in 1905
Jahresuhrenfabrik Catalog

27—Clock No. 272 in 1905
Jahresuhrenfabrik Catalog

28—Clock No. 231 in 1905
Jahresuhrenfabrik Catalog

29—Clock No. 281 in 1905
Jahresuhrenfabrik Catalog

30—Clock No. 217 in 1905
Jahresuhrenfabrik Catalog

31—Clock No. 219 in 1905
Jahresuhrenfabrik Catalog

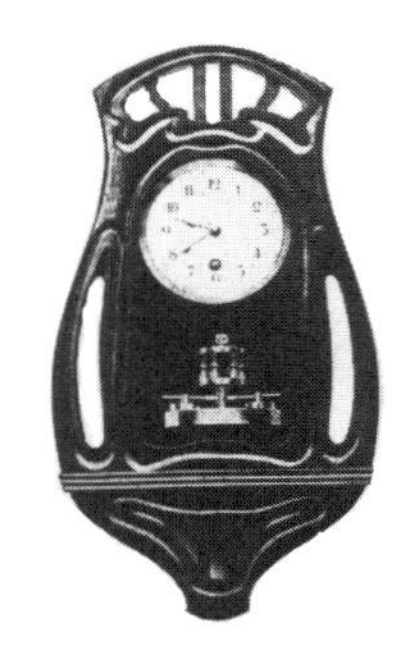

32—Clock No. 221 in 1905
Jahresuhrenfabrik Catalog

33—Clock No. 221a in 1905
Jahresuhrenfabrik Catalog

34—Clock No. 282 in 1905 Jahresuhrenfabrik Catalog

35—Clock No. 218 in 1905 Jahresuhrenfabrik Catalog

36—Clock No. 208 in 1905 Jahresuhrenfabrik Catalog

37—Clock No. 241 in 1905 Jahresuhrenfabrik Catalog

38—Clock No. 204 in 1905 Jahresuhrenfabrik Catalog

39—Clock No. 257 in 1905 Jahresuhrenfabrik Catalog

40—Clock No. 245 in 1905 Jahresuhrenfabrik Catalog

41—Clock No. 258 in 1905 Jahresuhrenfabrik Catalog

42—Clock No. 276a in 1905 Jahresuhrenfabrik Catalog

43—Clock No. 240 in 1905 Jahresuhrenfabrik Catalog

44—Clock No. 214 in 1905 Jahresuhrenfabrik Catalog

45—Clock No. 213 in 1905 Jahresuhrenfabrik Catalog

46—See Section 9, Plate 1439
Jahresuhrenfabrik

47—See Section 9, Plate 1595
Jahresuhrenfabrik

48—See Section 9, Plate 1439
Jahresuhrenfabrik

49—See Section 9, Plate 1623
Unknown

50—See Section 9, Plate 1171
Gebr. Junghans

51—See Section 9, Plate 1049
Jahresuhrenfabrik

52—See Section 9, Plate 1163
Kienzle Clock Factories

53—See Section 9, Plate 1056
Jahresuhrenfabrik

54—See Section 9, Plate 1191
Gustav Becker

55—See Section 9, Plate 1617
Jahresuhrenfabrik

56—See Section 9, Plate 1423
Kienzle Clock Factories

57—See Section 9, Plate 1437
Unknown

58—See Section 9, Plate 1419
Phillipp Haas

59—See Section 9, Plate 1051A
Jahresuhrenfabrik

60—See Section 9, Plate 1651A
Jahresuhrenfabrik

61—See Section 9, Plate 1423
Kienzle Clock Factories

62—See Section 9, Plate 1471A
Grivolas

63—See Section 9, Plate 1651A
Jahresuhrenfabrik

64—See Section 9, Plate 1718
Unknown

65—See Section 9, Plate 1471A
Grivolas

66—See Section 9, Plate 1415
Phillipp Haas

67—See Section 9, Plate 1207A
Gustav Becker

68—See Section 9, Plate 1007A
Jahresuhrenfabrik

69—See Section 9, Plate 1385
Badische Uhrenfabrik

70—See Section 9, Plate 1621
Unknown

71—See Section 9, Plate 1419
Phillipp Haas

72—See Section 9, Plate 1610
Jahresuhrenfabrik

73—See Section 9, Plate 1621
Unknown

74—See Section 9, Plate 1523
Kieninger & Obergfell

75—See Section 9, Plate 1317
Kieninger & Obergfell

76—See Section 9, Plate 1631
Jahresuhrenfabrik

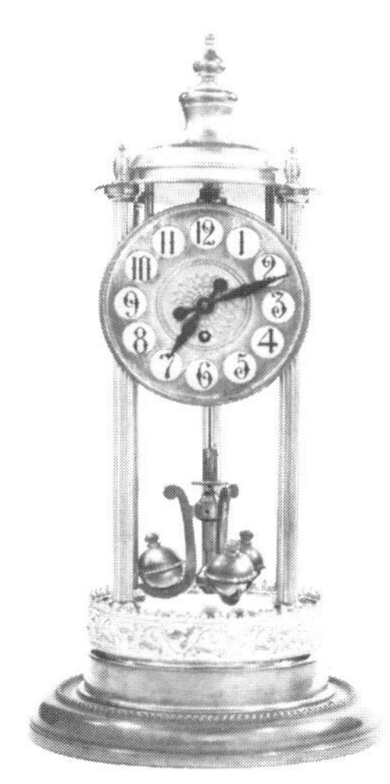

77—See Section 9, Plate 1145
Franz Vosseler

78—See Section 9, Plate 1257
Jahresuhrenfabrik

79—See Section 9, Plate 1257
Jahresuhrenfabrik

80—See Section 9, Plate 1145
Franz Vosseler

81—See Section 9, Plate 1043
Phillipp Haas

82—See Section 9, Plate 1175
Kienzle Clock Factories

83—See Section 9, Plate 1419
Phillipp Haas

84—See Section 9, Plate 1009
Jahresuhrenfabrik

85—See Section 9, Plate 1247
Kienzle Clock Factories

86—See Section 9, Plate 1614
Aug. Schatz & Sohne

87—See Section 9, Plate 1597
Jahresuhrenfabrik

88—See Section 9, Plate 1043
Phillipp Haas

89—See Section 9, Plate 1208A
Franz Vosseler

90—See Section 9, Plate 1504A
Jahresuhrenfabrik

91—See Section 9, Plate 1470E
Unknown

92—See Section 9, Plate 1687
Unknown

SECTION 7 Adjustment of the 400-Day Clock Escapement

We receive more questions about the adjustment and repair of 400-Day Clock escapements than about any other part of the clock. We sincerely hope that this chapter of the Guide will not only help you, but will also save you hours of time in making repairs.

We are fortunate to have Henry B. Fried as co-author of this chapter. Mr. Fried is a Certified Master Clockmaker and Certified Master Watchmaker, is a third generation watchmaker, who organized and taught the first horological class for the New York City Board of Education and continued to direct and teach classes in this field for the Board for thirty years.

Mr. Fried is Technical Director of the American Watchmakers Institute, Horological Editor of *The Jewelers Circular-Keystone,* and an accredited horological consultant-contributor to both the *Random House* and *Merriam-Webster Dictionaries.* His articles have appeared in almost every horological trade journal. He lectures extensively and has appeared on radio and television in connection with horology. He consults for the industry as well as for governmental agencies.

Mr. Fried has received the "Man of the Year Award" from the jewelery industry (WMDJAA). He is an honorary member and past president of The American Watchmakers Institute. He has been president of local, state and national horological trade associations.

A collector of all forms of horologia, Mr. Fried is a founder and Fellow (Silver Star) of The National Association of Watch and Clock Collectors, a Fellow of the British Horological Institute and Fellow of the American Watchmakers Institute.

Mr. Fried is author of numerous text books and articles on watch- and clockmaking, including *The Watch Repairers Manual, Bench Practices for Watch and Clockmakers* and many others. His latest books include *The Repair of Quartz Watches* and *Principles and Repair of Self-Winding Watches.*

Charles Terwilliger

INTRODUCTION

This section is devoted to a detailed study of the two types of escapements usually used in 400-Day Clocks: The *Graham* (dead beat) and the *pin pallet.* Early experimental clocks have been found with *verge* and *cylinder* escapements (See Section 6, Clocks 1-4), but the first successful production of clocks by Jahresuhrenfabrik (See Section 6, Clock 6) had movements with Graham escapements. Standard clocks made by Badische Uhrenfabrik during the early 1900's used pin pallet escapements. All other manufacturers, prior to World War I, used the Graham.

About 1954, Kundo introduced its miniature 400-Day Clock model with pin pallet escapement. It created unexpected problems, so Kundo retooled and offered the same clock with a Graham escapement. Subsequently, Kern, Schatz, Koma, Haller and Nisshindo used pin pallet escapements successfully in their miniature and midget movements.

While the principles of the Graham escapement apply to the pin pallet, the pin pallet is treated separately in this Section

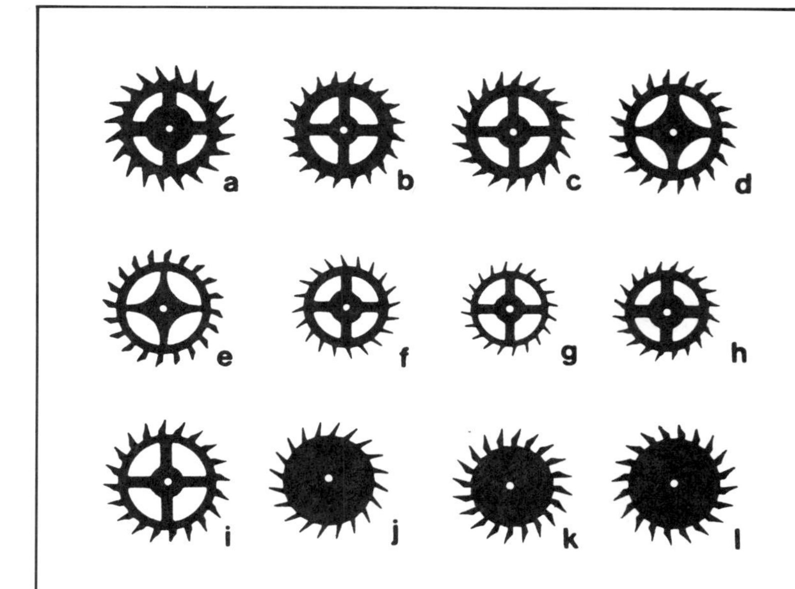

Figure 1 (20 Teeth)

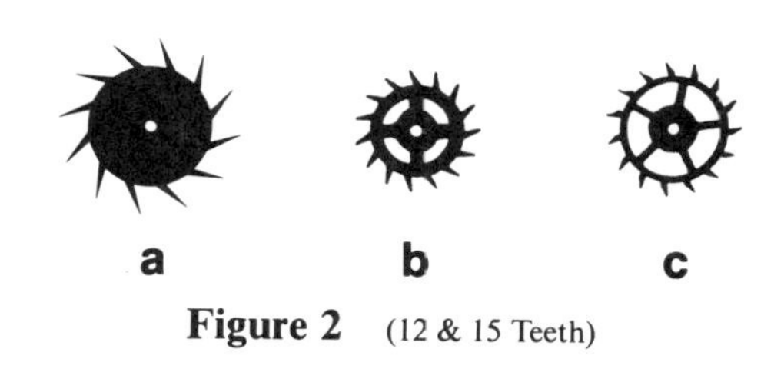

Figure 2 (12 & 15 Teeth)

because it differs in its action and repair. The locking and impulse surfaces of the Graham escapement are on the pallets, while the pin pallet escapement carries the same surfaces on the escape wheel teeth.

Graham (Dead Beat) Escapement

With the Graham escapement, there are only three types of escape wheels, those with 20 teeth (Figure 1) and those with 12 and 15 teeth (Figure 2). Differences in tooth shape are unimportant. Some escape wheels are solid; others are crossed out to lighten them thus minimizing inertia. There are two types of anchors, those with adjustable pallets shown in Figure 3 (drawings approximate) and solid (unadjustable) pallets (Figure 4). The first successful production of 400-Day Clocks, by Jahresuhrenfabrik, used anchors with solid pallets and Aug. Schatz & Sohne continued to use anchors with solid pallets during the firm's one hundred years of clock-making. One advantage of the solid pallet: it can't be

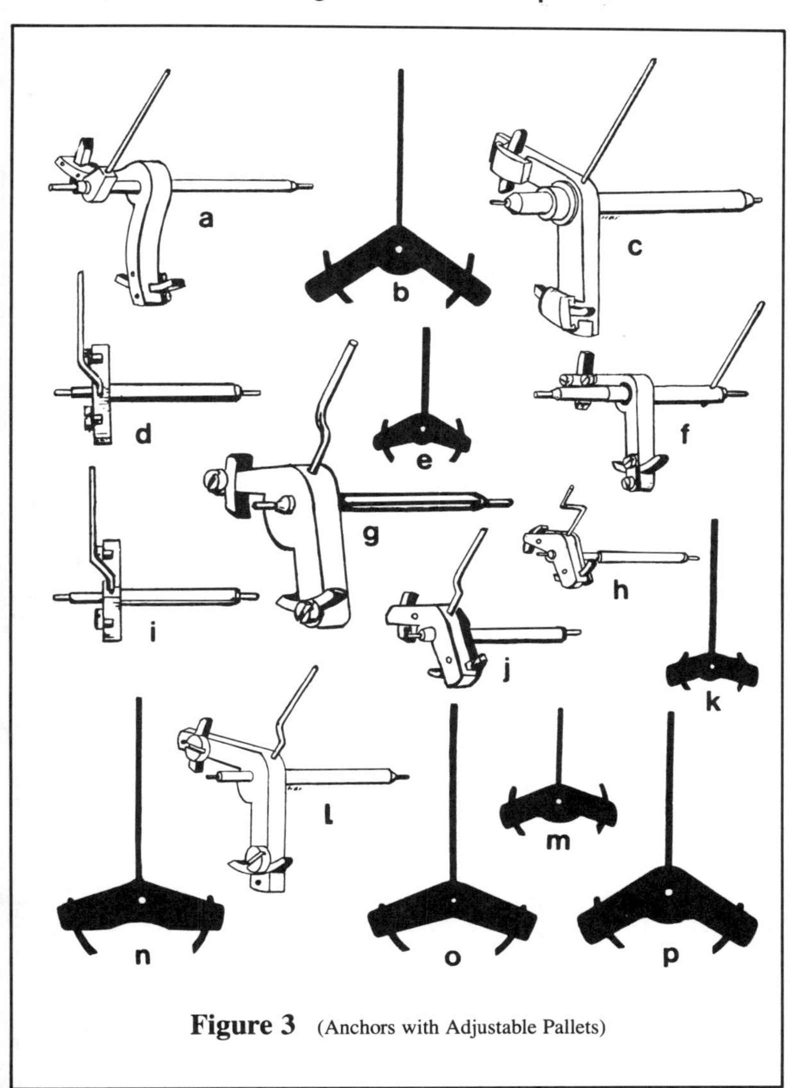

Figure 3 (Anchors with Adjustable Pallets)

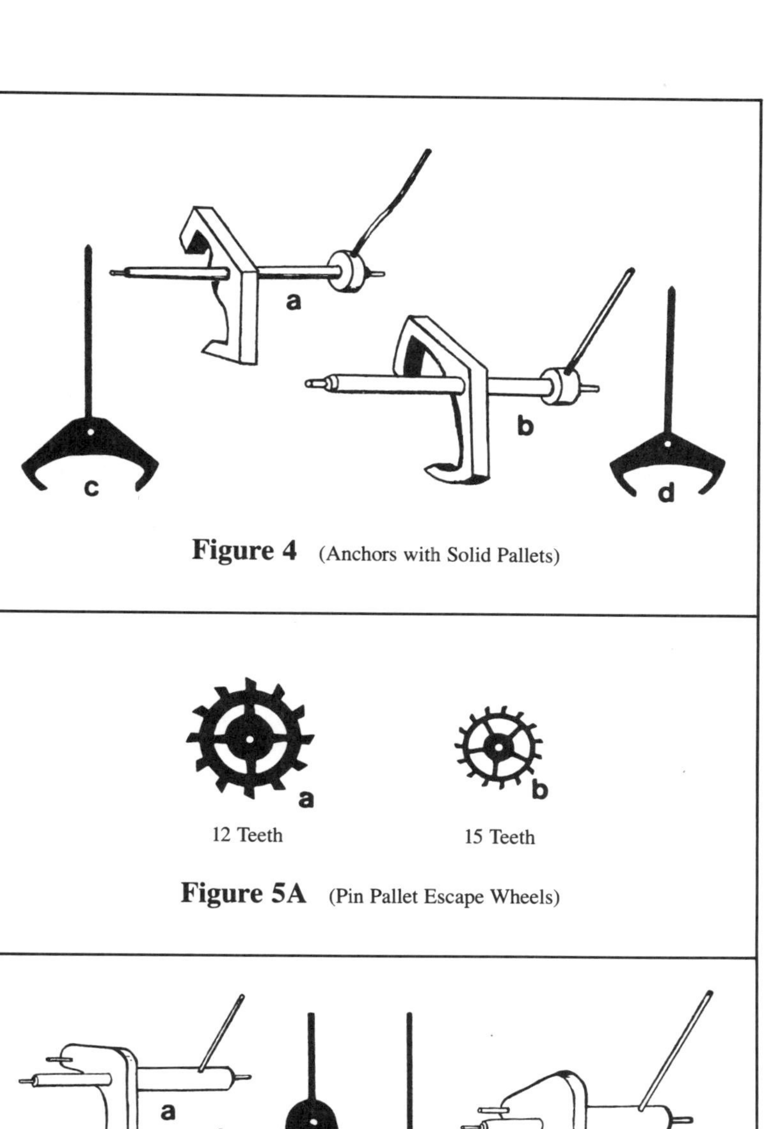

Figure 4 (Anchors with Solid Pallets)

Figure 5A (Pin Pallet Escape Wheels)

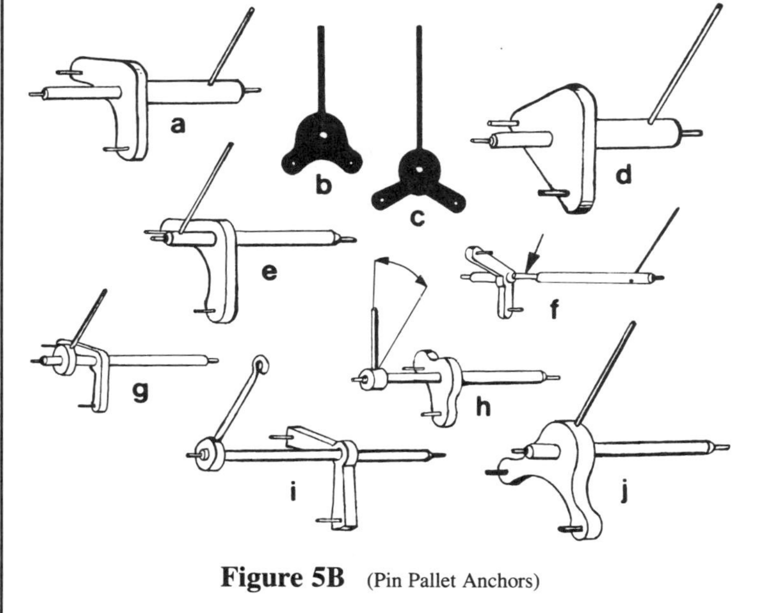

Figure 5B (Pin Pallet Anchors)

adjusted! In an attempt to correct a fault in the running of the clock, the inexperienced repairer will adjust everything! The majority of 400-Day Clock manufacturers who used Graham escapements equipped them with adjustable pallets.

Pin Pallet Escapement

With the pin pallet escapement, there are also two types of escape wheels, those with 12 teeth and 15 teeth (Figure 5a). There is only one form of pin pallet anchor shown in Figure 5b (drawings approximate) although the pins are set farther apart for the 15-tooth escape wheel than for the 12-tooth.

EXPLANATION OF THE GRAHAM (DEAD BEAT) ESCAPEMENT

Figure 6 shows the adjustable pallet escapement viewed from the back plate where the repairer can best observe the the escapement action through the especially made "peep holes" (See Figure 10) which appear in most movements. Ten of the escape wheel's twenty pointed teeth are shown here. The two hard-polished steel pallets, attached to the anchor, are shown in black. Each pallet is held by a clamp screwed to the anchor frame (on the back of the frame in the illustration shown) or by just the head of a screw.

The pallet on the left is called the "receiving" or "entrance" pallet because, in the escapement action, the escape teeth enter into the anchor following contact with it. The pallet on the right is called the "exit" or "discharging" pallet because the teeth leave the anchor following contact with it.

This type of escapement was invented by George Graham and is called *dead-beat* because the escape teeth remain motionless while locked on the pallets' curved surface despite any movement of the pallets. In other types of anchor escapements, there is a recoil, or backward movement of the escape wheel during such action.

A careful study of Figure 6, with the following observations, will aid in an understanding of the escapement action:

This particular escape wheel travels in a clockwise direction. (In clocks where the escape wheels turn in a counterclockwise direction, teeth and pallets are in reverse positions.) The 20 teeth are spaced 18° apart. The front of each tooth has a rake (slant) of 12° back from a radial line. This front slant of the tooth is necessary so that only its tip makes contact with the curved surface of the pallet during what is known as the *locking sequence* of the escapement action. If the tooth were not slanted, the corners of the pallet, especially the exit pallet, would scrape the front surface of the tooth causing a recoil of the wheel which would stop the action.

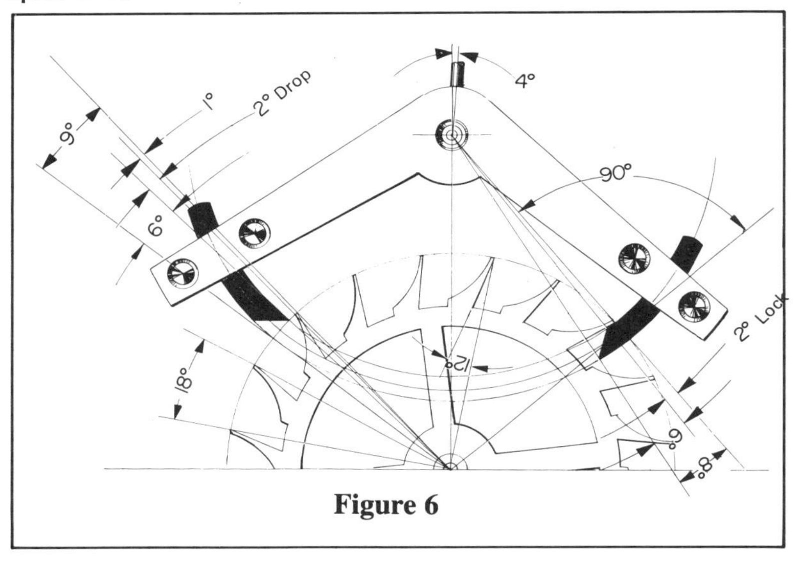

Figure 6

In Figure 6, an escape wheel tooth has just dropped off the receiving pallet while the fifth tooth in front of it has become locked on the exit pallet. The distance between the dropped tooth and the sharp *let-off* edge of the entrance pallet is 2° while the tooth thickness at its tip is 1°. The width or thickness of each pallet is 6°. Thus the combined widths of the tooth thickness, pallet and drop space equals 9° or the equivalent of half a tooth space. The 2° drop space represents the amount of free, unhampered movement of the escape wheel between contact with either pallet. Each pallet (see exit pallet) allows for an angular swing of 8°. Of this, 2° is on the circular locking surface and 6° is for the actual lifting action which takes place as the tooth slides over the slanted or impulse surface. Therefore, the 8° angular motion of the pallet causes the anchor pin at the top of the anchor to swing an initial 4° each side of the line-of-centers. Motion of the pin beyond this is caused by the momentum, or overswing, of the pendulum.

In fine regulator clocks, the teeth are more pointed and they lock closer to the impulse surface of the pallets. The lifting action is seldom more than 2°. These refinements are impracticable in the 400-Day Clock. First, the tooth tips must be made stronger to withstand bending if they should fail to drop off the pallet, and because of some mis-action or disturbance, be forced back on to the locking surface. Second, the lock has to be deep enough to prevent the escape wheel from unlocking prematurely should any undue vibration be given to the oscillating pendulum.

A drop space of 2° is necessary for the safe functioning of the escapement. Notice that both pallets swing along the same circular track. When one pallet is being lifted to unlock the escape tooth, the opposite pallet with its sharp edge is descending into the escape wheel. If the space is less than 2°, this sharp edge will dig into the back curve of the tooth thereby preventing any further movement of the pallets or of the escape wheel.

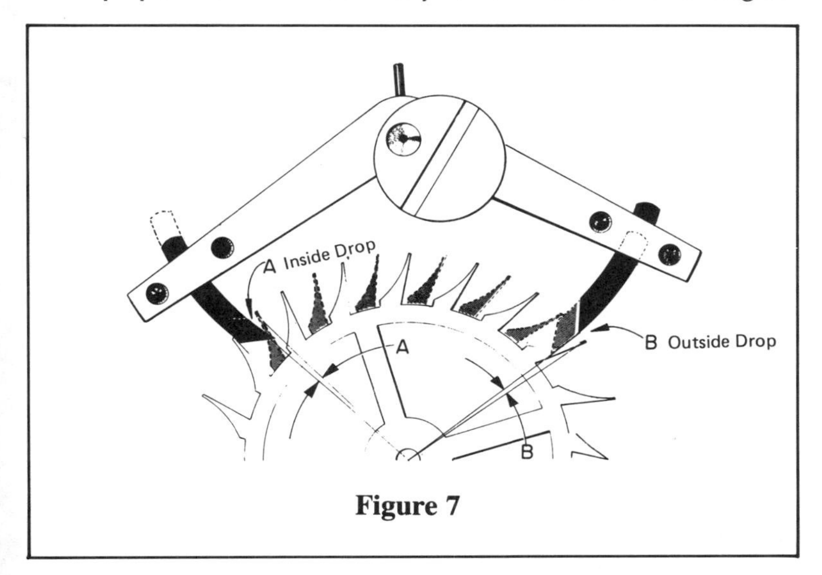

Figure 7

The impulse action must be relatively large, compared with the drop action, to provide a sufficient twisting of the suspension spring to maintain the pendulum motion.

Examining and Testing the Graham Escapement

First, remove the pendulum and the entire suspension unit. (With some clocks, it will also be helpful to remove the suspension bracket.) Next, place your finger on the tip of the anchor pin and move it back and forth slowly while checking all 20 escape teeth for correct locking and also observing whether they escape, or drop freely, from the entrance pallet. Repeat this observation on the exit pallet. Next, observe whether the drop space for all 20 teeth is the same off the entrance pallet as off the exit pallet. Figure 7 shows this test in application. When a tooth is locked on the entrance pallet, there are five teeth between the pallets. In this position, the drop space is at B and is called "outside drop" because the dropping of an escape tooth took place *outside* the anchor as did the locking. In the shaded view, which shows the alternate position, there are 6 teeth inside the anchor, a tooth locked on the exit pallet, and a tooth dropped from the entrance pallet. This is called "inside drop" shown at A, because both the lock and drop took place *inside* the anchor.

In a well adjusted escapement, the inside drop space A and outside drop space B must be equal *for all teeth.* If the drop spaces are unequal for any of the 20 teeth, it is an indication that at least one of these teeth may be bent. A smaller amount of inside drop space on one tooth may indicate that a tooth locked on the exit pallet is bent forward or the tooth just dropped (see A, Figure 11) is bent backward. If the outside drop space is smaller on one tooth, it may show that the dropped tooth is bent backward, or the tooth locked on the entrance pallet is bent forward. Straighten any bent teeth before making further adjustments (See Figure 12).

If the escapement action is being examined for the purpose of determining some fault in the clock, and it is found to be perfectly adjusted as described here, you can be certain that the fault lies elsewhere. ***Don't make adjustments to the escapement unless you are sure they are absolutely necessary.***

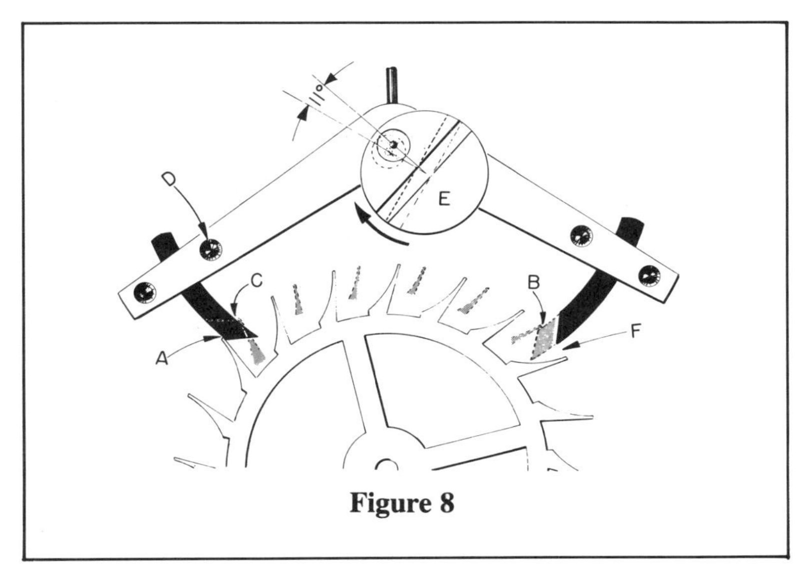

Figure 8

Making Adjustments to the Graham Escapement

Many repairers, watchmakers in particular, who are unfamiliar with the 400-Day Clock escapement, make the mistake of comparing the escapement action with that of a watch. They notice that the escape wheel teeth lock quite deeply on the 400-Day Clock pallets and they reason that, since this condition is detrimental to a watch's action, they should try to make the lock shallower by giving a clockwise adjustment of the eccentric nut E, Figure 8. Such an adjustment does carry the anchor upward, thereby lessening the locking, but the new position of the anchor causes multiple errors. (This short locking is shown in Figure 8 at A and in the shaded alternate view B.) First, the safe, deeper, lockings necessary for this type of escapement are disturbed. Second, the slot in the eccentric nut, twisted with a sharp screwdriver has probably been gouged making subsequent corrections difficult. (See Section 8, Question 2.) But the greatest upsetting of the adjustment occurs at F and at C, Figure 8. By turning the eccentric nut clockwise an angular distance of 11°, or a linear distance of only about 1/32 of an inch (approximately 1 millimeter) the outside drop space F is greatly increased at the expense of the inside drop space C shown in the alternate view.

With this maladjustment, the escapement is out of action and the clock will not run. The reason is that the escape tooth shown at B will not unlock because, when the opposite pallet descends during the unlocking attempt, it will dig into the back of the shaded tooth at C.

If the eccentric nut had been turned counterclockwise, the opposite effect would have been achieved (See Figure 9).

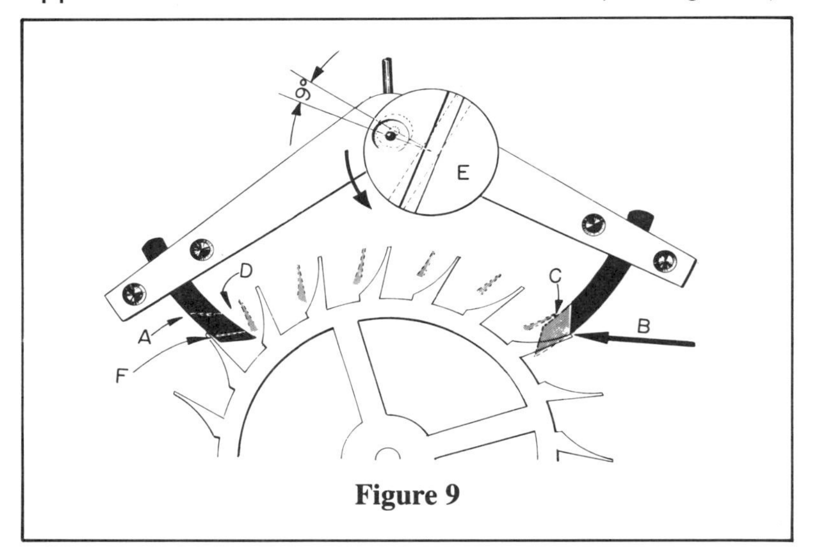

Figure 9

The locks are deepened. F shows the original position of the pallet before the eccentric nut E was turned. The distance below the dotted line at F indicates the increase in lock. C shows how the exit pallet lock is increased. A shows the position of the entrance pallet when the exit pallet is in the locking position. Turning the eccentric nut only 9° causes the outside drop space B to disappear, while the inside drop space D is greatly increased. The escapement in this condition again is inoperative.

About Adjusting the Pallets

Suppose the eccentric nut is left untouched, but one of the pallets is moved. What is the effect? Figure 10 illustrates what

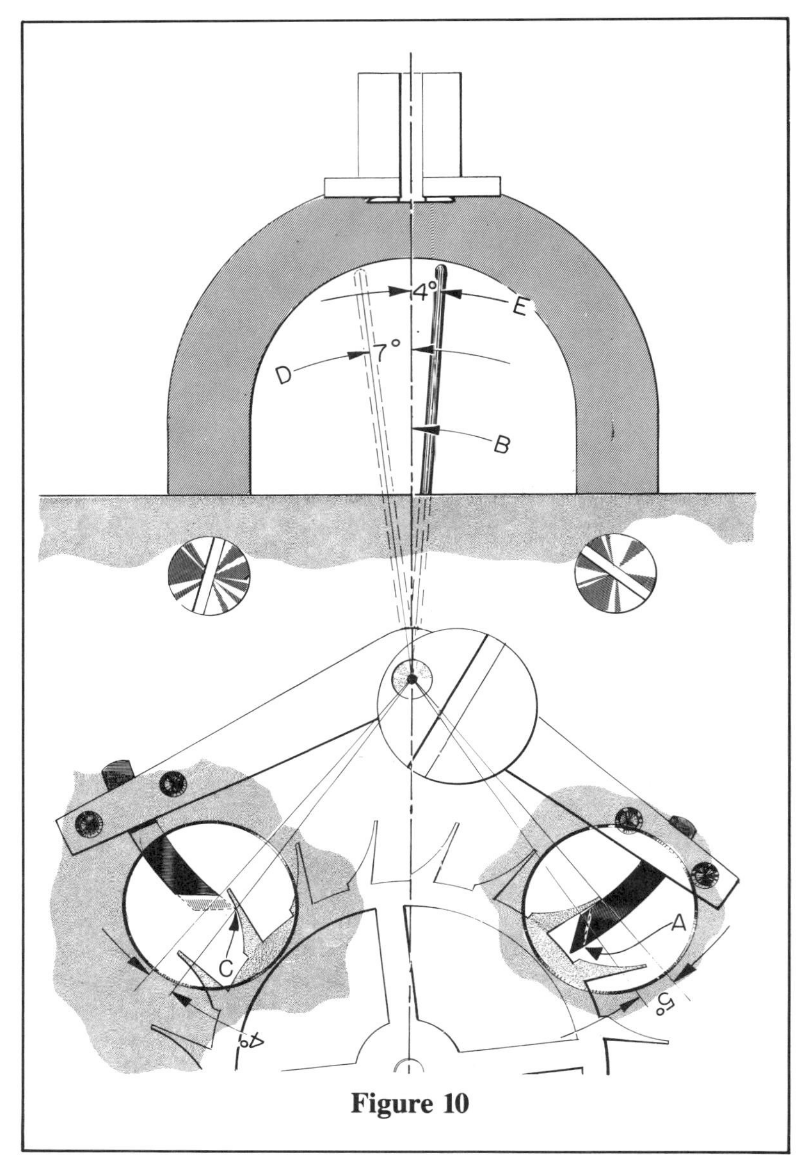

Figure 10

will take place. Here the exit pallet has been moved downward the distance from its original position shown by the dotted line A. This increased the lock from 2° to 5° (See Figure 6).

Because the pallet, to unlock, must now go up 3° further than it did previously, the opposite pallet will be required to descend a corresponding additional 3°. You should always remember that an increase of the lock on one pallet increases the lock on the opposite pallet a similar amount. (Increasing the lock on both pallets doubles the effect on *each* pallet.)

Increasing the lock to 5°, as illustrated in Figure 10, will make the escapement inoperative. Since the exit pallet must now rise 5° before the escape tooth will become unlocked, the opposite pallet, when it has gone down but 4°, will butt the curved back of the tooth shown at C, stopping the escapement and leaving a tooth hung up at 1° on the exit pallet's locking surface.

Furthermore, the anchor pin will move more from the line of centers on one side than on the other. This is shown in the upper section of Figure 10. Here the anchor pin at D travels 7° to the left of the center line B while the anchor pin at E still remains at the original 4°. (If it is possible that the escapement is still operative, a resetting of the suspension unit will be necessary to put the clock in beat again.)

In Figure 10, the "peep holes" found in most back plates are indicated with shading. It's through these holes that all observations of the teeth and pallets must be made. Some of the pre-World War I clocks did not have these holes and, on these clocks, the adjustments must be viewed from the sides, which is much less convenient (See Section 9, Plate 1475).

Summary of Adjustments

Up to this point, we have covered the general adjustments to the Graham escapement. To the right is a review in chart form of the basic adjustments that can be made. When the adjustments indicated in this table seem to have a combination of symptoms, treat each symptom separately.

Reparing Damaged Graham Escape Wheel Teeth

Sometimes the escapement will seem to be in good order, but at irregular intervals the escape wheel will trip off the pallet, instead of locking on it as it should. The clock may even run for a while in this condition, but its timekeeping would be most erratic. Or, sometimes a tooth will be hung up and fail to drop off the pallet, stopping the escapement, even though to casual inspection the teeth may seem to be in good shape.

Both of these conditions are most often caused by short, bent or broken escape wheel teeth. If you realize that it takes about five minutes for the escape wheel to make a single revolution, you can see why one such tooth can avoid immediate detection unless you make a thorough test of every tooth on each pallet.

Figure 11 shows an escape wheel with a tooth bent backward at A, one bent forward at B, and a broken tooth tip (short tooth) at C. In this Figure, although the tooth which is locked on the exit pallet is perfectly formed, the tooth which is bent backward at A has caused the escapement to stop operating, because the inside drop space no longer exists and the escape wheel cannot unlock.

When tooth B locks on the exit pallet, the same fault will be observed, because the forward bent tooth keeps the inside drop space (off the receiving pallet) below minimum requirements needed for clearance.

TROUBLE SHOOTING CHART
(Graham or Pin Pallet Escapement)

COMMON SYMPTOMS				CORRECTIONS			
Entrance Pallet Drop	Exit Pallet Drop	Entrance Pallet Lock	Exit Pallet Lock	Eccentric Nut*	Entrance Pallet	Exit Pallet	Pallet Lifting Surface Angles†
More	Less	Normal	Normal	Raise Pivot Hole			
Less	More	Normal	Normal	Lower Pivot Hole			
Normal	Normal	Deep	Deep		Move Up	Move Up	
Normal	Normal	Shallow	Shallow		Move Down	Move Down	
More	Less	Deep	Deep	Raise (§) Pivot Hole			
Normal	Normal	Deep	Shallow		Move Down		Make Entrance Pallet Steeper
Normal	Normal	Shallow	Deep			Move Down	Make Exit Pallet Steeper
Normal	Normal	Very Shallow	Deep			Move Down	Make Exit Pallet Steeper
Normal	Normal	Deep	Very Deep			Move Up	Make Exit Pallet Less Steep

* It is assumed that the eccentric nut carries the *anchor* pivot. If it carries the *escape wheel* pivot, correction should be made just the opposite to that indicated.

† Rather than attempting to grind and polish the pallets, you may find it more convenient to replace them.

§ If the locks become shallow in making this adjustment, consider this a new symptom and make further corrections as indicated.

Note these three basic rules of thumb:

(1) If drops are equal, do not move the eccentric nut.
(2) If drops are unequal, do not move the pallets.
(3) If locks are unequal, change the lifting surface angle of one or both pallets.

TABLE OF ADJUSTMENTS
(Graham or Pin Pallet Escapement)

WHEN YOU DO THIS:			THE RESULT IS THIS:			
Eccentric Nut	Entrance Pallet	Exit Pallet	Entrance Pallet Lock	Exit Pallet Lock	Entrance Pallet Drop	Exit Pallet Drop
Raise Pivot Hole			Decreases	Decreases	Decreases	Increases
Lower Pivot Hole			Increases	Increases	Increases	Decreases
	Move Up		Decreases	Decreases	No Change	No Change
	Move Down		Increases	Increases	No Change	No Change
		Move Up	Decreases	Decreases	No Change	No Change
		Move Down	Increases	Increases	No Change	No Change
	Move Up	Move Up	Decreases*	Decreases*	No Change	No Change
	Move Down	Move Down	Increases*	Increases*	No Change	No Change

**by the total of the combined distances both pallets are moved.*

If a short tooth (one with a broken tooth tip) drops off the receiving pallet, it will cause a light lock on the tooth which is about to drop onto the exit pallet. This tooth will also lock too lightly on either pallet.

To test for deformed teeth, hold your finger on the tip of the anchor pin and move it very slowly, observing the escapement action through the peep-holes in the back plate. Permit each escape tooth to move slowly across one pallet's lifting surface. When a tooth drops off the pallet, arrest the action of the anchor pin so that no further movement is permitted. Now inspect the lock on the opposite pallet and

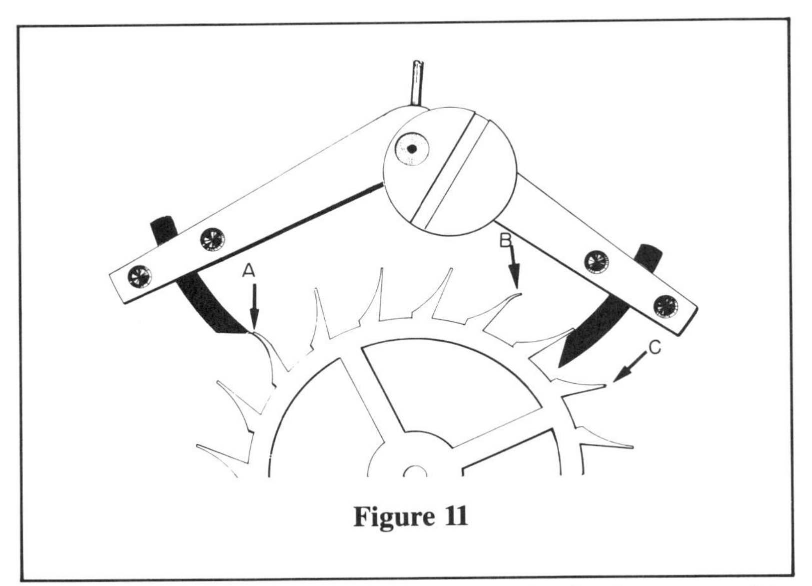

Figure 11

make a mental note of the amount of this lock. Repeat this on all 20 teeth for drop and lock. Any deformed or broken tooth can easily be located by observing obvious differences in the amount of a tooth's drop or lock.

Repairing Pallets in the Graham Escapement
Since the pallets are extremely hard tempered and move about 25 times slower than those on swinging pendulum clocks on the same size, they should wear much longer. Also, since the escape teeth are much broader than those of swinging pendulum clocks they distribute their wear over a broader area of the pallet's polished surfaces. Since they receive such little wear, the pallets should never need repair. Nor should their shape be changed in any way. It is best to leave the pallets alone. They should be raised or lowered only if it's absolutely necessary.

Repairing an Escape Wheel in the Graham Escapement
When an escape tooth is bent, you can often straighten it by grasping the bent tooth between the blades of a pivot-straightening tweezer (as illustrated in Figure 12) and pull-snapping it while squeezing the tweezers. Too much pressure on the tweezers is not recommended, for this might thin out the end of the soft tooth and raise a burr.

If you find an escape wheel with a short or broken tooth, it's best to obtain a new wheel, or a complete new wheel and pinion. If the wheel is from an old clock and you cannot match it, you can substitute one with the same number of teeth and approximately the same diameter. Sometimes it's possible to obtain a complete wheel and pinion which can be made to fit.

If you can get a complete wheel and pinion unit, check these four things, in addition to the wheel's diameter: (1) The escape wheel must be in the same vertical plane as the anchor, and aligned so that the wheel teeth strike the middle portion of the pallets. If they don't, it may be possible to shift the position of the anchor or the wheel on the arbor so that both will be in the same plane. (2) The diameter of the pinion should be as close to that of the original as possible. (3) The pinion should be positioned so that it makes full contact with the center wheel. (4) The pivots should be of the same diameters as those of the originals. If they are too large, either they should be reduced to fit, or you should carefully broach the pivot holes. If they are too small, you can bush the holes.

If a substitute wheel is not available, you can make a new one or have one made to the original sample. In cases where only the tip (less than 0.5mm) of the tooth is missing, you can shorten all of the other teeth on the escape wheel to the height of the shortest tooth. This operation, called "topping," is performed as follows: Place the escape wheel in the lathe as shown in Figure 13 and run the lathe, moderately fast, in the direction of the arrow. Steady a fine file on the adjusted T-rest, and bring it to bear gently on the wheel. This action cuts the teeth shorter without bending them, and without raising burrs on their front surfaces where the finest surface is needed for smooth operation.

Make frequent inspections during this operation to insure that the teeth are not cut too short. If they are cut too short, the 12° rake (slant) will be insufficient to provide clearance for the locking corner of the exit pallet. The pallet will then dig in to the front side of the tooth, causing the escape wheel to recoil and possibly causing the clock to stop.

When the teeth are shortened; their tips naturally become thicker, thereby reducing the available drop space. If they are found to be excessively thick, you can thin them by a

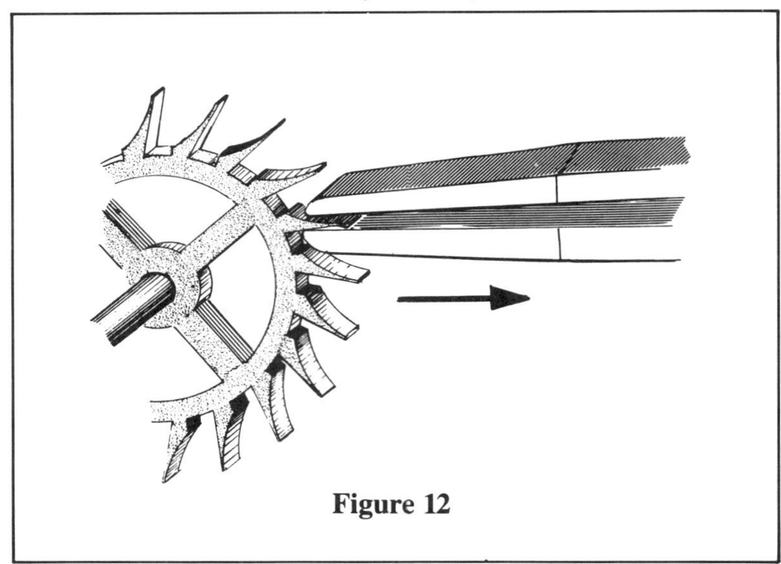

Figure 12

judicious filing of the back curve of each tooth, as shown in Figure 14, until the tooth-tips become as thin as the space shown by the arrow A. A curved file or an India oilstone slip can be used. This operation will also remove any burrs, if any remain from the topping shown in Figure 13.

When the wheel is finished in this manner, fit it to the movement. A trial manipulation of the anchor will reveal that the escape teeth, formerly high as that at A, Figure 15, now only reach to the pallets at dotted line B and C. In this condition, the escape wheel will cause the anchor to flutter like an alarm clock bell hammer. Therefore, both pallets will have to be extended to the amount indicated in black, *below* the dotted lines B and C. Extending both pallets equally will most likely provide the safe lock shown at B, and the alternate position D.

Here it should be pointed out that, in theory, the operation described above is not sound, since it upsets the geometric proportions of the original escapement design. In actual practice, however, the method has provided to be a workable repair expedient.

Do not expect to achieve the same result by lowering eccentric nut E. It will not work for the following reasons: The smaller the escape wheel, the closer its arbor must be to the anchor arbor and the smaller a circle must be scribed by the curved pallets. As mentioned above, if the escape wheel is made smaller than the original by topping, the geometric proportions in which the escapement was designed originally no longer exist. Turning down the eccentric nut will bring the escape wheel and anchor arbors closer, but since the circle scribed by the curved pallets cannot be made of a smaller diameter, because their shape is fixed, the disproportion remains uncorrected. Furthermore, turning the eccentric nut down causes the outside drop to disappear, as shown in Figure 9.

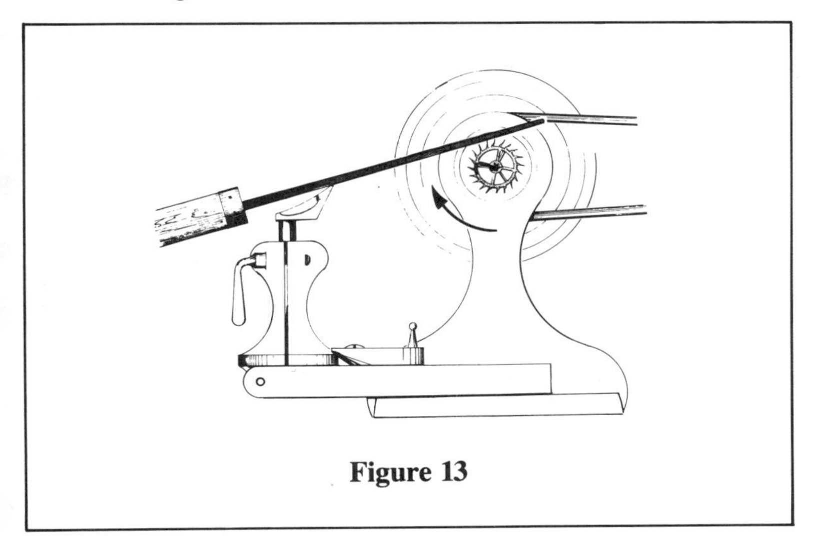

Figure 13

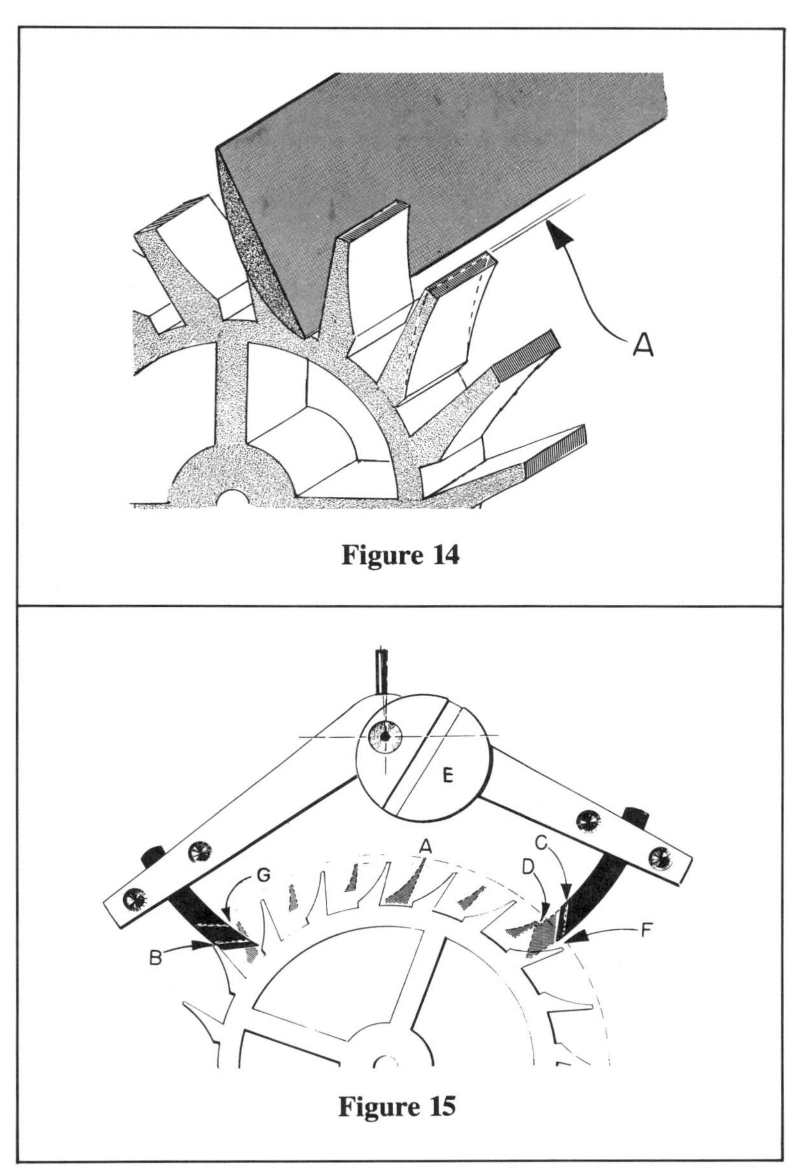

Figure 14

Figure 15

When the escape wheel was made smaller, the drop space was also slightly decreased. However, if the teeth are made thinner, the drop spaces will be sufficient as shown at G at inside drop, and the alternate position F at outside drop, in Figure 15.

EXPLANATION OF THE PIN PALLET ESCAPEMENT

Figure 16 is a geometric view of a typical 400-Day Clock pin pallet escapement. The escape wheel has 15 teeth and turns counterclockwise. The basic angular movement of the pallet is 8½°. This is based upon a lift angle of the escape teeth of 7°. The anchor pin's circular section provides an additional 1½° lift, half of the pin's diameter.

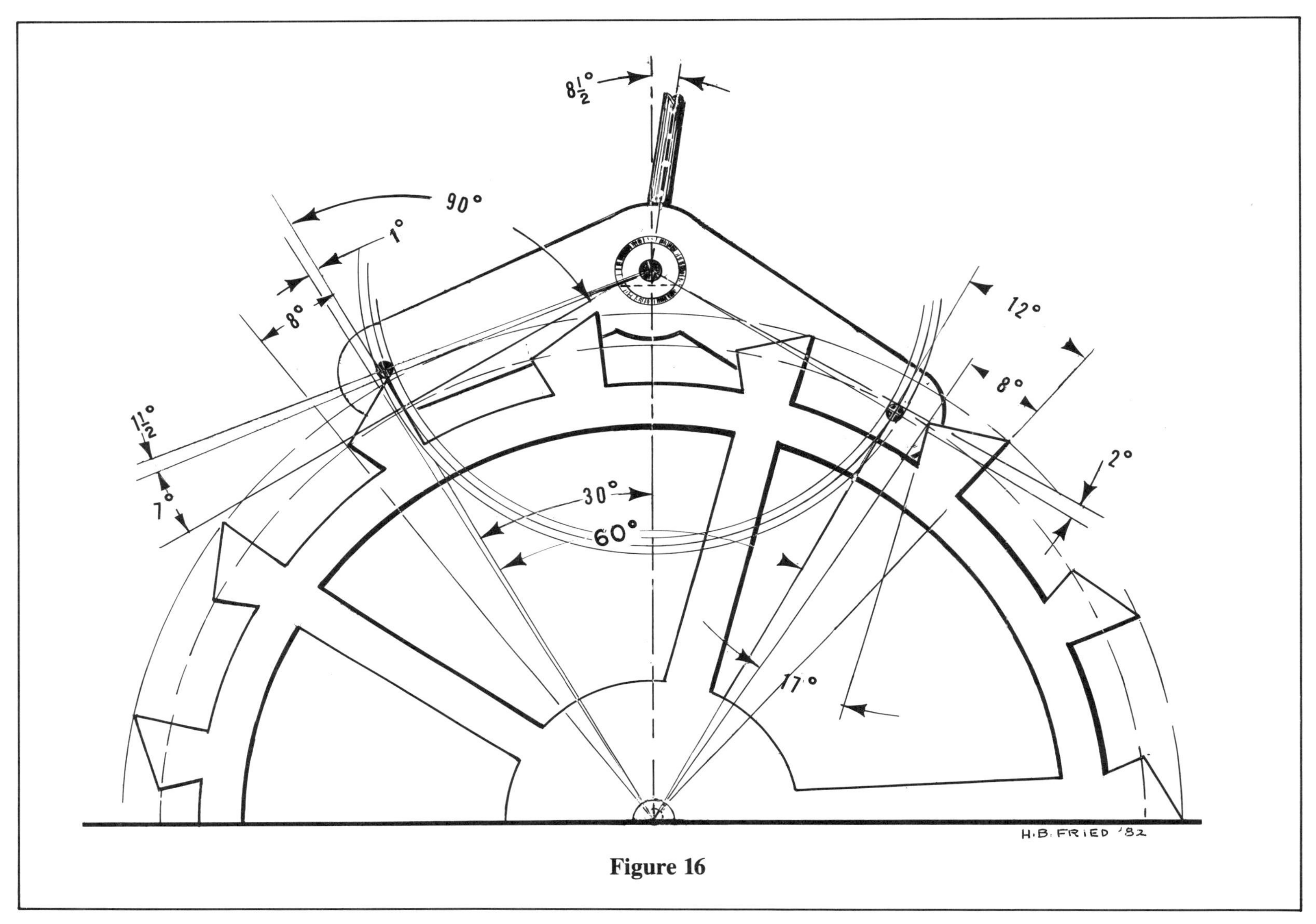

Figure 16

In this escapement, it is mainly the rake (the undercutting of the front of the escape teeth) that helps the pin to remain locked. In the drawing, the undercutting is shown as 17°, an exaggerated amount for the purpose of illustration. With many 400-Day Clock pin pallet escape wheels, the rake (or undercutting) has such a small angle (6° or less) that its effect is difficult to see. Nevertheless, when the pin of the pallet makes contact with the front of the tooth, not only does the rake of the locking surface draw the pin to the escape wheel rim, but the top of the round pin also supplies an additional draw which increases as the pin descends.

When unlocking takes place, the escape wheel is required to go backwards or recoil about 1°. This reverses the tension on the suspension spring.

The drop space, the distance the escape tooth moves without contact with the pallets, must also be greater than with the Graham (dead beat) escapement, because the back half of the pin must clear the tooth from which it has dropped.

The escapement is designed so that when in good condition, the initial lock, shown at the entrance pin on the right, is 2° measured from the escape tooth's entrance corner to the front center of the pallet pin.

Repairing the Pin Pallet Escapement

Unlike other 400-Day Clock escapements with pointed escape teeth and adjustable pallets, the pin pallet escapement has very sturdy escape wheel teeth. They could not conceivably become deformed unless by intention or ignorant manipulation. Therefore, repairs to teeth should never be necessary. The pins on these pallets are the thing to worry about. Some, particularly those with pre World War I clocks, may be dead hard. Others, made since 1949, are relatively soft and can be bent. However, if it becomes necessary to bend a pin, you may find that it will break off because it had been bent several times before. If a pin is broken, or becomes broken while being straightened, it may be possible to replace it by staking in a new pin of exactly the same diameter. However, this can be an extremely difficult operation and, if at all possible, avoid any attempt to adjust the pins. If the pin is loose, it should be staked tight.

Pivot holes for pin pallet anchors and escape wheels do not become worn. However, in some clocks, the anchor pivot hole is set in a peninsula of the plate which tempts bending and is an invitation to compound difficulties. It goes without saying that no attempt should ever be made to change the configuration of any part of the pin pallet escapement unless it is specifically called for.

Summary of Adjustments
The problem solving procedures suggested in the "Trouble Shooting Chart" and "Table of Adjustments," shown on a previous page, apply in pin pallet escapements in the same manner as they apply to Graham (dead beat) escapements.

Figure 17 shows the pallet pin as it makes contact with the oncoming escape tooth. It has become locked 2° below the corner of the escape tooth. Notice that the lift pin is above the let-off corner of the exiting tooth and there is about 4° of drop between it and the back corner of the tooth.

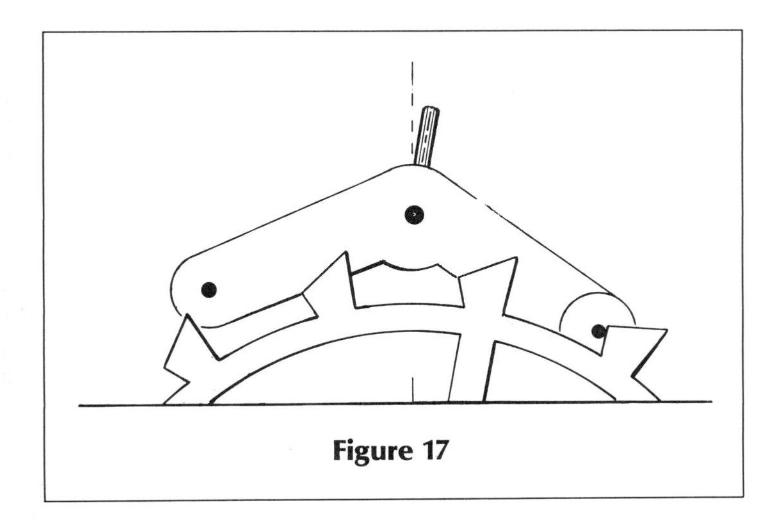

Figure 17

Figure 18 shows that the undercut locking surface of the tooth has caused the pin to be drawn to the escape wheel rim. Notice that because of the angle of the tooth's locking surface, the escape wheel has advanced from the position now shown in the dotted, phantom view. Also, the drop space at the exit pin A has increased and the pin also has ascended slightly compared with the phantom view.

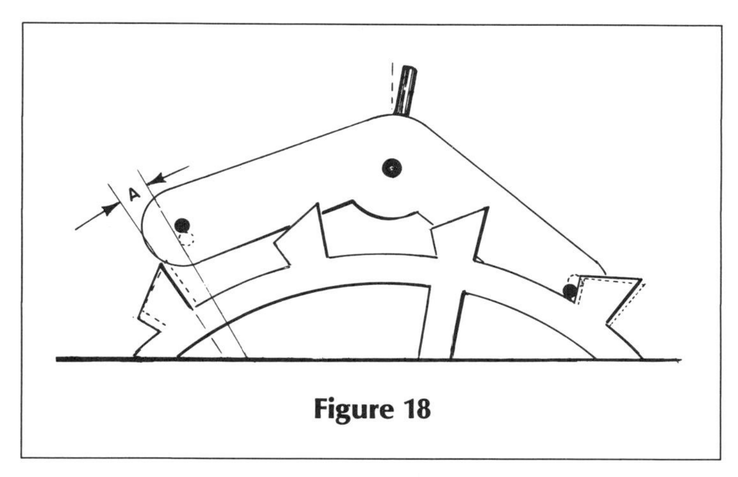

Figure 18

In Figure 19, the entrance pallet is unlocking and about to enter upon the tooth's lifting surface. This causes the escape wheel to recoil from its previous position as shown in the phantom view. Important also is that this recoil has reduced the drop between the exit pin and the tooth as shown at A. Notice as well, that this pin has descended below the tip of the tooth.

The design of this escapement does not waste space for unwanted, unproductive motion. Let us examine what happens to the venturesome few who would trample where horological engineers fear to tread; that is, bending pins.

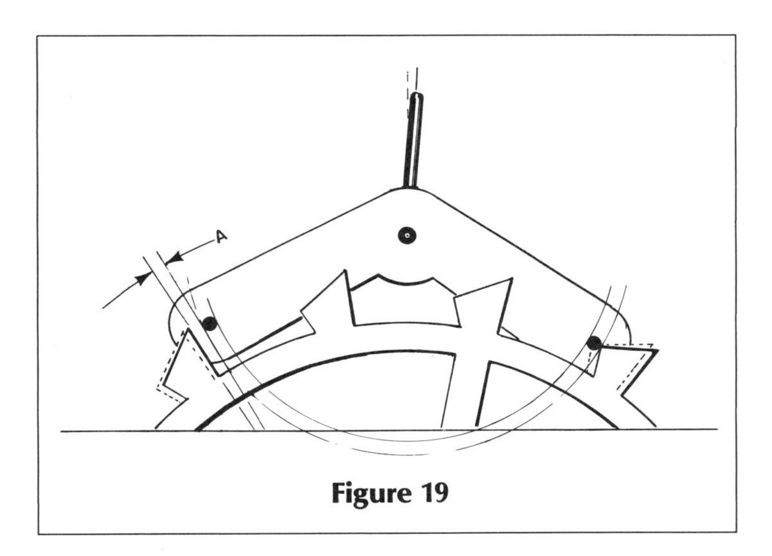

Figure 19

Bent Pins
Figure 20 shows an anchor in which the exit pin has been bent away from the pallet arbor as at A. The pin is now lodged on the lifting surface and the clock will stop with the escapement jammed. The outside or exit drop has vanished.

What if a pin is bent toward the pallet arbor as in Figure 21? Here the inside drop has disappeared. The exit pallet is locked, but the wheel cannot recoil and the escapement becomes jammed.

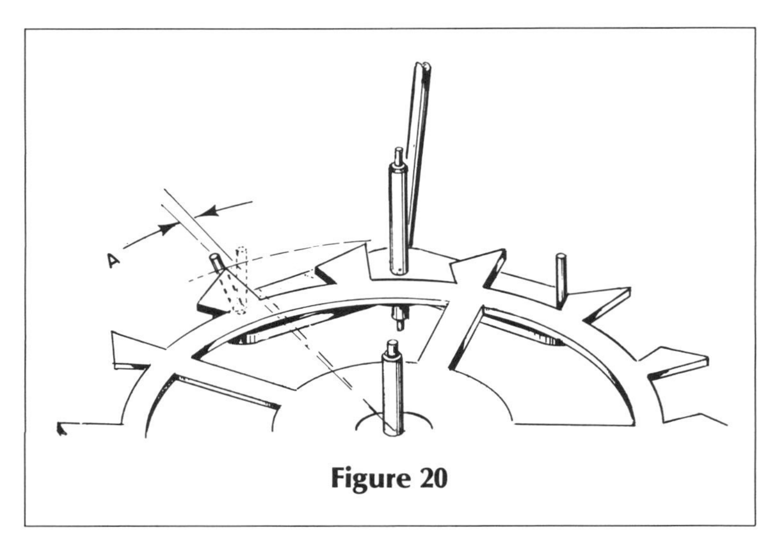

Figure 20

Bending pins in or out in relation to the escape wheel rim (if it can be done without their breaking) produces a similar effect. Bending the pins out will cause a tripping or rattling because of the lack of locking. Bending one or both pins in will result in too deep a lock, and when the escapement attempts to unlock, the opposite pin will "hang up", similar to the situation illustrated in Figure 21.

Bending one pin in or out affects the depth of lock of both pins an equal amount. Bending both pins in the same direction doubles the effect of bending only one pin.

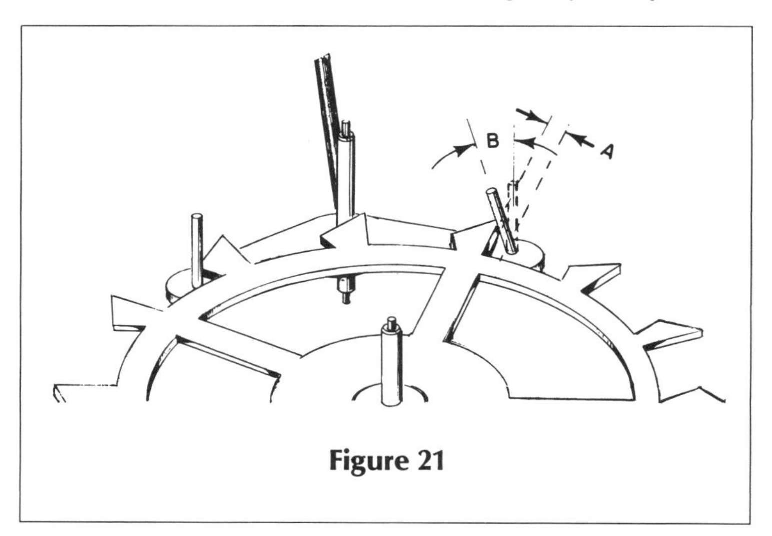

Figure 21

Bending or altering the original position of the peninsula, which holds the pallet arbor pivot, disrupts the precise geometry of the escapement. As shown in Figure 22, the pallet pins are planted exactly 60° apart. The pallet center (pivot) is at the intersection of right angles drawn from tangents on the escape wheel pitch circle, which determines the circular paths of the entrance corners of the escape teeth. Altering the position of the pallet pivot will change the inside and outside drop spaces, alter the lifting angles of the escape teeth in relation to the pins, and affect the locking possibilities. Even if an escapement action were possible, the efficiency of the escapement will be adversely affected. **In short, the pin pallet escapement is not designed to be altered or adjusted after the clock leaves the factory.**

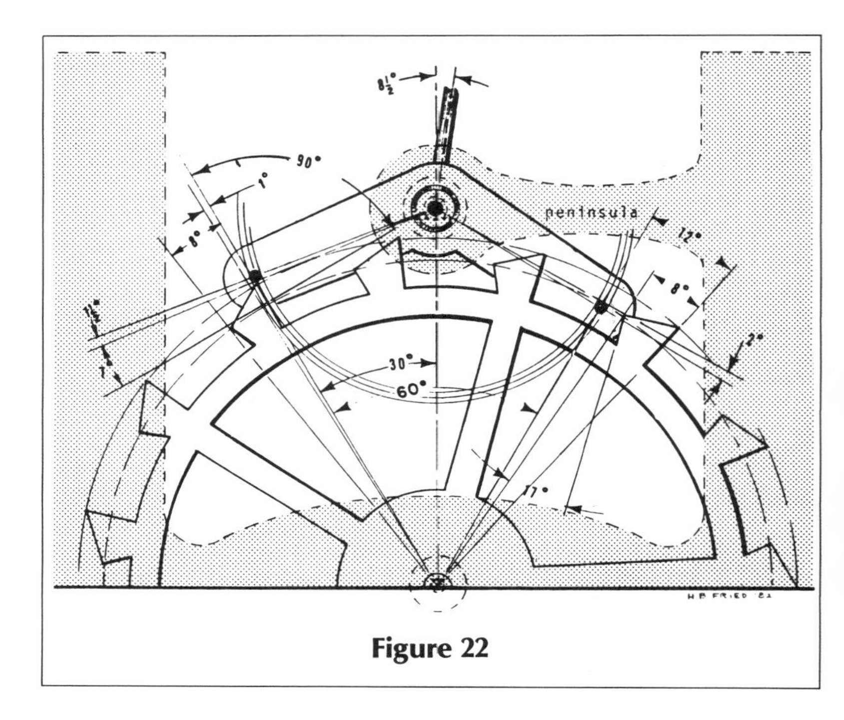

Figure 22

CONCLUSION

This section of the Guide has detailed the construction and repair of the two types of escapements commonly used in 400-Day Clocks. It should be emphasized again that any adjustment of the escapement of a 400-Day Clock should be undertaken only as a last resort and only after the initial set-up has been thoroughly studied as outlined in this Section. When adjustments are required, reference to the "Trouble Shooting Chart" and "Table of Adjustments" should help reduce the time required to return the clock to its original, factory settings.

SECTION 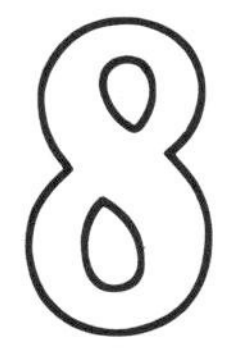 Answers to the Most Frequently Asked Questions about 400-Day Clock Repair

The following questions and answers appear to cover most of the problems which will confront you. However, if you are faced with a specific repair problem not covered somewhere in this Guide, or if you have other questions about the clock, we'll gladly try to help. Write: Question Department, The Horolovar Company, Box 264, St. Clair Shores, Michigan 48080. If the question is about repair, give as many details of the symptoms, and the steps you have taken, as you possibly can. And...*be sure to enclose an addressed, stamped return envelope.*

QUESTION (1): How do you adjust a 400-Day Clock escapement?

ANSWER:
Questions about the 400-Day Clock escapement have been asked so frequently that there appeared to be a great need for a thorough explanation of this part of the clock as well as for details covering its adjustment and repair. Therefore, Section 7 of this Guide, written with the help of Henry B. Fried, one of the country's foremost authorities on watch and clock escapements, has been devoted entirely to this subject.

A thorough reading of Section 7 should give you full knowledge of the fundamental principles of the 400 Day-Clock escapement and also answer any questions about its adjustment and repair.

QUESTION (2): How can you turn a tight eccentric nut without chewing up the sides of the slot?

ANSWER:
If the eccentric nut obviously has never been turned before, it is more than likely that it doesn't need turning now. (Be sure to read Section 7 "Adjustment and Repair of the 400-Day Clock Escapement" before you start moving the nut.) If you find that it is necessary to move it, first make a tiny scratch across the rim of the nut so you will not only know how far you move it, but also in what position it was originally. You may wish to put it back exactly where it was!

There seems to have been no standard among manufacturers for either the size or tightness of the eccentric nut. Some nuts can be turned easily with a screwdriver; others offer so much resistance that it is almost impossible to move them without injuring the sides of the slot.

Actually, it is best not to use a regular screwdriver, not even to test the tightness. Since the slot sides are parallel, and a screwdriver's end is tapered, all pressure put on the screwdriver is directed only upon the top edge of the slot. It's bound to leave an ugly scar.

At almost any good hardware store, you can buy an inexpensive tool known as an offset screwdriver. It is all steel, has two usable ends, and gives better leverage than a regular screwdriver. Buy two of them and you will have four different sized nut-turning ends. Grind one end so that it fits into the slot, making sure that the flat sides are parallel. With this tool, there is little likelihood that you will damage the sides of the slot while you are turning. The three other ends of the screwdrivers may be ground to fit different sizes slots in other eccentric nuts.

Since 1949, several manufacturers have replaced eccentric nuts with a U-shaped, die stamped area around the pivot hole leaving the hole on an arm. Here adjustment is made merely by bending the arm slightly up or down. It makes adjustment simpler, but its disadvantage is that if it is moved too often the arm will break off.

QUESTION (3): How can you stop the anchor from fluttering?

ANSWER:
This question is often worded in other ways. For example: "I replaced a suspension spring with a HOROLOVAR spring of the proper strength indicated, but the clock now gains an hour every hour. I used a thinner spring, but it still gains several hours in a day. Is there something wrong with the spring?"

What has happened is that, unnoticed, the anchor fluttered. That is, the escape wheel skipped several teeth during one turn of the pendulum. Actually, for the clock to gain an hour every hour, the pendulum would have to rotate back and forth at a speed of sixteen times a minute!

Fluttering usually takes place when the fork on the suspension spring unit is positioned too low on the suspension spring. If the back plate of your clock is illustrated in Section 9 of this Guide, and if a unit number is indicated below the plate, check the unit geometry in Section 10 to see whether the fork is in a position lower than that indicated. (If you are using a Horolovar Suspension Unit, this check is not necessary.) If no unit number is indicated under the plate, the clock was manufactured prior to 1949, and you may be able to solve the problem by raising the fork slightly on the suspension spring.

But if the anchor still flutters, even with the fork slightly raised, there may be insufficient lock to the escape wheel teeth on the pallets. (See "Trouble Shooting Chart" in Section 7.) This can be remedied by lowering *one* pallet very slightly.

QUESTION (4): How much clearance should there be between the anchor pin and the fork tines?

ANSWER:
There should be *only* enough clearance to ensure that the tines do not bind the pin at any point in its movement. If the fork is spread open too wide, much of the power of the pin's impulse will be lost. If it is too tight, it will bind the pin and all power will be lost.

You should check the amount of clearance when the anchor pin is at its extreme angle, because when the pin is at its farthest point from center (in either direction) the clearance between pin and fork tines is at the minimum.

Although most forks are designed to minimize this condition, by having tines with rounded or knife edges or a closed end so the width of the opening cannot be changed, it is a good idea to check the clearance of the fork on the anchor pin every time the suspension unit is attached—especially since it is very easy to bend the tines without noticing it. This happens most often when a new suspension spring is being inserted into the unit.

One quick, practical test of the setting of the fork tines is to start the pendulum rotating and then rock the clock slightly front to back. This will cause the pendulum to swing back and forth as it rotates. By observing the suspension spring near the fork, you can tell if binding between the anchor pin and the fork tines is occurring.

QUESTION (5): How should a broken anchor pin be replaced?

ANSWER:
Anchor pins on most newer clocks are made of soft steel, but on many old clocks, the pins were hardened. When these older types break off, repairs are more difficult to make. If a soft pin was already bent when you received the clock, and you wish to straighten it, you can minimize the possibility of breakage by grasping the pin with pliers just above the anchor and making the bend above this point.

If the pin is broken off so that a piece remains inside the anchor, the expedient (and most expensive) thing to do is to replace the entire anchor unit. However, if the piece is not hardened, it is possible to drill it out. Taper the end of a new steel pin, approximately 1 millimeter or less in diameter, and drive it tightly into the hole.

If a stub of the broken pin remains, try to remove it by unscrewing it, for some pins are threaded into the anchor. But whether they're threaded or not, the simplest way to replace a pin after the stub has been removed is to drive a new pin into the hole so that it is friction tight. A more skillful job can be done by tapping the hole and threading the new pin to fit.

Some repairmen replace broken pins by soft-soldering them back onto the anchor. Although this method can be used to perform a functional repair, and might conceivably be resorted to if a broken tempered pin cannot be removed from the anchor, generally it should be discouraged. The use of solder in any visible place on the clock, no matter how carefully it's applied, is poor horological workmanship.

If a soldered joint is to be made, it's best to use a soldering iron, rather than an alcohol lamp or a torch. With a lamp, and particularly with a torch, it's all too easy to reach a temperature which will seriously affect the temper in the pallets.

QUESTION (6): How do you put the pendulum in beat?

ANSWER:
First, it helps to know exactly what is meant by the pendulum being "in beat". This can best be understood by the following instructions with a reference to Figure 1.

Once the clock is started, it continues to run because the power, originating with the wound mainspring, is converted mechanically into a series of little impulses which individually push the rotating pendulum back and forth. This pendulum action originates with the escape wheel teeth (ewt) which, through the anchor pallets (p), pass alternate impulses on to the suspension fork (f) via the anchor pin (ap). The fork, in turn, transfers the back and forth motion of the anchor pin to a twisting motion of the suspension spring (sp). The twist ultimately passes down the length of the spring to the bottom suspension block (bb) where it ends as a little push of sufficient intensity to maintain the oscillation of the pendulum attached to it.

To keep this action continuous, the push must be given to the pendulum at an equal distance from its dead position in either direction. Otherwise the clock will soon stop. The adjustment by which the push is made equal in both directions of rotation is know as putting the pendulum "in beat".

Adjustment of the pendulum beat is made by turning the entire suspension unit with the pendulum attached on its axis (a-x). This is accomplished by one of several different methods depending upon the design of the clock. In all clocks, the suspension unit is held by its top block (tb) in some form of saddle (s). (See Section 15 for various types of suspension saddles.) Except in one or two of the very early models, provision has been made for this saddle to be turned. It may be necessary to loosen a set screw (ss) to move it, or the saddle may just be friction tight (ft).

In some old clocks, when the saddle is fixed, (for instance, Section 6, Clock 50; Section 9, Plate 1171), the adjustment can

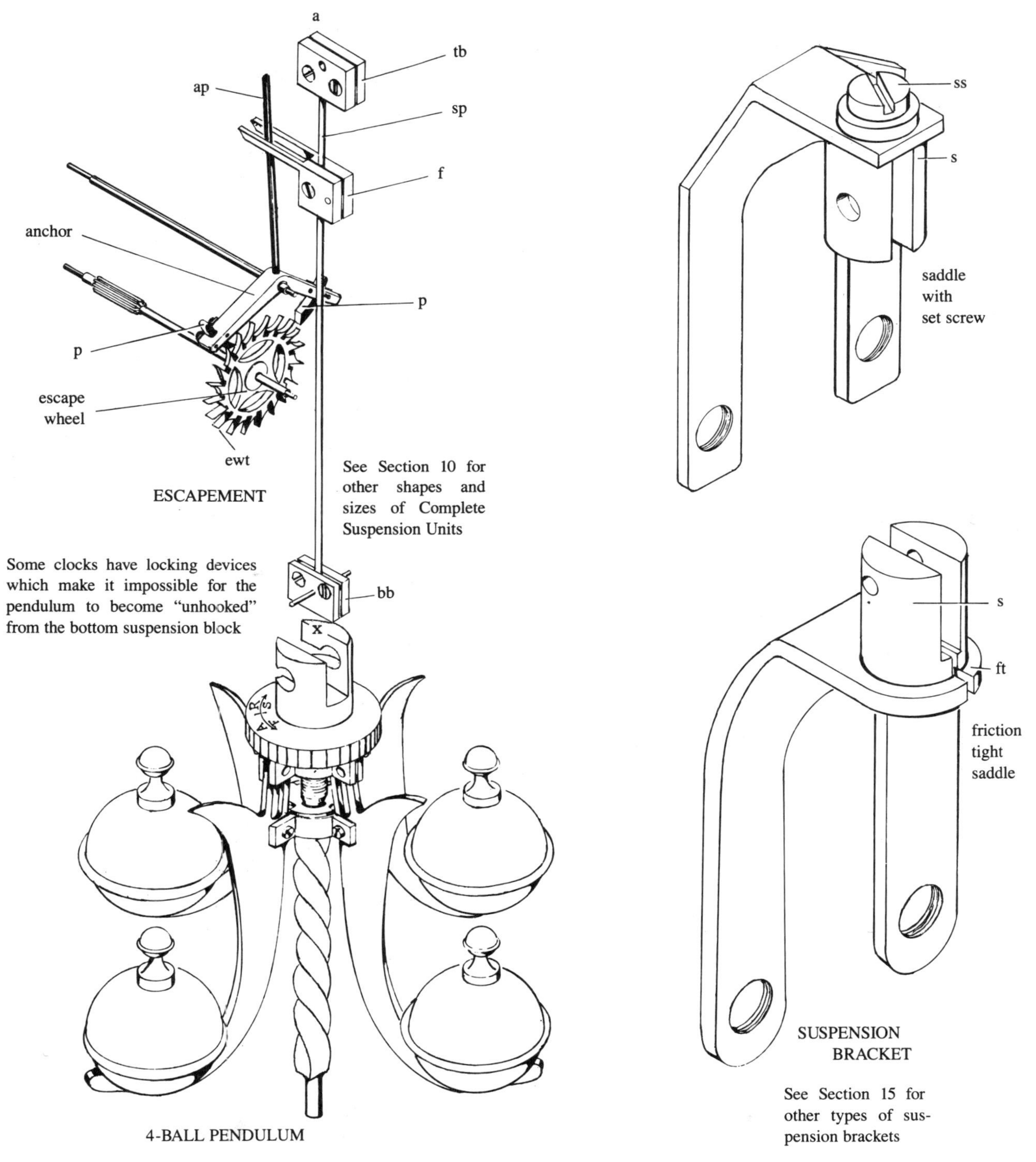
a
tb
ap
sp
f
anchor
p
p
escape wheel
ewt
ESCAPEMENT
See Section 10 for other shapes and sizes of Complete Suspension Units
Some clocks have locking devices which make it impossible for the pendulum to become "unhooked" from the bottom suspension block
bb
x
4-BALL PENDULUM
ss
s
saddle with set screw
s
ft
friction tight saddle
SUSPENSION BRACKET
See Section 15 for other types of suspension brackets

Figure 2

be made only by giving the suspension spring (sp) a slight but permanent twist at a point near the top suspension block (tb), a rather delicate operation best accomplished with tweezers.

Here's how to find the correct position of the suspension unit which will put the pendulum in beat:

Check the clock to be sure that it's wound and that it is in a level position. Also, check the suspension spring (sp) to see that it is absolutely straight and that the two blocks (tb and bb) and fork (f) *are in the same vertical plane as the spring.* (If one of them is slightly out of line, it can be put back into position by twisting the spring a little. However, if the spring is badly bent or kinked, it should be replaced. If the spring is bent between the fork and the top suspension block, it should be replaced since it is nearly impossible to sufficiently straighten the spring in this area.

Next, copy the protractor shown in Figure 2 and cut it out so that it will fit on your clock base. Working from the back of the clock, allow the pendulum to come to rest and mark the pendulum ball farthest from you using a small piece of masking tape. Align the protractor so that the three marks lie directly beneath the pendulum ball opposite the marked ball. The three lines represent the zero position.

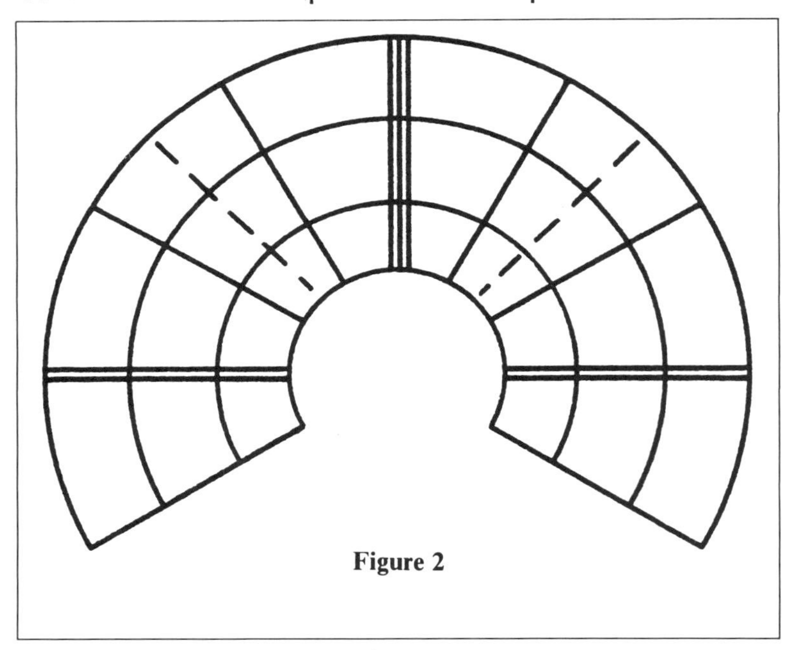

Figure 2

Start the pendulum rotating in either direction by giving it one full turn, no more. Listen for the sound of the escape wheel striking the anchor pallet. The use of a beat amplifier (See Section 17) makes it easier to hear this sound—the tic and toc. Note the position of the marked pendulum ball with respect to the protractor. When the beat is perfectly set, the sound of the escape wheel striking the anchor pallet will occur equally distant from the zero mark on the protactor (the three marks). If the beat is not set properly, adjust the saddle by turning it slightly. It takes very little motion of the saddle to greatly effect the beat of a 400-Day Clock. The graduations on the protractor are intended to help determine the location of the marked ball when the sound occurs. These marks are not intended as limits.

It is easiest to determine the direction that the saddle should be turned by visualing a line between the locations of the marked ball when the sound occurs on each side of the zero line. For example, after starting the pendulum rotating, you note that the sound occurs above the last mark on the right side of the zero mark and above the third mark on the left side of the clock. This means that it is necessary to rotate the "locations" of the sounds clockwise. In order to do this, rotate the saddle in the counter-clockwise direction. While only a small adjustment is ordinarily necessary, it may take several tries before the saddle is correctly positioned and the beat is properly set.

On clocks having saddles with set screws (ss), it is always a good idea to check the beat after the set screw has been tightened. The saddle may have been turned in tightening without your noticing it. This step is unnecessary if a beat setting tool is used. (See Section 17.) The use of a beat amplifier and a beat setting tool makes beat setting easier, faster, and more accurate.

For the time being, leave the protractor in place as it can be used to check the action of the clock and is very helpful in regulating the clock.

QUESTION (7): How much should the pendulum rotate?

ANSWER:

The minimum number of degrees that a pendulum of the standard size clock should rotate is about 270° or ¾ of a turn. If you are using the protractor shown in Figure 2, the dashed lines indicate 270°. Anything less, particularly if there is a little overswing, usually indicates a loss of power somewhere in the train.

It is sometimes possible to increase the turn of the pendulum by raising the fork slightly on the suspension spring. However, if this doesn't help, or if it impairs the functioning of the escapement, check the escapement action. If that is in order, then try to locate the source of the power loss. Look for too much play in the fork, a loose top block, a distorted mainspring or gummy oil in pivot holes or mainspring barrel.

There's considerably more leeway in the *maximum* amount that a pendulum will turn. In the standard clock, a 360° to 450° turn (1 to 1¼ turns) is not unusual. As a matter of fact, some miniature movements have pendulums that make as much as two turns.

In general then, if the pendulum of a clock makes at least three quarters of a turn, and if there is an overswing of one half inch or more, you can be reasonably sure that the clock will run at least 400 days.

QUESTION (8): Is there any trick to regulating the pendulum so that the clock will keep good time?

ANSWER:
Actually no, but before turning the regulating nut, it is well to determine which way to turn it to make the desired correction. For most ball pendulums, a clockwise turn of the nut will expand the position of the weights, thereby making the clock run slower. Sometimes arrow markings on the nut will also show you which direction to turn. Turning toward A (Advance) or F (Fast) will speed the clock up; toward R (Retard) or S (Slow) will slow it down. However, don't blindly depend upon a clockwise turn of the nut to slow the clock. Some new clocks have pendulums that turn just the opposite!

If you have left the beat setting protractor in place you can quickly bring the clock very close to proper regulation by using a stop watch and counting the number of pendulum beats or rotations. Begin by looking up the number of Beats per Minute for the suspension unit that is installed in your clock. This information is shown in Section 10 (Also see Question 11 in this Section). Allow the pendulum to continue rotating or give it a very gentle start and allow it to settle into its normal cycle. Choose one of the lines on the protractor and start the stop watch as the marked pendulum ball passes over that line. Count each time the marked ball passes over the line (both directions of rotation) and stop the stop watch when the number of beats per minute is reached. If your suspension unit shows 8 beats per minute, 8 rotations of the pendulum should take exactly 60 seconds.

If the stop watch does not read 60 seconds, adjustment of the pendulum is required. As an example, if the reading on the stop watch is 58 seconds, the clock is running fast and the pendulum should be adjusted towards the R or S. If the stop watch reads 61 seconds, adjust the pendulum in the A or F direction. Using the approach of counting the number of Beats per Minute will enable you to quickly come very close to the proper regulation setting. However, you will probably have to allow the clock to run at least 24 hours and use the dial and hands for final regulation.

For final regulation, it is best to keep a record of the day and hour when the clock is set, the amount of the change made and whether it was to make the pendulum go fast or slow. At first, you can make an adjustment every day or two, moving the pendulum regulating nut perhaps only a quarter of an inch at a time. Then, if the clock seems to show no great gain or loss of time, make your check once a week on the same day for two or three weeks. When the time comes that a shift of only 1/16 of an inch of the regulating nut (on a ball pendulum) determines whether the clock is fast or slow, you know that you have it regulated as finely as possible.

The actual regulating operation should be performed on the pendulum regulating nut just as the pendulum is about to reverse its direction. If you wait for the pendulum to reach this point, you will be able to start it rotating again at this same point by merely releasing it from your fingers. If you swing the pendulum just a few degrees past reversal point, no great harm will be done. People who swing the pendulum so that it will rotate much more than its normal cycle after each regulation immediately inject an error in the rate of the clock, for the clock will start keeping time again only after the pendulum has returned to its normal cycle. In the meantime, the movement of the hands, having been slowed down somewhat by the long "overswing" of the pendulum, will indicate a loss of time. Although this would normally suggest that a *fast* regulation of the pendulum nut be made, such an adjustment may be unnecessary.

Some clock repairmen, with a knowledge of this phenomenon, reset the hands with their master timepiece only after the clock has run a few hours. This is, in fact, an excellent idea but, unfortunately, time consuming.

Often people ask why no one has made a chart which would relate the number of minutes a day that a clock could be made to gain or lose to a certain amount of turn of the pendulum regulating nut. As a matter of fact, such a chart would not be practical. The relationship would vary from model to model with differences in the type of pendulum, as well as in the diameters of the regulating nuts, being just two variables. A chart would have to be made for each clock.

QUESTION (9): Do all 400-Day Clock pendulums make 8 turns per minute?

ANSWER:
Until about 1953, practically all 400-Day Clock pendulums did make 8 turns or beats per minute. However, several of the miniature movements made since that time have pendulums making 6, 8, 10, 12, or 15 beats per minute. In order to help the repair person, the number of Beats per Minute are shown for each key-wind suspension unit in Section 10. On older 400-Day Clocks, for which suspension units are not shown, it is best to assume that the pendulum will make 8 turns per minute. If the approach of counting the number of beats per minute does not seem to be working, then install the dial and hands and follow the procedure described in Question 8 for the final regulation of a clock.

QUESTION (10): How long should the suspension spring be?

ANSWER:
The proper length of the suspension spring varies with the type of clock.

Most clocks imported since 1949 have a pendulum guide cup on the base into which an extended axis of the pendulum is supposed to hang. The purpose of this cup is twofold: (1) It shows whether or not the clock is sufficiently level to operate properly, and (2) it prevents the pendulum from swinging freely when the clock is picked up, thereby guarding against a bent suspension spring or a broken dome.

The spring for clocks having these guide cups should be long enough to insure that the pendulum will remain suspended within the confines of the cup, but not so long that it will touch the bottom. Since the cup is seldom more than 3/16" deep, the spring should be measured carefully especially if it is being cut from a longer piece.

The length of spring for clocks equipped with pendulum locking devices is even more critical. For some, a difference of 1/16" may make either the clock or the locking device inoperative. After installing a new suspension spring, check the locking of the pendulum and the safe clearance between the pendulum and the locking bracket or socket before putting the pendulum in beat and regulating.

Requirements are less precise for clocks having ball pendulums without guide cups or locking devices, or for those having old style disc pendulums. In these cases, the clock can be made to run properly even if the pendulum is as much as an inch or more above the base. However, the shorter a spring is the thinner it will have to be if the clock is to be regulated properly.

Spring lengths indicated for the suspension units illustrated in Section 10 are correct for clocks with guide cups and pendulum locking devices. For clocks made prior to 1949, the spring should be cut so the pendulum is approximately one quarter of an inch above the base. If a higher pendulum is desired, a somewhat thinner spring will have to be used. While discussing suspension spring length, it might be well to read the answer to Question 20.

QUESTION (11): What is the best way to determine whether a replacement suspension spring is of the correct strength for the clock?

ANSWER:
If the correct suspension strength for the clock is not known (or if the back plate of the clock is not illustrated in this Guide), you will have to make a guess trying one of the two following approaches. The first approach is to find a back plate illustration in the guide which appears to be close to the back plate on your clock and start with the thickness recommended for this clock. The second approach is to select a spring of middle strength, such as .004" (.102mm) for a Standard size movement, or a .0025" (.064mm) for a miniature movement, and try it. Counting the number of Beats per Minute as described in Question 8 will enable you to quickly determine if the spring thickness chosen is close to what is required.

If you are comfortable with the Beats per Minute concept, some rules of thumb concerning the regulation of 400-Day Clocks may be helpful. Generally, the turning of the regulation nut on the pendulum will change the time required for the specified number of Beats per Minute approximately 4 seconds when adjusting the pendulum from full Slow to full Fast. Thus, if the pendulum adjustment is centered in its range, the time required for the specified number of Beats will be plus or minus 2 seconds. The second rule of thumb is that a change of .0001" (.0025mm) in spring thickness will change the time required for the specified number of Beats per Minute by approximately 3.7 seconds for a standard sized clock and 4.1 seconds for a miniature sized clock.

As an example, let us assume that you are working on a standard sized clock manufactured before World War I, for which no suspension spring thickness is shown in the Guide. The age of the clock indicates that you can expect that the pendulum will have 8 Beats per Minute. You install a .0040" thick suspension spring and find that, with the pendulum adjustment in its center position, the clock requires 54 seconds to complete 8 beats. This result indicates that you need to slow the clock. The rule of thumb indicates that you can expect to change the time required for 8 beats by approximately 3.7 seconds for each change in suspension spring thickness of .0001". Thus, you need to use a thinner spring and, using the second rule of thumb, a change to the spring thickness of .0002" to .0038" should be the next try. Since you began your work by centering the pendulum in its adjustment range, you should have no difficulty in bringing the clock into proper regulation with the .0038" thick spring.

QUESTION (12): How do you replace a suspension spring in a clock when the suspension blocks have no screws, but are just pressed together?

ANSWER:
The most satisfactory way to replace the suspension spring is to replace the existing cheaper top and bottom blocks for ones with screws. Competition among the German 400-Day Clock manufacturers became so keen for the U.S. dollar in 1952-1953 that several manufacturers reduced their cost without thought to possible future repair problems. Most of these manufacturers evidently saw the error of their ways, but in the meantime possibly 50,000 or more clocks with suspension springs of the type you have described found their way into U.S. homes. This is a sufficiently large number of clocks to make it worth your while to glance at the suspension block of every clock before estimating on its repair. What may appear to be a simple suspension spring replacement may turn out to be a suspension *unit* replacement with probably more than normal time needed to obtain the new unit or to repair the old one.

Usually the fork is attached to the spring in the customary way with a screw so it can be reused.

It is possible to replace a suspension spring using the old blocks by prying them open, replacing the spring, and closing them again tightly with brass rivets. Here is the way it's done: drill two small holes through the blocks, one on either side of the spring. Countersink them slightly so that the rivets will hold after their ends are filed flush. *Be sure to remove any broken pieces of spring from the old blocks* before inserting the new spring and, of course, make sure that the new spring is of the correct strength and length and is correctly seated in the blocks before you rivet them.

The chief disadvantage at attempting to rivet the old blocks is that it's hard to know whether the spring is being held tightly enough. If the spring hasn't been held tightly and loosens when you test it, the riveting job has to be done over.

Although it is preferable to use the blocks made by the manufacturer of a clock, those of other manufacturers may be used. Just keep in mind that the top block must be able to move freely in the saddle so the suspension spring will always pull from the pivoting point rather that from the bottom of the block, as it will if the fit is too tight. Also, the bottom block must fit into the pendulum hook easily so the spring will not be bent when the pendulum is being attached or removed. To free a tight block, use a fine file to remove burrs for screw heads or ends. It is possible, of course, to increase the opening in the saddle or pendulum hook with a file, but this method is to be discouraged since it irreversibly changes the original design. Never attempt to open these by bending as they may break.

QUESTION (13): How tight should the hands be?

ANSWER:
The hands should turn very easily. If they are properly adjusted, you should be able to turn the minute hand with your little finger without exerting much pressure.

The minute hand on some clocks imported in the 1950's was set so tightly that it is difficult to turn it without bending it. The pressure needed to turn it is so great that it forces the anchor to flutter, no matter what position of the cycle the pendulum is in. (It is normal for the anchor to flutter when the hands are being turned while the pendulum is in the *middle* of its cycle. In fact, this is a test you can make to prove that the escapement is in order.)

Hands that turn stiffly are usually found in clocks that do not have a tension washer behind the cannon pinion. (See Question 14.) Tension is obtained from either a purposely crimped cannon pinion, or from a hand nut which presses tightly against the minute hand bushing, or from both.

A pinched cannon pinion is the result of poor design. It is to be found along with other ill-advised economies, in clocks which have been assembled with price rather than quality in mind. There is nothing wrong with a design which allows hand tension to be obtained directly from the cannon pinion. Such an arrangement is found in many of the best clock movements. In these cases, the cannon pinion if fitted closely to the centerwheel arbor along a considerable length. Tension control is obtained from a split section of its tube, a distance of as much as a third of its length. With this design, it it possible to obtain a positive, yet easy movement of the hands.

The tube of the pinched 400-Day Clock cannon pinion is usually turned to a smaller diameter for about ⅛th of an inch at its center to make it thinner and thus easy to pinch. An examination of the cannon pinion will reveal that the actual pinching was probably done with a pliers-like tool which left only a small indentation where its two jaws bit into the brass. Since the entire tension control of the hands is limited to these two little indentations, it is easy to understand why this control is uneven, why it varies from clock to clock—too loose in some, too tight in others. Inevitably, the wear on just these two points is bound to cause trouble.

If tension is obtained just by pressure of the hand nut on the hand bushing, i.e., by squeezing the cannon pinion from end to end (without a tension washer), the clock will also inevitably give trouble.

The permanent cure for these conditions is to redesigning the cannon pinion so that a tension washer can be put behind it. This means that the tube must be smoothed on the inside and shortened in length by about 1.5 to 2.0 millimeters by turning down the thickness of the pinion proper. There is usually enough room so that the thinner pinion still will engage the intermediate wheel without interference. However, you should always examine the relative thickness of the wheel and pinion, as well as the play in the intermediate wheel, to be sure that the operation can be performed succesfully.

Unfortunately, it is not always possible to take the time required to make permanent repair. The practical action to take therefore, is to readjust the pinch in the cannon pinion so that the hands will turn with ease. This repair should be good for many years.

The difference between hands which are too loose or too tight may be the difference of a few thousandths of a millimeter in the tightness of the hand nut on the end of the arbor. The nut is supposed to be tightened until it rests against a shoulder at the end of the threaded portion of the arbor. (See Section 18, No. 109.) It frequently happens, however, that even though it appears to be tight, the nut doesn't actually reach this point. The result is that when the hands are turned backward, the nut backs off and the hands get too loose. When the hands are turned forward, they get too tight. Not infrequently, when you try to manipulate this poor arrangement, you either strip the threads in the hand nut or the threaded end of the arbor breaks off in the nut.

To correct the faulty condition of this design, it is often necessary to do some adroit filing or to add some shims. If the hands are always too tight, or if you cannot make the nut stay tight, the minute hand bushing can be thinned just enough to allow the nut to be drawn tightly on the arbor while at the same time providing a smooth and easy turning of the hands. It might even be necessary to shorten the length of the cannon pinion if the hand bushing is already thin. If the hands are too loose, a shim in the form of a bent washer can be inserted between the hand nut and the minute hand bushing thereby providing a tension washer in front of the cannon pinion rather than behind it. Some manufacturers have, in fact, used this expedient. (Section 18, No. 109.) However, here again the permanent repair should be to shorten the cannon pinion, as described above, so that a tension washer can be put behind it.

QUESTION (14): How important is the tension washer behind the cannon pinion?

ANSWER:
The purpose of the tension washer is to provide a sufficient amount of friction to the motion train, so that the drive from the center wheel arbor will be positive and yet will allow for an easy manual "setting" of the hands. Not all clocks have a tension washer behind the cannon pinion, but clocks that do operate with one usually cause less trouble than those that do not.

Often when a clock is dismantled, this little washer is overlooked and lost. When the motion train is reassembled without it, the hands fall loosely, even when the hand nut (or pin) is tightened. In the repairperson's attempt to make the hands tighter, the nut is often turned too hard. As a result, its threads are stripped or the threaded end of the center wheel arbor breaks off.

Tension washers are of various designs: round, oval, triangular, etc. The shape is unimportant and it is not difficult to make a washer if one is needed. Take a piece of steel from a broken mantel clock suspension spring, or from any similar thin, steel stock. Through it, drill or punch a hole just large enough to fit loosely against the centerwheel arbor hub which is close to the front plate. Trim the thin steel around this hole so that you have a washer no larger in diameter than the portion of the cannon pinion which is not cut into by the pinion teeth. This procedure will assure that points of contact will not interfere with the ends of the intermediate wheelteeth. When this is done, curve the washer slightly so that it will require a slight pressure from the cannon pinion to flatten it. This little pressure, or tension, is sufficient to assure a positive operation of the motion train and still allow the hands to turn freely and easily. (Section 18, No. 109.) (See also Question 13: How tight should the hands be?)

QUESTION (15): How is it possible for the hour hand to lose time even though it is tight on the hour wheel pipe? How can this fault be corrected.?

ANSWER:
Occasionally, incredible as it may seem, intermediate wheels and hour wheels from some clocks have been found to have one too many teeth...an error which would have been discovered at the manufacturers if the clock had been tested. This extra tooth will cause a lack of synchronization between the minute and hour hands. Although wheel cutting machines were corrected before many clocks having this fault got into production, the number of clocks that have reached the U.S. have been sufficient to cause many owners concern and to baffle many repairmen.

There is only one way to make the correction. Locate the faulty wheel and replace it with one with the correct number of teeth.

QUESTION (16): When you clean a clock, is it always necessary to remove the mainspring?

ANSWER:
Definitely yes. Many repairmen are inclined to skip this important operation either because they are too timid to remove the mainspring from the barrel, or because they realize that if they did remove it, they would be unable to replace it properly without a mainspring winder. To cover this omission, they usually just add a little fresh oil and hope for the best.

If the clock is of very recent manufacture and its mainspring was inserted at the factory, you might get by using this obviously slipshod method. However, it is unfair to the customer to neglect this important step. For three very good reasons, a neglected mainspring may be the source of considerable trouble for you:

1. The old oil may be so gummy that the mainspring coils actually stick to each other or to the barrel sides. This condition reduces power to such an extent that the expedient of adding fresh oil will result in only a temporary improvement. Even though the clock may seem to run satisfactorily at the moment, it will probably be returned to you in only a few months. During this period, because its power will be erratic, the clock will not keep good time.

2. One of the mainspring eyes may be torn. When the mainspring is wound, particularly if gummy oil is causing extra resistance, the eye may break. If it does, the resulting shock will usually cause major damage to the teeth of both barrel and first wheel. Should this happen while the clock is in your shop, you'll probably lose your profit on the repair.

3. The coils of the mainspring may be distorted because a previous repairman inserted it in the barrel by hand, an almost impossible job to do without distorting the coils. If you have been making a regular practice of inserting the mainspring by hand, you'd better check your own work in this way: Remove the spring after you've inserted it and see whether the edges of the coils are parallel when you hold the spring by its end. The odds are better than ten to one that the coil will be distorted. When a distorted mainspring uncoils, much of its energy is lost in the pressure that the twisted coils exert on the sides of the barrel. This energy loss weakens the entire power train to a point where the clock will no longer run.

Since a mainspring winder is relatively expensive, and hardly worth the investment unless a specialty is being made of clock repair, it is a fair question to ask how, then, can the mainspring be properly replaced? There are two answers: (1) Send the mainspring, barrel, and arbor to someone who has a winder and will do the job for you, or (2) plan to replace the mainspring with a new one which will come to you already coiled to a diameter slightly smaller than the inside diameter of the barrel. (See Section 17). Work the binding wire that's around the coil over to the edge of the spring, which will allow you to insert the spring in the barrel over the top of the barrel hook. You can then pry off the wire letting the coils open in the barrel.

If the outer eye of the spring has not caught on to the barrel hook, do this: With the exposed part of the mainspring facing down, press the barrel firmly against the bench with the palm of your hand until the spring is in as far as it will go. Insert the arbor, place the cover in position, and tap it around its edges with a fiber hammer until it falls into place. (You can use the padded jaws of a vise to give extra, but gentle, pressure against the lower half of the cover while you tap the upper half with a fiber hammer.) One complete turn of the arbor with a key is then usually enough to slip the spring around inside the barrel until the outer eye of the mainspring engages with the barrel hook. Remove the cover again and check to make sure that the inner coil and eye fit snugly around the arbor and arbor hook. It must not be possible for the arbor hook to free itself from the mainspring eye. Then oil (See Question 18), and replace the cover.

A word of caution about inserting a mainspring into a barrel, particularly if you don't own a winder. Check the *direction* in which the arbor hook is to fasten itself to the inner eye of the spring by placing the arbor into the barrel. Practically all 400-Day Clock mainsprings are coiled into the barrel counter-clockwise. However, some of the pre-World War I clocks had barrels with mainsprings coiled *clockwise.*

QUESTION (17): What is the purpose of the little extra barrel and mainspring attached to the third wheel in some clocks? How should it be set?

ANSWER:
This unusual device appeared only in some Standard size clocks manufactured by Schatz. It was designed to function in the same way as a fuzee. That is, it was supposed to cause an even distribution of power throughout the entire unwinding of the mainspring, correcting for the fact that the power is slightly greater as the mainspring starts to unwind than it is at the end of the wind. In theory, there is probably some justification for the device, but the fact that the manufacturer discontinued it probably indicates that it was not successful. It's also just one more thing to get out of order. The little mainspring in the barrel is subject to breakage and, unless the cover has been put on carefully, the arbor will bind against it.

If the little mainspring breaks, it (1) can be replaced, or (2) a new wheel and pinion can be obtained *without* the barrel unit, or (3) the arbor can be fixed to the barrel. There will be no loss in the timekeeping of the clock if the barrel is eliminated. If you make the little mainspring replacement, you'll notice that the bent hook on its outer end does not attach directly to the barrel hook. An extra, loose and straight piece of spring, about ⅜" long, rests between the barrel hook and the loop mainspring hook. It's quite a trick to wind the little mainspring in the barrel by hand and still keep your sense of humor!

Since the inner mainspring eye easily bends away from the arbor hook be careful not to turn the arbor backward before you replace the entire barrel unit in the movement. When the clock is wound, the little mainspring will automatically wind itself up to the proper tension.

QUESTION (18): Where should you oil the clock?

ANSWER:
The *mainspring* should be oiled with a good mainspring lubricant. The coils unwind very slowly and if the clock is to keep good time over a period of several years, it is important that you use a lubricant that will not oxidize or become gummy. Wet the edges of the coils. A small sized artist's brush makes a good applicator and the oil that remains on it is sufficient to oil around the inner mainspring eye, the arbor and on the inner surface of the barrel. Don't try to "soak" the mainspring. The oil will pour out of the barrel cover when the slot becomes turned to the bottom.

Use a good grade of clock oil on all *pivots.* Remember that too much oil is almost as bad as too little. Under no circumstances should the pivot hole oil cups be completely filled because eventually the oil will flow all over the plate. If you can see that there is oil just around the pivot in the hole, you've used the correct amount. Begin with the arbor holes in the *barrel* and *barrel cover.* Often ignored, these are oiled most easily before the barrel is inserted into the movement. Oil all *pivots* in both front and back plates. Place a small drop behind the *ratchet wheel,* in the *hole of the click,* and at the point where the click spring touches the click. The *anchor pin* should be oiled only slightly at the place where it touches the fork tines. If so much oil is put on that it joins the space between the pin and the tines, it will run down the pin and all over the anchor. Finally, barely touch two or three of the *escape wheel teeth tips* with a very small amount of oil. Eventually, a thin film will become distributed over every tooth tip as well as on the locking and impulse surfaces of the pallets. Here, too, a small sized artist's brush is a useful tool for oiling the anchor pin and escapement. The hairs of the brush should be barely "wet".

Do not oil any of the wheels and pinions in the movement. None is needed.

Do not oil the wheels and pinions of the motion train unless the intermediate wheel has a steel arbor and pinion with one pivot in the plate and the other in a bridge. These two pivots should be oiled.

Ball pendulums should not be oiled anywhere. Disc pendulums should be oiled in the places where the regulating rod enters the two little weights.

QUESTION (19): Can 400-Day Clock parts of different manufacture be interchanged?

ANSWER:
Certain 400-Day Clock parts are interchangeable and others can be made interchangeable with only minor alteration. However, it would be practically impossible to prepare a list of such parts which would be of any value to the repairer. There have been more than a dozen German manufacturers and assemblers of 400-Day Clocks, many of whom made more than one size movement. There are also hundreds of thousands of old clocks still in running order made by manufacturers no longer in existence.

QUESTION (20): What do you do when in a brand new, correctly set up clock (1) the pendulum touches the base, (2) you've shortened the suspension spring to correct for this, and (3) you still can't get the pendulum to swing slowly enough? Should you then try a thinner spring?

ANSWER:
No. You would just be piling error upon error. Such a clock is probably one of several on the market which had either (1) thin, brass rings at the top and bottom of the columns holding the movement platform, or (2) a thin movement holding platform. In shipping, the column rings may have telescoped thus lowering the movement just enough to make the pendulum touch the base; or, if the platform is thin and has become bent, the movement may be leaning backward, a condition which will also lower the pendulum.

If the rings have telescoped, they should be replaced, for it is practically impossible to straighten them and still have them look right. You can prevent thin rings from telescoping again by inserting some common iron washers inside of them so that the pressure from the tightened bolt will act upon the washers rather than upon the rings. In later clocks, the rings were made of heavier gauge brass. These will not telescope.

Usually you can straighten a slightly bent platform with hand pressure using the movement as a lever. Be sure that your effort is directed on the plate, not on the dial or arch as these may bend or break. Also, be sure that all of your effort is directed at the platform and not at the bottom of the columns where they join the base. When the movement is straight again, tighten the nuts under the base.

If you have shortened the suspension spring too much, you will have to replace it with one of correct strength and length.

QUESTION (21): How can you diagnose the fault when an apparently well-ordered clock still fails to run?

ANSWER:
If the movement has been well cleaned and oiled, and otherwise appears to be in order, make these eight tests:

1. Check the pendulum beat. (See Question 6.) Perhaps the suspension spring has become slightly twisted without your noticing it.

2. Examine the fork to see whether it is too tight or, perhaps, too loose on the anchor pin. (See Question 4.) The tines may be bent, and although they're apparently the right distance apart at the end, they may close up at the point where they touch the pin. If the tines are too open, much of the power from the mainspring will be dissipated. There should be *only* enough clearance to ensure that the tines will never grip the pin *in any position* of its movement.

3. Examine the anchor pin to see that it is tight. (See Question 5.) Some pins are threaded into the anchor; others are driven in and held by friction. If it is not obvious which condition prevails, try screwing the pin home before attempting to drive it in.

If the pin is of the type that's purposely offset to bring the top portion *nearer* to the suspension spring, it will probably be found to be threaded. However, the tightening process may leave the pin in an off-center position and throw the pendulum off-beat. If it's only a few degrees off, compensate for it by adjusting the pendulum beat. However, if it's way off center, it would be best to stake the pin in its correct position.

4. Remove the suspension unit to see if the ends of the suspension spring have been set *exactly* in the center of both the top and bottom block. If the spring is kinked between the fork and top block, you should replace it; if kinked between fork and bottom block, try to straighten it.

5. Check the top suspension block in its saddle to be sure that it doesn't bind or is not too loose. The block should be free so that the pendulum will hang from the holding screw or pin rather than from the bottom of the block. If the block is so loose that it moves when the anchor pin activates the suspension fork, it will absorb a part of the power, all of which should be directed to the pendulum.

6. Note whether the suspension spring is touching the guard holding screws or suspension guard at any point. Also note whether the fork is touching the suspension bracket.

7. Look through the movement, particularly at the area directly behind the back plate, to see whether the ends of any screws are touching moving parts. In some movements, the screw holding the ratchet wheel bridge has the same head and thread as the pillar screws but is shorter. The longer screw may have been interchanged and is pressing against the barrel. Also, the lower guard holding screw, if tightened without the guard on, may be pressing against the barrel.

8. Wait until the clock stops of its own accord. Then, carefully start the pendulum rotating again *without touching the hands.* (The slightest movement of the hands may temporarily release a "catch".) Then note whether:

(A) The pallets hang between two teeth without touching them (which means that there is no power from the escape wheel). If they do, remove the hands, dial and motion train (hour wheel, intermediate wheel, and cannon pinion). If the clock will run properly without them, the problem can be located somewhere within the motion train. (See Question 13 and 14.) Make sure that the hour hand is far enough onto the hour wheel pipe so that its front cannot touch the minute hand bushing. If the clock still won't run with the motion train removed, the problem must lie somewhere between the plates.

Actually there can be only about three causes for a hung escape wheel: (1) one or more teeth or pinion leaves are bent somewhere in the train, (2) an arbor is bent, or (3) a pivot is bent.

(B) A pallet is locked to an escape wheel tooth, thus preventing the tooth from escaping. Because it may take some time before a pallet actually does lock, the clock may run satisfactorily for a period of time.

A locked pallet is usually the result of a bent escape wheel tooth, a bent pivot or an escape wheel that is out of round. These are among the more tantalizing problems that arise, because the locking may occur only infrequently and rarely at a time when you are actually observing the clock.

One way to check on this condition is to turn the movement upside down and, with your finger pressing upward on the anchor arbor, allow the anchor to flip back and forth so that at least two or three complete turns of the escape wheel are made. This procedure brings the escape teeth and pallets as closely together as possible. If they are going to catch at all, you'll feel it with your finger. This is also a good test to check on a possible bent escape wheel tooth.

(C) The escapement seems to function properly, while at the same time, the escape wheel teeth drop sluggishly onto the pallets with no audible snap. This condition is customarily found in clocks that have not been cleaned for a while and are loaded with gummy oil. However, since we are presuming that the clock has been thoroughly cleaned and oiled, the problem is to locate the cause of the power loss.

If the mainspring was inserted into the barrel by hand, the chances are that its coils were seriously distorted. It is almost impossible to insert a 400-Day Clock mainspring in its barrel by hand without distorting its coils. Such distortion causes the edges of the mainspring coils to scrape the sides of the barrel as it unwinds, thus using a large portion of its power. This can be checked by removing the mainspring and letting it hang by its outer end. If the sides of the spring are not exactly parallel, it has been distorted. The only sure way of inserting the spring into the barrel properly is with a mainspring winder. (See Question 16.)

Loss of power also can be caused by binding of a wheel and pinion. An easy check on this is to insert the arbors between the plates two at a time, first the barrel and first wheel, then the first and second wheels, etc. Be sure to tighten all four pillars before testing each pair. Finally, insert everything between the plates except the anchor, attach the ratchet wheel, click, and click spring and, with a key, give the winding arbor one quarter turn. The escape wheel should start spinning immediately. If it doesn't, look again carefully at all wheel teeth, pinion leaves and pivots. One of them has to be out of alignment.

QUESTION (22): I've been in the repair business over 20 years and can make a nice profit on the more expensive clocks, but I've given up taking in 400-Day Clocks. How can I make a profit when I have to cut my repair charges so much?

ANSWER:
You should not let the retail price of the clock have any influence on what your repair charge should be. In fact, you are losing a great profit opportunity by not openly soliciting 400-Day Clock repair in order to get as much of this business as possible in your area.

If you have been repairing 400-Day Clocks for a long time, you know that it takes you much less time to repair a clock now than it did when you started. You also now have your most important timesaver, the ***Horolovar 400-Day Clock Repair Guide,*** on your bench within easy reach. Therefore, you should have a good idea about how much time it will take to make a repair.

QUESTION (23): Does the eccentric nut always carry the anchor pivot hole?

ANSWER:
Until about 1957, the eccentric nut in all 400-Day Clocks, even those made before World War I, controlled the position of the anchor pivot. However, when the very small 400-Day Clock movements began to appear, some movements were made with the eccentric nut designed to carry the escape wheel pivot. In using the "Trouble-Shooting Chart" in Section 7 for making adjustments, the instruction "raise (or lower) pivot hole" refers to the *anchor* pivot hole. If the eccentric nut carries the *escape wheel* pivot, the correction made should be just *opposite* to that indicated.

QUESTION (24): I've always turned the regulating nut on 400-Day Clock pendulums clockwise to slow the clocks down. *However, I've noticed that with some of the newer clocks, I have to turn counterclockwise to slow down. This is very confusing. Can you list the plate number of the clocks in the Guide with pendulums having regulation nuts that must be turned* counterclockwise *to make them go* slower?

ANSWER:
It is impossible to furnish such a list because many manufacturers made both types of pendulums yet gave no corresponding indication of the change on the back plates. Although the pendulum in each case was redesigned to eliminate play in the regulation, the regulating nut could just as well have been designed to turn in the usual way: *clockwise to make the clock go slower.* Why it wasn't is one of those mysteries of foreign manufacturing.

Incidentally, it's always a good idea to explain to the customer that, if he expects the clock to keep good time, he must give it final regulation in his home. Show him the pendulum regulating nut and tell him which way it should be turned to make the clock go faster or slower. Make note on the repair ticket of the correct direction at the time you are regulating the clock.

QUESTION (25): Are there any special instructions for the repair of the Schatz 1000-Day Clock?

ANSWER:
No. The clock should be treated exactly the same as the Schatz miniature 400-Day Clock. Actually, the only major differences in basic construction between these two Schatz movements are that the 1000-Day Clock has a larger barrel, a longer, heavier mainspring and a heavier first wheel. Also, the anchor in the Schatz 1000-Day movement has a longer arbor because the distance between the front and black plates is greater than with the miniature 400-Day movement. Otherwise, most of the parts in the two types of Schatz movements are interchangeable.

QUESTION (26): What strength Horolovar suspension spring should be used for the little 8-Day Staiger torsion pendulum clock?

ANSWER:
The correct strength Horolovar spring is .0017" (.043mm). Since this strength is not available, the .0018" (.046mm) spring can be used if thinned down a little. It's not easy to make a suspension spring replacement in this clock because the top and bottom suspension fittings are pinched onto the ends of the spring rather than attached with screws.

QUESTION (27): I have a clock in the shop (Section 9, Plate 1087), which I have cleaned and checked over carefully. It runs perfectly when the hands are off, but with the hands on, the clock always stops when both minute and hour hands are being raised. Can you suggest where the loss of power might be?

ANSWER:
The clock you have is probably the one made by Edgar Henn Company which is covered with a glass sided case. One lot of this model clock, probably numbering several thousand, was originally made with hands which were too heavy for the movement. You can either replace the hands with lighter ones, or substitute a shorter pair without harming the appearance of the clock. Unfortunately, the Henn company went out of business in 1956, and Henn substitute hands are not available. You'll just have to find a lighter weight pair that will fit.

QUESTION (28): My biggest problem is not with the clocks. It's with their owners! They take the clock home without knowing how to set it up, they bend the spring which throws the pendulum out of beat, then they bring it back complaining that it wasn't repaired right! How can I convince them that it was their fault?

ANSWER:
You can be sure that your customers are not any happier to have to return the clocks to you than you are to see the clocks come back! So, the problem is not one of how to convince these people that it's their fault. It's a matter of giving the owners some basic instruction on how to handle their clocks properly and, of equal importance, how *not* to abuse them.

Here is our suggestion for a solution: Before turning a 400-Day Clock over to its owner, ask him if he knows how to handle it and regulate it properly. If he says "yes", tell him you'd like to review a few points with him to be sure he knows what your guarantee includes and what it doesn't. Most likely he'll welcome anything you can tell him because there's a good chance that the only information he's ever had about the clock came from the printed instructions he received with it. Unfortunately, these were not too complete and may have been lost.

Here, then, is a set of general instructions which you can use with every repair that you are prepared to guarantee.

Note that there are several places in which alternate instructions are given, a procedure made necessary because of the many different models in existence. You should assist your customer by checking the paragraphs which specifically apply to the clock in question.

Copies of this Repair Warranty are available at cost from The Horolovar Company. (A sample is reproduced on this and the following pages.)

QUESTION (29): How can you identify a genuine Horolovar suspension spring?

ANSWER:
There are many suspension springs on the market, some of which upon casual examination appear to be exactly the same as Horolovar. Unfortunately, it is almost impossible to identify the genuine product without a chemical analysis. Therefore, it it is best to rely on the identifying package. Horolovar suspension springs are never sold in bulk and there are no "seconds" sold in unlabeled packages. All Horolovar springs are sold in the familiar red and white envelopes with the name HOROLOVAR TEMPERATURE COMPENSATING prominently displayed. Envelopes contain either three springs of a single strength or a dozen springs of assorted strengths.

It is important to remember that *if you are following suspension spring strength (thickness) recommendations in the Horolovar Repair Guide, Horolovar suspension springs are the only ones which will allow you to regulate the clock within the regulating limits of the pendulum.*

QUESTION (30): Is it possible to tighten the two movement holding screws too much?

ANSWER:
Yes, it is possible to tighten the screws on some clocks so much that the movement pillars will bend. This in turn will pull the lower part of the plates together which may eliminate the end shake required for the barrel and first wheel arbors and, consequently, stop the clock. Thus, care should be taken not to turn the movement holding screws too tightly. Since the only function of these two screws is to keep the movement from falling off of its support, a firm turn with pliers is sufficient. Some manufacturers have placed spacer washers under the pillars to prevent them from becoming distorted when the movement holding screws are tightened. This is an acceptable refinement, but is not really a necessary one.

QUESTION (31): How can I tell where to put the fork on a suspension spring unit that is not illustrated in the Guide?

ANSWER:
When the position of the fork on the suspension spring is not known for an old clock, you will have to find the correct position by trial and error. A good starting is to attach the

400-DAY CLOCK REPAIR WARRANTY

With Operating Instructions

Repair Ticket No. ________ Date __________

IDENTIFICATION

This is a ______________________________ Key-wind 400-Day Clock
(Fill in name of manufacturer if known)

WARRANTY

This clock has been expertly repaired by

and is guaranteed against any mechanical failure, *not caused by abuse*, for a period of ____________ following the above date.

Per ____________

OWNER PLEASE NOTE

The most vital part of this clock is the thin wire "spring" from which the pendulum is suspended. It is only a little thicker than a human hair and must be treated with great care so that it does not become twisted or bent. Your clock has been equipped with a HOROLOVAR TEMPERATURE COMPENSATING SUSPENSION SPRING which is the best and toughest spring available. It should last forever under normal service and strains, but *it will not stand abuse.*

To get the most satisfaction and enjoyment from your clock, the person in charge of it should carefully review the operating instructions as given on the following pages. The paragraphs that apply to this clock have been checked.

OPERATING INSTRUCTIONS FOR YOUR 400-DAY CLOCK

(Checked paragraphs apply to *your* clock)

☑ To Lock and Unlock the Pendulum Suspension

(Read checked paragraphs *A*, *B* or *C*) Before the pendulum is put into motion, which is the way the clock is started, it is necessary to get it into a free-suspended position.

Check *A*, *B*, or *C*

A **Clocks with Detached Pendulums and Without Suspension Spring Guards** (or with ring guards attached to the movement back plate)

The pendulum should be *very carefully* hooked on to the pin block on the suspension spring. Attaching the pendulum to the pin block (and detaching it when necessary) is a most important operation, and great care must be taken not to bend the spring even the least bit! Don't try to do it with one hand. It's best to hold the little pin block in position with the finger of one hand while "hooking" the pendulum on it with the other. (When removing the pendulum from the pin block, remember that the pendulum "hook" has to be raised slightly before the pin can be eased out.)

B **Clocks with Detached Pendulums and with Suspension Spring Guards that Can Be Locked**

These spring protection pieces, or guards, all located at the back of the movement, have different designs. Some just cover the pin block attached to the lower end of the suspension spring; others actually clamp the pin block so it can't be moved. The guard itself may be held in place friction-tight, or by a thumb screw. Whatever the design, this protection device must be raised up and away from the pin block and kept away.

With the pin block on the suspension spring thus exposed, the pendulum should then be very carefully hooked on. Attaching the pendulum to the suspension spring (and detaching it when necessary) is a most important operation, and great care must be taken not to bend the spring even the least bit! Don't try to do it with one hand. It's best to hold the little pin block in position with the finger of one hand while "hooking" the pendulum on it with the other. (When removing the pendulum from the pin block, remember that the pendulum "hook" has to be raised slightly before the pin can be eased out.)

C **Clocks with Attached Pendulums and with Pendulum Locking Devices** (Read checked paragraphs D, E or F)

If *C* is checked, check *D*, *E*, or *F*

D **Kundo Clock**
The pendulum locking lever is in the back, under the movement. A "safety" cotter pin may be in position (to prevent the lever from opening accidentally during shipping) and this pin must be removed before the locking lever can be opened. (If the clock has been factory packed, a brass clamp may be attached to the upper part of the pendulum. It should be pulled off and discarded.) The pendulum should be carefully held, when the locking lever is pushed to the side, allowing it to be lowered gently into the free position: a drop of about ⅛".

Caution: When this action is reversed; that is, when the pendulum is to be locked again for moving or shipping, remember that it must be *carefully raised (about ⅛") into its socket* so that the locking lever will hold it in a firm position. If you pull the locking lever over *without* first raising the pendulum into its socket, no locking takes place. In such a case, the suspension spring is quite likely to be seriously bent if the clock is tipped.

If the clock is to be shipped, it is advisable to replace the "safety" pin in the hole by the lever. If the original cotter pin that came with the clock is not available, a satisfactory pin can be made from a slightly opened paper clip. (If no hole for a "safety" pin is provided, the locking lever should be held in the locked position with a rubber band.)

E **Schatz Clock**
The pendulum of this clock is locked and unlocked by a lever which is located in front, or in back, of the clock at the bottom of the base. When the lever is moved to an extreme right position, the pendulum is raised and locked; when it is moved to the extreme left position, the pendulum is lowered and is "free."

A second and more secure pendulum lock is provided for use when the clock is shipped. This lock is at the back of the movement, located in the lower piece of the tubular suspension spring guard. This lower guard can be adjusted—either locked (lowered) or unlocked (raised)—by means of a thumb screw. Except when the clock is being shipped, this adjustable guard should be raised as high as it will go, and held there by tightening the thumb screw.

To lock the pendulum for shipping, loosen the thumb screw and lower the piece in such a way that the two prongs provided not only go through the two holes in the pendulum locking bracket, but also through two holes in the disc at the top of the pendulum. (It may be necessary to rotate the pendulum slightly to find a pair of matching holes through which the prongs will go.) When the two prongs of the piece have thus been fully seated, the thumb screw should be tightened to hold it in place.

F **Other Clocks**
There are many different types of pendulum locking mechanisms, most of which provide some means of raising the pendulum against a bracket, at the rear of the movement, thereby supporting the pendulum at top and bottom during shipping. If the method of operating the mechanism on your clock is not clear, read the instructions in the paragraphs above which explain other types of locking devices. If they don't help you to determine how yours works, take the clock to your local clock repairman who will be glad to explain the operation to you.

☑ To Place the Clock

The clock should be placed in a position *free from vibration*. Television sets, pianos, unsteady book cases or tables are all *unsatisfactory* places. A mantel, a steady piece of furniture, or even a wall bracket, is to be preferred. It will not help its timekeeping qualities if the clock is placed directly over a radiator, or in a window where it will be subjected to direct sunlight or to drafts.

☑ To Level the Clock

The surface on which the clock rests does not have to be completely level. But the clock itself *must* be level. You don't need a leveling tool if the clock is provided with a pendulum guide cup on the base. The clock is level when the tip of the pendulum is directly over, or inside, the guide cup.

If the base of the clock is provided with leveling screws, any leveling correction can be made by raising or lowering one or more of these screws. If leveling screws are not provided, the correction can be made by inserting one or more pieces of stiff cardboard in two places (about 4 inches apart to prevent wobbling), under the base.

☑ To Start the Clock—Very Important!

Carefully rotate the pendulum, in either direction, so that it is about one complete turn from dead center, then release it. This will start the pendulum rotating more than is usually required. Depending upon the clock, the normal rotation may be as little as ¾ turn or, for some miniature or midget clocks, as much as a turn and a half. After the pendulum has continued to rotate for a half-hour or so, it will have "settled down" to its normal cycle. Once you know how far the pendulum normally rotates in one direction, always give it just a little more than this amount whenever you have to start it in the future. But *never rotate the pendulum more than one and one half turns from dead center.* If you do, a permanent twist may be left in the suspension spring which will prevent the clock from running. (No clock repairman can be expected to replace a suspension spring, free of charge, if the spring has been permanently twisted or otherwise bent).

✔ To Set the Hands

Set the time by moving the minute hand in either direction. The hour hand will automatically follow. If the hour hand does not point exactly to the hour when the minute hand is on 12, shift its position, in either direction (it's on only friction tight) until it does point to the hour. Don't be alarmed if the escapement "flutters" when the hands are being moved.

✔ To Regulate the Clock (Read checked paragraphs *G* or *J*)

Check *G* or *J* — If *G*, check *H* or *I*

G On 3-ball and 4-ball pendulums, a round, knurled "regulating disc" (about the size of a nickel) is at the top. When turned, it will make the balls go toward, or away from, the center of the pendulum. The direction in which the regulating disc should be turned to make the clock go faster or slower depends upon the design of the pendulum. Look for the letters F (fast) and S (slow) on top of, or near, the regulating disc.

To make *this* clock go SLOWER

H Turn regulating disc *clockwise*

I Turn regulating disc *counterclockwise*

J On disc pendulums of pre-World War I clocks, there are two little disc weights which can be moved by turning a threaded steel rod with a key. They will go *toward* the center of the pendulum (turn key toward *F* or *A* to make the clock go *faster*) or away from the center (turn key toward *S* or *R* to make the clock go *slower*). A double-end key for both regulating and winding is available for this clock. It can be obtained from your clockmaker.

✔ To Adjust for Accurate Timekeeping

If the clock is to be used as a timekeeper, the final regulating has to be done when it is in its position and leveled. The most practical way of regulating the clock is as follows: At least a half-hour after the clock has first been set up and started, at which time the pendulum should be rotating back and forth at its normal rate, set the hands to some accurate time source such as a quartz watch or electric clock. Cover the clock with its dome or case and don't touch it again for a week. Start keeping a record with a table similar to this:

DATE	RIGHT TIME	CLOCK TIME	MINUTES DIFFERENCE		AMOUNT OF REGULATION AND DIRECTION
			FAST	SLOW	
			XXXX	XXXX	XXXXXXXXXXXX

At the end of the week (or sooner, if the clock is gaining or losing time rapidly), compare the time with the accurate source and note the number of minutes that the clock has run fast or slow. To correct its rate of gain or loss, stop the pendulum at the point where it reverses its cycle, and adjust the regulating disc on the pendulum (or regulating rod on disc pendulum clocks) as described above. (A half turn will make a significant change in the timekeeping, so if the clock is only 5 or 10 minutes fast or slow, start with a ¼ or ⅛ turn.) Then, release the pendulum at, or just a little beyond, the place where you stopped it. (By making the regulating adjustment while holding the pendulum at one end of its cycle, you avoid overswinging it when you start it again. An overswing prevents accurate regulation.) Reset the hands to the correct time and let the clock run without further correction for another week.

Be sure to make a note of the amount the clock ran fast or slow, and the amount of turn you gave to the pendulum regulating disc, i.e., ½ turn, ¼ turn, etc. After two or three weeks, you will reach a point when a ⅛ turn of the outside edge of the regulating disc will change the timing from fast to slow (or from slow to fast). You have then reached the maximum possible regulation. Do not make any further regulating adjustments unless the clock is moved to another position.

Don't expect the clock to keep accurate time for 400 days! Every two weeks, or whenever necessary, remove the dome and nudge the minute hand to the correct time. With this attention, the clock will keep as good time as an 8-day mantel clock.

✔ To Wind the Clock

When fully wound, the 400-Day Clock will usually run for 500 or more days. However, many people like to make a ceremony of the winding operation by doing it once a year on a birthday or wedding anniversary. All 400-Day Clocks that are wound from the back of the movement wind to the left—counter-clockwise. The mainspring is a heavy one and a little pressure is needed to turn the key. It is impossible to overwind the clock, so keep turning the key until the spring is obviously wound tight. (Hint: The clock will keep more accurate time throughout the year if it is wound fully once every two months.)

✔ To Keep the Key Handy

One of the best places to keep the key, so that it will be available when the winding takes place, is right under the clock!

✔ To Clean Glass or Plastic

Most 400-Day Clocks are covered with glass or plastic domes, or by glass or plastic sided cases. The clocks are a lot more attractive when the glass or plastic is kept clean and bright.

Glass Domes Wash occasionally in very hot water and wipe off with a clean, dry towel. The dome can be kept bright over long periods by just wiping it off with a clean, damp, paper towel while it is on the clock. When removing the dome from the clock, always grip it as near to the bottom as possible. This will keep your fingerprints in an area where they won't show.

Glass Sided Cases Wipe both sides with a soft cloth, wrung out with hot water. Wipe off with a clean, soft, dry towel. Do not dip the case in water. The sides can be kept bright over long periods by wiping occasionally with a clean, dry cloth while the case is on the clock. Avoid putting your fingers on the glass when removing the case.

Plastic Domes or Cases Follow the same suggestions as above, but avoid extremely hot water. Always use a *very* soft cloth when wiping. Plastic is relatively soft and will eventually become dull if wiped repeatedly with a rough cloth or paper towel.

fork so that there is about ¼" between the top of the fork and the bottom of the top block. On most clocks, this location will result in the fork being centered in the suspension bracket opening. If, with the fork in this position the anchor "flutters", raise the fork slightly. If the anchor moves with a "jerky" motion, lower the fork slightly. If the escapement is in proper adjustment you will find a position between these extremes that will be correct.

QUESTION (32): I have a Kundo Midget clock and used Horolovar Unit 5F as recommended. However, with the regulating nut turned to maximum slow, the clock runs very fast. Shouldn't I use a lighter spring?

ANSWER:

No. The 5F Unit is correct for this clock. You are probably

turning the regulating nut clockwise to slow it down. This is one of several pendulums designed to do exactly the opposite. (See answer to Question 24.)

QUESTION (33): I need to thin a suspension spring slightly. What is the best way to do this?

ANSWER:
Suspension springs can be thinned easily by making a few strokes with a fine emery cloth. The thinning is best accomplished by gluing the emery cloth to two short pieces of wood such as Popsicle sticks. You will be thinning the suspension spring in the area between the fork and the bottom block. Begin by installing the suspension spring in the clock. Then pinch the suspension spring between the pieces of backed emery cloth using your thumb and forefinger. Use your other hand to steady the clock and to hold the pendulum down with a gentle pull! Make your thinning strokes begin at the fork and stop when you reach the bottom block. Release the spring and take another stroke beginning again at the fork.

Make only a few strokes at a time to avoid thinning the suspension spring too much. The emery cloth removes the spring material very quickly so work carefully. Check the alignment of the fork and the bottom block to make sure that you have not given the spring a slight twist during the thinning process.

QUESTION (34): A customer brought in a pre-World War I 400-Day Clock with 4-ball pendulum for repair. Can this pendulum be original with the clock? About what year did 4-ball pendulums replace the old disc pendulums?

ANSWER:
The first ball pendulums were 3-ball and appeared about 1890. 4-ball pendulums were first used about 1900. However, they didn't generally replace the disc pendulums until after World War I. In the meantime, some factories made certain models of clocks with both disc and 4-ball pendulums, some even providing dealers with both types, leaving it up to the buyer to decide which he wanted.

QUESTION (35): I am at the end of the pendulum adjustment and the clock is running too fast. How can I slow the clock?

ANSWER:
You can make a clock run slower by using a thinner suspension spring or by using a longer suspension spring, if there is room for a longer spring. A thicker suspension spring or a shorter suspension spring will make the clock run faster.

QUESTION (36): I find it impossible to adjust the escapement on some clocks so that the anchor pin will go smoothly from one side to the other. How can a jerk be eliminated?

ANSWER:
If the jerk is very slight, don't try to eliminate it. This minor backlash is perfectly normal with many clocks. Here's what's happening:

Near the end of the pendulum's rotation, the suspension spring is twisted sufficiently to cause the fork to move the anchor far enough to allow an escape tooth to be released from a pallet. The tooth drops high on the opposite pallet and almost immediately thereafter the pendulum reverses its rotation. During this reverse cycle, the spring untwists, and the fork guides the anchor back. The pallet pulls away from the escape tooth until the tooth engages the pallet's impulse surface. The resulting impulse given to the anchor causes a slight forward twist to the spring which is what keeps the pendulum in motion. When the escape tooth finally drops off of the pallet, the fork immediately assumes its position on the spring without interference. It is at this point that there may be a slight, momentary backlash of the anchor pin.

In short, the minor backlash is merely a repositioning of the fork by the spring after the escape tooth has ceased its impulse action. Don't confuse this action with "fluttering", which is covered in the answer to Question 3.

QUESTION (37): What is the best way to lacquer the brass parts of the clock after they have been cleaned and polished?

ANSWER:
Best results will be obtained by having both the polishing and lacquering done by a professional. Firms that specialize in silver polishing will usually work on brass. However, these firms may be available only in larger cities. Also, many musical instrument repairers are used to lacquering brass. Check your Yellow Pages.

It is not necessary to lacquer the plates. A good polishing by buff stick or wheel, followed by careful handling to avoid fingerprints, will keep the plate attractive for as long as four or five years. If you lacquer the plates, all traces of lacquer in the pivot holes and oil sinks must be carefully cleaned out before the clock is assembled.

QUESTION (38): When viewed from the end of the anchor, the anchor pin is angled to one side. Is this proper?

ANSWER:
No, the anchor pin should be centered directly above the anchor when viewed from the end of the anchor. Occasionally, you will find a clock where the anchor pin has been bent in order to bring the clock into beat. Setting the beat in this manner makes an easy job difficult and may result in a broken anchor pin when attempts are made to straighten the anchor pin. You can minimize the possibility of breaking the anchor pin by gripping it with needle nose pliers just above the anchor and making the bend above this point. Please note that, when viewed from the side of the anchor, many anchor pins are designed to be bent so that the top of the anchor pin is nearer the fork. In these instances, of course, the pin should not be straightened.

QUESTION (39): Why doesn't the old suspension spring in my clock match the dimensions of the recommended suspension spring shown in Section 9 of the Guide?

ANSWER:
The dimensions of the spring in the Repair Guide are developed for the particular alloy that is used to make Horolovar Suspenion Springs. Springs made of other alloys require different dimensions for the springs. It is almost impossible to correlate spring characteristics between different alloys without extremely sophisticated laboratory equipment. Thus, it is best to rely on the recommended spring sizes shown in the Rapair Guide and to use Horolovar Suspension Springs.

QUESTION (40): Do you have any suggestions for installing a Horolovar Mainspring in a mainspring barrel?

ANSWER:
The following procedures have worked well. As a first step, make note of the direction that the spring is wound in the barrel. Then remove the spring and clean the barrel. Some barrels are relatively thin and may have a tendency to bulge when the mainspring is released in the barrel. To protect against this bulging, tighten a worm gear hose clamp (available in automotive parts stores) around the outside of the barrel before inserting the mainspring. After the hose clamp is tightened, check to see that the spring is wound in the proper direction and then push the mainspring into the barrel. The edge of the barrel will cause the wire keeper to come off as the mainspring is pushed in place. Remove the hose clamp and install the arbor. It may be necessary to turn the arbor until the hole in the outer coil of the mainspring catches on the barrel hook. Lubricate the mainspring by using a good quality clock oil applied to the exposed ends of the mainspring coils (mainspring greases are too thick). Finally, install the barrel cover.

SECTION 9 400-Day Clock Back Plate Illustrations

Back plate markings serve as "fingerprints" to help to identify many clocks and manufacturers. Prior to October 1913, 400-Day Clocks could have completely blank back plates or plates with serial numbers on them or the words "Germany" or "Made in Germany". But after that date, the U.S. Customs required the manufacturer to mark either his own name or the name of the U.S. importer on the plate as well as the words "Germany" or "Made in Germany". Because the name of the importer often appears above the words "Made in Germany," the importer is often mistakenly thought to be the manufacturer.

Round plate movements stamped "Germany" are occasionally seen in cases marked "France". The back plate from one of these movements is illustrated in Plate 1179 in this Section. In the case of unmarked movements in unmarked cases that appear to be French, there is reason to believe that both movement and case were made in Germany even though in some cases the wheels and pinions have French characteristics.

There is also confusion about 400-Day Clock *round plate* movements not carrying "Germany" or "Made in Germany" markings. Some sources believe these movements, especially those in four-glass cases, were made in France. This seems unlikely, as French 400-Day Clock production was extremely small and the round plates, quite different from the round plates in Germany, are marked with the maker's monogram (Plate 1471A). To date, no 400-Day Clock movements with the words "France" or "Made in France" stamped on the back plate have been found.

Many clocks manufactured during the first quarter of the century had serial numbers stamped on their back plates. Since each manufacturer chose a different style or size of numeral, and since he usually stamped his serial number in about the same position on every plate, if serial numbers appear, they sometimes provide a clue to manufacturer identity. It is not known exactly when serialization of 400-Day Clock movements was stopped, but probably about 1930.

The patent numbers D.R.P. No. 144688 and U.S.P. 751686 which appear on the back plates of several different manufacturers (Plates 1043 and 1047) refer to a specially designed "twin loop" pendulum (Section 13, Pendulum 10) that was thought to help compensate for the effect of temperature changes on the suspension spring. This patent does not refer to any part of the movement. Prior to the issuance of the patent, the back plates carried the words "Patent Angemeldet/Patents Applied" (Plate 1471). The German patent for this pendulum was issued to Andreas Huber of Munich on 5 April 1902 and the U.S. patent to Henrich Sattler of Munich on 9 February 1904. The patent numbers D.R.P. 144687 which appear on the back plates of some Jahresuhrenfabrik clocks (Plate 1049) is for a similar pendulum of different design (Section 13, Pendulum 7). To add to the confusion, however, these pendulums were not always used on every clock carrying the pendulum's patent numbers. The number D.R.G.M. 484408 and 502714 have not been identified.

The back plates in this section represent an accumulation of over forty years, many having been supplied by cooperative repairers. The Horolovar Company is always happy to receive approximate drawings of plates not illustrated, particularly when the variations from plates in this section are significant.

HOW TO USE THE BACKPLATE ILLUSTRATIONS

How Plates are Arranged
Plates are arranged alphabetically where there is at least one letter in the marking. Following them, at random, are plates having serial numbers only. Finally, blank plates with no names, letters or numbers are at the end.

Note that alphabetization is by the *first letter* of a name or trade mark. Thus Gustav Becker will be found in G, The National Silver Company will be in T, etc.

If there is no name or trade mark, other lettered markings are alphabetized. For instance, Made in Germany will be in M, Germany will be in G, the letters D.R.G.M. in D, etc.

About the Manufacturers
If the name of the manufacturer is known, it appears at the top left of the plate.

About the Date of Manufacture
The date of manufacture, i.e., circa 1900, appears above the top right of the plate. It is an *approximate* date, sometimes based upon sketchy information.

About Horolovar Spring Strengths
The Horolovar spring strengths shown under each back plate are based on the assumption that the clock has its original pendulum, and that the pendulum's shape, size or weight has not been altered.

An asterisk (*) following a spring size indicates that there is some question about whether the clock tested has the original pendulum or whether the pendulum has been altered.

A minus sign (-) following a spring size indicates that the required spring falls between two *Horolovar* stock strengths, and the strength shown is the heavier and must be thinned down. (See Section 8, Question 33 for instructions for thinning a suspension spring.)

About Pendulums
The type of pendulum most often found with the plate is listed beneath the lower right corner of the plate. However, it should be noted that manufacturers often made other styles of pendulums available. Therefore, a four ball pendulum found on a clock with a plate which shows "Disc Pendulum" may not necessarily be incorrect.

In a few instances, manufacturers used miniature movements with standard sized dials, columns, and bases. In these cases, the pendulum may be identified as a "Standard Pendulum". This identification refers to the size of the pendulum.

About Horolovar Mainspring Sizes
Where it has been possible, the size of the Horolovar Mainspring to be used with the clock is shown under each plate: 19 x 36, 19 x 38, etc. (See Section 17 for details and dimensions of mainspring sizes; also Section 18, No. 75.)

Illustrations
The *outlines* of the plates are approximate. The holes are *for position only* and particularly in the case of pivot holes, are generally larger than the actual holes in the plates.

All *threaded* holes are shown in solid black.

About the Suspension Units
Complete Suspension Units for clocks imported since 1949 are illustrated in Section 10. The Horolovar Numbers that designate these Units are indicated under each pack plate illustration: Units 1, 3A; Units 6, 7, 8, 9; Unit 13B; etc. All Units which have been identified are included in order to maintain a historical record of the Suspension Units used with the post World War II 400-Day Clocks.

A choice of Units for a particular clock may be indicated for one or more of several reasons. In some instances, different units have slightly different forks or blocks. These differences allow a more precise dating of some clocks. In other instances, the length of the Unit may be different for use with movements which are located at different heights above the base. When a choice of Units is indicated, it is advisable to refer to the illustrations and descriptions in Section 10 in order to determine which one is appropriate and recommended for a particular clock.

Some recommended Units have slightly different shaped blocks and forks from the Unit originally installed in the clock. Functionally, there is no difference between the original Unit and the recommended Unit.

Kieninger & Obergfell c 1930

A

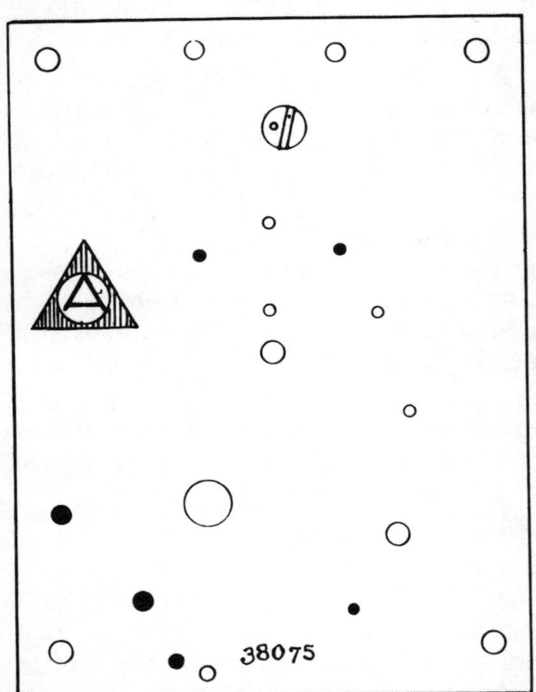

Plate 1001 4-Ball Pendulum
USE .0032" (.081mm) HOROLOVAR
Unit 1 (19 x 38)

Kern & Link c 1932

A.C. BECKEN CO.

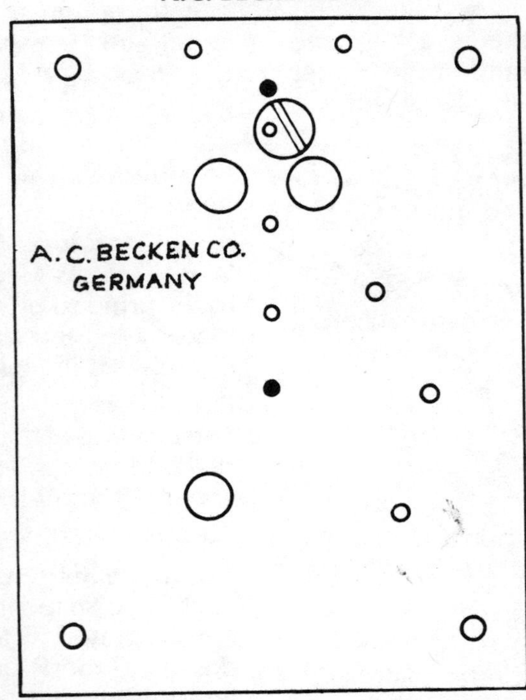

Plate 1002 Disc Pendulum
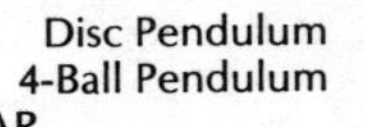
4-Ball Pendulum
USE .0036" (.091mm) HOROLOVAR
Units 11A, 11B (18 x 38)

Wintermantel Uhrenfabrik c 1900

ALB. WINTERMANTEL

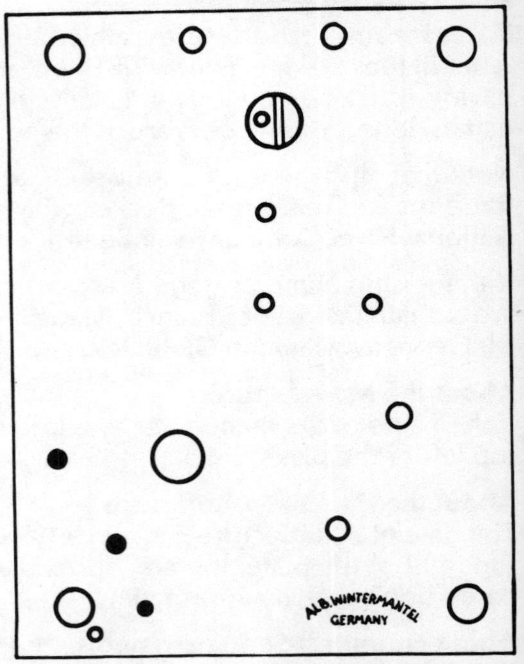

Plate 1003 Disc Pendulum
USE .004"* (.102mm*) HOROLOVAR
See Appendix 107 (19 x 36)

Kieninger & Obergfell c 1953

AMC

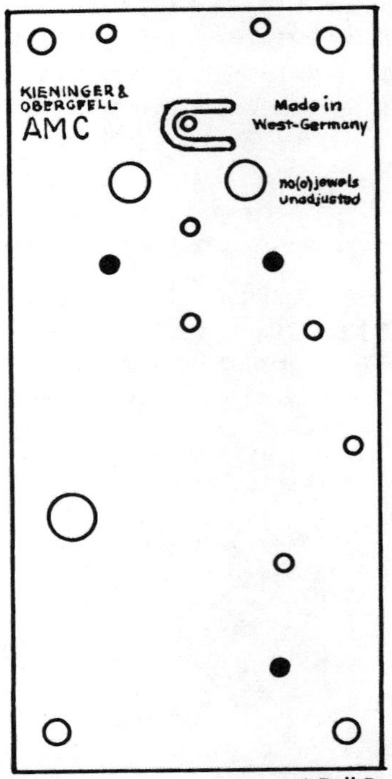

Plate 1004 4-Ball Pendulum
USE .0032" (.081mm) HOROLOVAR
Unit 3C
See Appendix 76 (19 x 38)

Kieninger & Obergfell c 1965

AMC

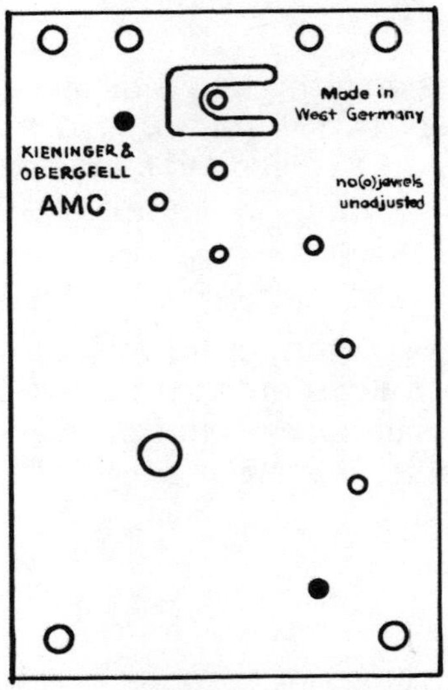

Plate 1004A 4-Ball Pendulum
Miniature Clock
USE .0023" (.058mm) HOROLOVAR
Unit 5E
See Appendix 77 (14 x 30)

Konrad Mauch c 1950

AMERICAN CLOCK CO.

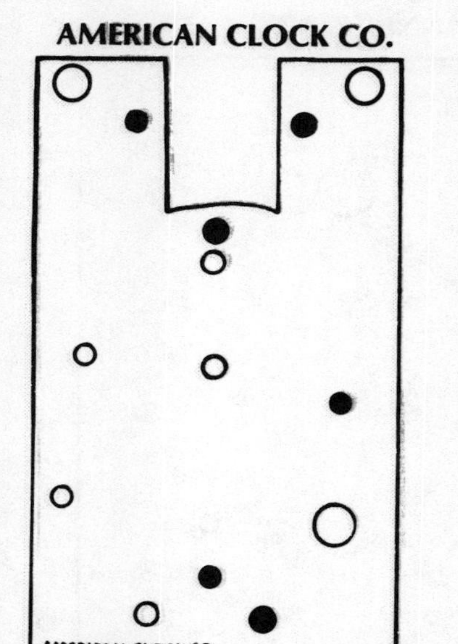

Plate 1005 4-Ball Pendulum
USE .0035" (.089mm) HOROLOVAR
Units 13A, 13B (20 x 38)
See Appendix 63, 64, 84

Konrad Mauch c 1954

AMERICAN CLOCK CO.

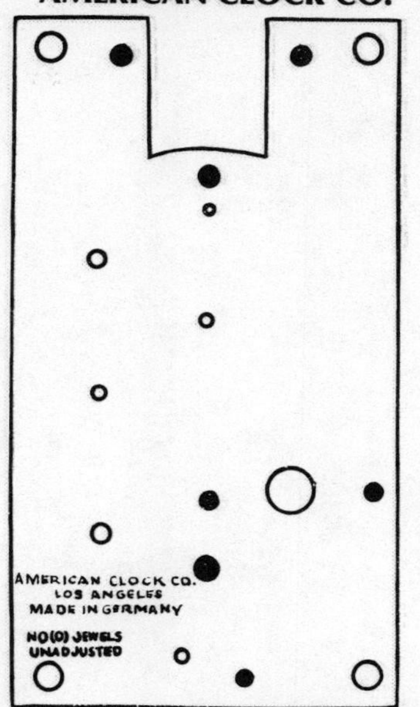

Plate 1006 4-Ball Pendulum Miniature Clock
USE .0032" (.081mm) HOROLOVAR
Unit 14A
USE .003" (.076mm) HOROLOVAR
Unit 14B Original (13 x 32)
See Appendix 52, 63, 64, 85 Revised (15 x 32)

Phillipp Haas c 1906

ANNIVERSARY

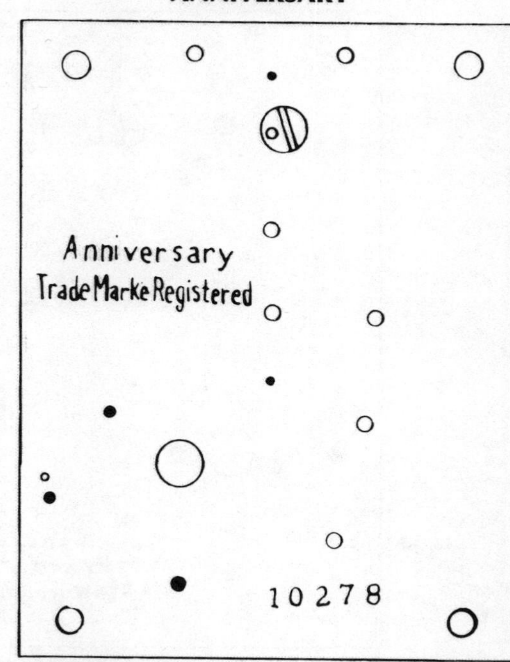

Plate 1007 Disc Pendulum
USE .0028"* (.071mm*) HOROLOVAR
See Appendix 1, 49, 89, 118 (19 x 36)

Jahresuhrenfabrik c 1905

ANNIVERSARY

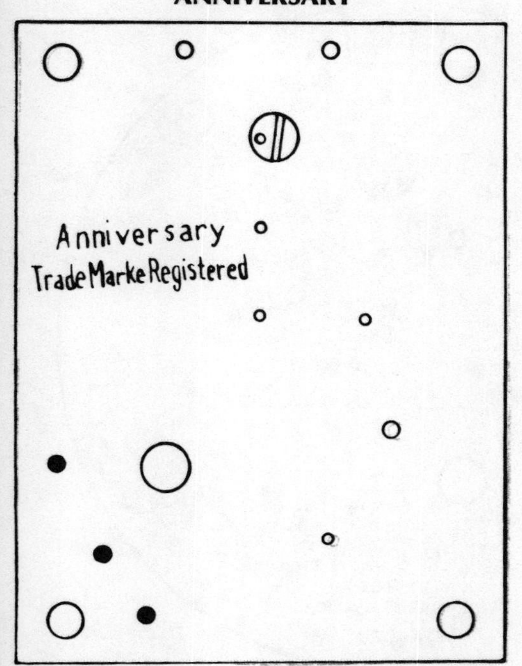

Plate 1007A Disc Pendulum
USE .004" (.102mm) HOROLOVAR
See Appendix 1 (19 x 36)

Kienzle Clock Factories c 1906

ANNIVERSARY

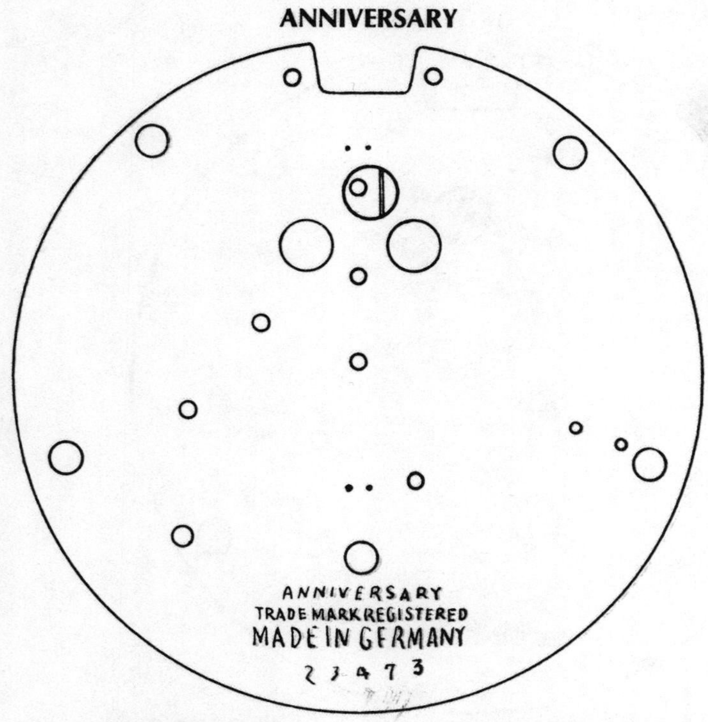

Plate 1008 Disc Pendulum
USE .0038"* (.097mm*) HOROLOVAR
See Appendix 1, 73 (19 x 36)

Jahresuhrenfabrik c 1906

ANNIVERSARY

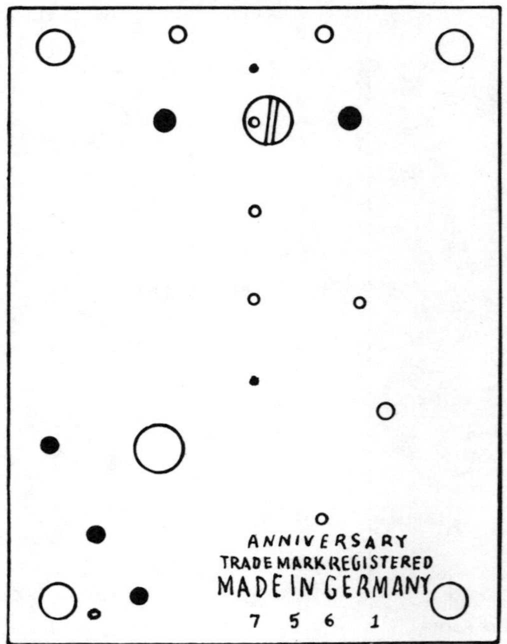

Plate 1008A Disc Pendulum
USE .004" (.102mm) HOROLOVAR
See Appendix 1, 90 (19 x 36)

Jahresuhrenfabrik c 1905

ANNIVERSARY

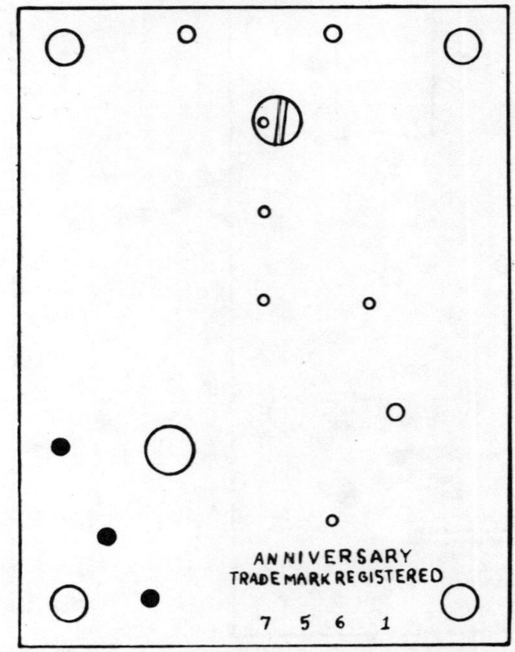

Plate 1009 Disc Pendulum
USE .004" (.102mm) HOROLOVAR
See Appendix 1 (19 x 36)

Manufacturer Not Known c 1904

ANNIVERSARY

made in Germany

Plate 1009A Disc Pendulum
USE .004"* (.102mm*) HOROLOVAR
See Appendix 1, 73, 88 (19 x 36)

Manufacturer Not Known c 1904

ANNIVERSARY

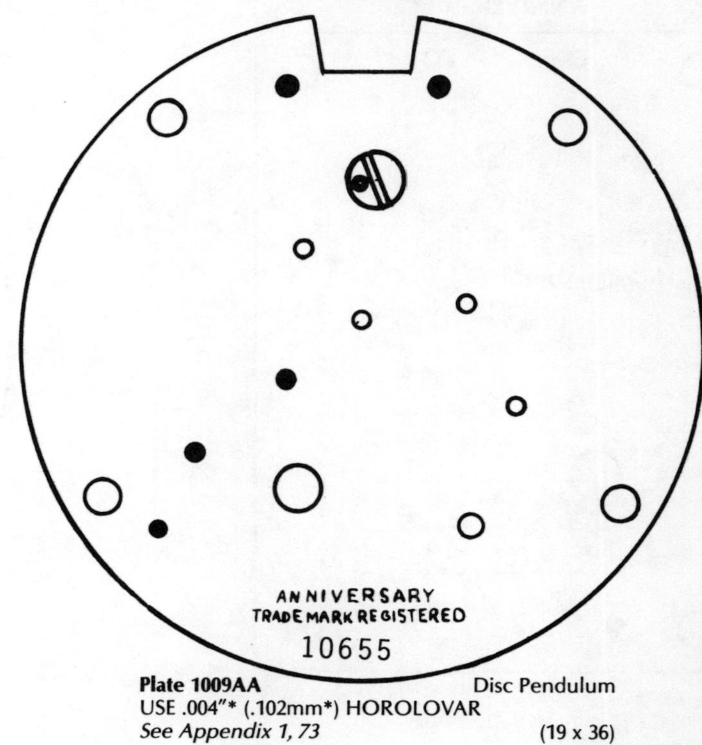

Plate 1009AA Disc Pendulum
USE .004"* (.102mm*) HOROLOVAR
See Appendix 1, 73 (19 x 36)

Aug. Schatz & Sohne c 1954

A. SCHATZ & SONS

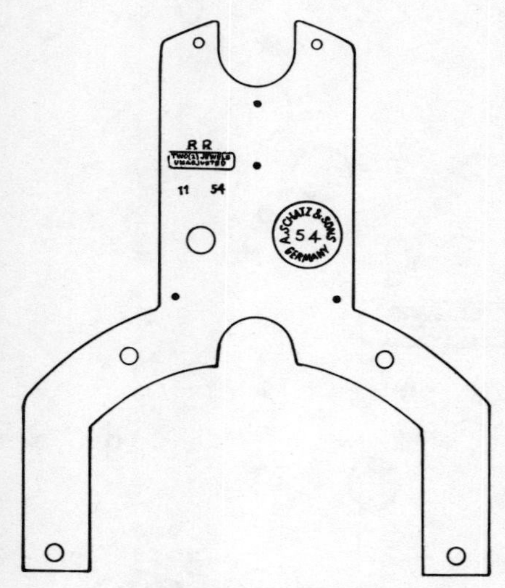

(½ actual size)

Plate 1010 4-Ball Pendulum 1000-Day Clock

USE .0024" (.061mm) HOROLOVAR
Unit 10B (21 x 50)
See Appendix 4, 18, 19, 63, 66, 80, 113

Aug. Schatz & Sohne c 1955

A. SCHATZ & SONS

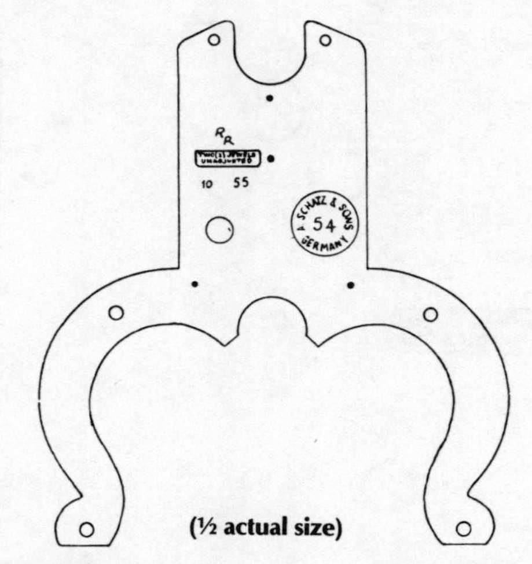

(½ actual size)

Plate 1010A 4-Ball Pendulum 1000-Day Clock

USE .0024" (.061mm) HOROLOVAR
Unit 10B (21 x 50)
See Appendix 4, 18, 19, 63, 66, 80, 113

Uhrenfabrik Herr c 1953

ASSOCIATED MERCHANDISING CORP.

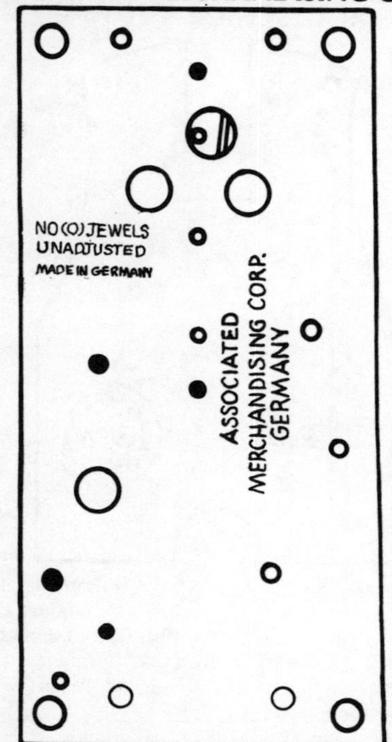

Plate 1012 4-Ball Pendulum

USE .0035" (.089mm) HOROLOVAR
Units 19, 20A, 21

USE .0038" (.097mm) HOROLOVAR
Unit 27B

Square Clock
USE .0033" (.084mm) HOROLOVAR
Unit 20B (19 x 36)
See Appendix 106, 109, 110

Aug. Schatz & Sohne c 1954

AUG. SCHATZ & SOHNE

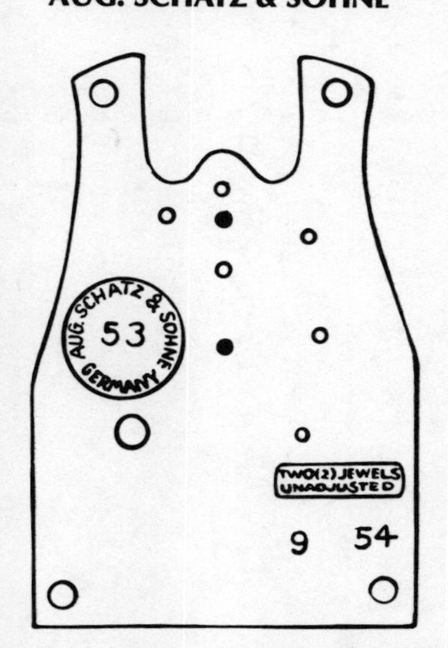

Plate 1013 Dome Clock—3-Ball Pendulum
Coach Clock—4-Ball Pendulum

USE .0023" (.058mm) HOROLOVAR
Unit 10A (13 x 30)
See Appendix 4, 5, 18, 63, 66, 79

Aug. Schatz & Sohne c 1954

AUG. SCHATZ & SOHNE

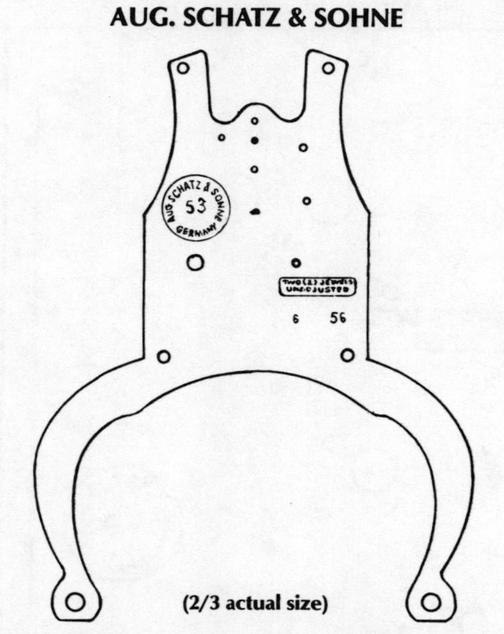

(2/3 actual size)

Plate 1013A Dome Clock—3-Ball Pendulum
Coach Clock—4-Ball Pendulum

USE .0023" (.058mm) HOROLOVAR
Unit 10A (13 x 30)
See Appendix 4, 5, 18, 63, 66, 79

Aug. Schatz & Sohne c 1957

AUG. SCHATZ & SOHNE

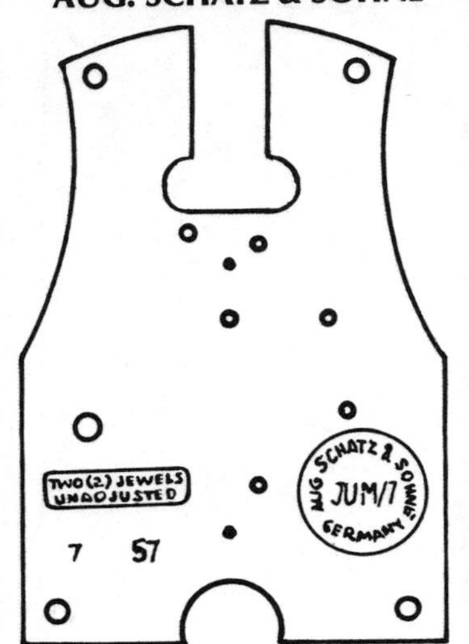

Plate 1013B
4-Ball Pendulum
Midget Clock
Pin Pallet Escapement
USE .0022" (.056mm) HOROLOVAR
Unit 10C (13 x 30)
See Appendix 4, 18, 40, 63, 66, 79

Aug. Schatz & Sohne c 1955

AUG. SCHATZ & SOHNE

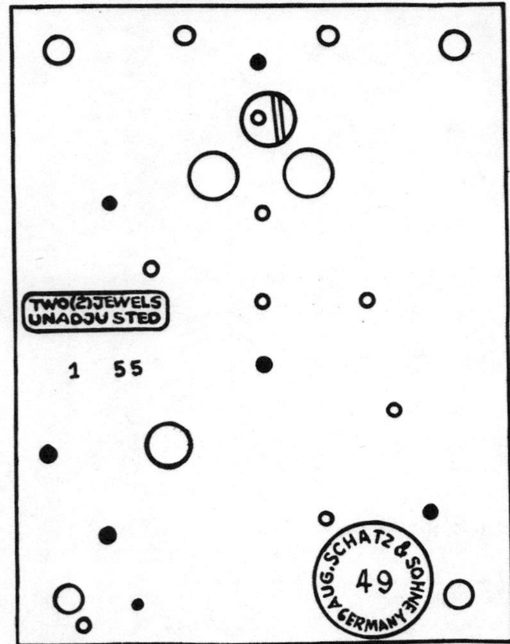

Plate 1014 4-Ball Pendulum
USE .004" (.102mm) HOROLOVAR
Units 6, 7, 8, 9 (19 x 36)
See Appendix 17, 18, 78

Aug. Schatz & Sohne c 1950

AUG. SCHATZ & SOHNE

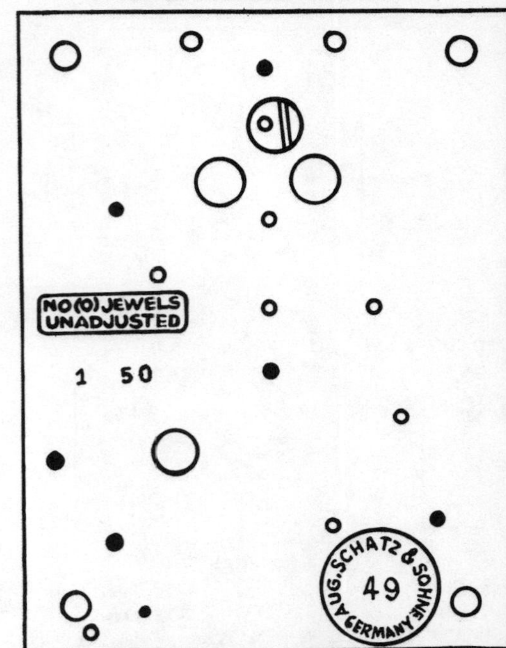

Plate 1014A 4-Ball Pendulum
USE .004" (.102mm) HOROLOVAR
Units 6, 7, 8, 9 (19 x 36)
See Appendix 18, 78

Aug. Schatz & Sohne c 1952

AUG. SCHATZ & SOHNE

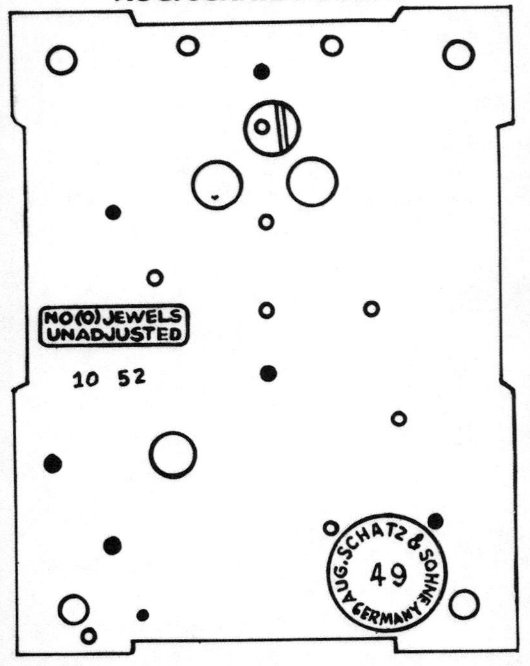

Plate 1014B 4-Ball Pendulum
USE .004" (.102mm) HOROLOVAR
Units 6, 7, 8, 9 (19 x 36)
See Appendix 18, 62, 78

Aug. Schatz & Sohne c 1963

AUG. SCHATZ & SOHNE

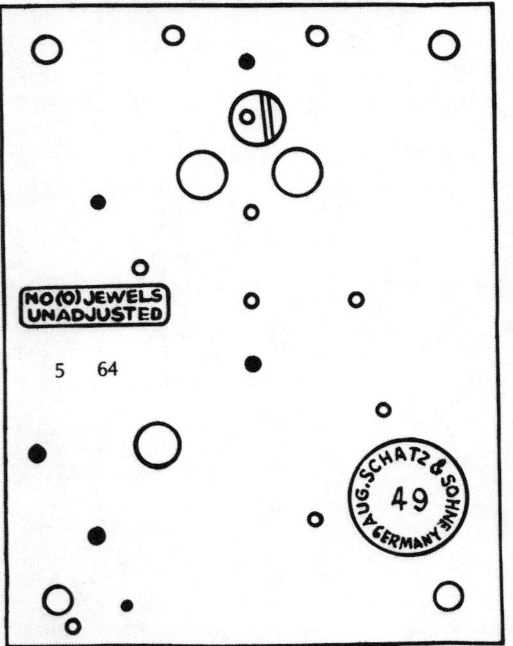

Plate 1014C 4-Ball Pendulum
USE .004" (.102mm) HOROLOVAR
Units 6, 7, 8, 9 (19 x 36)
See Appendix 18, 78

Badische Uhrenfabrik c 1902

B

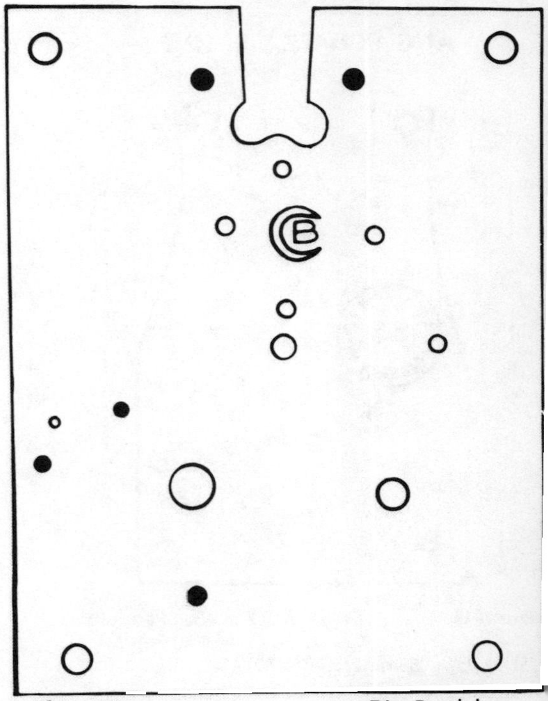

Plate 1015
Disc Pendulum
3-Ball Pendulum
Lantern Pinions
USE .0045"-* (.114mm-*) HOROLOVAR
See Appendix 63 (20 x 38)

Uhrenfabrik M. Reiner c 1954

EURAMCA TRADING CORP.

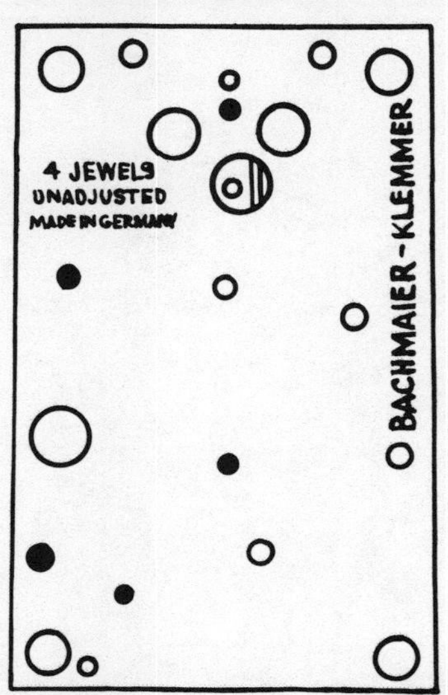

Plate 1015A 4-Ball Pendulum Miniature Clock

USE .0028" (.071mm) HOROLOVAR
Unit 28C (16 x 36)
See Appendix 25, 55, 60, 109

Uhrenfabrik Herr c 1952

BACHMAIER-KLEMMER

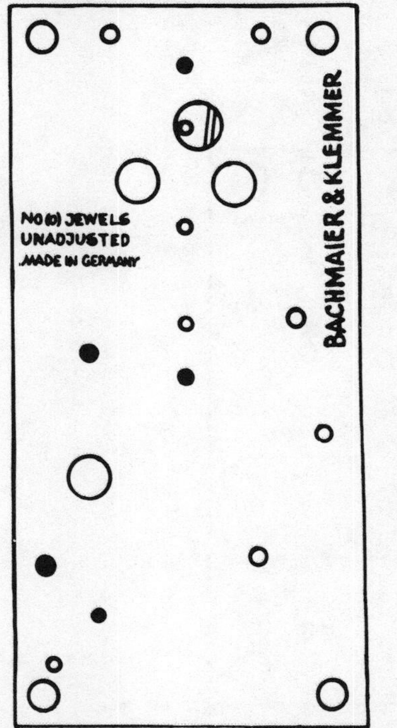

Plate 1016AAA 4-Ball Pendulum

Dome Clock
USE .0035" (.089mm) HOROLOVAR
Units 19, 20A, 21

Square Clock
USE .0033" (.084mm) HOROLOVAR
Unit 20B (19 x 36)
See Appendix 109, 110

Uhrenfabrik M. Reiner c 1954

BACHMAIER-KLEMMER

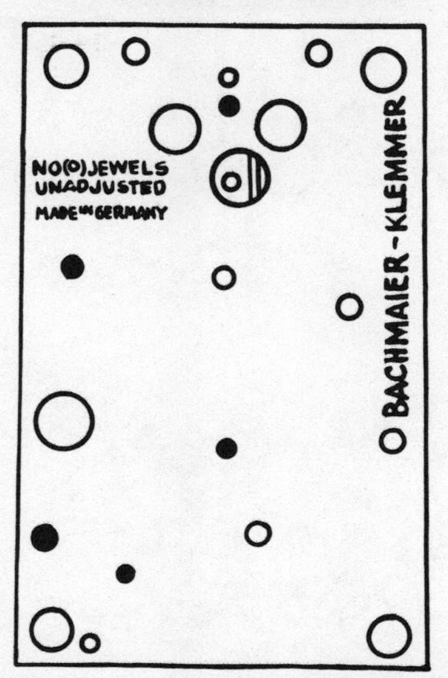

Plate 1016 Standard 4-Ball Pendulum Miniature Movement

USE .0032" (.081mm) HOROLOVAR
Unit 23B (16 x 36)
See Appendix 25, 55, 109

Uhrenfabrik Herr c 1952

BACHMAIER-KLEMMER

GERMANY
NO(0) JEWELS
UNADJUSTED
MOHERTUS TRADING CO.
BACHMAIER & KLEMMER

Plate 1016AA 4-Ball Pendulum

USE .004" (.102mm) HOROLOVAR
Unit 27A (19 x 36)
See Appendix 130

Badische Uhrenfabrik c 1890

BADENIA REGD.

BADENIA
REGD.
D.R.G.M.
484408
D.R.G.M.
502714
made in Germany

Plate 1016A Disc Pendulum Pin Pallet Escapement Lantern Pinions

USE .0038"* (.097mm*) HOROLOVAR
See Appendix 12, 63, 96 (20 x 38)

W. Petersen c 1954

BADUF G.m.b.H.

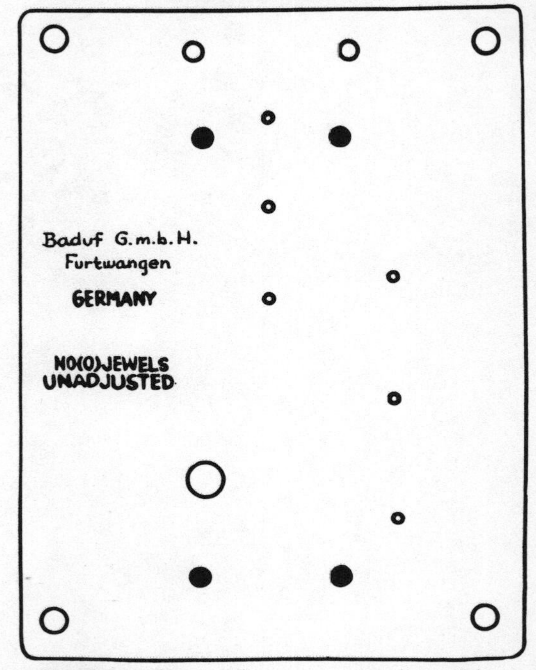

Plate 1017 4-Ball Pendulum

USE .0033" (.084mm) HOROLOVAR
Units 18B, 18C (19 x 38)
See Appendix 9, 59, 65

Badische Uhrenfabrik c 1902

BAD UHRENFABRIK

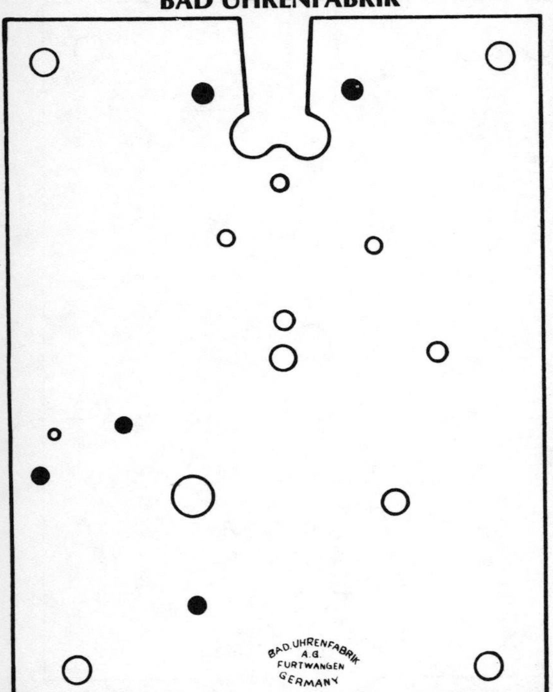

Plate 1019 Disc Pendulum
3-Ball Pendulum
Lantern Pinions
USE .0035"* (.089mm*) HOROLOVAR
See Appendix 63 (20 x 38)

Badische Uhrenfabrik c 1902

BAD UHRENFABRIK

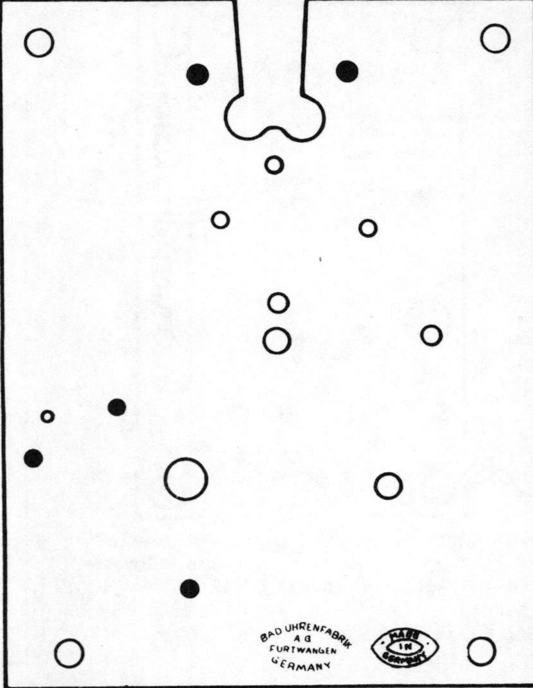

Plate 1019A Disc Pendulum
3-Ball Pendulum
Lantern Pinions
USE .0035"* (.089mm*) HOROLOVAR
See Appendix 63 (20 x 38)

Kieninger & Obergfell c 1950

BECKEN

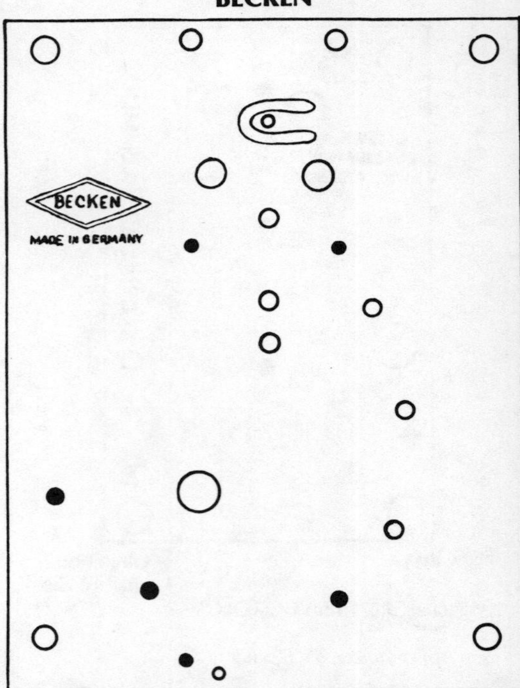

Plate 1023 4-Ball Pendulum
USE .0032" (.081mm) HOROLOVAR
Unit 1 (19 x 38)
See Appendix 76

Kieninger & Obergfell c 1952

BECKEN

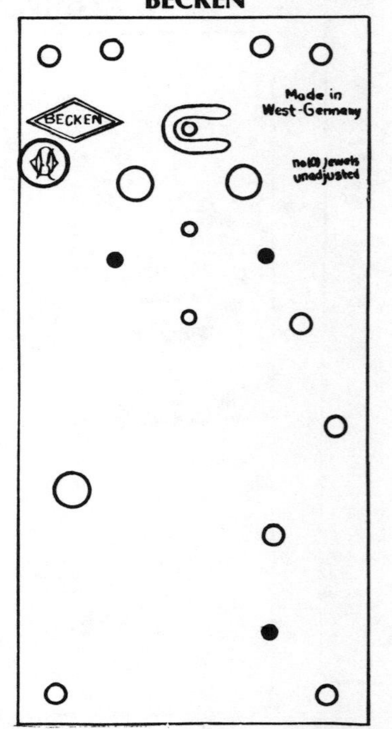

Plate 1027 4-Ball Pendulum
USE .0032" (.081mm) HOROLOVAR
Units 1, 3A, 3B, 3C
See Appendix 76 (19 x 38)

Kieninger & Obergfell c 1954

BECKEN

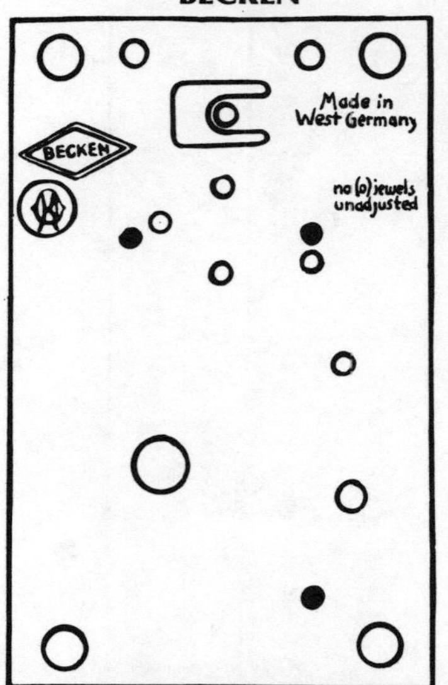

Plate 1031 4-Ball Pendulum
Miniature Clock
USE .0023" (.058mm) HOROLOVAR
Units 5A, 5B, 5C, 5D, 5E (14 x 30)
See Appendix 77, 111

Gustav Becker c 1900

BHA

BHA
1461151

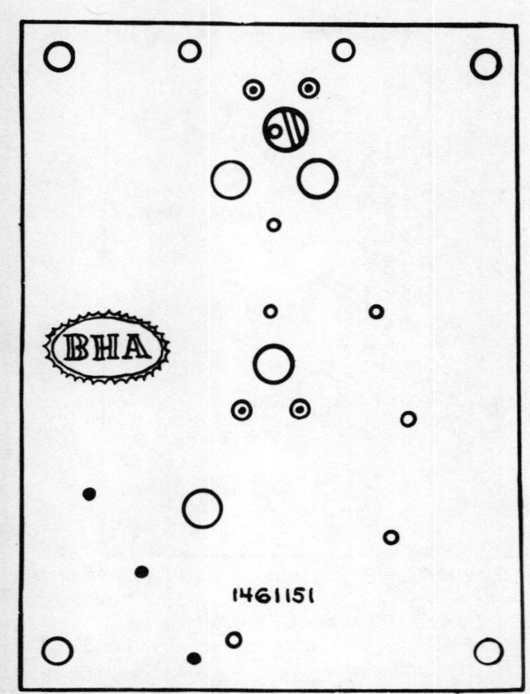

Plate 1032 Disc Pendulum
USE .004″ (.102mm) HOROLOVAR
See Appendix 11, 70, 74, 87 (19 x 38)

Gustav Becker c 1905

BHA

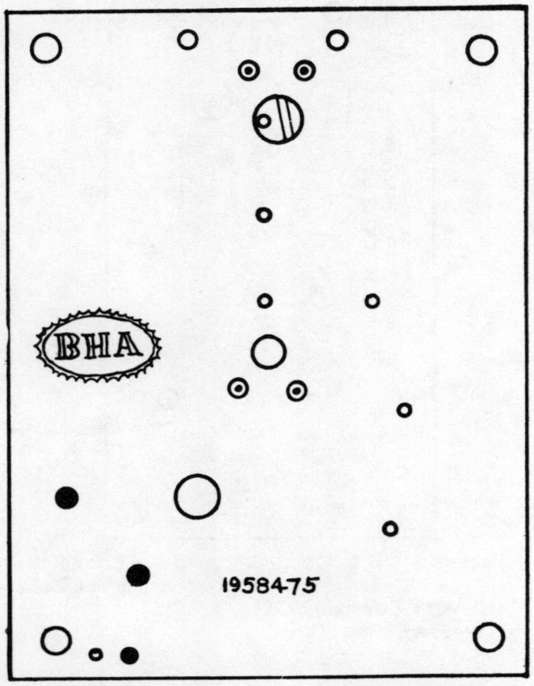

BHA
1958475

Plate 1032A Disc Pendulum
USE .004″ (.102mm) HOROLOVAR
See Appendix 11, 70, 74, 87 (19 x 38)

Gustav Becker c 1907

BHA

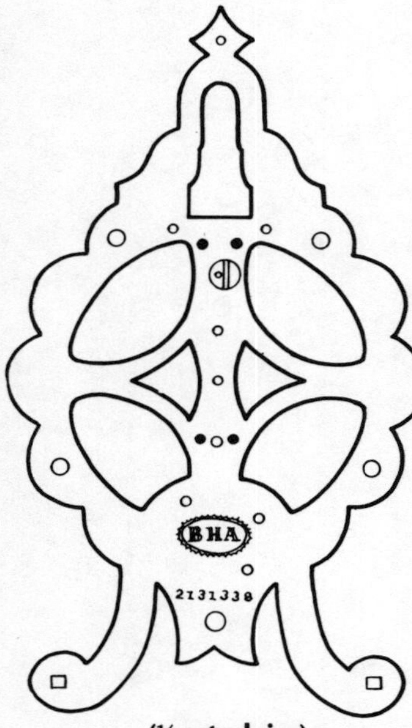

BHA
2131338

(½ actual size)

Plate 1033 Skeleton Disc Pendulum
Skeleton Clock
USE .004″ (.102mm) HOROLOVAR
See Appendix 69, 70, 74 (19 x 38)

Kern & Link c 1949

CHARLES NUNNENMAN

CHARLES NUNNENMAN
IMPORTER
GERMANY

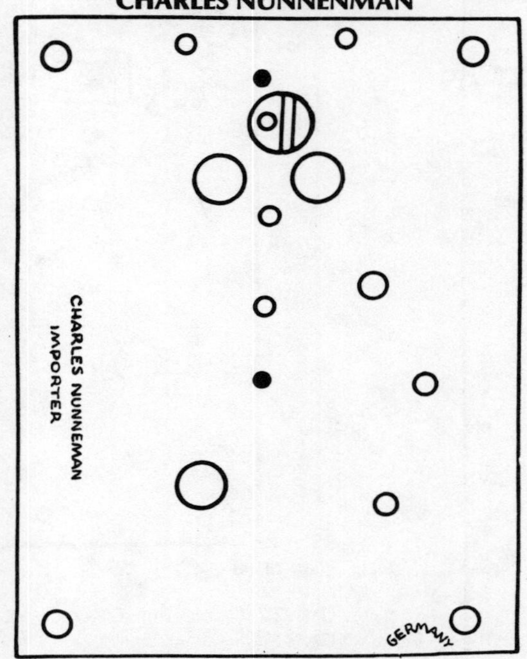

Plate 1034 4-Ball Pendulum
USE .0036″ (.091mm) HOROLOVAR
Unit 11A (18 x 38)

Kieninger & Obergfell c 1950

COOP

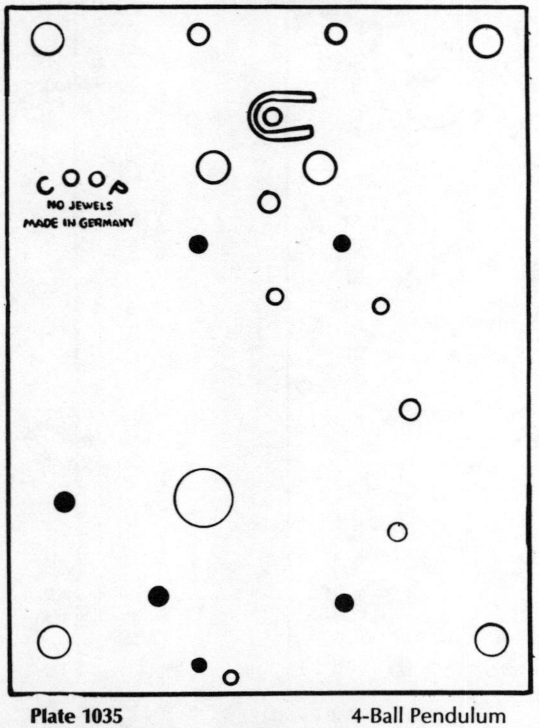

COOP
NO JEWELS
MADE IN GERMANY

Plate 1035 4-Ball Pendulum
USE .0032″ (.081mm) HOROLOVAR
Units 1, 3A (19 x 38)
See Appendix 76

Kieninger & Obergfell c 1950

CUCKOO CLOCK MFG. CO.

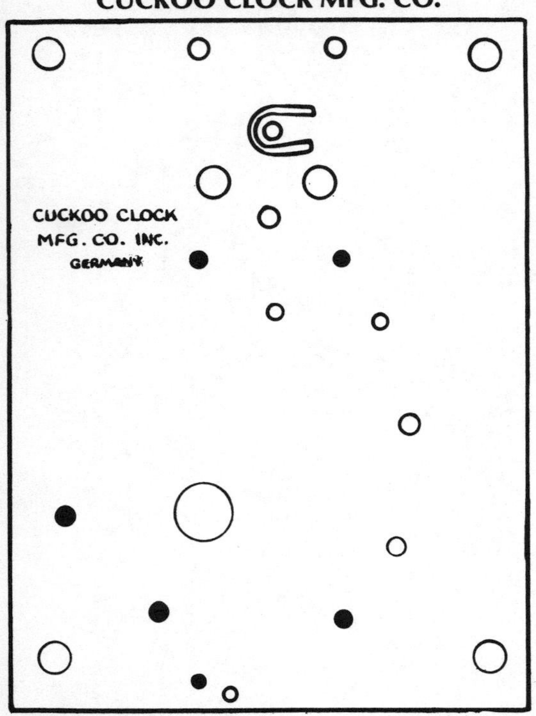

Plate 1035E 4-Ball Pendulum
USE .0032" (.081mm) HOROLOVAR
Units 1, 3A (19 x 38)
See Appendix 76

Edgar Henn c 1952

CUCKOO CLOCK MFG. CO.

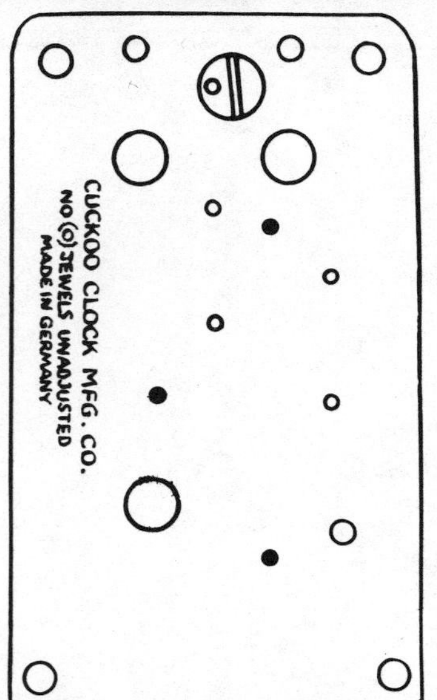

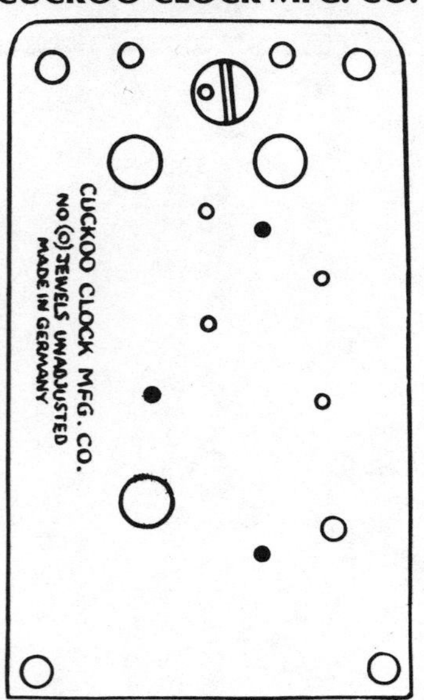

Plate 1036 4-Ball Pendulum
Miniature Clock
USE .0023" (.058mm) HOROLOVAR
Units 29A, 29B (19 x 32)

Edgar Henn c 1953

CUCKOO CLOCK MFG. CO.

CUCKOO CLOCK MFG. CO.
NO (0) JEWELS UNADJUSTED
MADE IN GERMANY

Plate 1037 4-Ball Pendulum
Miniature Clock
USE .0023" (.058mm) HOROLOVAR
Unit 29C (19 x 32)

Edgar Henn c 1953

CUCKOO CLOCK MFG. CO. INC.

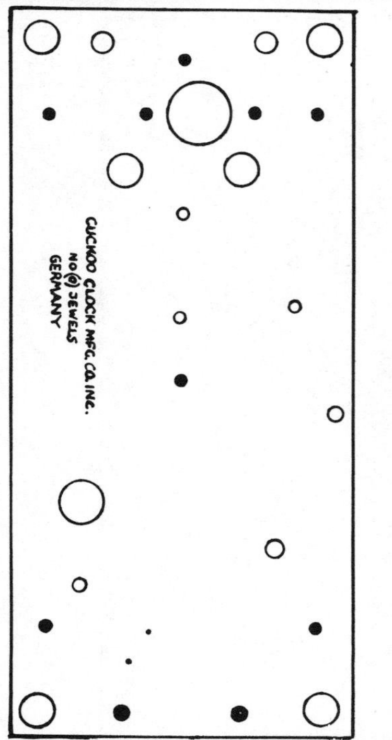

Plate 1038 4-Ball Pendulum
USE .0037" (.094mm) HOROLOVAR
Unit 32 (19 x 38)
See Appendix 39

Kieninger & Obergfell c 1953

CUCKOO CLOCK MFG. CO. INC.

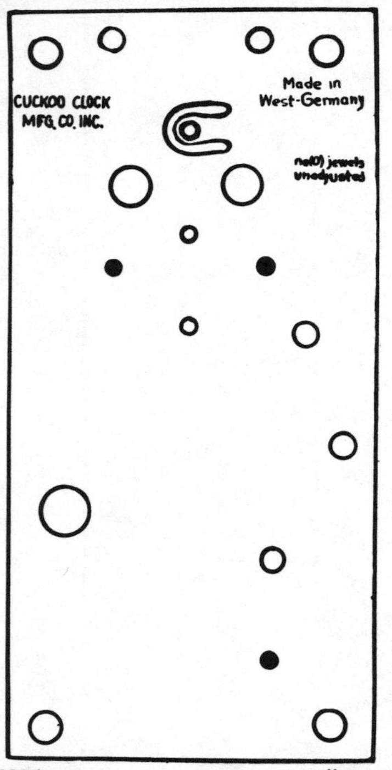

Plate 1039A 4-Ball Pendulum
USE .0032" (.081mm) HOROLOVAR
Unit 3C (19 x 38)
See Appendix 76

Kieninger & Obergfell c 1956

CUCKOO CLOCK MFG. CO. INC.

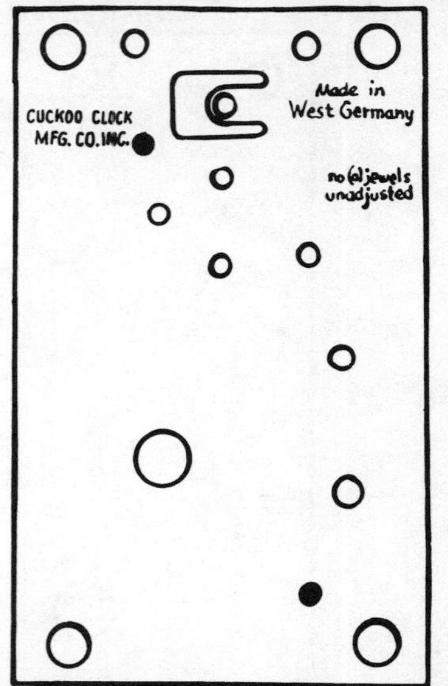

Plate 1039B 4-Ball Pendulum
Miniature Clock
USE .0023" (.058mm) HOROLOVAR
Units 5C, 5D, 5E (14 x 30)
See Appendix 77

Kieninger & Obergfell c 1953

CUCKOO CLOCK MFG. CO. INC.

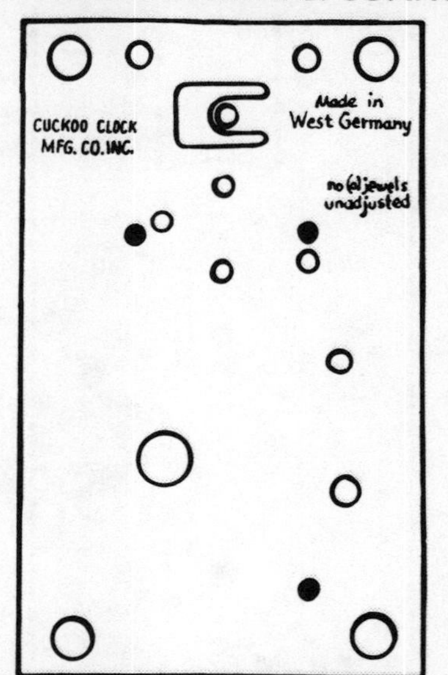

Plate 1040 4-Ball Pendulum Miniature Clock

USE .0023" (.058mm) HOROLOVAR
Units 5C, 5D, 5E (14 x 30)
See Appendix 77, 111

Uhrenfabrik Herr c 1951

CUCKOO CLOCK MFG. CO. INC.

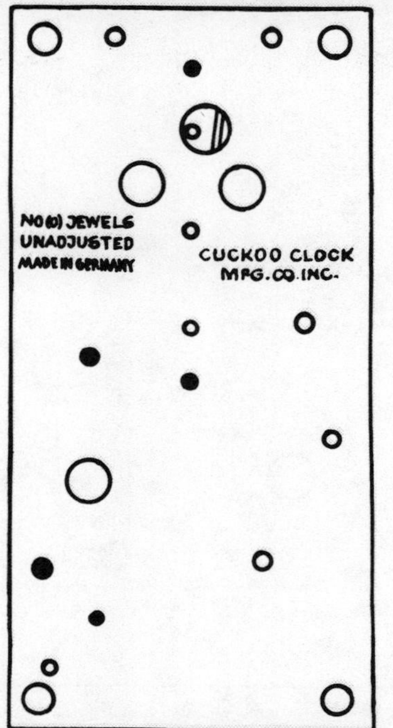

Plate 1040A 4-Ball Pendulum

USE .0035" (.089mm) HOROLOVAR
Units 19, 20A, 21

USE .0038" (.097mm) HOROLOVAR
Unit 27B
See Appendix 106, 109, 110 (19 x 36)

Uhrenfabrik Herr c 1951

CUCKOO CLOCK MFG. CO. INC.

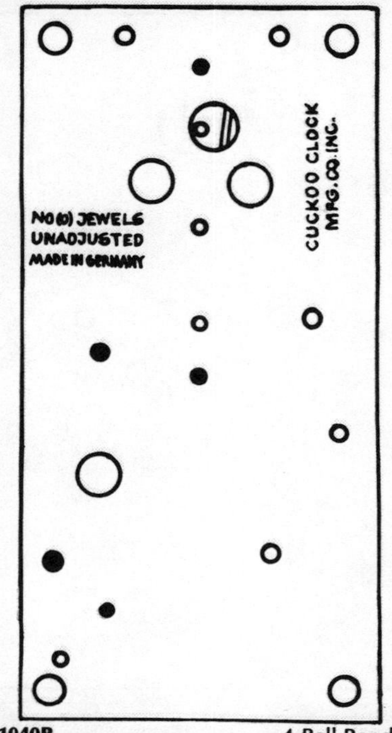

Plate 1040B 4-Ball Pendulum

USE .0035" (.089mm) HOROLOVAR
Units 19, 20A, 21

USE .0038" (.097mm) HOROLOVAR
Unit 27B
See Appendix 106, 109, 110 (19 x 36)

Uhrenfabrik M. Reiner c 1954

CUCKOO CLOCK MFG. CO.

NO (0) JEWELS
UNADJUSTED
MADE IN GERMANY
CUCKOO CLOCK
MFG. CO. INC.

Plate 1040BB 4-Ball Pendulum Miniature Clock

USE .0028" (.071mm) HOROLOVAR
Unit 28C (16 x 36)
See Appendix 25, 55, 109

Kieninger & Obergfell c 1965

DE BRUCE WATCH CO., INC.

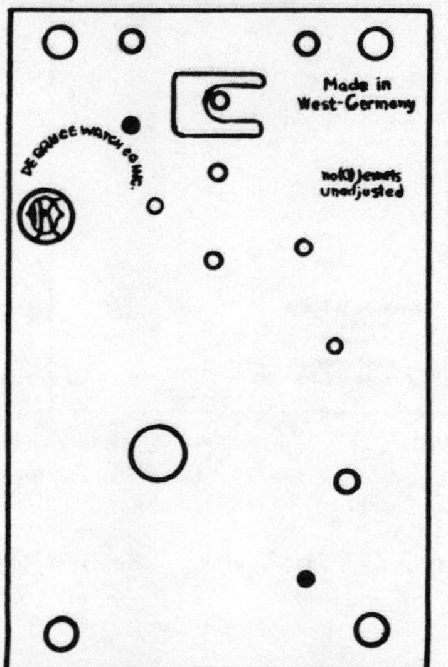

Plate 1040D 4-Ball Pendulum Miniature Clock

USE .0023" (.058mm) HOROLOVAR
Unit 5E (14 x 30)
See Appendix 77

Kieninger & Obergfell c 1952

DE BRUCE WATCH CO., INC.

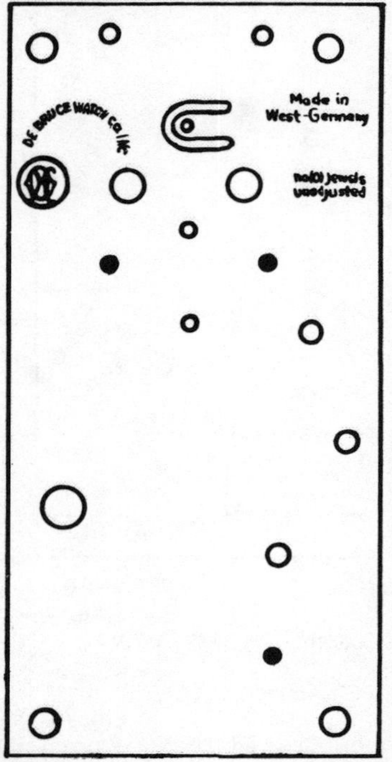

Plate 1040C 4-Ball Pendulum

USE .0032" (.081mm) HOROLOVAR
Unit 3C (19 x 38)
See Appendix 76

Jahresuhrenfabrik c 1910

D

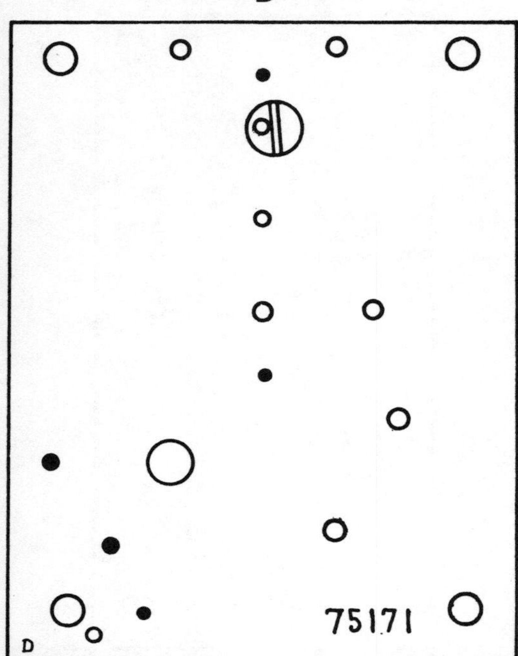

Plate 1040E Disc Pendulum
USE .004"* (.102mm*) HOROLOVAR
(19 x 36)

Uhrenfabrik Herr c 1952

DEJOLIE JEWELERS

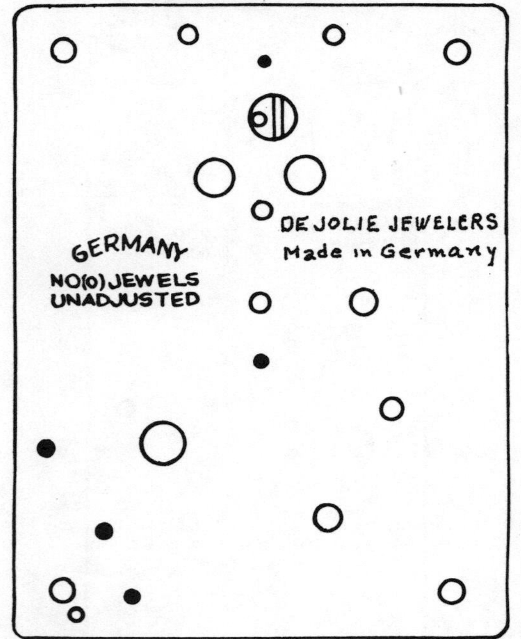

Plate 1040F 4-Ball Pendulum
USE .004" (.102mm) HOROLOVAR
Unit 27A (19 x 36)
See Appendix 130

Konrad Mauch c 1950

DEMMLER IMPORTS

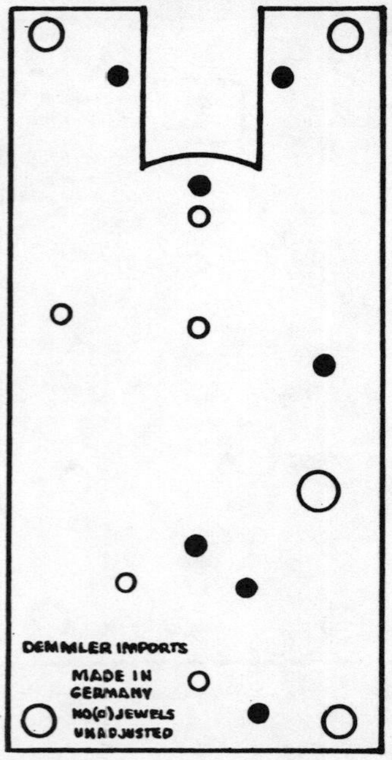

Plate 1040H 4-Ball Pendulum
USE .0035" (.089mm) HOROLOVAR
Units 13A, 13B, 13C (20 x 38)
See Appendix 63, 64, 84

Konrad Mauch c 1954

DEMMLER IMPORTS

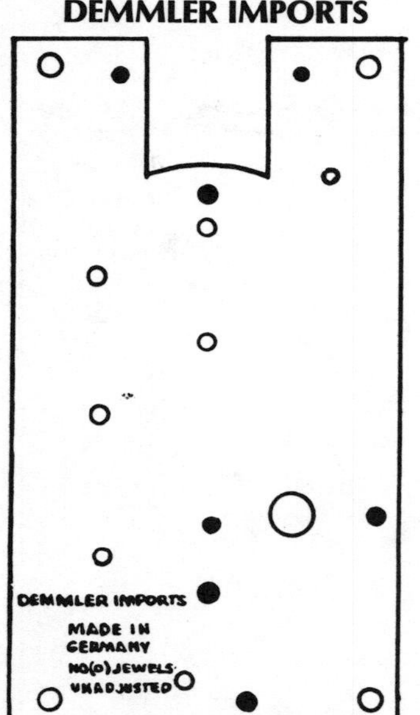

Plate 1040J 4-Ball Pendulum
Miniature Clock
USE .0032" (.081mm) HOROLOVAR
Unit 14A
USE .003" (.076mm) HOROLOVAR
Unit 14B Original (13 x 32)
See Appendix 52, 63, 64, 85 Revised (15 x 32)

Konrad Mauch c 1957

DEMMLER IMPORTS

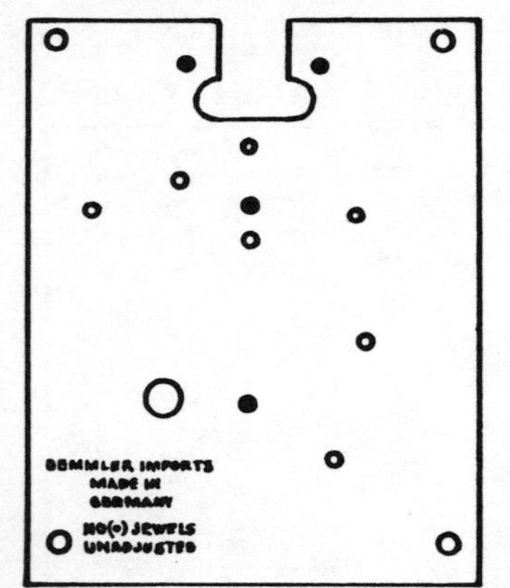

Plate 1040K 3-Ball Pendulum
Pin Pallet Escapement
USE .0022" (.056mm) HOROLOVAR
Units 38, 38A Original (12 x 25)
See Appendix 53, 63, 64, 86 Revised (14 x 25)

Badische Uhrenfabrik c 1890

D.R.G.M.

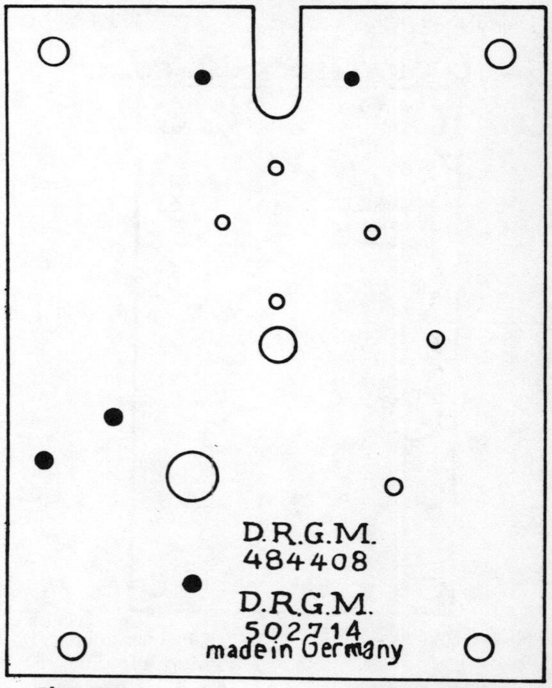

Plate 1041 Disc Pendulum
Pin Pallet Escapement
Lantern Pinions
USE .0038"* (.097mm*) HOROLOVAR
See Appendix 12, 63, 96 (20 x 38)

Phillipp Haas c 1907

D.R.P.

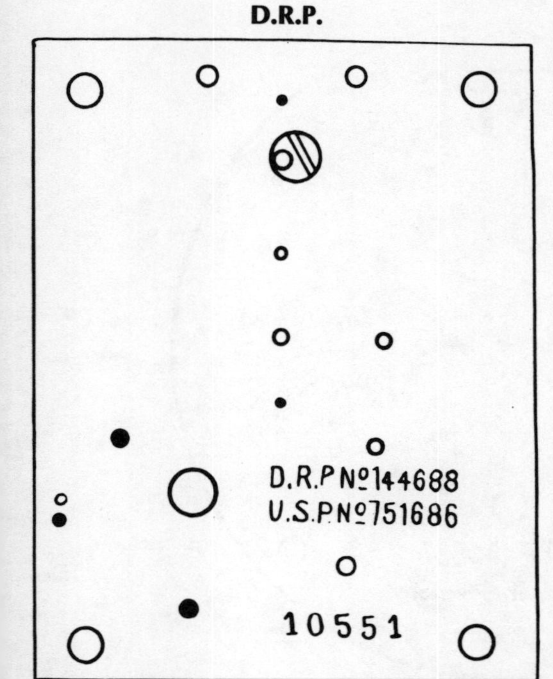

Plate 1043 Disc Pendulum
USE .004"* (.102mm*) HOROLOVAR
See Appendix 13, 89, 118, 127 (19 x 36)

Jahresuhrenfabrik c 1908

D.R.P.

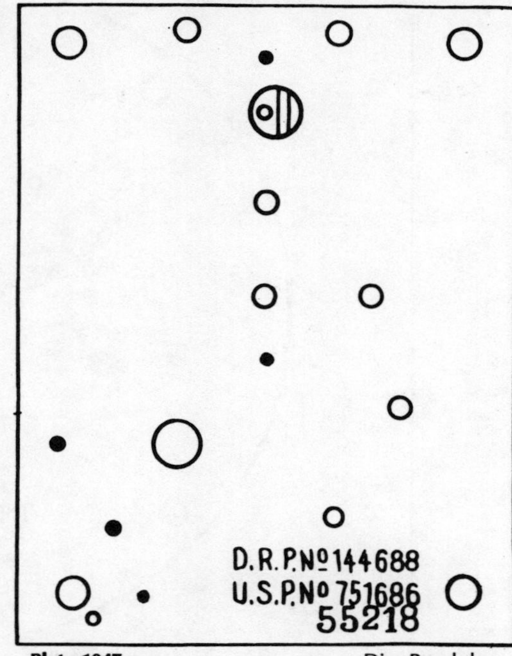

Plate 1047 Disc Pendulum
USE .0037"* (.094mm*) HOROLOVAR
USE .0045"* (.114mm*) HOROLOVAR
6½" Suspension Spring
See Appendix 13 (19 x 36)

Jahresuhrenfabrik c 1906

D.R.P.

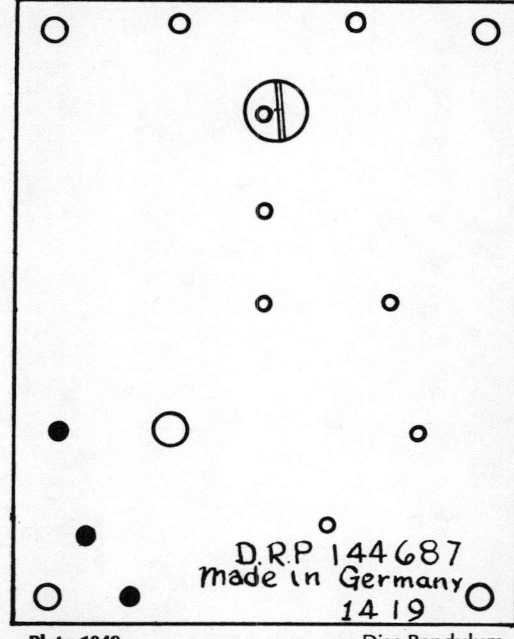

Plate 1049 Disc Pendulum
USE .004" (.102mm) HOROLOVAR
See Appendix 14 (19 x 36)

Jahresuhrenfabrik c 1906

D.R.P.

Plate 1049A Disc Pendulum
USE .004" (.102mm) HOROLOVAR
See Appendix 14, 88 (19 x 36)

Jahresuhrenfabrik c 1905

D.R.P.

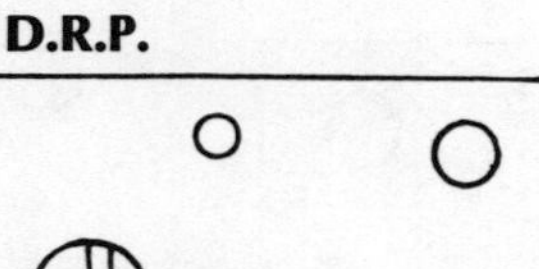
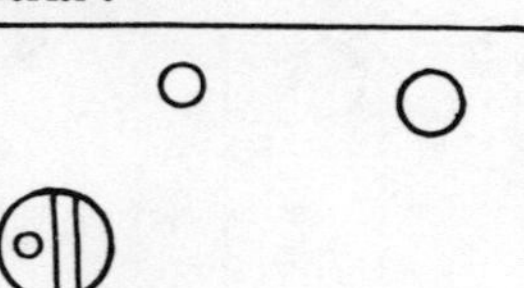
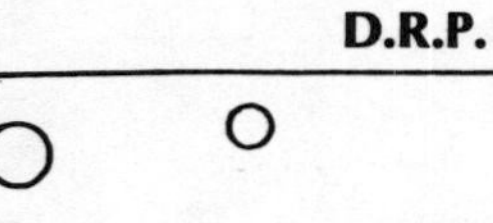
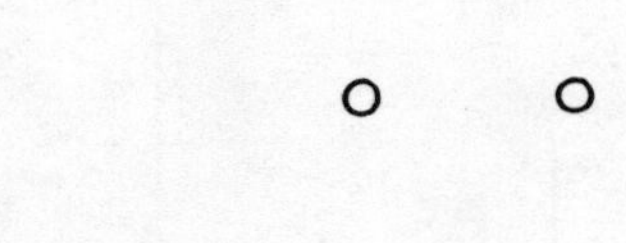

Plate 1051 Disc Pendulum
USE .005"* (.127mm*) HOROLOVAR
6¾" Suspension Spring
See Appendix 13 (19 x 36)

Jahresuhrenfabrik c 1908

D.R.P.

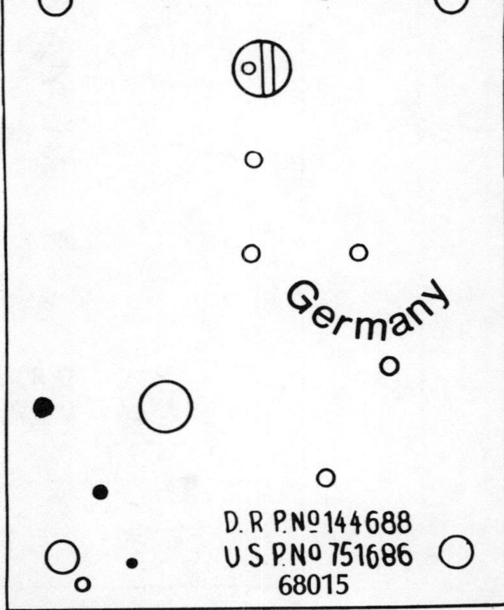

Plate 1051AA Disc Pendulum
USE .005"* (.127mm*) HOROLOVAR
6¾" Suspension Spring
See Appendix 13 (19 x 36)

Jahresuhrenfabrik c 1908

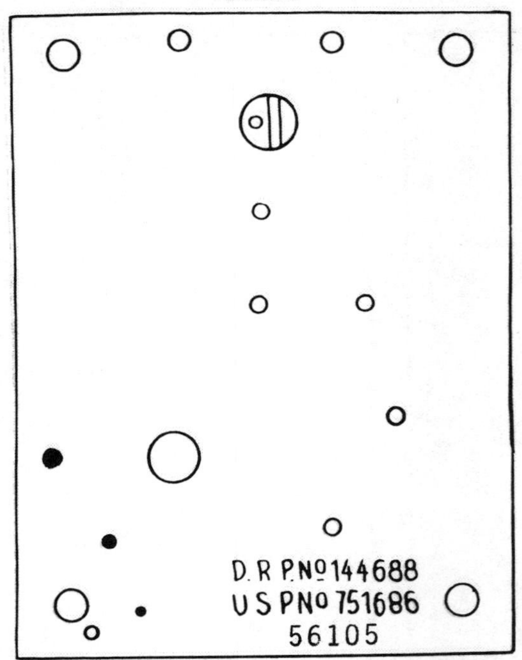

Plate 1051A Disc Pendulum
USE .005"* (.127mm*) HOROLOVAR
6¾" Suspension Spring
See Appendix 13 (19 x 36)

Manufacturer Not Known c 1900

D.R.P.

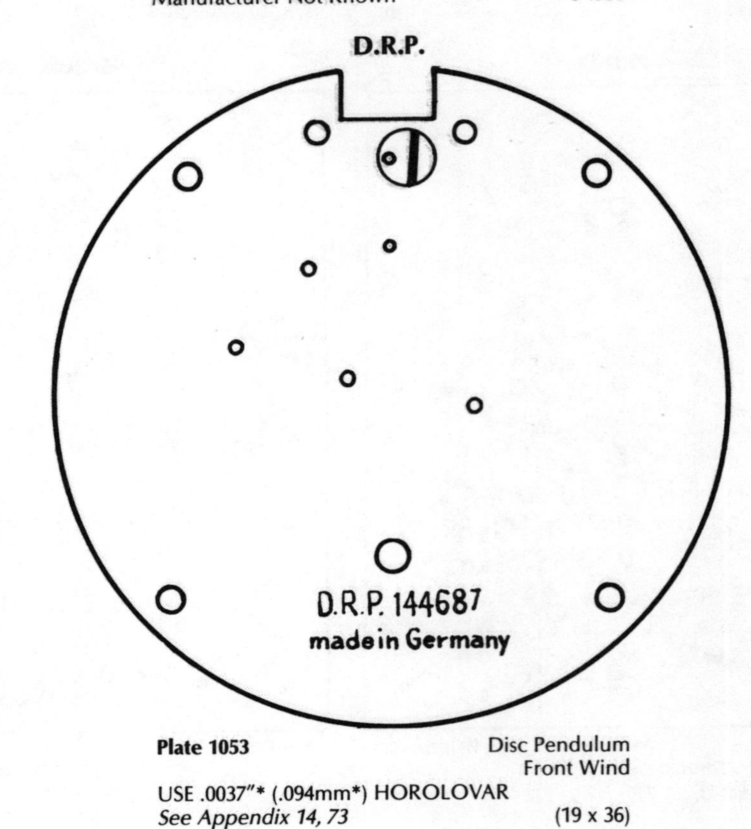

Plate 1053 Disc Pendulum
Front Wind
USE .0037"* (.094mm*) HOROLOVAR
See Appendix 14, 73 (19 x 36)

Jahresuhrenfabrik c 1904

D.R.P.

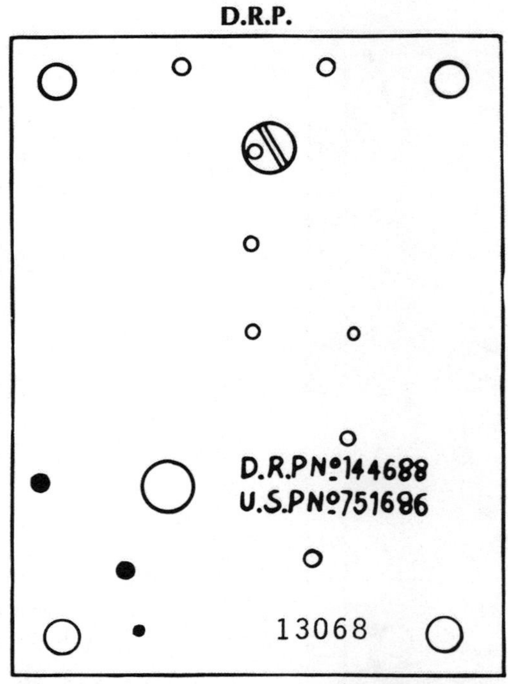

Plate 1055 Disc Pendulum
USE .0038"* (.097mm*) HOROLOVAR
See Appendix 13, 127 (19 x 36)

Jahresuhrenfabrik c 1906

D.R.P.

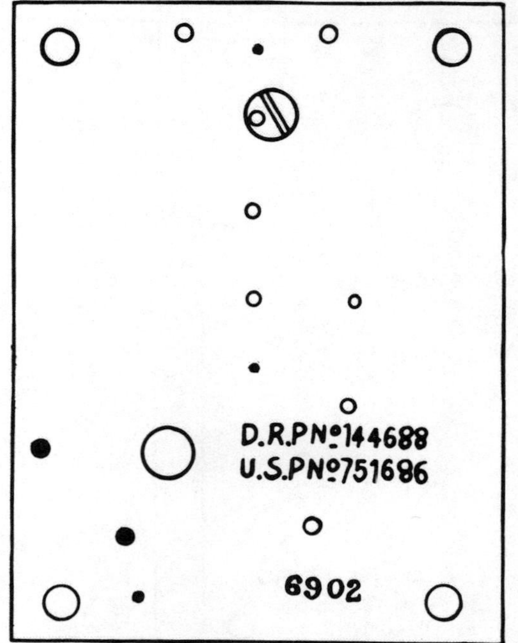

Plate 1056 Disc Pendulum
USE .0038"* (.097mm*) HOROLOVAR
See Appendix 13 (19 x 36)

Edgar Henn c 1953

EDGAR HENN

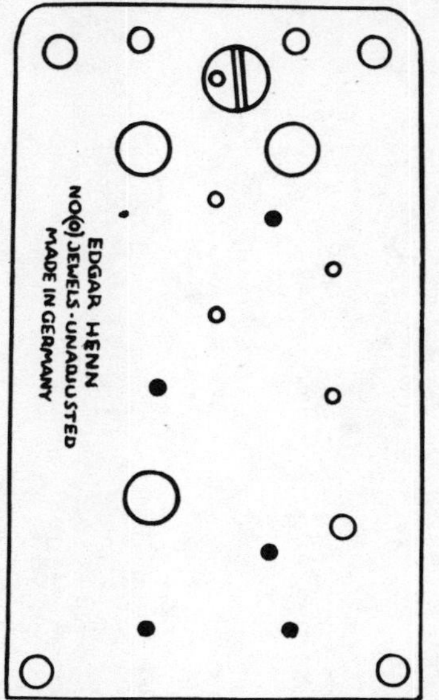

Plate 1057 4-Weight Pendulum Miniature Clock
USE .0023" (.058mm) HOROLOVAR
Unit 29C (19 x 32)
See Appendix 95

Edgar Henn c 1953

EDGAR HENN

Plate 1057AA 4-Weight Pendulum Miniature Clock
USE .0023" (.058mm) HOROLOVAR
Unit 29A, 29B (19 x 32)
See Appendix 95

Edgar Henn c 1952

EDGAR HENN

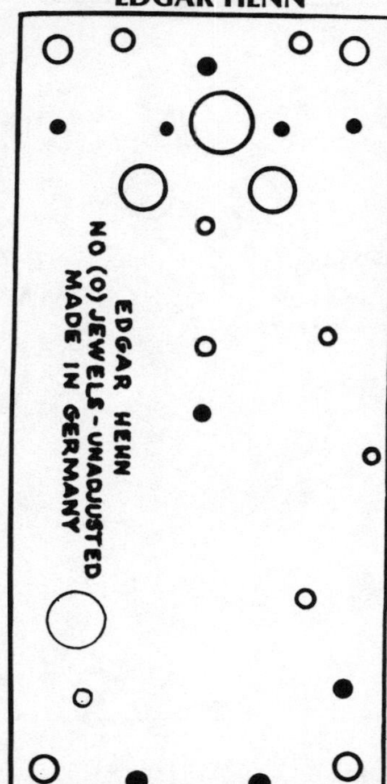

Plate 1057A 4-Ball Pendulum
USE .0037" (.094mm) HOROLOVAR
Unit 32 (19 x 38)
See Appendix 39

Edgar Henn c 1951

EHF (E. F. HENN)

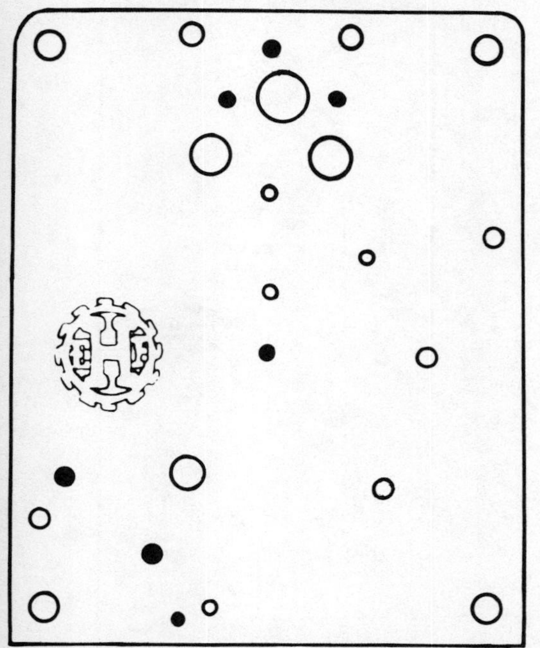

Plate 1059 4-Ball Pendulum
USE .0035" (.089mm) HOROLOVAR
Unit 31 (19 x 38)
See Appendix 39

Uhrenfabrik Herr c 1953

EHF (E. F. HENN)

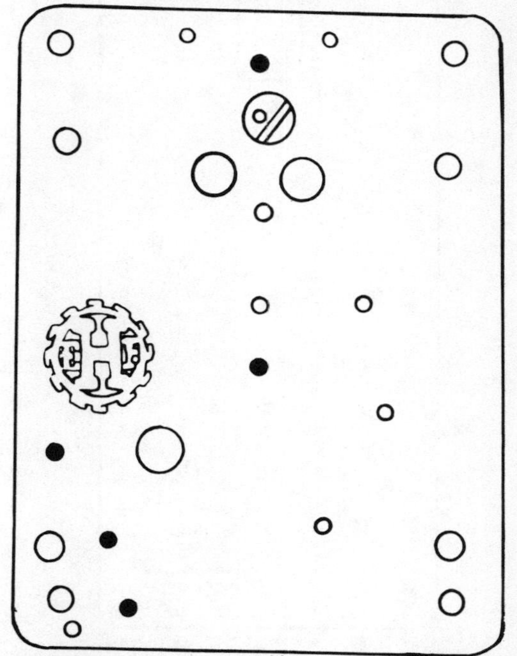

Plate 1063 4-Ball Pendulum
USE .004" (.102mm) HOROLOVAR
Unit 27A (19 x 36)
See Appendix 8, 117, 130

Uhrenfabrik Herr c 1952

EMO WATCH CO.

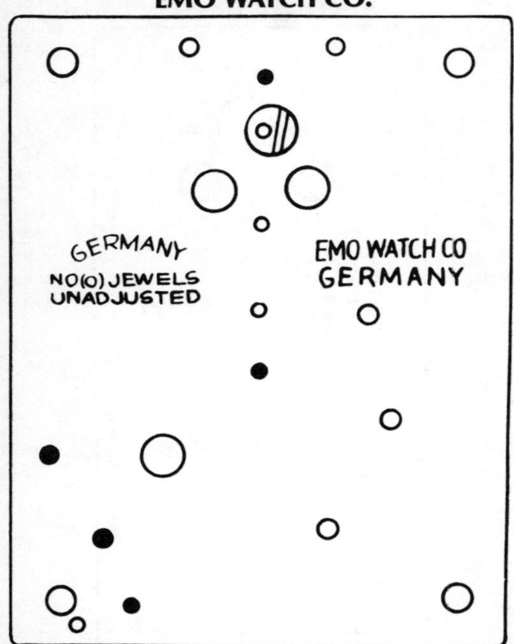

Plate 1067 4-Ball Pendulum
USE .004" (.102mm) HOROLOVAR
Unit 27A (19 x 36)
See Appendix 130

Uhrenfabrik Herr c 1952

EMO WATCH CO.

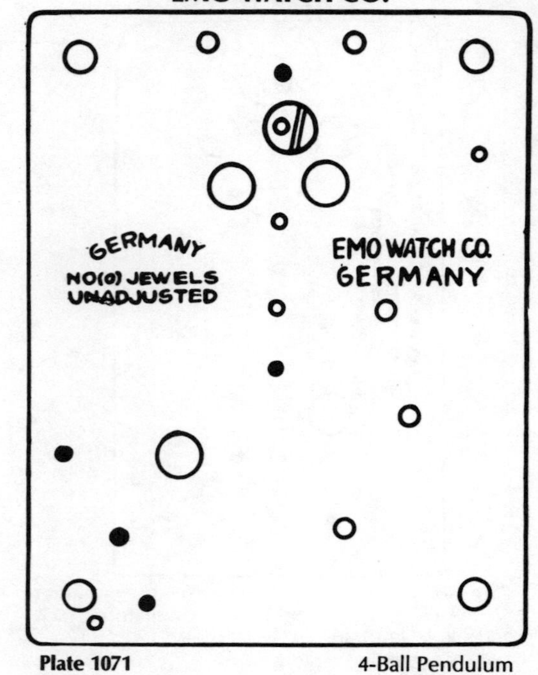

Plate 1071 4-Ball Pendulum
Hour Striker
USE .004" (.102mm) HOROLOVAR
Unit 27A (19 x 36)
See Appendix 56

Kieninger & Obergfell c 1930

E.R.

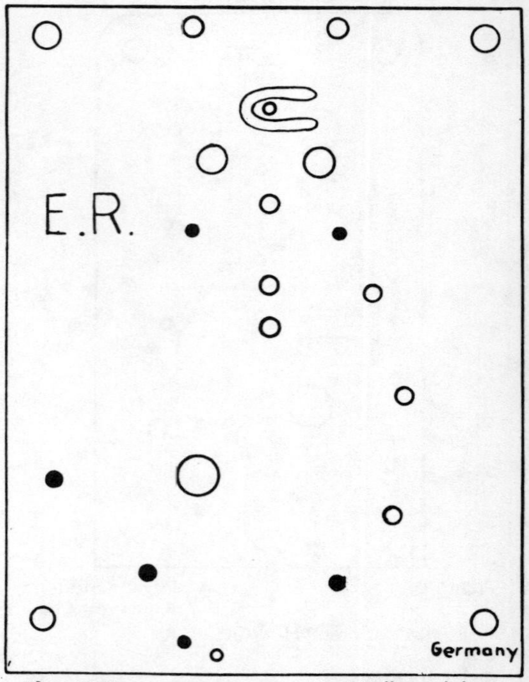

Plate 1075 4-Ball Pendulum
USE .0032" (.081mm) HOROLOVAR
Unit 1 (19 x 38)
See Appendix 76

Jauch & Haller c 1952

EULE

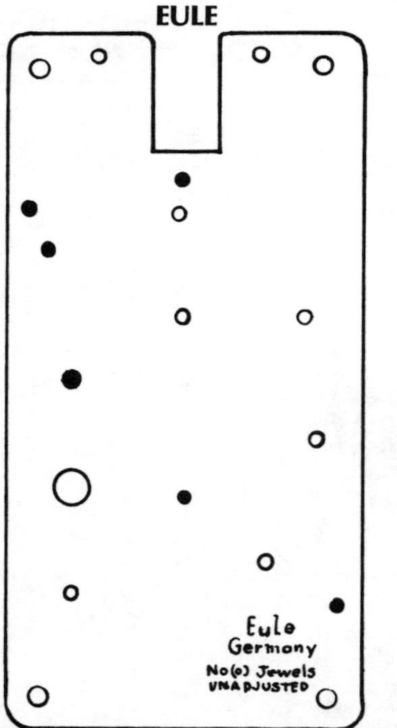

Plate 1076 4-Ball Pendulum
USE .0038" (.097mm) HOROLOVAR
Unit 33 (19 x 38)
See Appendix 63, 64

J. Link & Co. c 1952

EURAMCA TRADING CO.

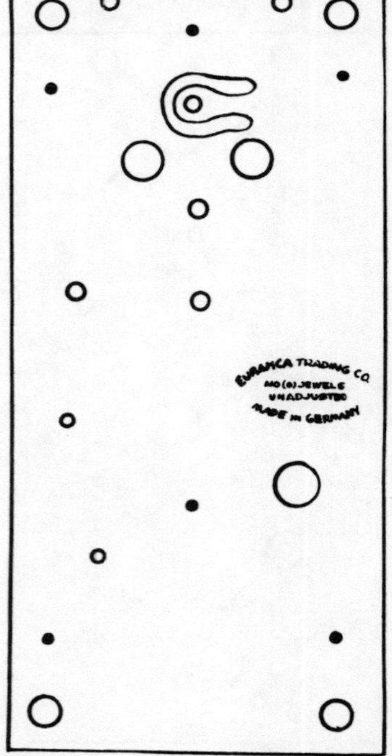

Plate 1077 4-Ball Pendulum
USE .0037" (.094mm) HOROLOVAR
Units 16, 17 (19 x 38)
See Appendix 119

Edgar Henn c 1951

EURAMCA TRADING CORP.

Plate 1079 4-Ball Pendulum
USE .0035" (.089mm) HOROLOVAR
Unit 31 (19 x 38)
See Appendix 39

Edgar Henn c 1952

EURAMCA TRADING CORP.

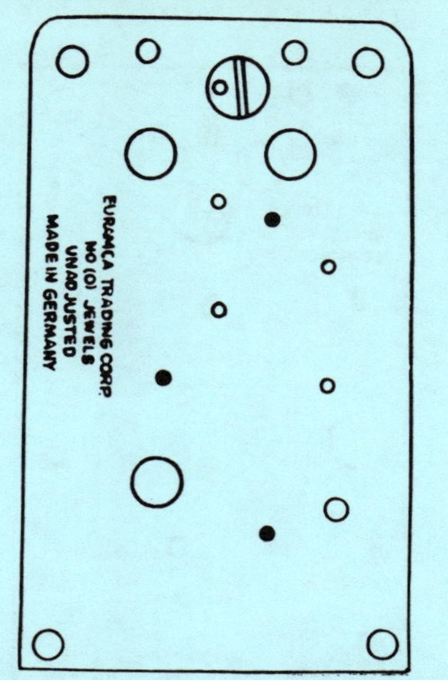

Plate 1087 4-Ball Pendulum Miniature Clock
USE .0023" (.058mm) HOROLOVAR
Units 29A, 29B (19 x 32)
See Appendix 95

Edgar Henn c 1953

EURAMCA TRADING CORP.

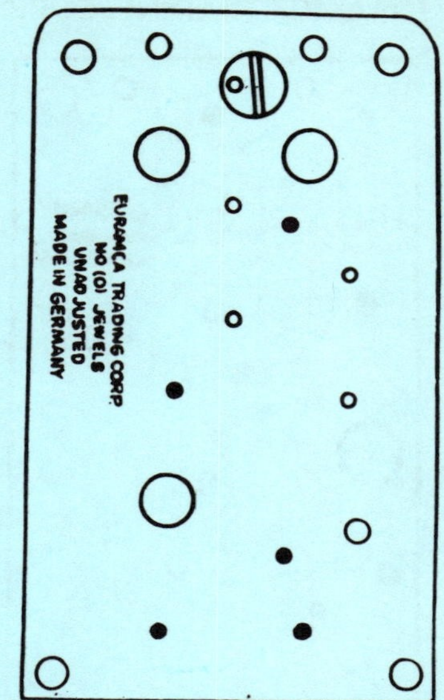

Plate 1088 4-Ball Pendulum Miniature Clock
USE .0023" (.058mm) HOROLOVAR
Units 29A, 29B (19 x 32)
See Appendix 95

Edgar Henn c 1953

EURAMCA TRADING CORP.

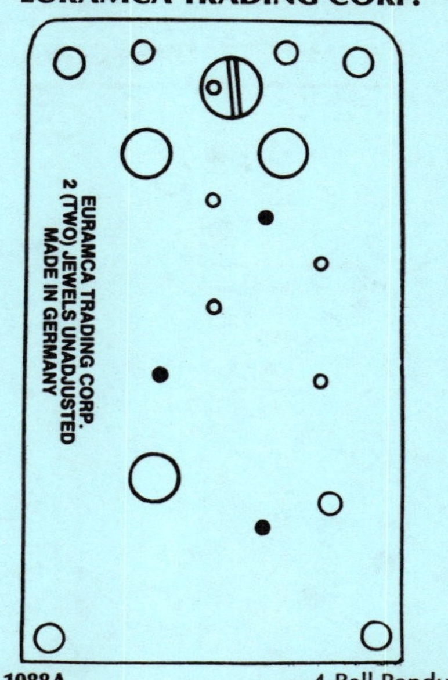

Plate 1088A 4-Ball Pendulum Miniature Clock
USE .0023" (.058mm) HOROLOVAR
Units 29A, 29B (19 x 32)
See Appendix 95

Kieninger & Obergfell c 1955

EURAMCA TRADING CORP.

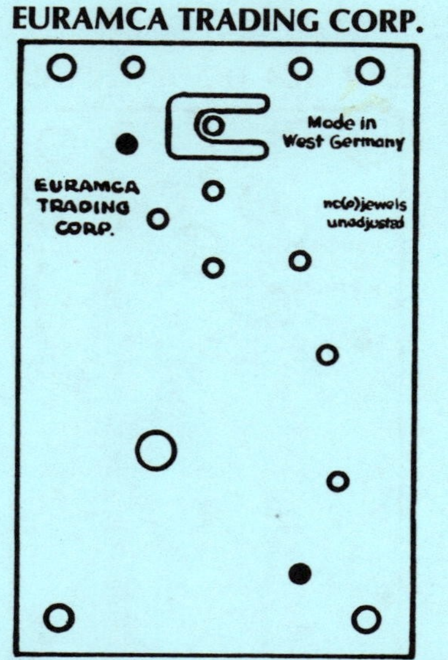

Plate 1089 4-Ball Pendulum
USE .0023" (.058mm) HOROLOVAR
Unit 5E (14 x 30)
See Appendix 77

Kieninger & Obergfell c 1951

EURAMCA TRADING CORP.

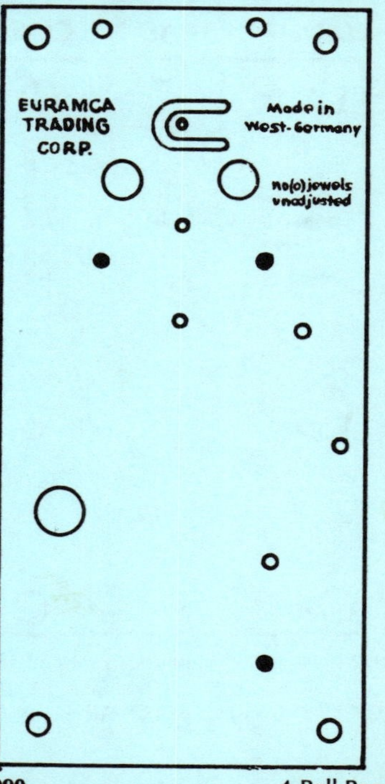

Plate 1090 4-Ball Pendulum
USE .0032" (.081mm) HOROLOVAR
Unit 3C (19 x 38)
See Appendix 76

Uhrenfabrik M. Reiner c 1951

EURAMCA TRADING CORP.

EURAMCA
TRADING CORP.
GERMANY
NO(0) JEWELS
UNADJUSTED

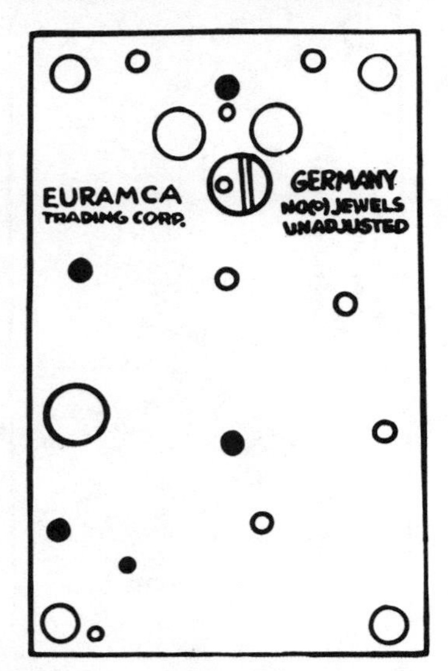

Plate 1091 Standard 4-Ball Pendulum Miniature Movement
USE .003" (.076mm) HOROLOVAR
Units 22, 23A, 24 (16 x 36)
See Appendix 25, 54, 109

Uhrenfabrik M. Reiner c 1954

EURAMCA TRADING CORP.

4 JEWELS
UNADJUSTED
MADE IN GERMANY
EURAMCA
TRADING CORP.

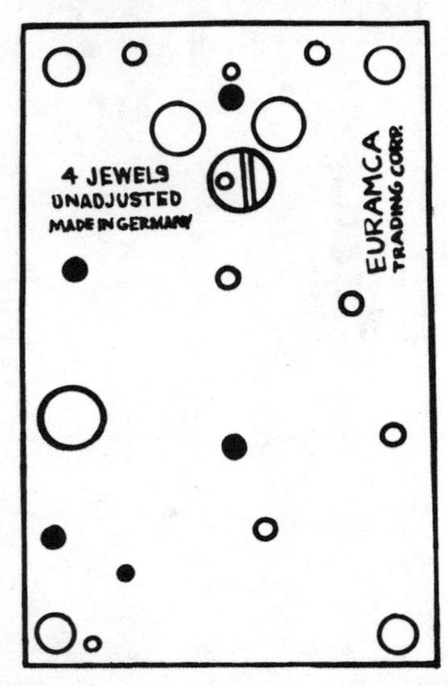

Plate 1091A 4-Ball Pendulum Miniature Clock
USE .0028" (.071mm) HOROLOVAR
Unit 28C (16 x 36)
See Appendix 25, 55, 60, 109

Uhrenfabrik M. Reiner c 1954

EURAMCA TRADING CORP.

EURAMCA
TRADING CORP.
GERMANY
NO(0) JEWELS
UNADJUSTED

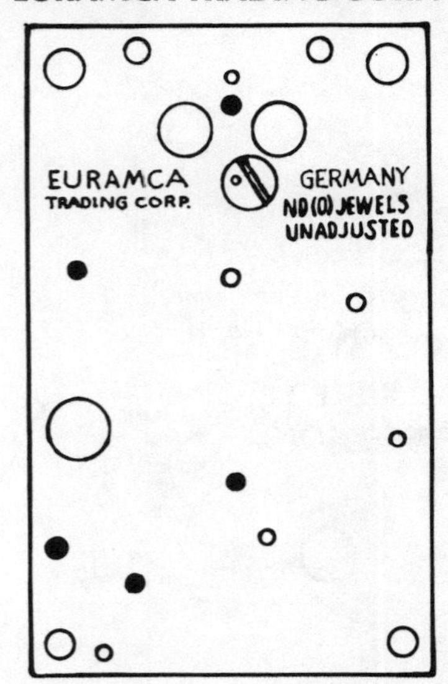

Plate 1092 Standard 4-Ball Pendulum Miniature Movement
USE .0032" (.081mm) HOROLOVAR
Units 23B (16 x 36)
See Appendix 25, 55, 109

Uhrenfabrik M. Reiner c 1953

EURAMCA TRADING CORP.

GERMANY
NO(0) JEWELS
UNADJUSTED
EURAMCA
TRADING CORP.

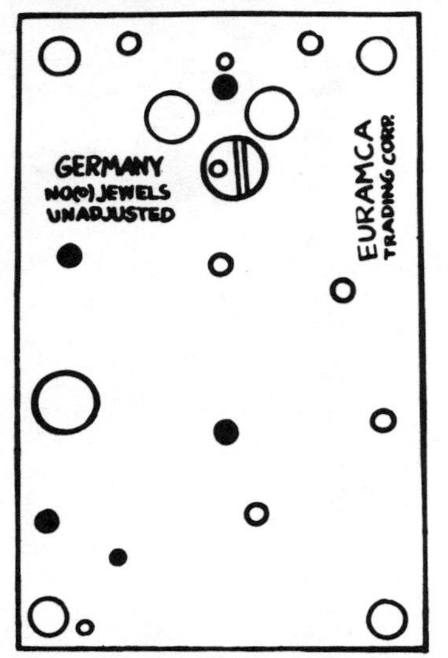

Plate 1092A Standard 4-Ball Pendulum Miniature Movement
USE .0032" (.081mm) HOROLOVAR
Units 23B (16 x 36)
See Appendix 25, 55, 109

Uhrenfabrik M. Reiner c 1953

EURAMCA TRADING CORP.

NO(0) JEWELS
UNADJUSTED
MADE IN GERMANY
EURAMCA
TRADING CORP.

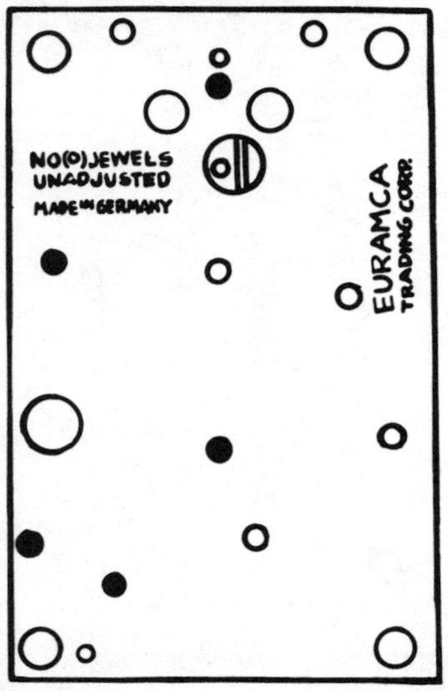

Plate 1093 4-Ball Pendulum Miniature Clock
USE .0028" (.071mm) HOROLOVAR
Units 28C (16 x 36)
See Appendix 25, 55, 109

Uhrenfabrik M. Reiner c 1954

EURAMCA TRADING CORP.

4 Jewels Unadjusted
Made in Germany
Euramca
Trading Corp.

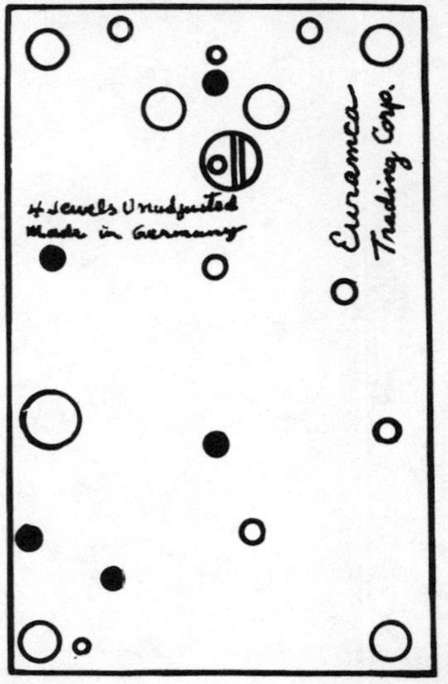

Plate 1093A 4-Ball Pendulum Miniature Clock
USE .0028" (.071mm) HOROLOVAR
Unit 28C (16 x 36)
See Appendix 25, 55, 60, 109

Uhrenfabrik M. Reiner c 1954

EURAMCA TRADING CORP.

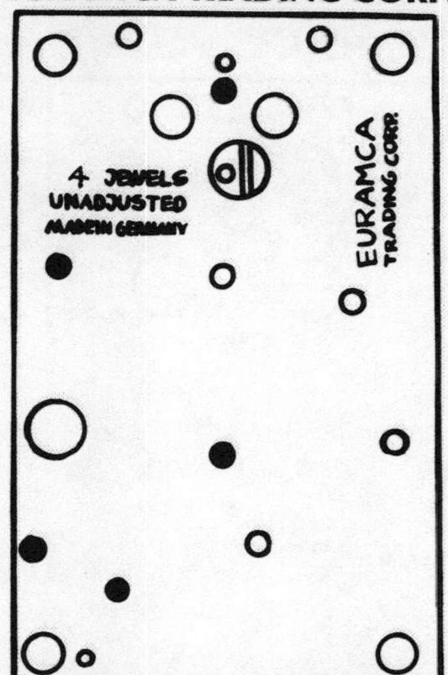

Plate 1093B 4-Ball Pendulum Miniature Clock
USE .0028" (.071mm) HOROLOVAR
Unit 28C (16 x 36)
See Appendix 25, 55, 60, 109

Uhrenfabrik Herr c 1952

EURAMCA TRADING CORP.

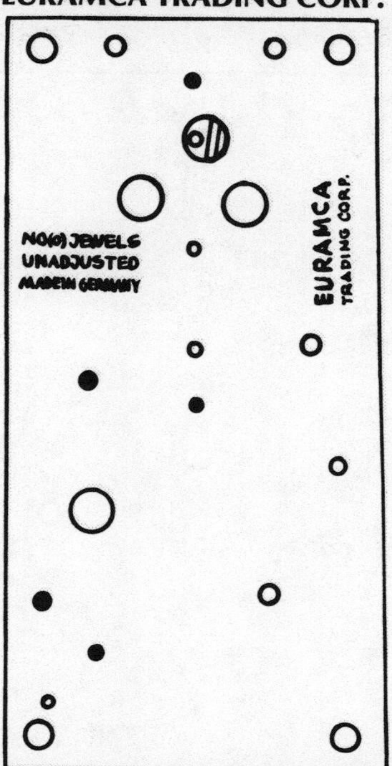

Plate 1096 4-Ball Pendulum
USE .0035" (.089mm) HOROLOVAR
Units 19, 20A, 21
USE .0038" (.097mm) HOROLOVAR
Unit 27B (19 x 36)
See Appendix 106, 109, 110

Uhrenfabrik Herr c 1952

EURAMCA TRADING CORP.

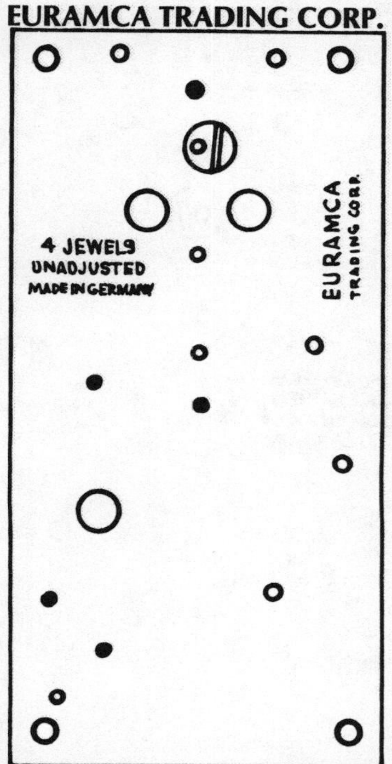

Plate 1097 4-Ball Pendulum
USE .0035" (.089mm) HOROLOVAR
Units 19, 20A, 21
USE .0038" (.097mm) HOROLOVAR
Unit 27B (19 x 36)
See Appendix 60, 106, 109, 110

Edgar Henn c 1952

EURAMCA TRADING CORP.

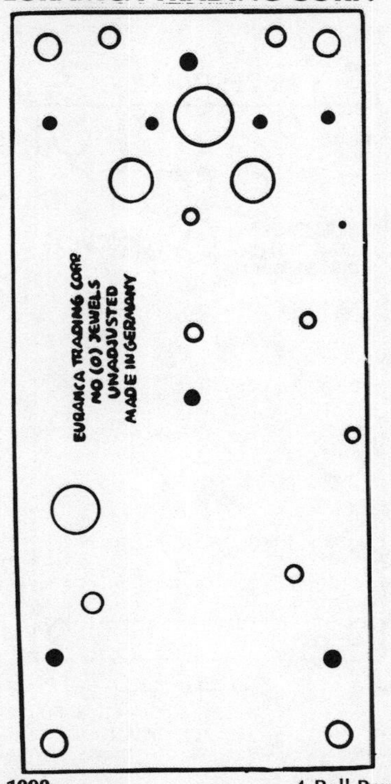

Plate 1098 4-Ball Pendulum
USE .0035" (.089mm) HOROLOVAR
Unit 31 (19 x 38)
See Appendix 39

Edgar Henn c 1953

EURAMCA TRADING CORP.

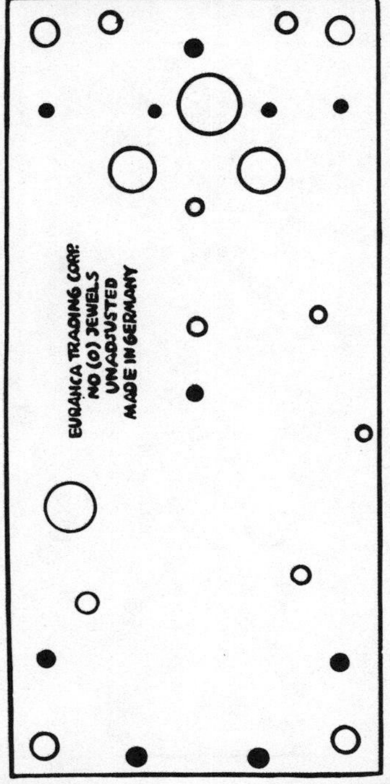

Plate 1099 4-Ball Pendulum
USE .0037" (.094mm) HOROLOVAR
Unit 32 (19 x 38)
See Appendix 39

Uhrenfabrik Herr c 1950

EYH

GERMANY
NO(0) JEWELS
UNADJUSTED
H. EICKMANN & CO.
MADE IN GERMANY
EYH

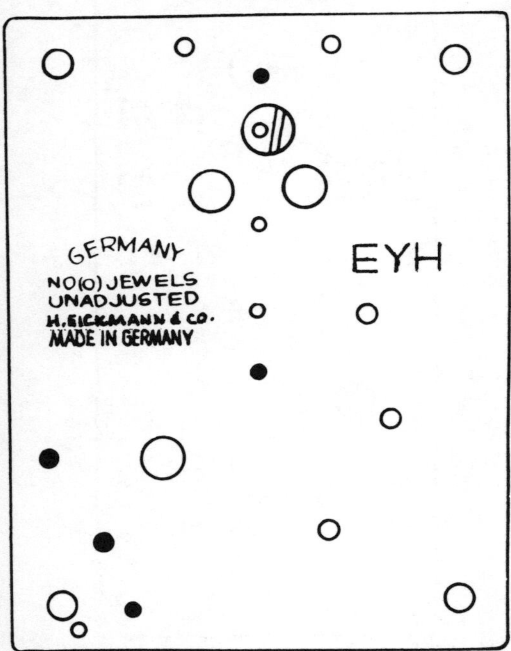

Plate 1100 4-Ball Pendulum
USE .004" (.102mm) HOROLOVAR
Unit 27A (19 x 36)
See Appendix 130

Jahresuhrenfabrik c 1910

F

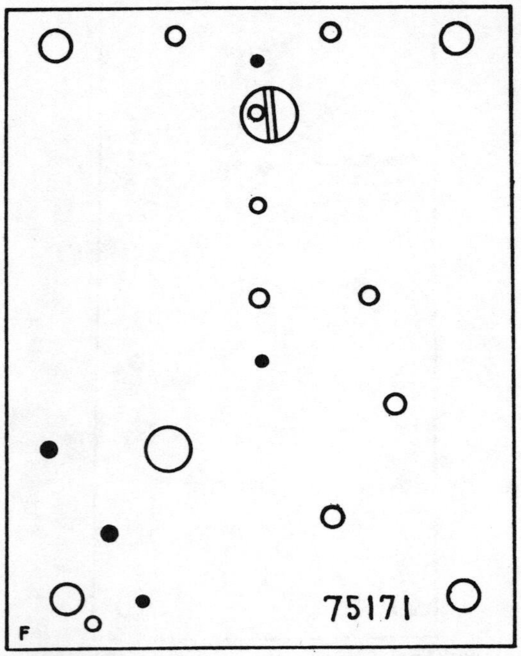

75171
F

Plate 1101 Disc Pendulum
USE .004"* (.102mm*) HOROLOVAR
(19 x 36)

Franz Hermle c 1951

FHS (F. S. HERMLE)

FHS
GERMANY

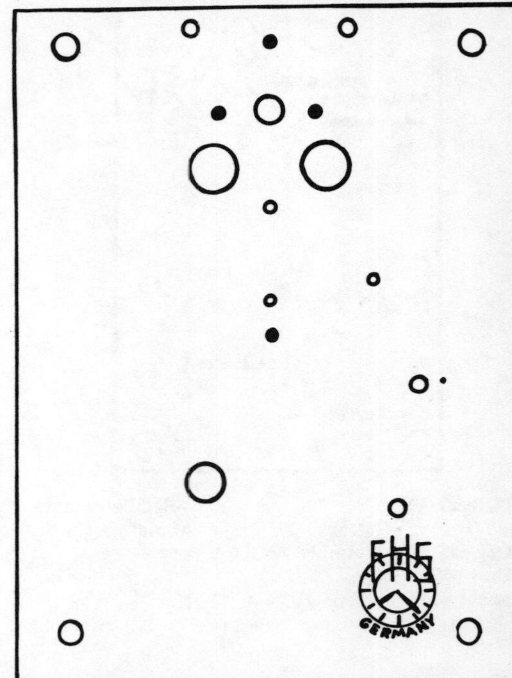

Plate 1103 4-Ball Pendulum
USE .0033" (.084mm) HOROLOVAR
Unit 25A (18 x 38)
See Appendix 71

Franz Hermle c 1951

FHS (F. S. HERMLE)

70
FHS
911-000

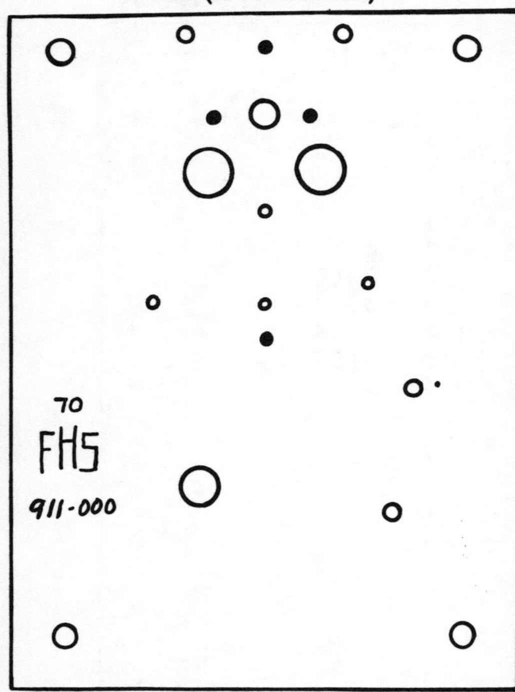

Plate 1103A 4-Ball Pendulum
USE .0033" (.084mm) HOROLOVAR
Unit 25A (18 x 38)
See Appendix 71

Franz Hermle c 1954

FHS (F. S. HERMLE)

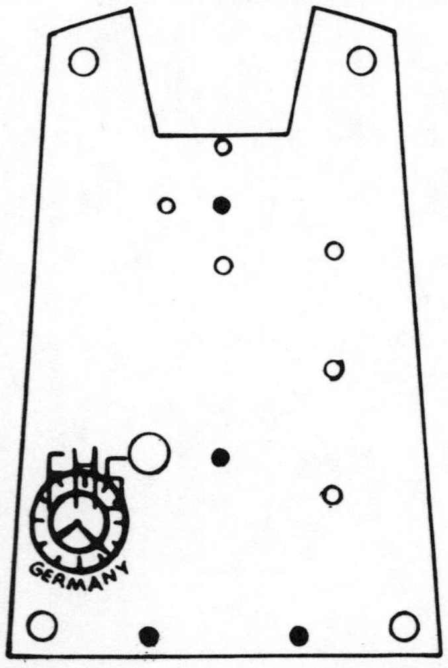

FHS
GERMANY

Plate 1104 4-Ball Pendulum Miniature Clock
USE .0022" (.056mm) HOROLOVAR
Unit 25B (13 x 30)
See Appendix 63

Uhrenfabrik M. Reiner c 1954

FORESTVILLE

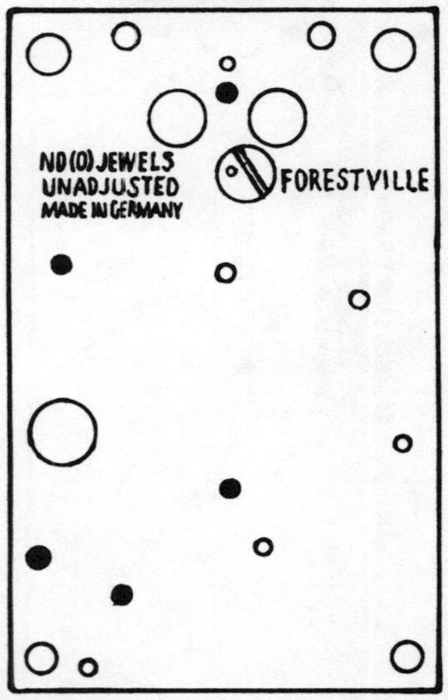

NO(0) JEWELS
UNADJUSTED
MADE IN GERMANY
FORESTVILLE

Plate 1105 Standard 4-Ball Pendulum Miniature Movement
USE .0032" (.081mm) HOROLOVAR
Unit 23B (16 x 36)
See Appendix 25, 55, 109

Uhrenfabrik Herr c 1953

FORESTVILLE

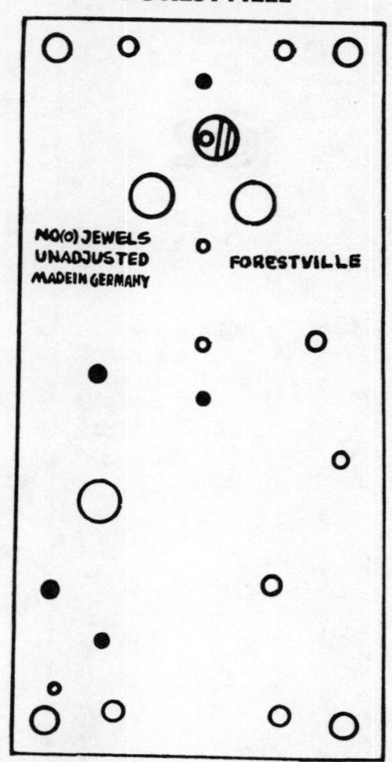

Plate 1105A 4-Ball Pendulum
USE .0035" (.089mm) HOROLOVAR
Units 19, 20A, 21
USE .0038" (.097mm) HOROLOVAR
Unit 27B (19 x 36)
See Appendix 106, 109, 110

Kieninger & Obergfell c 1949

FORESTVILLE CLOCK CO.

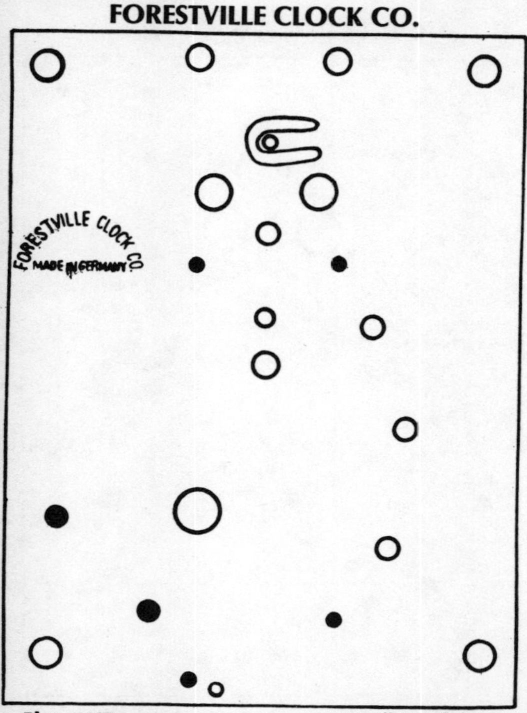

Plate 1107 4-Ball Pendulum
USE .0032" (.081mm) HOROLOVAR
Units 1, 3A (19 x 38)
See Appendix 76

Uhrenfabrik M. Reiner c 1953

FORESTVILLE CLOCK CO.

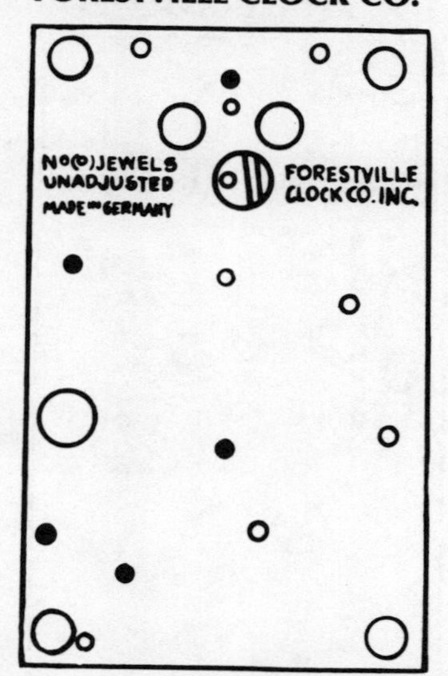

Plate 1106 Standard 4-Ball Pendulum Miniature Movement
USE .003" (.076mm) HOROLOVAR
Units 22, 23A, 24 (16 x 36)
See Appendix 25, 54, 109

J. Link & Co. c 1952

FORESTVILLE CLOCK CO.

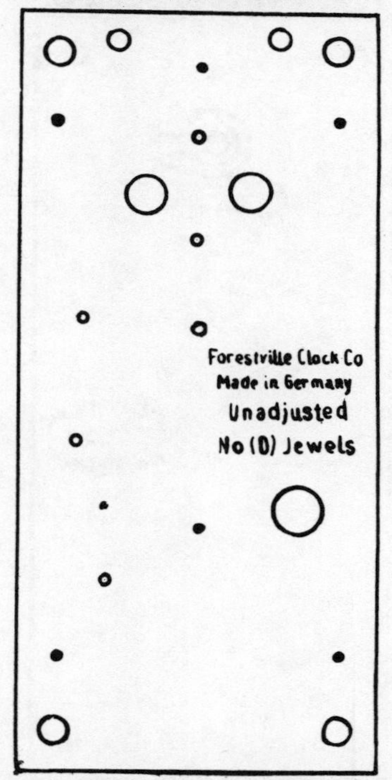

Plate 1111 4-Ball Pendulum
USE .0037" (.094mm) HOROLOVAR
Units 16, 17 (19 x 38)
See Appendix 24

Edgar Henn c 1953

FORESTVILLE CLOCK CO.

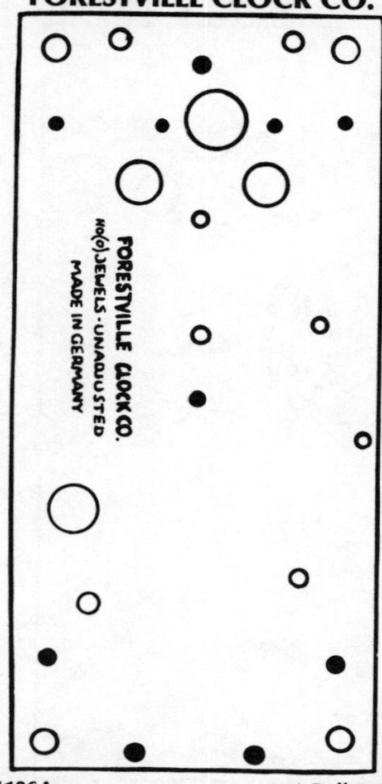

Plate 1106A 4-Ball Pendulum
USE .0037" (.094mm) HOROLOVAR
Unit 32 (19 x 38)
See Appendix 39

Uhrenfabrik M. Reiner c 1951

FORESTVILLE CLOCK CO.

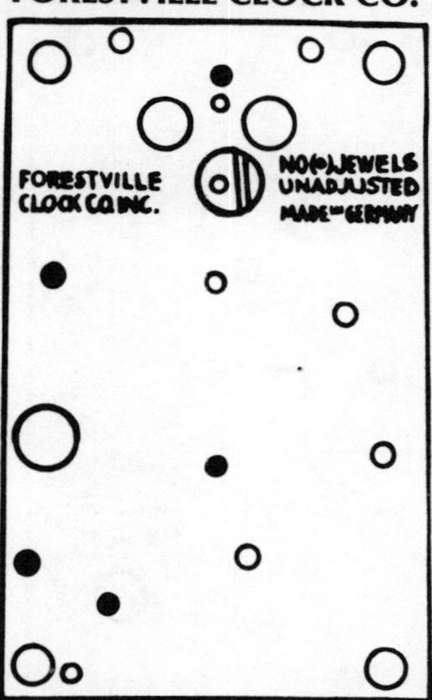

Plate 1115 Standard 4-Ball Pendulum Miniature Movement
USE .003" (.076mm) HOROLOVAR
Units 22, 23A, 24 (16 x 36)
See Appendix 25, 54, 109

Edgar Henn c 1953

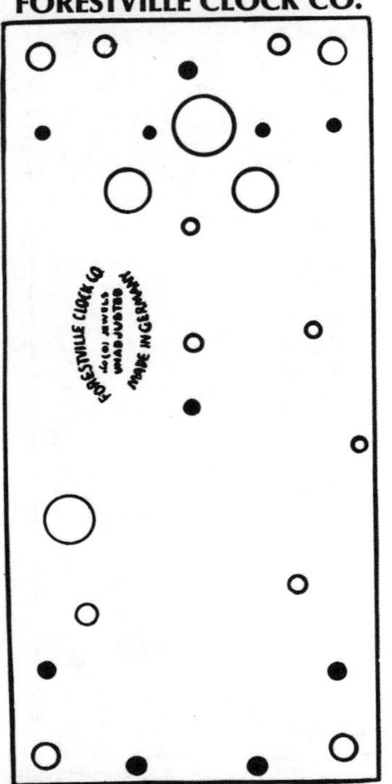

Plate 1116 4-Ball Pendulum
USE .0037" (.094mm) HOROLOVAR
Unit 32 (19 x 38)
See Appendix 39

Edgar Henn c 1952

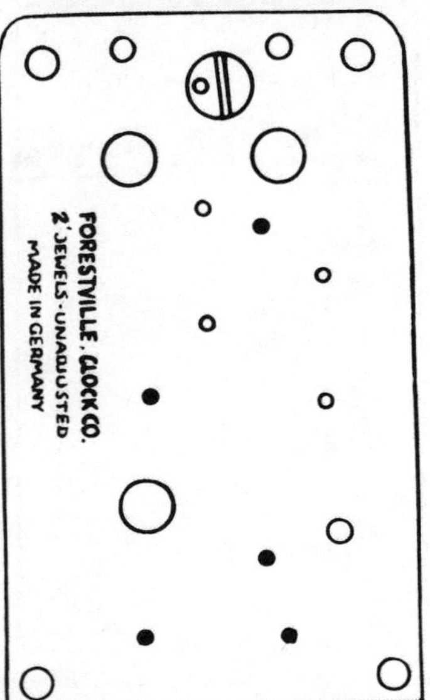

Plate 1117 4-Weight Pendulum Miniature Clock
USE .0023" (.058mm) HOROLOVAR
Units 29A, 29B (19 x 32)
See Appendix 95

J. Link & Co. c 1952

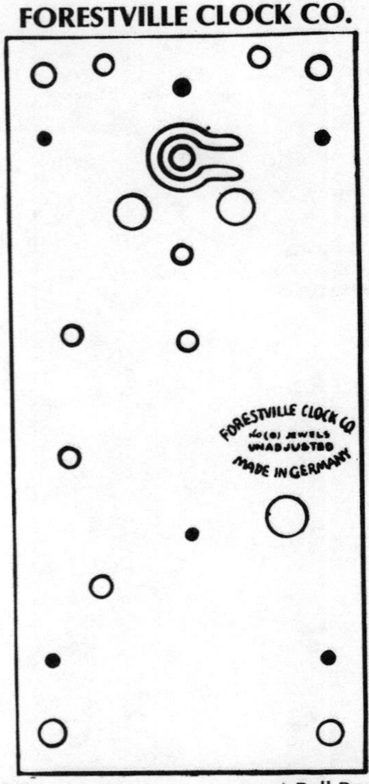

Plate 1119 4-Ball Pendulum
USE .0037" (.094mm) HOROLOVAR
Units 16, 17 (19 x 38)
See Appendix 38, 119

J. Link & Co. c 1953

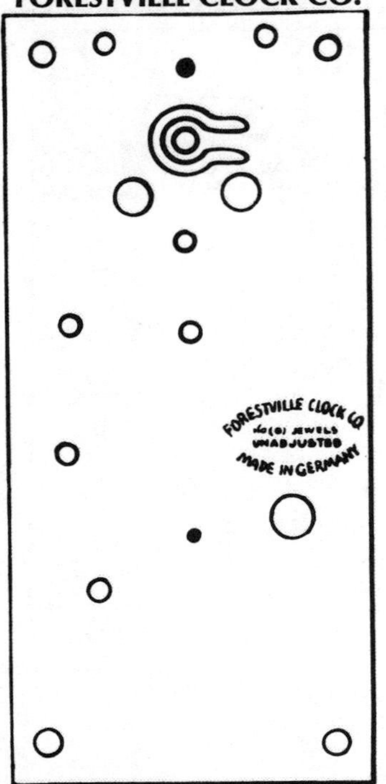

Plate 1120 4-Ball Pendulum
USE .0037" (.094mm) HOROLOVAR
Units 16, 17 (19 x 38)
See Appendix 38, 119

J. Link & Co. c 1952

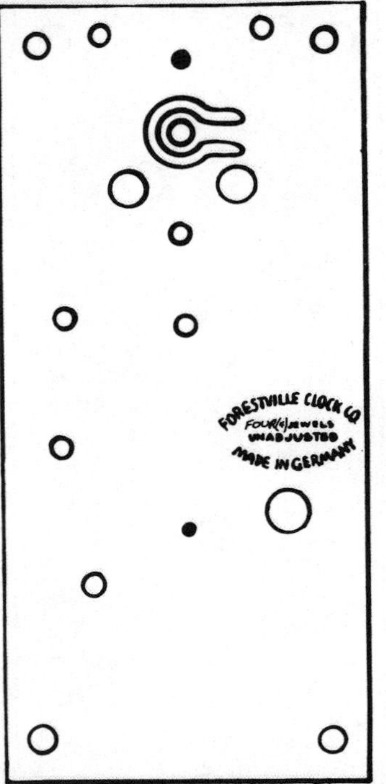

Plate 1120A 4-Ball Pendulum
USE .0037" (.094mm) HOROLOVAR
Units 16, 17 (19 x 38)
See Appendix 38, 119

Uhrenfabrik Herr c 1952

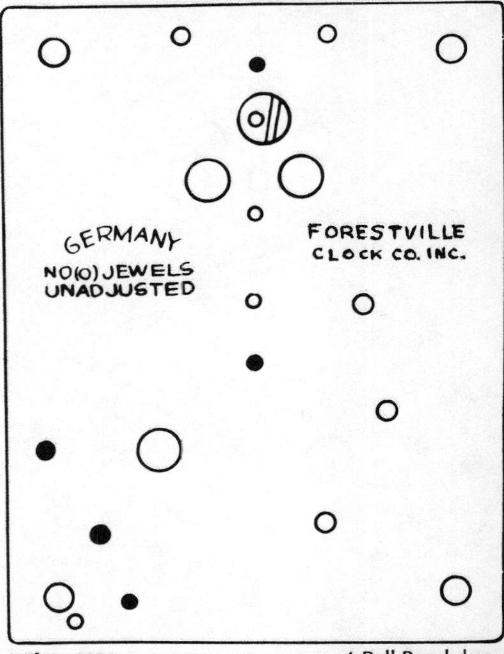

Plate 1121 4-Ball Pendulum
USE .004" (.102mm) HOROLOVAR
Unit 27A (19 x 36)
See Appendix 130

Franz Hermle c 1954

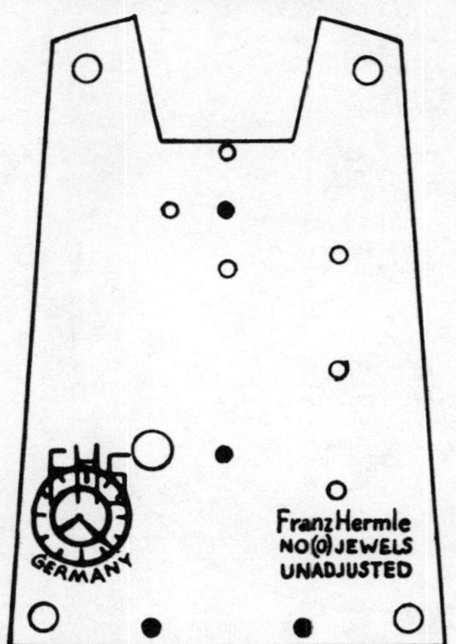

Plate 1124 4-Ball Pendulum Miniature Clock
USE .0022" (.056mm) HOROLOVAR
Unit 25B (13 x 30)
See Appendix 63

Franz Hermle c 1954

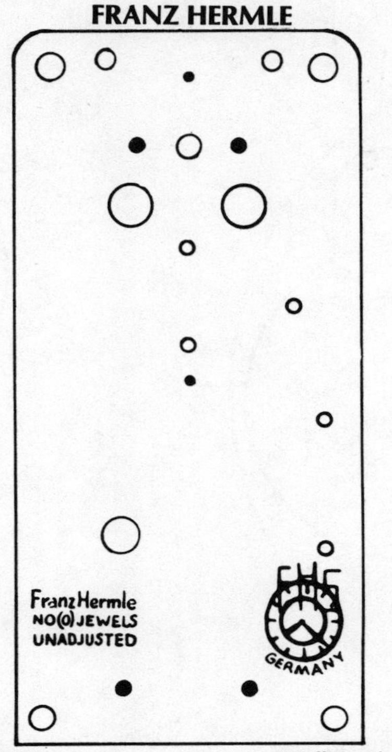

Plate 1126 4-Ball Pendulum
USE .0033" (.084mm) HOROLOVAR
Unit 25A (18 x 38)
See Appendix 71

Kieninger & Obergfell c 1951

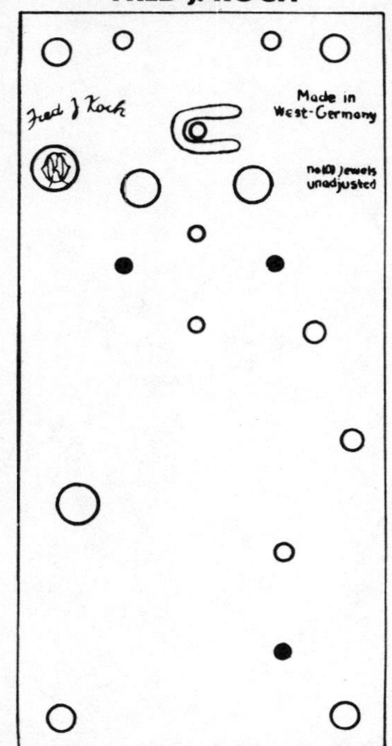

Plate 1135 4-Ball Pendulum
USE .0032" (.081mm) HOROLOVAR
Units 1, 3A, 3B, 3C (19 x 38)
See Appendix 76

Kieninger & Obergfell c 1954

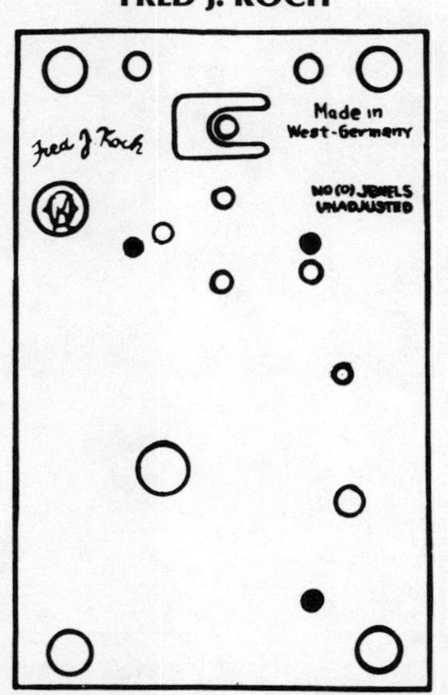

Plate 1139 4-Ball Pendulum Miniature Clock
USE .0023" (.058mm) HOROLOVAR
Units 5A, 5B, 5C, 5D, 5E (14 x 30)
See Appendix 77, 111

Kieninger & Obergfell c 1952

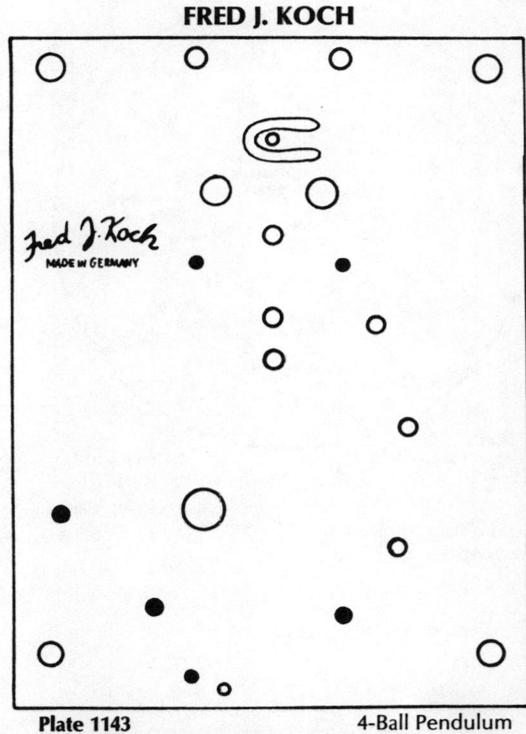

Plate 1143 4-Ball Pendulum
USE .0032" (.081mm) HOROLOVAR
Units 1, 3A (19 x 38)
See Appendix 76

Franz Vosseler c 1908

FR. VOSSELER
FR. VOSSELER
SCHWENNINGEN
GERMANY

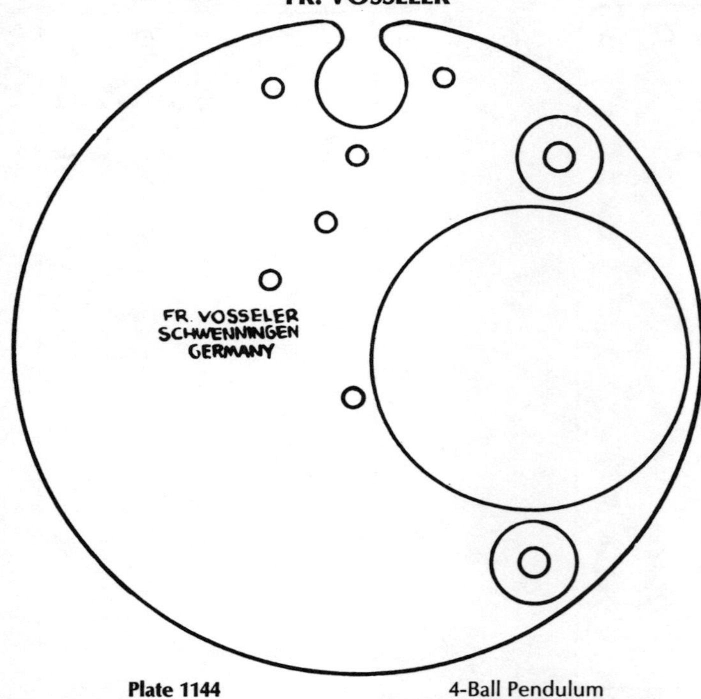

Plate 1144 4-Ball Pendulum
Underslung Suspension
USE .0028"* (.071mm*) HOROLOVAR
See Appendix 63 (19 x 36)

Franz Vosseler c 1910

FR. VOSSELER
FR. VOSSELER
SCHWENNINGEN
GERMANY
D.R.G.M. MADE IN GERMANY

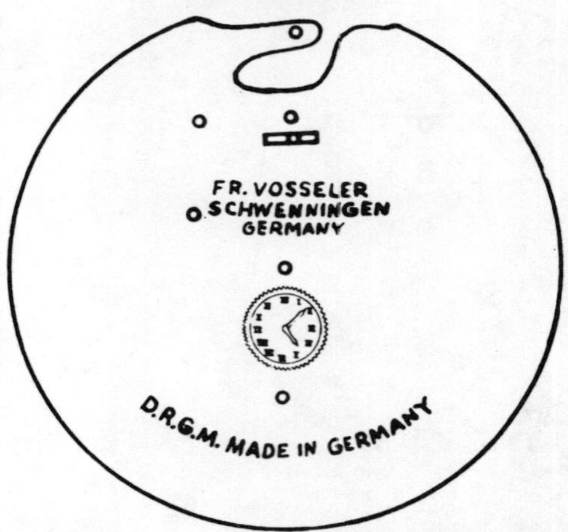

Plate 1145 3-Ball Pendulum
Miniature "Louvre" Clock
Underslung Suspension
USE .003"* (.076mm*) HOROLOVAR
See Appendix 26, 93

Franz Vosseler c 1910

FR. VOSSELER
FR. VOSSELER
SCHWENNINGEN
GERMANY

Plate 1145A 3-Ball Pendulum
Miniature "Louvre" Clock
Underslung Suspension
USE .003"* (.076mm*) HOROLOVAR
See Appendix 26, 93

Jahresuhrenfabrik c 1908

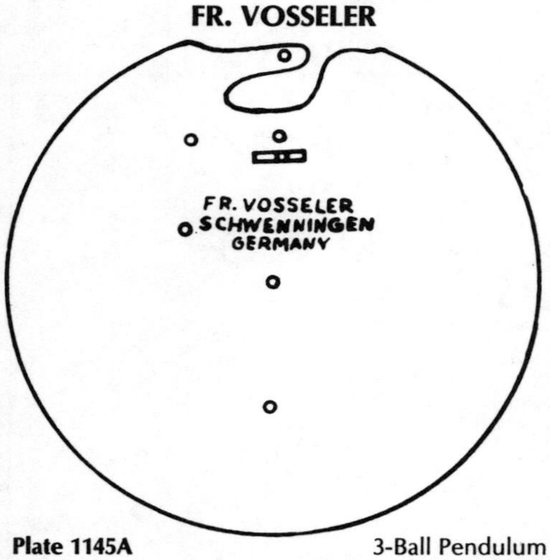

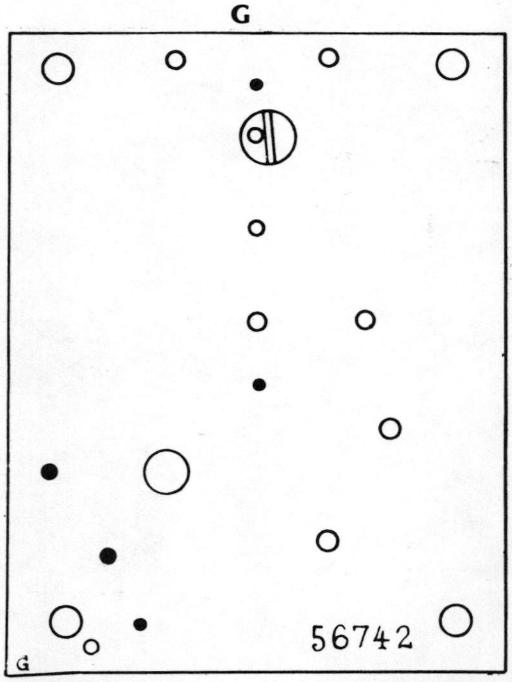

G
56742
G

Plate 1146 Disc Pendulum
4-Ball Pendulum
USE .005"* (.127mm*) HOROLOVAR
7¼" Suspension Spring
(19 x 36)

Jahresuhrenfabrik c 1907

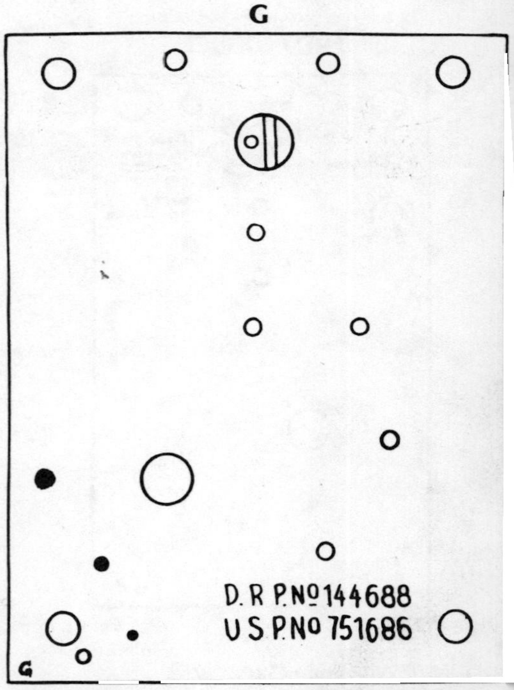

G
D.R P.No 144688
U S P.No 751686
G

Plate 1146A Disc Pendulum
USE .005"* (.127mm*) HOROLOVAR
7¼" Suspension Spring
(19 x 36)

Gustav Becker c 1900

G B

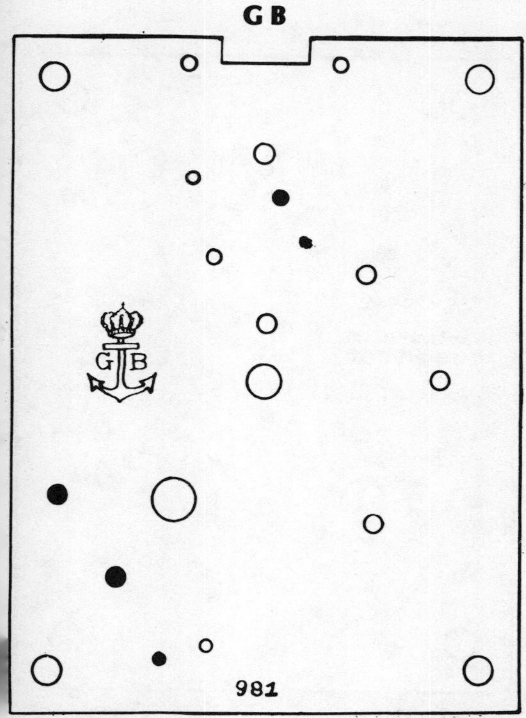

Plate 1147 Disc Pendulum
4-Ball Pendulum
Pin Pallet Escapement
Lantern Pinions
USE .004" (.102mm) HOROLOVAR
See Appendix 70, 74 (19 x 38)

Kieninger & Obergfell c 1930

GEO. BORGFELDT & CO.

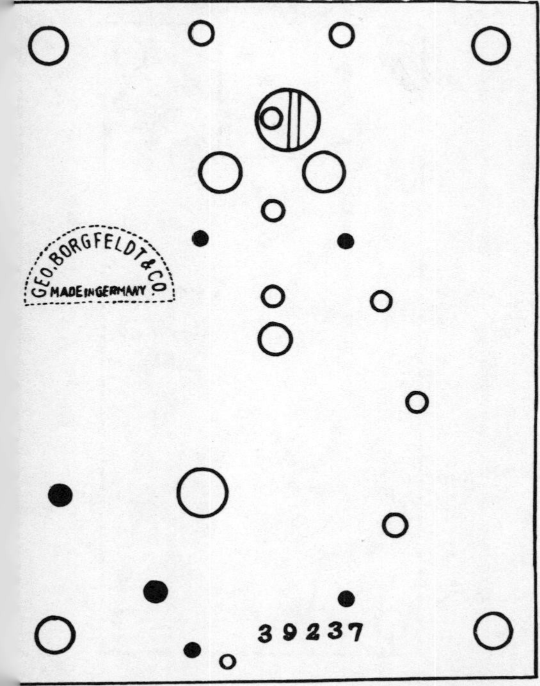

Plate 1151 4-Ball Pendulum
USE .0032" (.081mm) HOROLOVAR
Unit 1 (19 x 38)
See Appendix 76

Gustav Becker c 1907

G B

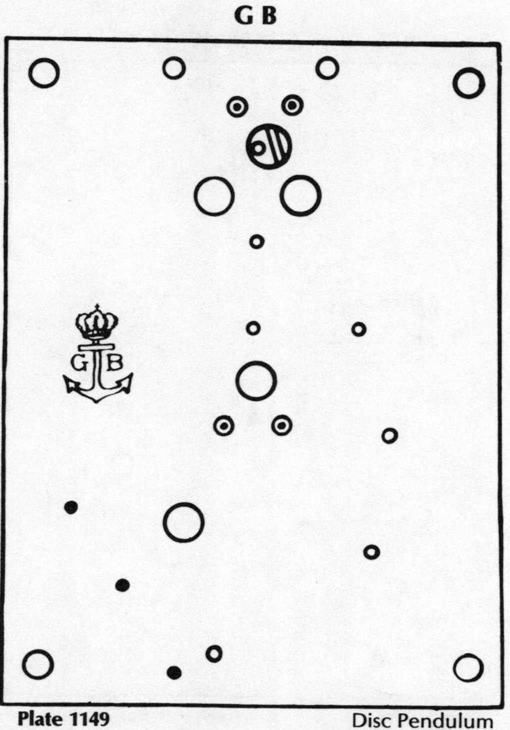

Plate 1149 Disc Pendulum
4-Ball Pendulum
USE .004"* (.102mm*) HOROLOVAR
See Appendix 11, 74 (19 x 36)

Kieninger & Obergfell c 1930

GEO. BORGFELDT & CO.

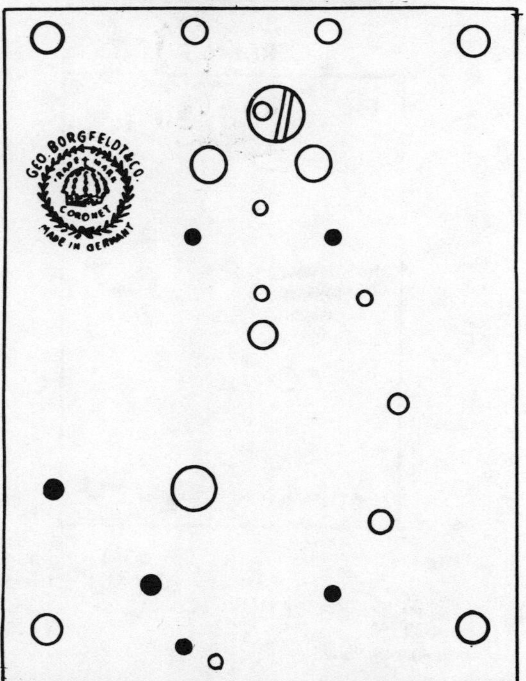

Plate 1155 4-Ball Pendulum
USE .0032" (.081mm) HOROLOVAR
Unit 1 (19 x 38)
See Appendix 76

Manufacturer Not Known c 1881

G.W. & CIE

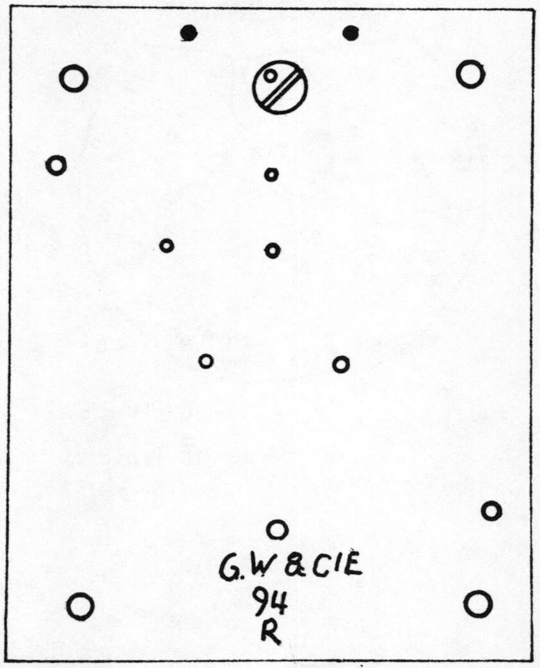

G.W &CIE
94
R

Plate 1150 Special Disc Pendulum
Front Wind
See Appendix 73 (19 x 36)

Georg Wurthner c 1957

GEORG WURTHNER

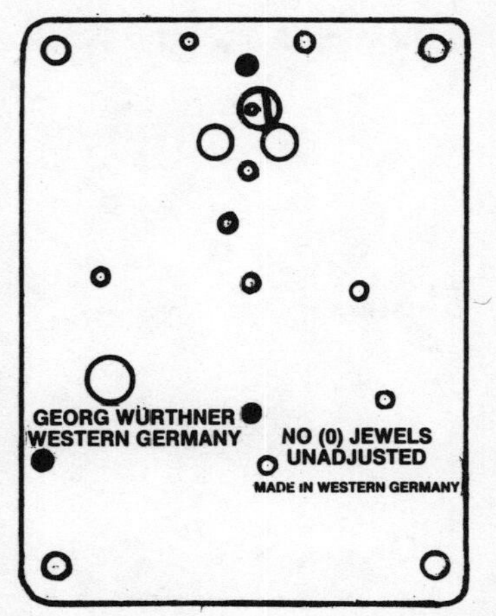

Plate 1157 4-Ball Pendulum
Midget Clock
USE .002" (.051mm) HOROLOVAR
Unit 37 (9.5 x 27)
See Appendix 27

Georg Wurthner c 1957

GEORG WURTHNER

NO (0) JEWELS
UNADJUSTED
MADE IN WESTERN GERMANY
GEORG WURTHNER
WESTERN GERMANY

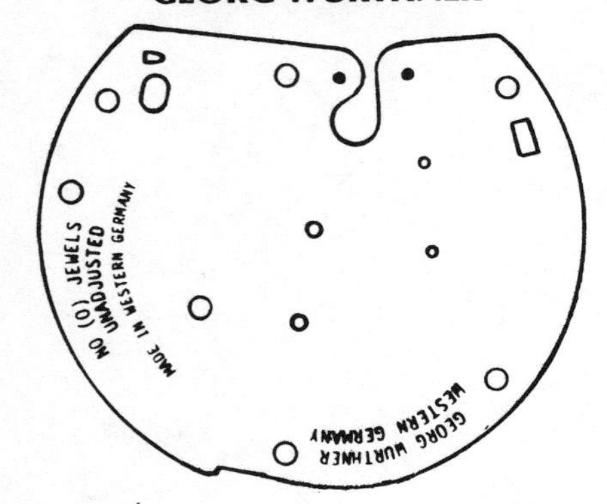

Plate 1157A 4-Ball Pendulum
Sub-Midget
14 Day Clock

See Appendix 104

Uhrenfabrik Herr c 1951

GERMANY

GERMANY
NO(0) JEWELS
UNADJUSTED

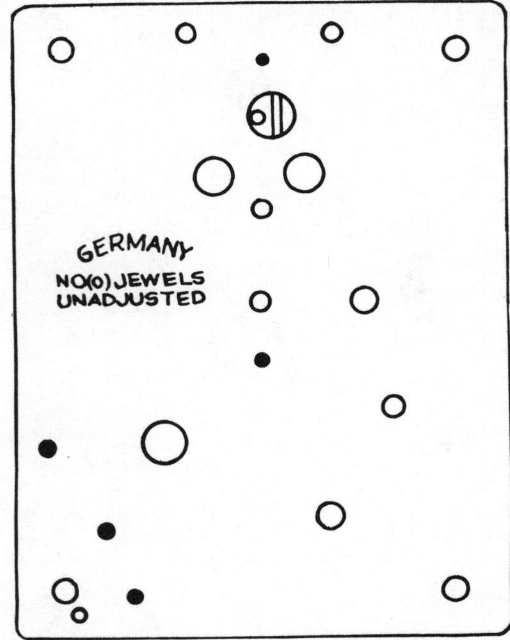

Plate 1159 4-Ball Pendulum
USE .004" (.102mm) HOROLOVAR
Unit 27A (19 x 36)
See Appendix 130

Uhrenfabrik Herr c 1952

GERMANY

GERMANY
NO(0) JEWELS
UNADJUSTED

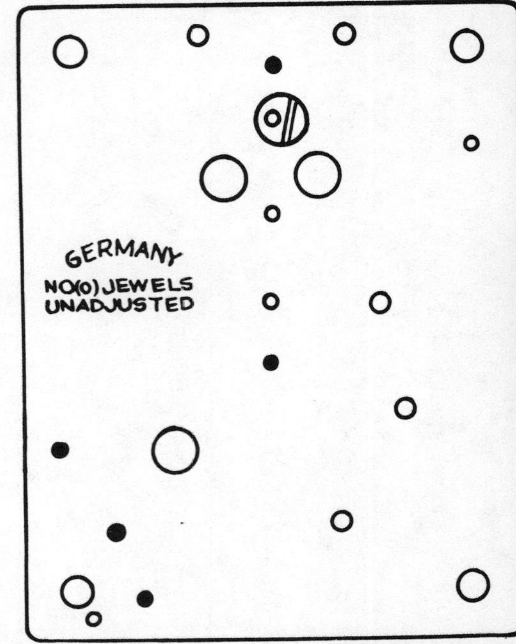

Plate 1159A 4-Ball Pendulum
Hour Striker
USE .004" (.102mm) HOROLOVAR
Unit 27A (19 x 36)
See Appendix 56

Sigfried Haller c 1968

GERMANY

GERMANY

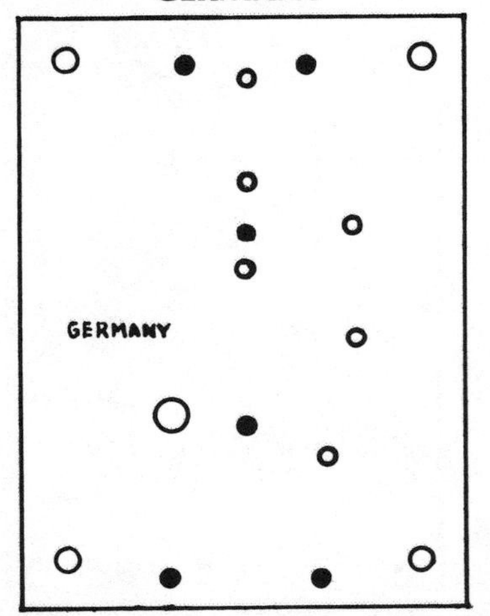

Plate 1160 4-Ball Pendulum
Miniature Clock
USE .0019" (.048mm) HOROLOVAR
Unit 39 (12 x 25)
See Appendix 24

Sigfried Haller c 1970

GERMANY

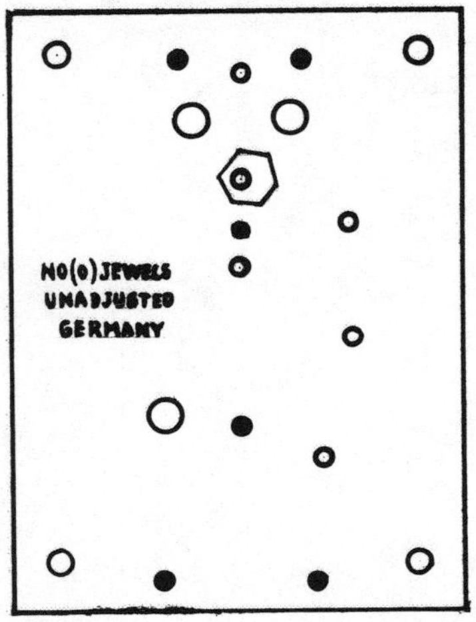

NO(0) JEWELS
UNADJUSTED
GERMANY

Plate 1161 4-Ball Pendulum
Midget Clock
USE .0019" (.048mm) HOROLOVAR
Unit 39 (12 x 25)
See Appendix 25

Sigfried Haller c 1951

GERMANY

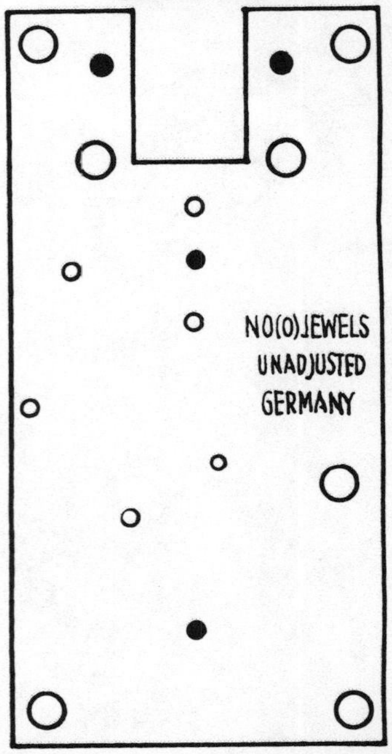

NO(0)JEWELS
UNADJUSTED
GERMANY

Plate 1162 4-Ball Pendulum
USE .0038" (.097mm) HOROLOVAR
Unit 33 (18 x 38)
See Appendix 63, 64

Sigfried Haller c 1952

GERMANY
NO(O)JEWELS
UNADJUSTED
GERMANY

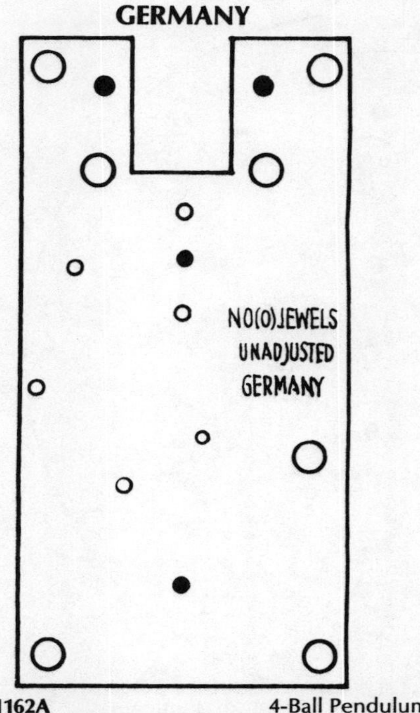

Plate 1162A 4-Ball Pendulum
USE .0038″ (.097mm) HOROLOVAR
Unit 33 (18 x 38)
See Appendix 63, 64

Kienzle Clock Factories c 1908

GERMANY
108518
Germany

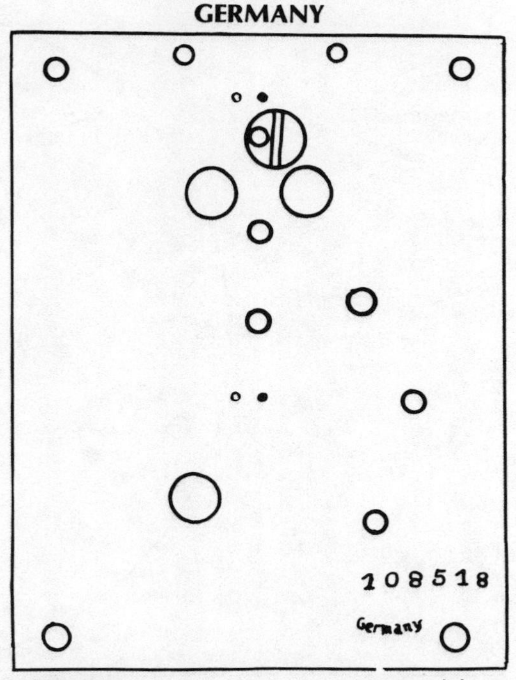

Plate 1163 Disc Pendulum
USE .0035″* (.089mm*) HOROLOVAR
(18 x 38)

Kienzle Clock Factories c 1909

GERMANY
Germany
126 880

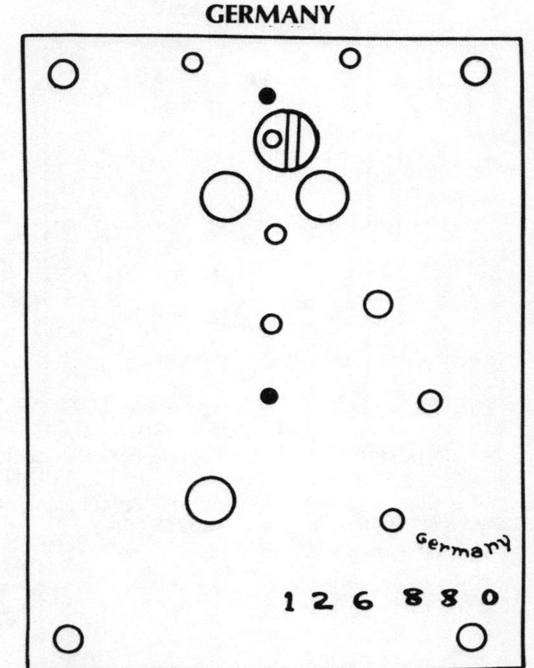

Plate 1167 Disc Pendulum
USE .0035″* (.089mm*) HOROLOVAR
See Appendix 41 (18 x 38)

Gebr. Junghans c 1910

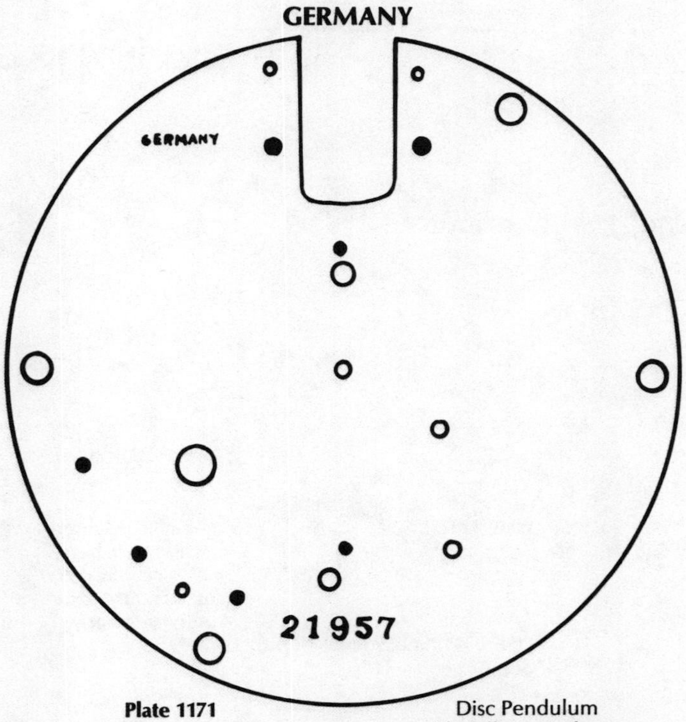

GERMANY
GERMANY
21957

Plate 1171 Disc Pendulum with Balls on Scales
USE .004″* (.102mm*) HOROLOVAR
See Appendix 29, 63 (19 x 36)

Manufacturer Not Known

GERMANY
GERMANY
74

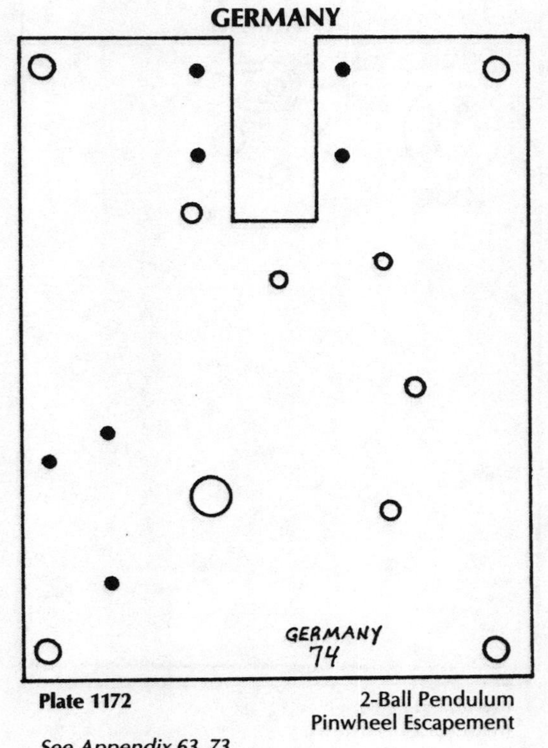

Plate 1172 2-Ball Pendulum Pinwheel Escapement
See Appendix 63, 73

Kienzle Clock Factories c 1911

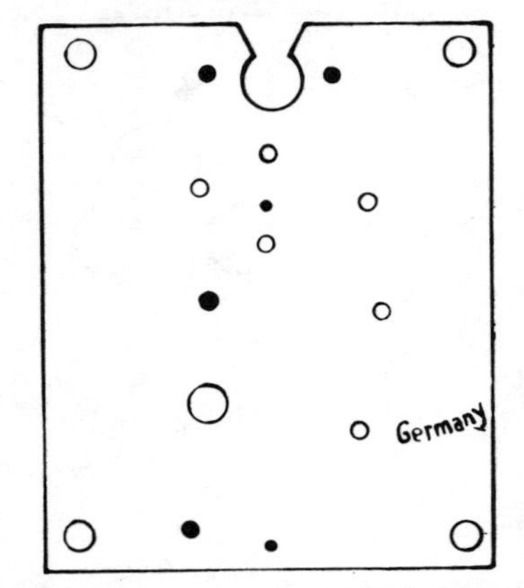

Plate 1175 Semispherical Pendulum
Miniature Clock: 5" High
Pin Pallet Escapement
USE .002"— (.051mm—) HOROLOVAR
See Appendix 30, 63

Manufacturer Not Known c 1904

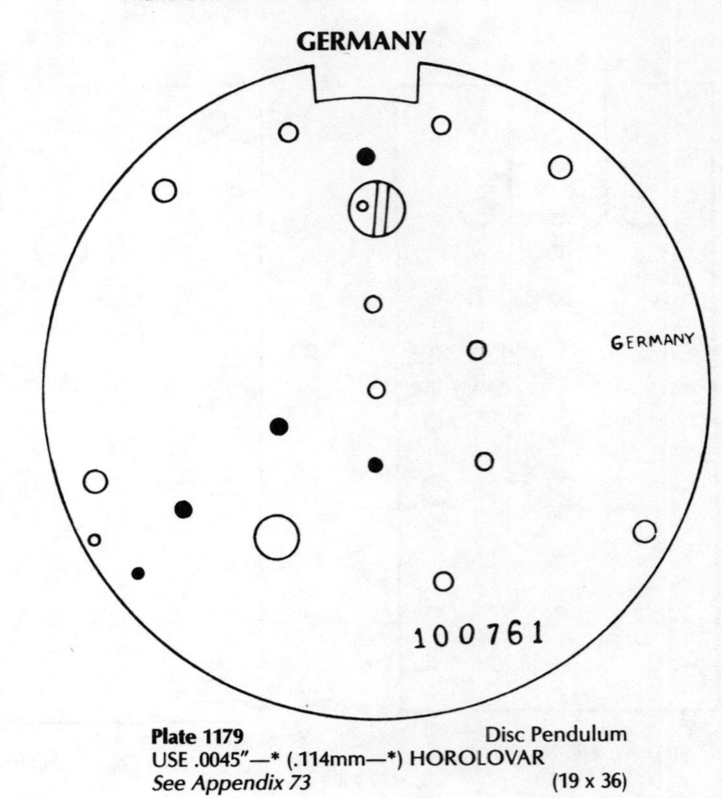

Plate 1179 Disc Pendulum
USE .0045"—* (.114mm—*) HOROLOVAR
See Appendix 73 (19 x 36)

Kieninger & Obergfell c 1950

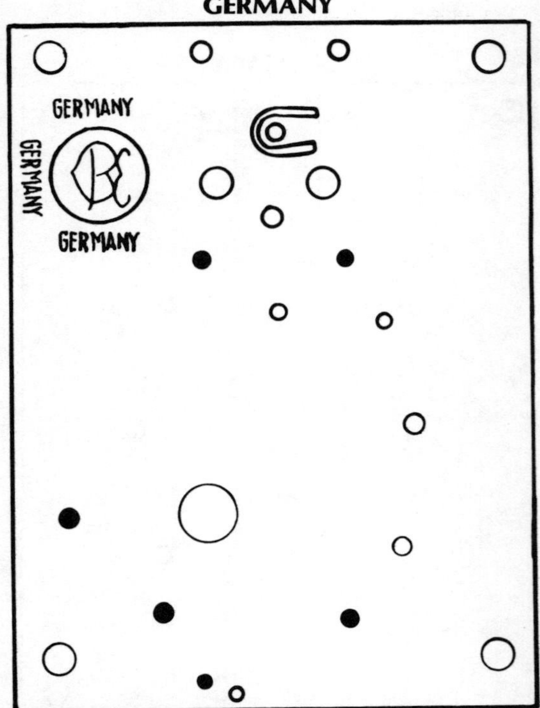

Plate 1180 4-Ball Pendulum
USE .0032" (.081mm) HOROLOVAR
Units 1, 3A (19 x 38)
See Appendix 76

Jahresuhrenfabrik c 1905

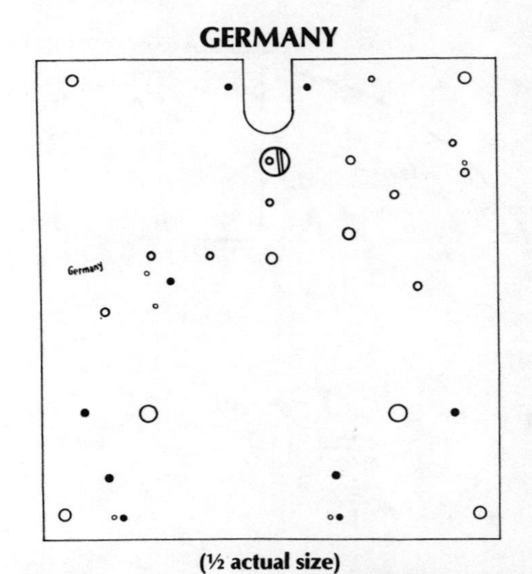

(½ actual size)

Plate 1181 4-Ball Pendulum
Hour and Half-hour
Year Striker
Graham Escapement
7⅛" Suspension Spring
USE .0035"* (.089mm*) HOROLOVAR
See Appendix 31

Kienzle Clock Factories c 1906

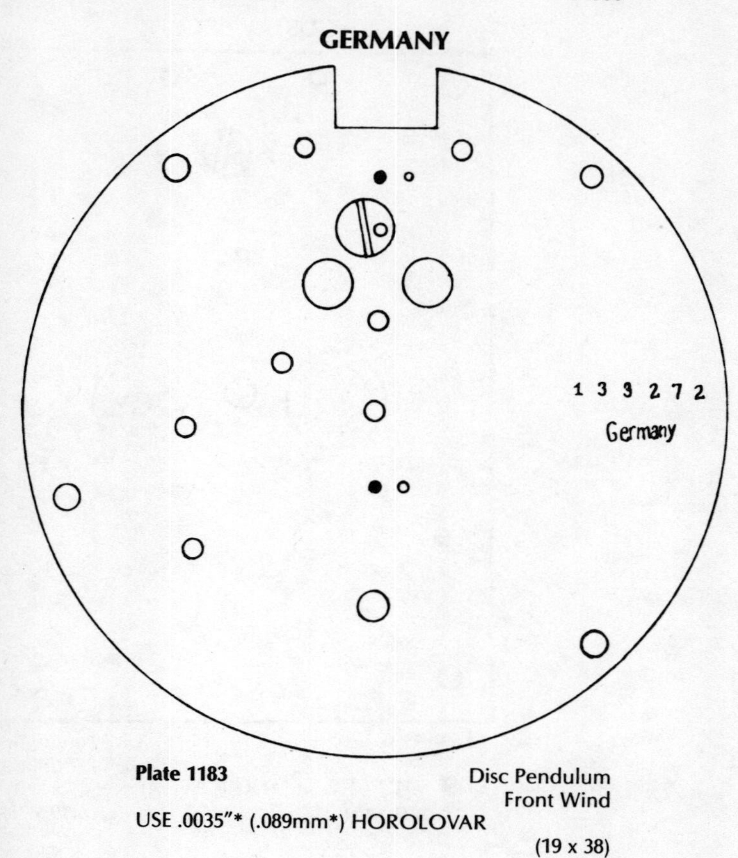

Plate 1183 Disc Pendulum
Front Wind
USE .0035"* (.089mm*) HOROLOVAR
(19 x 38)

Sigfried Haller c 1960

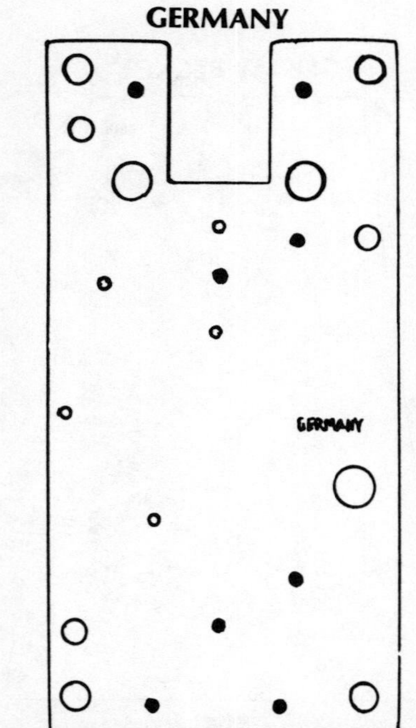

Plate 1185A 4-Ball Pendulum
USE .003" (.076mm) HOROLOVAR
Units 42, 42A (18 x 38)
See Appendix 63, 64

Gustav Becker c 1907

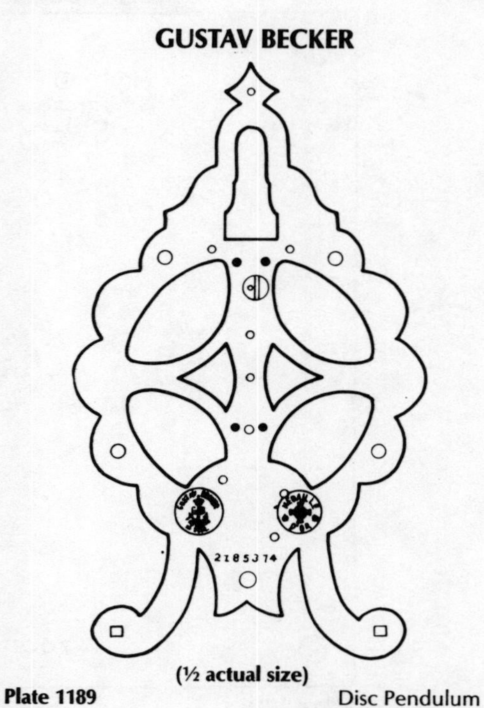

(½ actual size)

Plate 1189 Disc Pendulum
Skeleton Clock
USE .004"* (.102mm*) HOROLOVAR
See Appendix 69, 70, 74 (19 x 38)

Gustav Becker c 1902

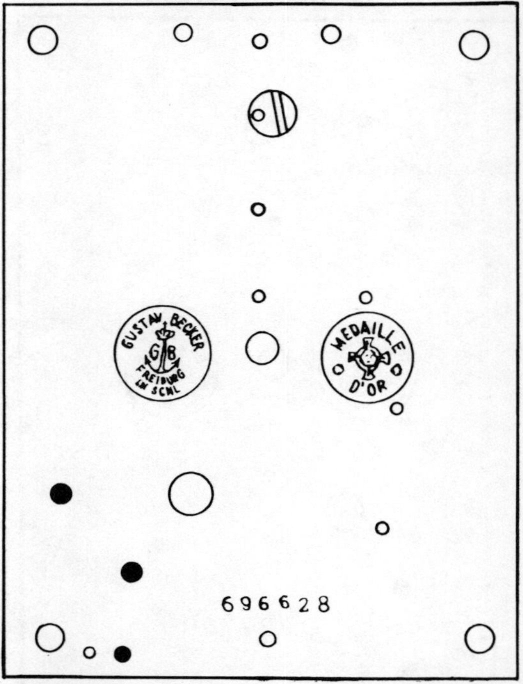

Plate 1191 Disc Pendulum
4-Ball Pendulum
USE .004"* (.102mm*) HOROLOVAR
See Appendix 70, 74 (19 x 38)

Gustav Becker c 1903

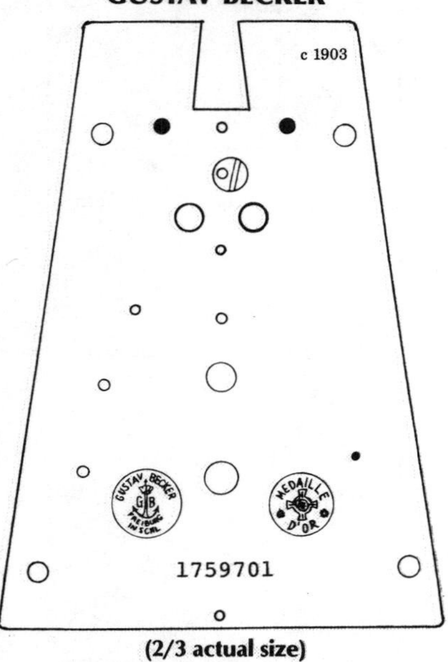

(2/3 actual size)

Plate 1193 Special Pendulum Wall Clock

See Appendix 3, 70, 74 (19 x 38)

Gustav Becker c 1902

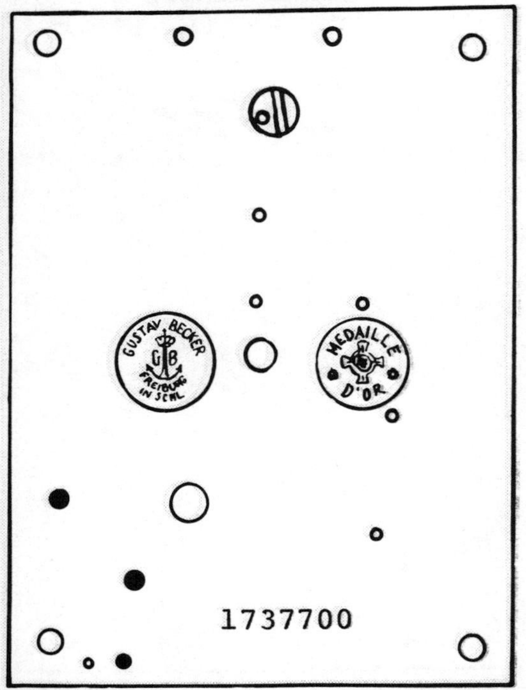

Plate 1195A Disc Pendulum 4-Ball Pendulum

USE .004"* (.102mm*) HOROLOVAR

See Appendix 70, 74 (19 x 38)

Gustav Becker c 1902

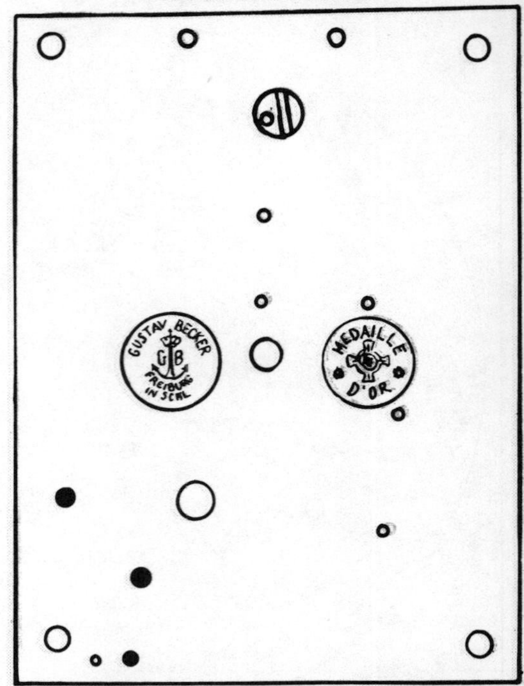

Plate 1195 Disc Pendulum 4-Ball Pendulum

USE .004"* (.102mm*) HOROLOVAR

See Appendix 74 (19 x 38)

Gustav Becker c 1906

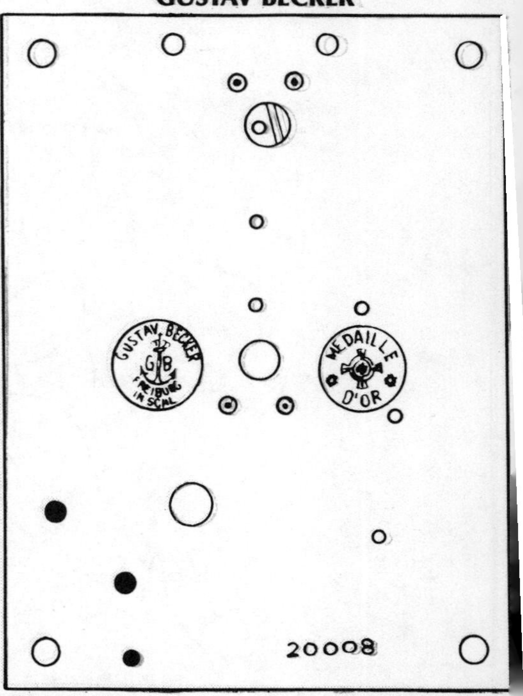

Plate 1199 Disc Pendulum 4-Ball Pendulum

USE .004"* (.102mm*) HOROLOVAR

See Appendix 11, 70, 74, 87 (19 x 38)

Gustav Becker c 1905

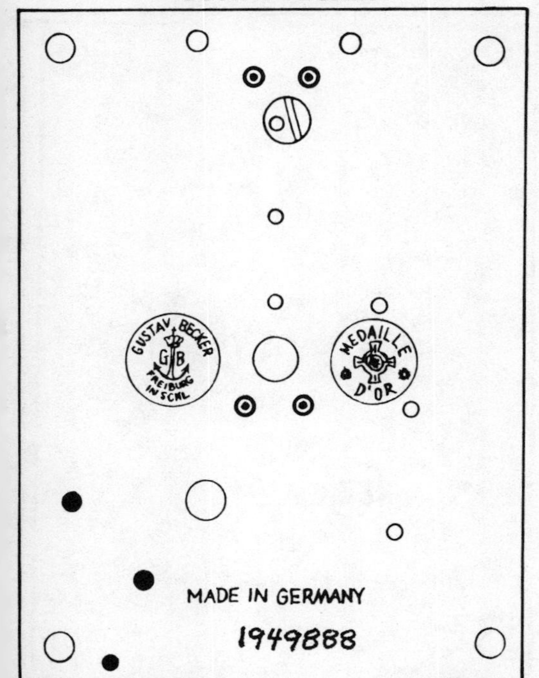

Plate 1201
Disc Pendulum
4-Ball Pendulum

USE .004"* (.102mm*) HOROLOVAR
See Appendix 11, 70, 74, 87 (19 x 38)

Gustav Becker c 1903

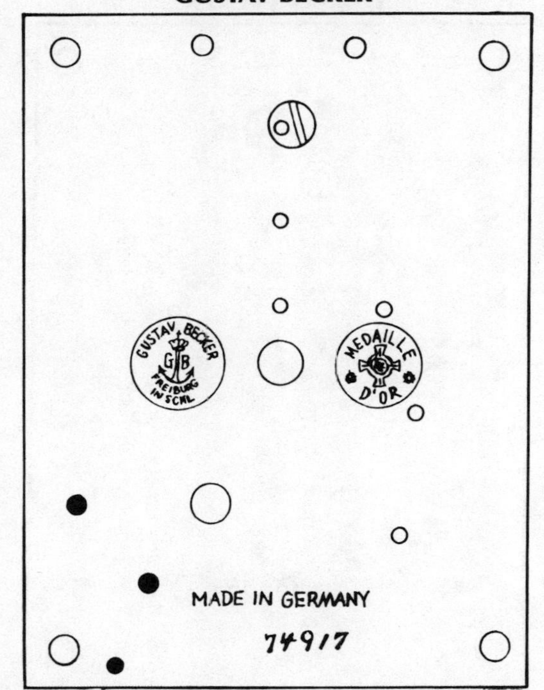

Plate 1203
Disc Pendulum
4-Ball Pendulum

USE .004"* (.102mm*) HOROLOVAR
See Appendix 70, 74, 87 (19 x 38)

Gustav Becker c 1910

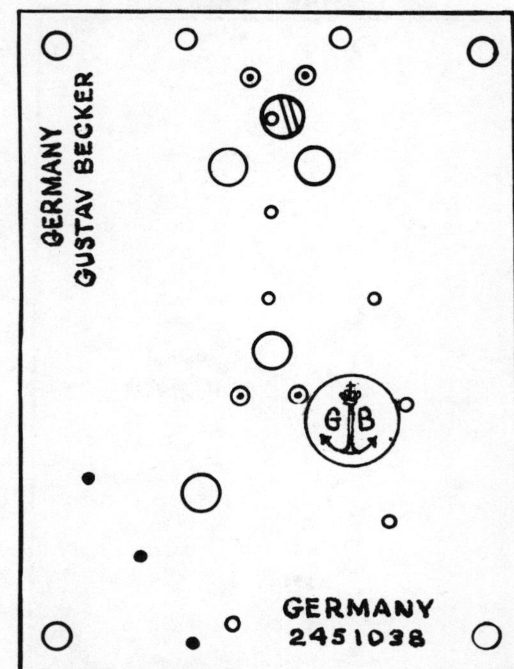

Plate 1206
Disc Pendulum
4-Ball Pendulum

USE .004"* (.102mm*) HOROLOVAR
See Appendix 11, 70, 74, 87 (19 x 38)

Gustav Becker c 1910

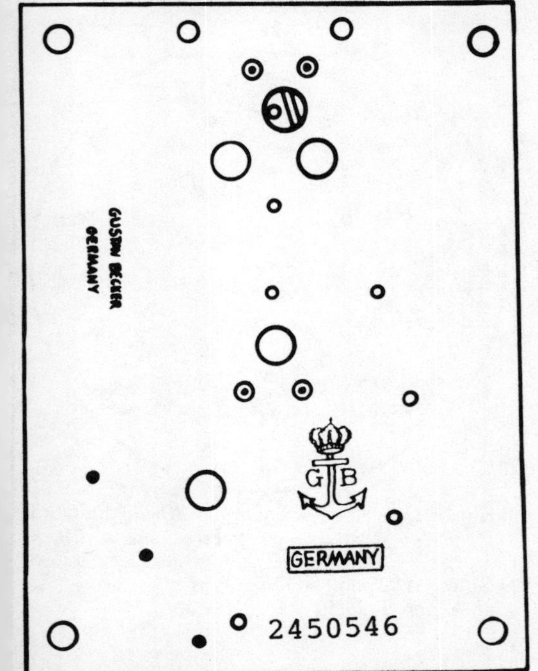

Plate 1206A
Disc Pendulum
4-Ball Pendulum

USE .004"* (.102mm*) HOROLOVAR
See Appendix 11, 70, 74, 87 (19 x 38)

Gustav Becker c 1910

GUSTAV BECKER

GB

2465214

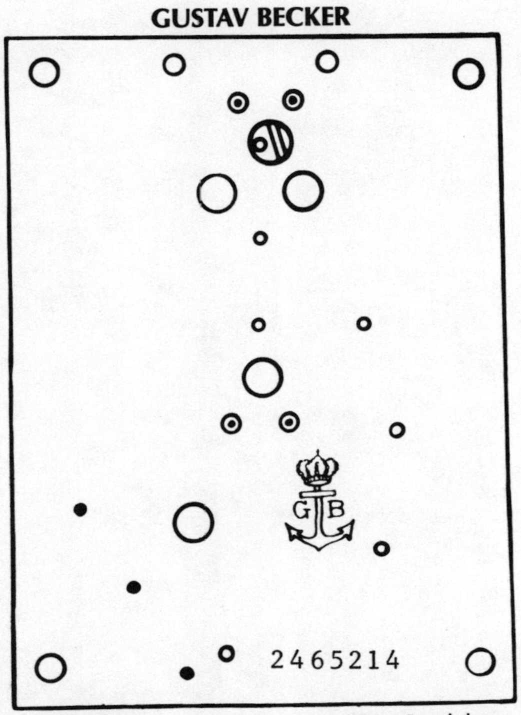

Plate 1206B
Disc Pendulum
4-Ball Pendulum

USE .004"* (.102mm*) HOROLOVAR
See Appendix 11, 70, 74, 87 (19 x 38)

Gustav Becker c 1910

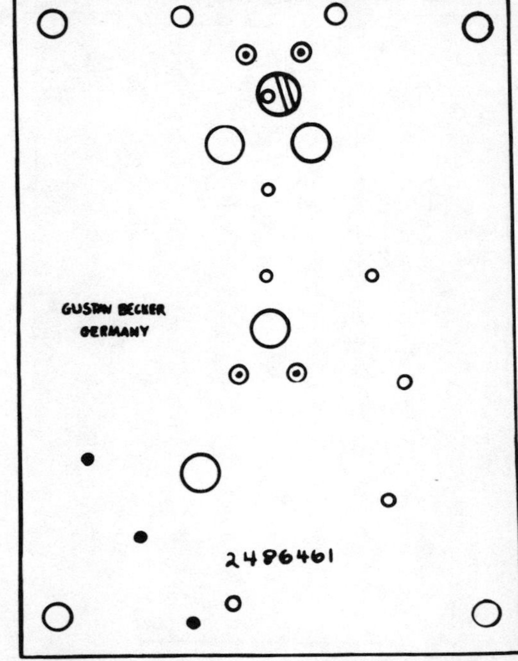

Plate 1207
Disc Pendulum
4-Ball Pendulum

USE .004"* (.102mm*) HOROLOVAR
See Appendix 11, 70, 74, 87 (19 x 38)

Gustav Becker c 1908

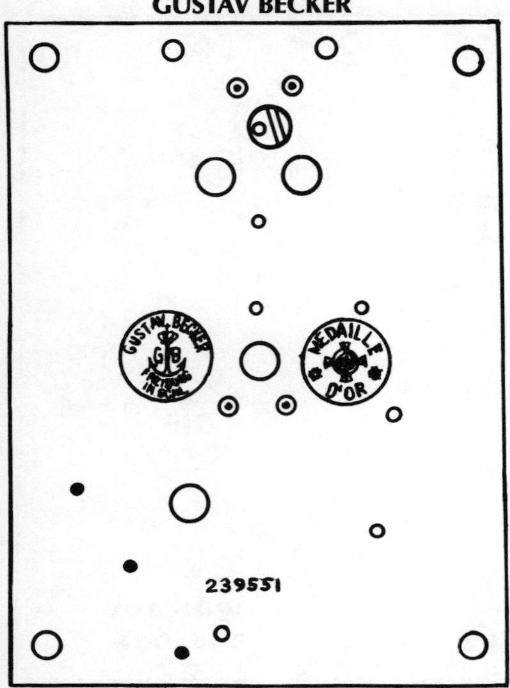

Plate 1207A Disc Pendulum
4-Ball Pendulum
USE .004"* (.102mm*) HOROLOVAR
See Appendix 11, 70, 74, 87 (19 x 36)

Gustav Becker c 1909

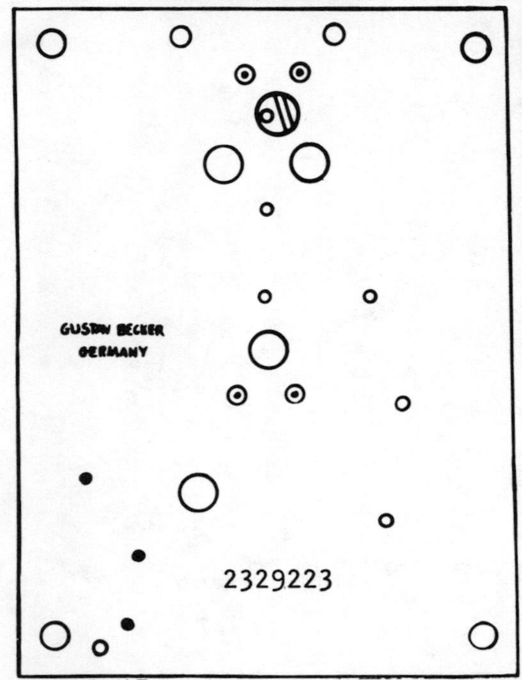

Plate 1207AA Disc Pendulum
4-Ball Pendulum
USE .004"* (.102mm*) HOROLOVAR
See Appendix 11, 70, 74, 87 (19 x 38)

Gustav Becker c 1906

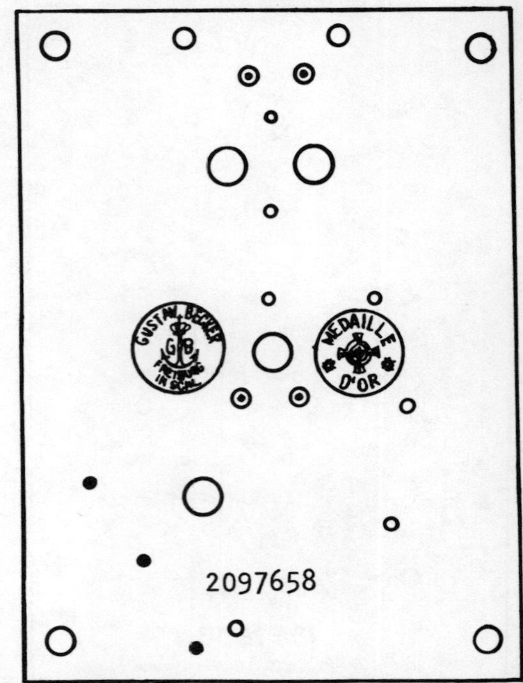

Plate 1207B Disc Pendulum
4-Ball Pendulum
USE .004"* (.102mm*) HOROLOVAR
See Appendix 11, 70, 74, 87, 97 (19 x 38)

Gutenbacher Uhrenfabrik c 1952

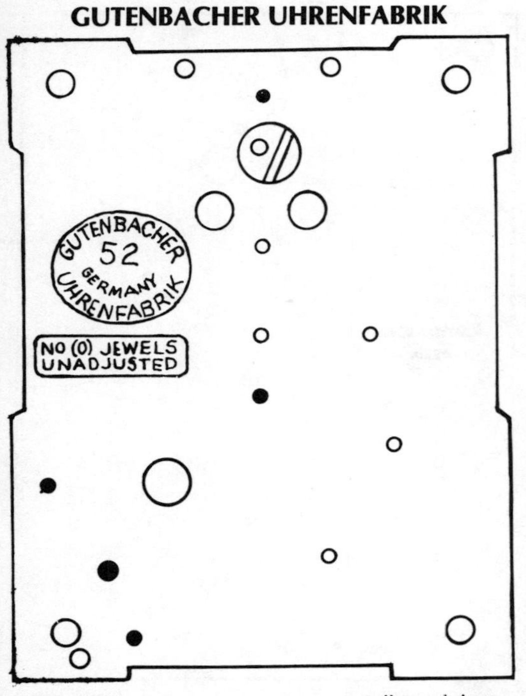

Plate 1208 4-Ball Pendulum
USE .004"* (.102mm*) HOROLOVAR
See Appendix 62 (19 x 36)

Franz Vosseler c 1910

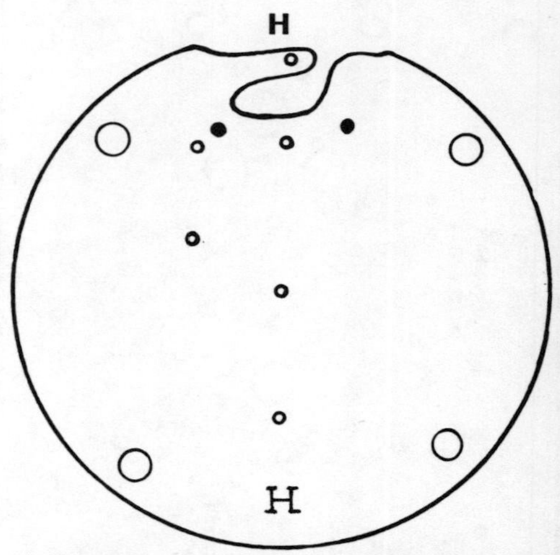

Plate 1208A 3-Ball Pendulum
Miniature "Louvre" Clock
Underslung Suspension
USE .003" (.076mm) HOROLOVAR
See Appendix 26, 93

Jahresuhrenfabrik c 1909

H

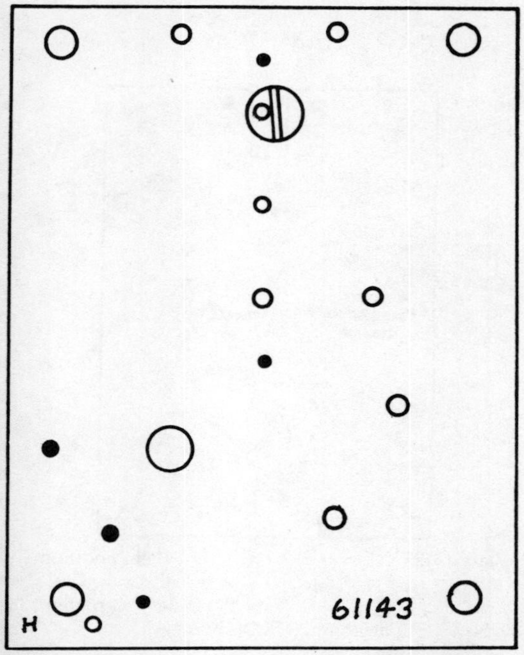

H

61143

Plate 1208B Disc Pendulum
USE .004"* (.102mm*) HOROLOVAR
(19 x 36)

Uhrenfabrik M. Reiner c 1954

HALL CRAFT CORP.

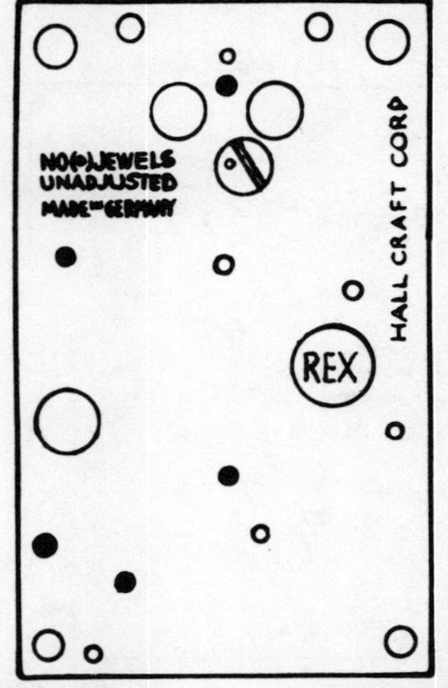

Plate 1210A 4-Ball Pendulum Miniature Clock
USE .0028" (.071mm) HOROLOVAR
Unit 28C (16 x 36)
See Appendix 25, 55, 109

Uhrenfabrik M. Reiner c 1951

HALL CRAFT CORP.

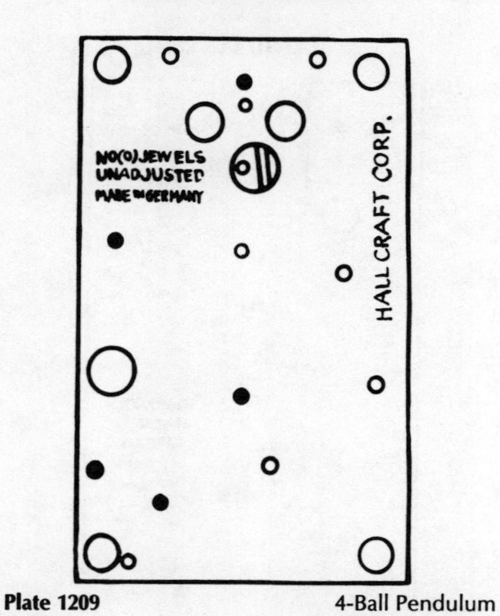

Plate 1209 4-Ball Pendulum Miniature Clock
USE .0025" (.064mm) HOROLOVAR
Units 28A, 28B (16 x 36)
See Appendix 25, 54, 109

Uhrenfabrik M. Reiner c 1954

HALL CRAFT CORP.

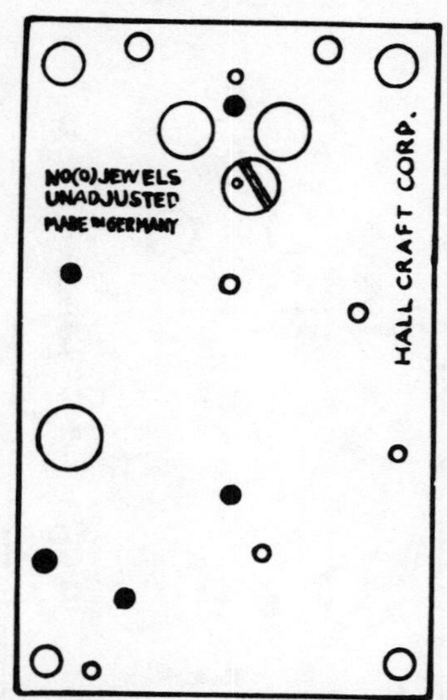

Plate 1210 4-Ball Pendulum Miniature Clock
USE .0028" (.071mm) HOROLOVAR
Unit 28C (16 x 36)
See Appendix 25, 55, 109

Uhrenfabrik Herr c 1951

HALL CRAFT CORP.

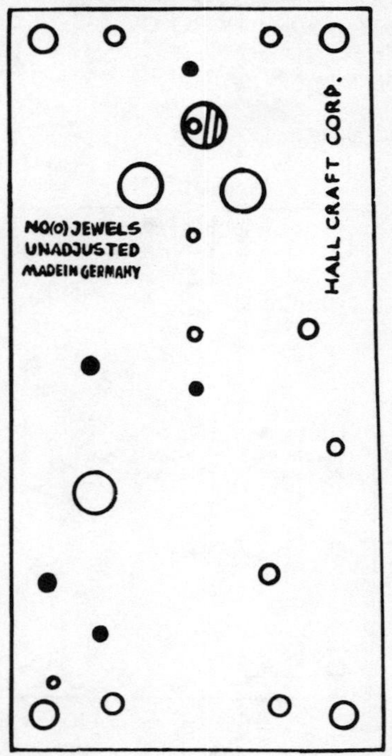

Plate 1211A 4-Ball Pendulum
USE .0035" (.089mm) HOROLOVAR
Units 19, 20A, 21
USE .0038" (.097mm) HOROLOVAR
Unit 27B (19 x 36)
See Appendix 106, 109, 110

Uhrenfabrik Herr c 1951

HALL CRAFT CORP.

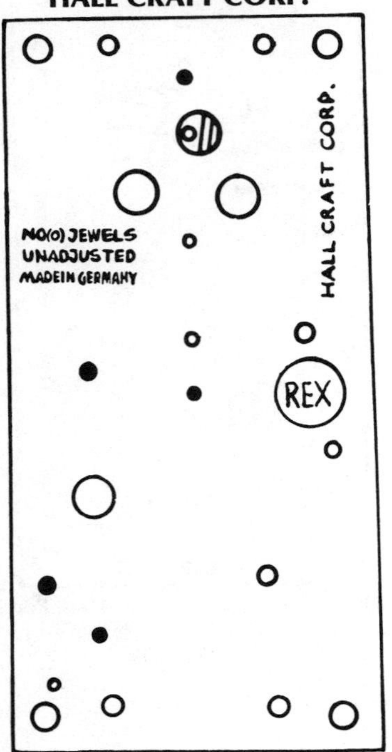

Plate 1212 4-Ball Pendulum
USE .0035″ (.089mm) HOROLOVAR
Units 19, 20A, 21
USE .0038″ (.097mm) HOROLOVAR
Unit 27B (19 x 36)
See Appendix 106, 109, 110

Sigfried Haller c 1960

HAMILTON

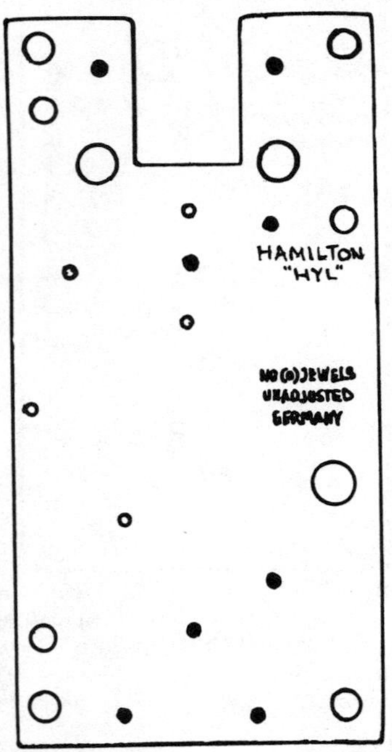

Plate 1212B 4-Ball Pendulum
USE .003″ (.076mm) HOROLOVAR
Units 42, 42A (18 x 38)
See Appendix 63, 64, 108

Sigfried Haller c 1972

HAMILTON

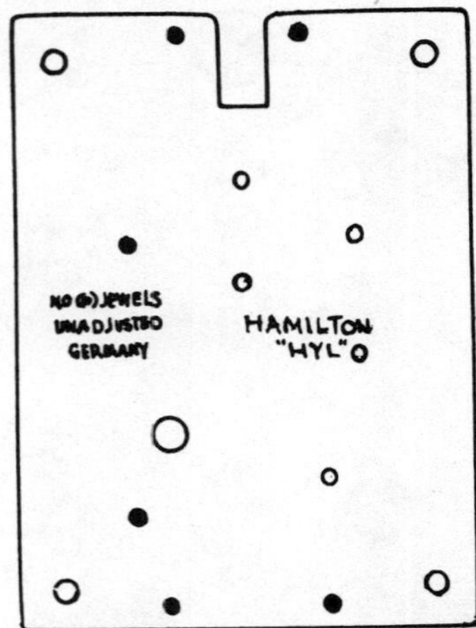

Plate 1212C 4-Ball Pendulum
Miniature Clock
Pin Pallet Escapement
USE .0022″ (.056mm) HOROLOVAR
Units 43, 43A (12 x 24)
See Appendix 63, 64

Sigfried Haller c 1973

HAMILTON

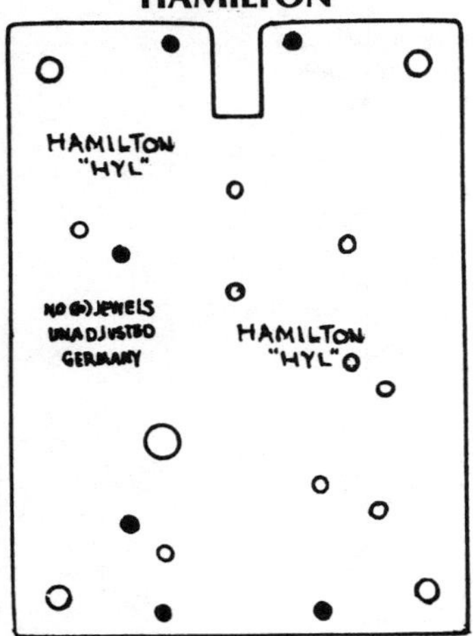

Plate 1212D 4-Ball Pendulum
Miniature Clock
USE .0022″ (.056mm) HOROLOVAR
Units 43, 43A (12 x 24)
See Appendix 63, 64, 108

Sigfried Haller c 1974

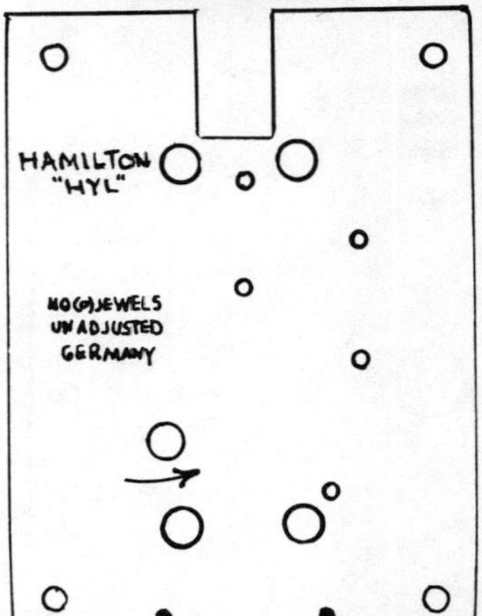

Plate 1212E 4-Ball Pendulum
Midget Clock
Pin Pallet Escapement
USE .0018″ (.046mm) HOROLOVAR
Units 44, 44A (12 x 24)
See Appendix 63, 64, 108

Sigfried Haller c 1930

HASI

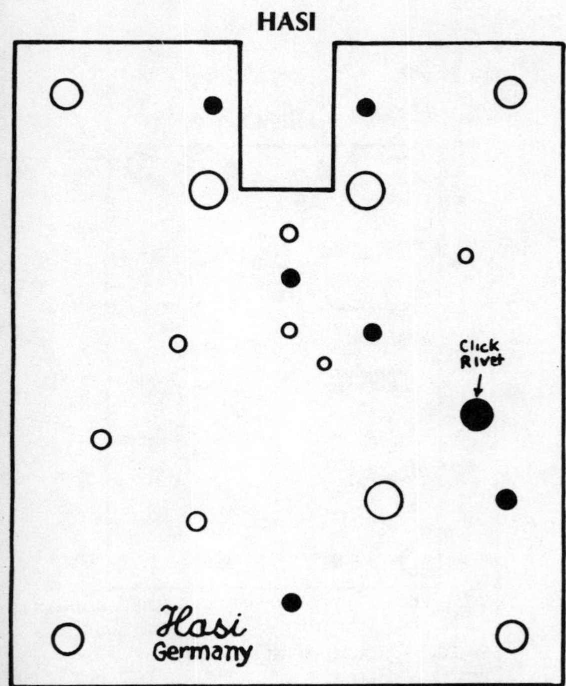

Plate 1213 4-Ball Pendulum
USE .0038" (.097mm) HOROLOVAR
Unit 33A (19 x 38)
See Appendix 63, 64

Sigfried Haller c 1951

HASI

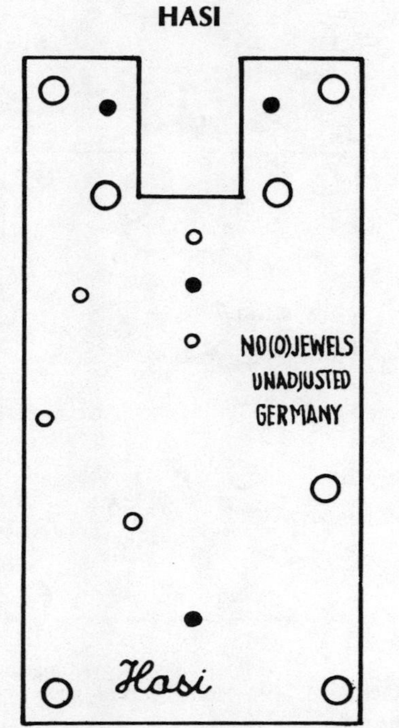

Plate 1213A 4-Ball Pendulum
USE .0038" (.097mm) HOROLOVAR
Unit 33A (19 x 38)
See Appendix 63, 64

Aug. Schatz & Sohne c 1952

H. COEHLER CO.

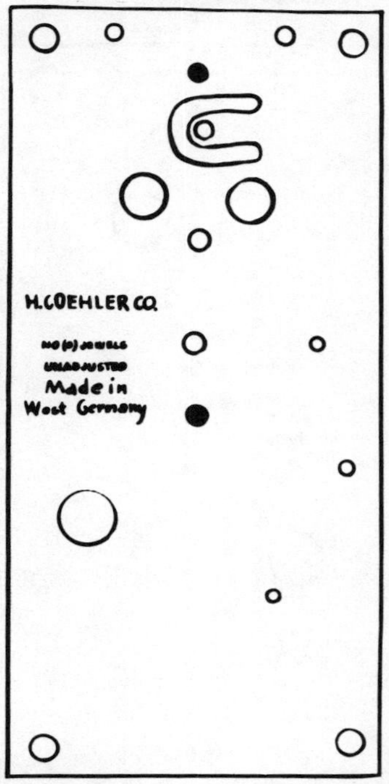

Plate 1214 4-Ball Pendulum
USE .004" (.102mm) HOROLOVAR
Units 6, 7, 8, 9 (19 x 36)
See Appendix 6, 78

Aug. Schatz & Sohne c 1954

H. COEHLER CO. INC.

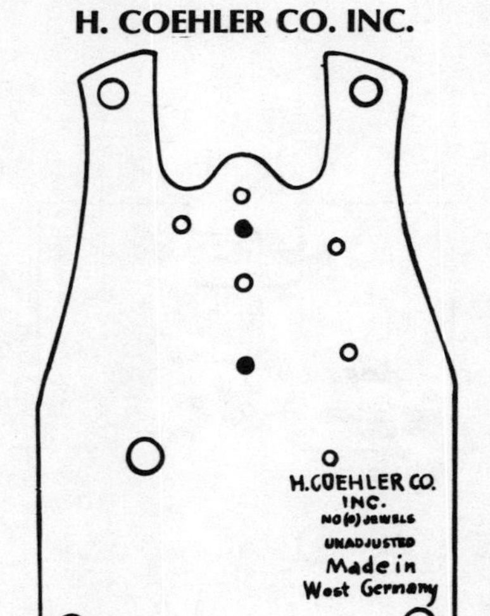

Plate 1214A Dome Clock—3-Ball Pendulum
Coach Clock—4-Ball Pendulum
USE .0023" (.058mm) HOROLOVAR
Unit 10A (13 x 30)
See Appendix 4, 5, 63, 66, 79

Kieninger & Obergfell c 1952

H. COEHLER CO. INC.

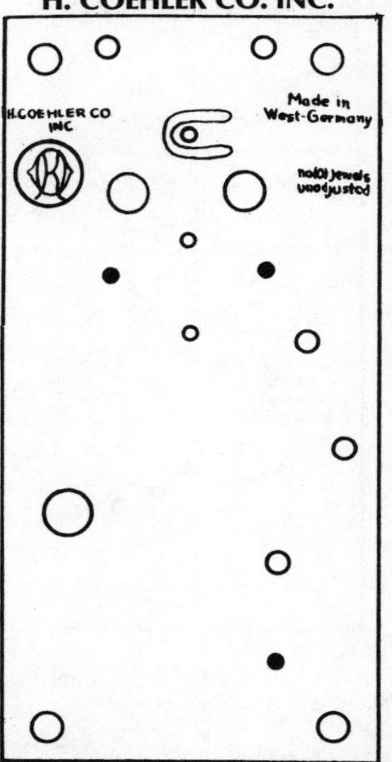

Plate 1214C 4-Ball Pendulum
USE .0032" (.081mm) HOROLOVAR
Units 1, 3A, 3B, 3C (19 x 38)
See Appendix 76

Kern & Sohne c 1953

HECO

Heco
HENRY COEHLER CO. INC.
GERMANY
NO(0) JEWELS
UNADJUSTED
KS

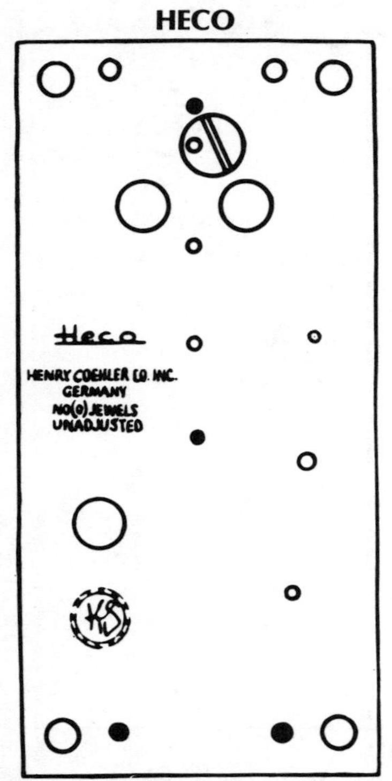

Plate 1216 4-Ball Pendulum
USE .0036" (.091mm) HOROLOVAR
Unit 11B (18 x 38)

Kern & Sohne c 1958

HECO

Heco
HENRY COEHLER CO. INC.
GERMANY
NO(0) JEWELS
UNADJUSTED

Plate 1217 4-Ball Pendulum Miniature Clock
USE .002" (.051mm) HOROLOVAR
Units 12A, 12B, 12C
USE .0019" (.048mm) HOROLOVAR
Unit 12D (12 x 25)
See Appendix 63

Kern & Sohne c 1959

HECO

Heco
HENRY COEHLER CO. INC.
GERMANY
NO(0) JEWELS
UNADJUSTED
KS
GERMANY

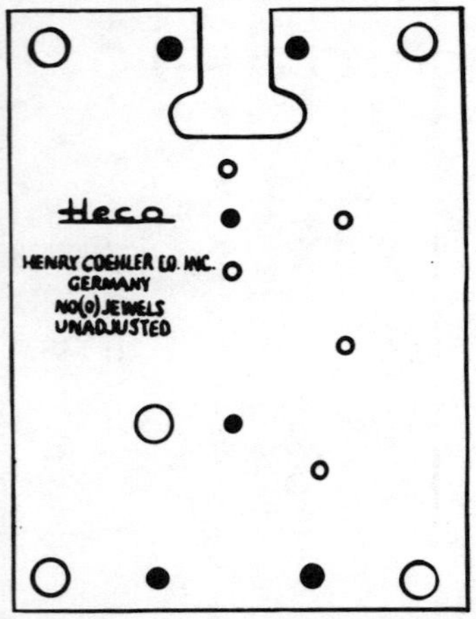

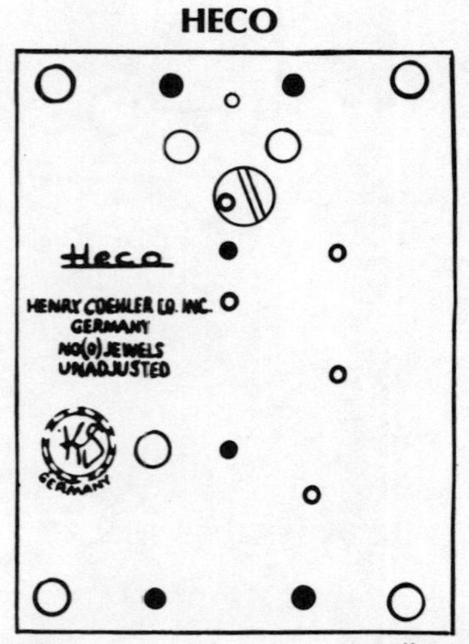

Plate 1217A 4-Ball Pendulum Miniature Clock
USE .002" (.051mm) HOROLOVAR
Units 12A, 12B, 12C
USE .0019" (.048mm) HOROLOVAR
Unit 12D (12 x 25)
See Appendix 25

Kern & Sohne c 1960

HECO

Heco
HENRY COEHLER CO. INC.
GERMANY
NO(0) JEWELS
UNADJUSTED
KS

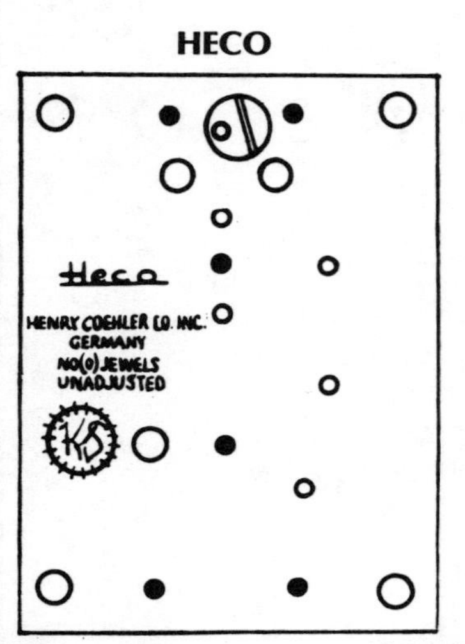

Plate 1217B 4-Ball Pendulum Miniature Clock
USE .002" (.051mm) HOROLOVAR
Units 12A, 12B, 12C
USE .0019" (.048mm) HOROLOVAR
Unit 12D (12 x 25)

Kern & Sohne c 1958

HECO

Heco
GERMANY
KS

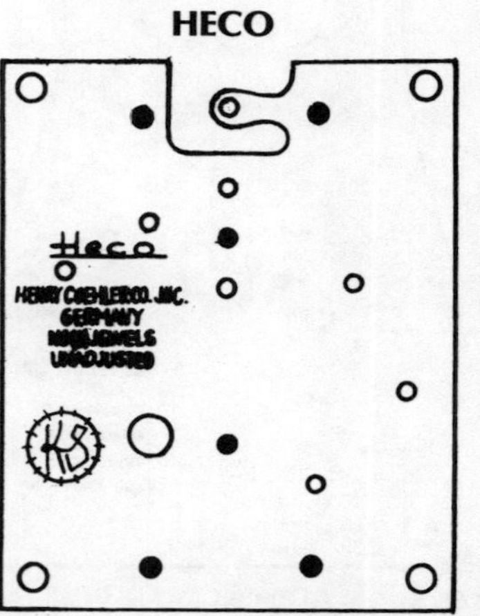

Plate 1217C 4-Ball or Disc Pendulum Midget Clock Pin Pallet Escapement
USE .0023" (.058mm) HOROLOVAR
Unit 12E (14 x 25)

Kern & Sohne c 1961

HECO

Heco
HENRY COEHLER CO. INC.
GERMANY
NO(0) JEWELS
UNADJUSTED

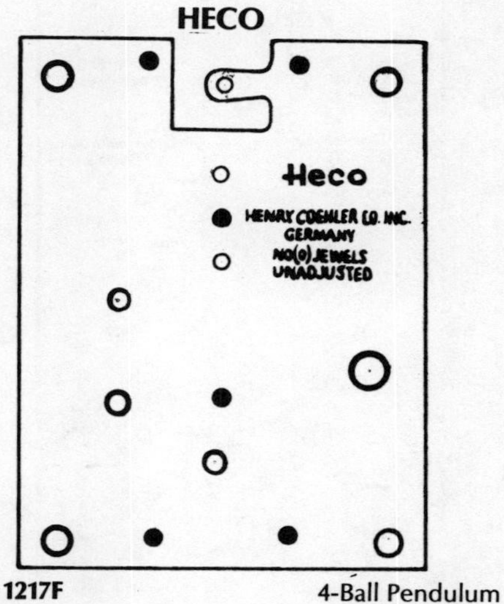

Plate 1217F 4-Ball Pendulum Miniature Clock
USE .0023" (.058mm) HOROLOVAR
Unit 12F (14 x 30)

Kern & Sohne c 1953

HECO

Heco
HENRY COEHLER CO. INC.
GERMANY
NO(0) JEWELS
UNADJUSTED
K.U.S. SIV
Germany

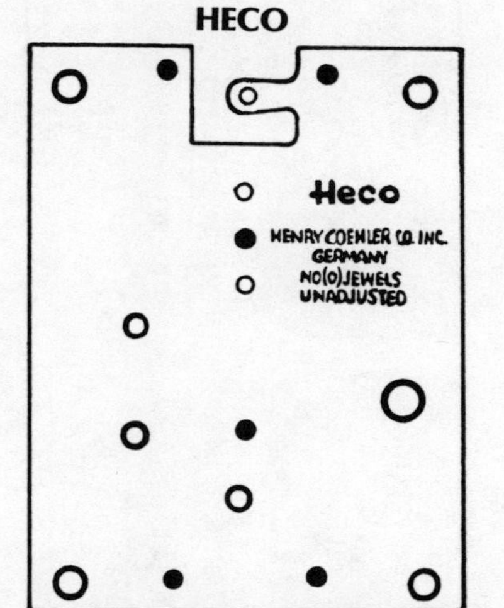

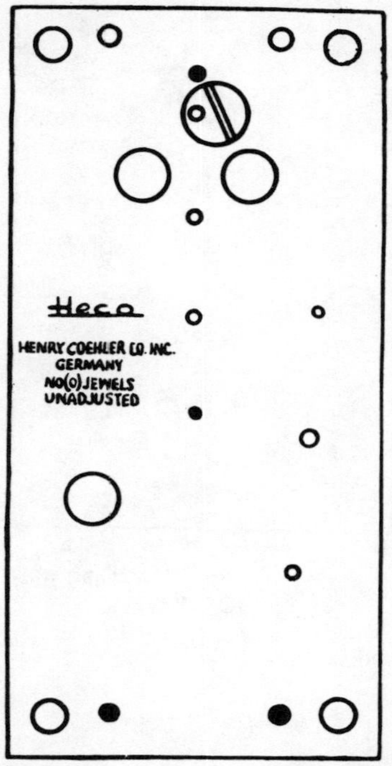

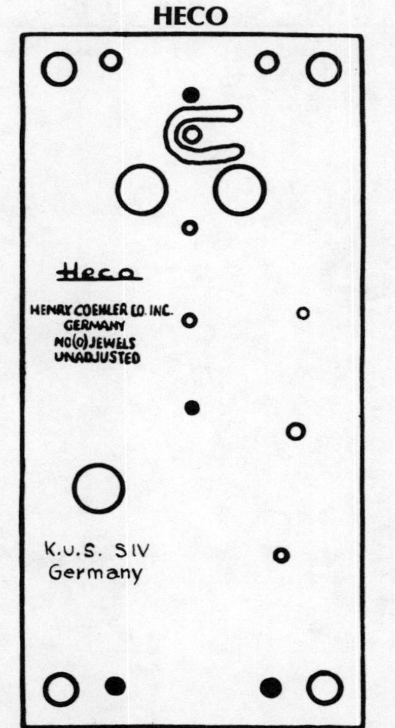

Plate 1218A 4-Ball Pendulum
USE .0036" (.091mm) HOROLOVAR
Unit 11B (18 x 38)

Kern & Sohne c 1961

HECO

Heco
HENRY COEHLER CO. INC.
GERMANY
NO(0) JEWELS
UNADJUSTED

Plate 1217G 4-Ball Pendulum Miniature Clock
USE .0023" (.058mm) HOROLOVAR
Unit 12F (14 x 30)

Kern & Sohne c 1953

HECO

Heco
HENRY COEHLER CO. INC.
GERMANY
NO(0) JEWELS
UNADJUSTED

Plate 1218 4-Ball Pendulum
USE .0036" (.091mm) HOROLOVAR
Unit 11B (18 x 38)

Uhrenfabrik Herr c 1952

H. EICKMANN & CO.

NO(0) JEWELS
UNADJUSTED
H. EICKMANN & CO.
MADE IN GERMANY

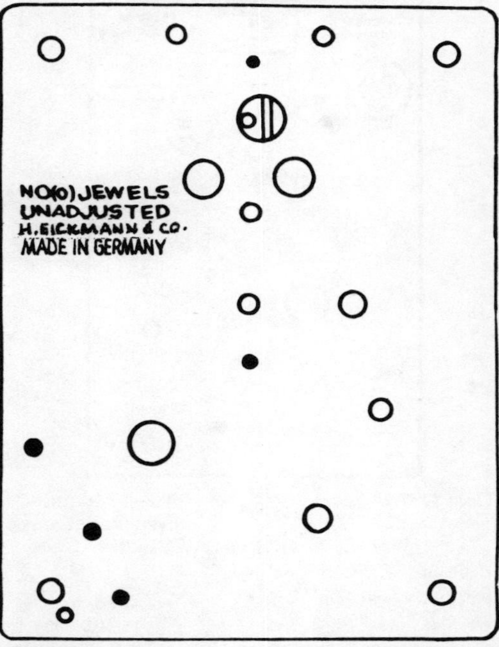

Plate 1221 4-Ball Pendulum
USE .004" (.102mm) HOROLOVAR
Unit 27A (19 x 36)
See Appendix 130

Uhrenfabrik M. Reiner c 1951

HENRY COEHLER

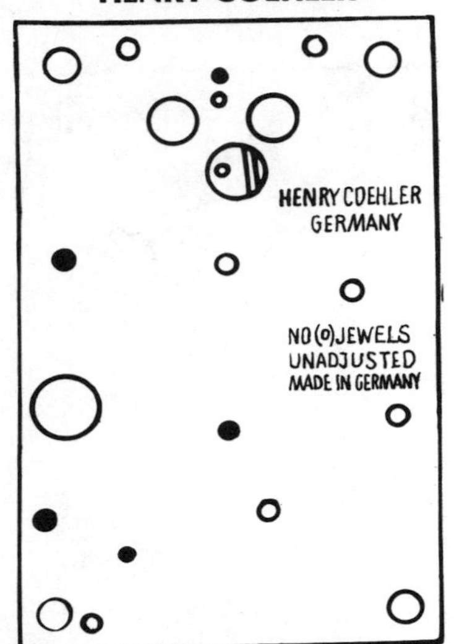

Plate 1222 4-Ball Pendulum Miniature Clock
USE .0025" (.064mm) HOROLOVAR
Units 28A, 28B (16 x 36)
See Appendix 25, 54, 109

Kieninger & Obergfell c 1952

HENRY COEHLER CO. INC.

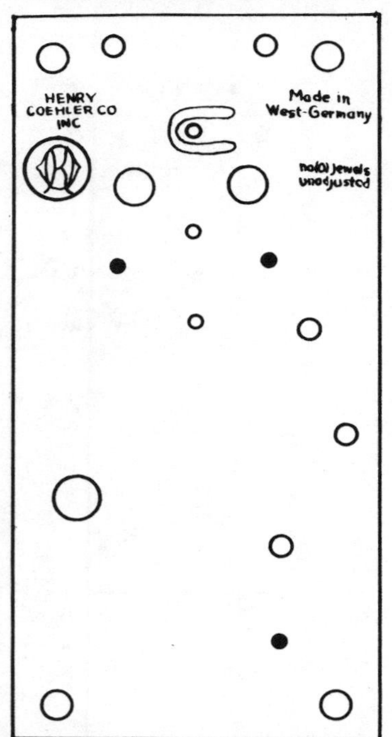

Plate 1227 4-Ball Pendulum
USE .0032" (.081mm) HOROLOVAR
Units 1, 3A, 3B, 3C (19 x 38)
See Appendix 76

Kieninger & Obergfell c 1952

HENRY COEHLER CO. INC.

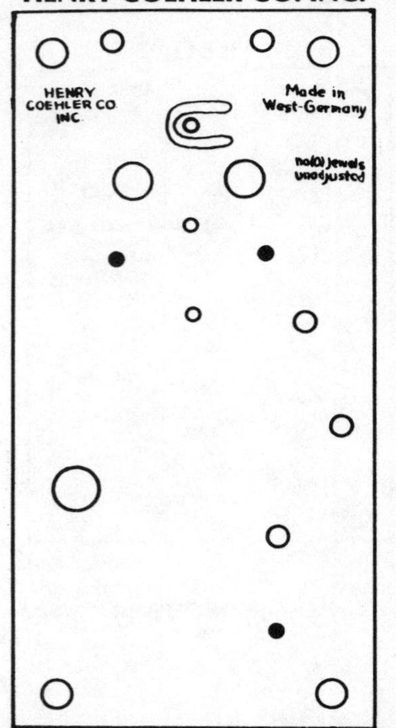

Plate 1227A 4-Ball Pendulum
USE .0032" (.081mm) HOROLOVAR
Units 1, 3A, 3B, 3C (19 x 38)
See Appendix 76

Kieninger & Obergfell c 1954

HENRY COEHLER CO. INC.

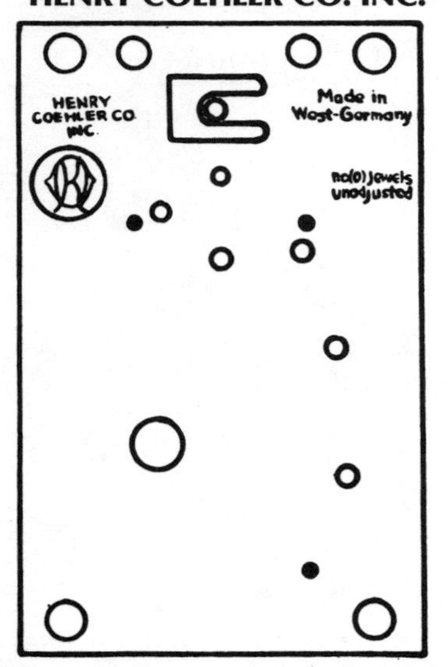

Plate 1231 4-Ball Pendulum Miniature Clock
USE .0023" (.058mm) HOROLOVAR
Units 5A, 5B, 5C, 5D, 5E (14 x 30)
See Appendix 77, 111

Sigfried Haller c 1952

HENRY COEHLER CO. INC.

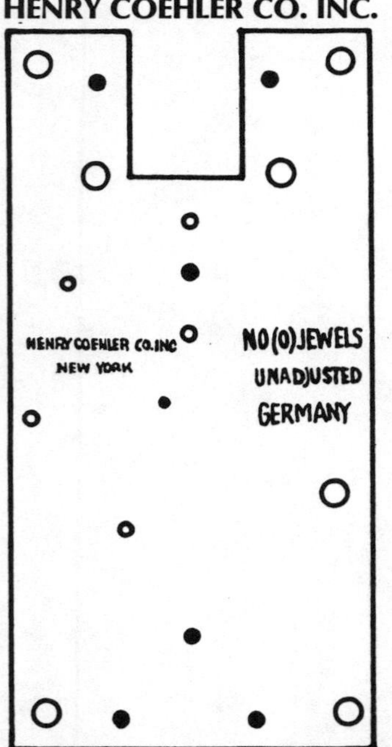

Plate 1232 4-Ball Pendulum
USE .0038" (.097mm) HOROLOVAR
Unit 33 (18 x 38)
See Appendix 63, 64

Kieninger & Obergfell c 1952

HENRY COEHLER CO. INC.

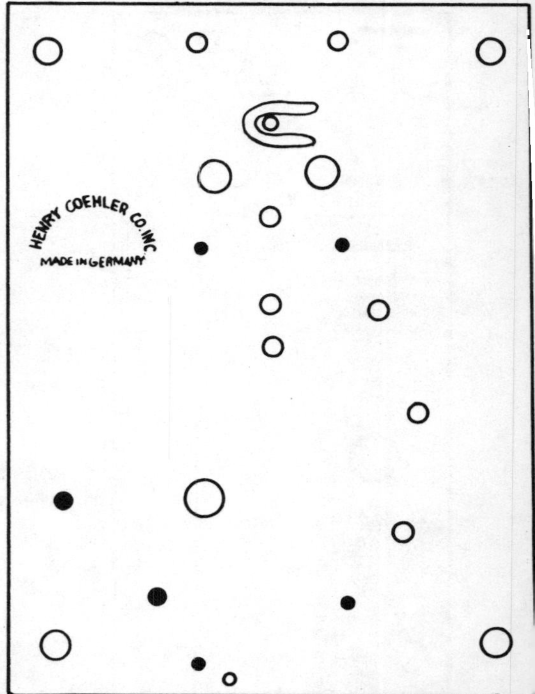

Plate 1239 4-Ball Pendulum
USE .0032" (.081mm) HOROLOVAR
Units 1, 3A (19 x 38)
See Appendix 76

Uhrenfabrik Herr c 1951

HERR

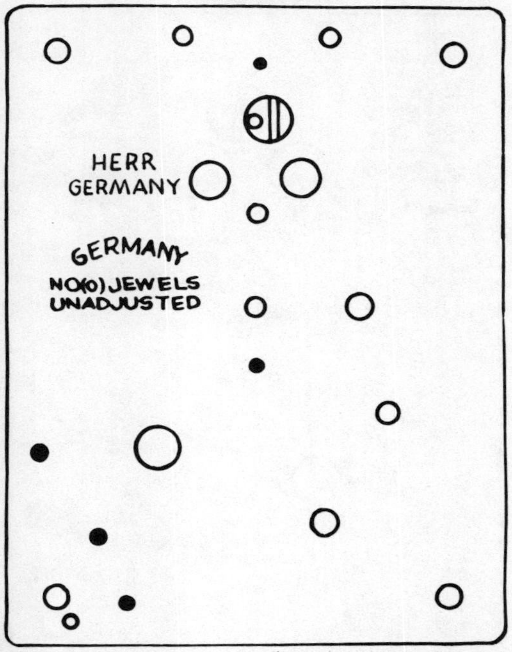

HERR
GERMANY
GERMANY
NO(0) JEWELS
UNADJUSTED

Plate 1241 4-Ball Pendulum
USE .004" (.102mm) HOROLOVAR
Unit 27A (19 x 36)
See Appendix 130

Uhrenfabrik M. Reiner c 1951

HERR

HERR
GERMANY
NO(0) JEWELS
UNADJUSTED

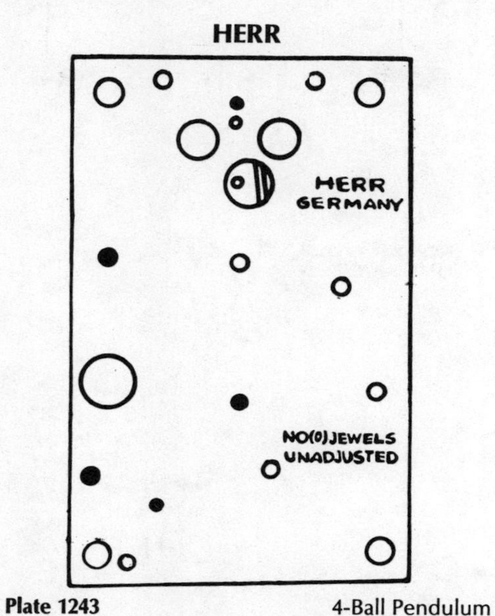

Plate 1243 4-Ball Pendulum
Miniature Clock
USE .0025" (.064mm) HOROLOVAR
Units 28A, 28B (16 x 36)
See Appendix 25, 54, 109

Uhrenfabrik M. Reiner c 1954

HERR

HERR
GERMANY
NO(0) JEWELS
UNADJUSTED

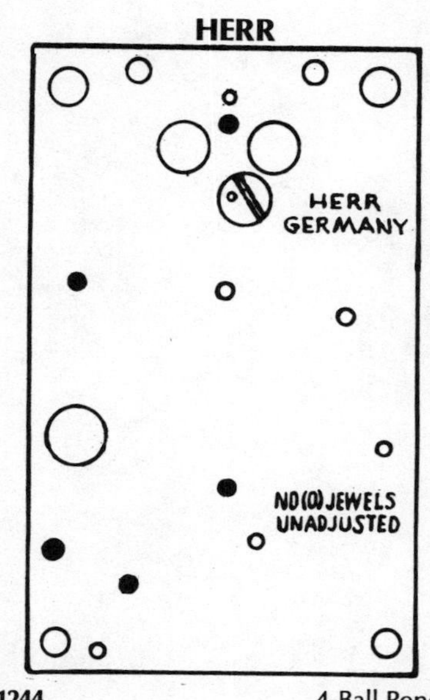

Plate 1244 4-Ball Pendulum
Miniature Clock
USE .0028" (.071mm) HOROLOVAR
Unit 28C (16 x 36)
See Appendix 25, 55, 109

Georg Wurthner c 1952

H. EULE

NO(0) JEWELS
UNADJUSTED
MADE IN
GERMANY
H. EULE

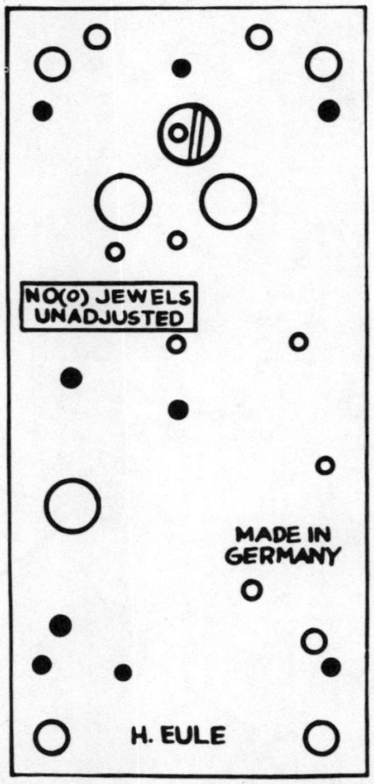

Plate 1245 4-Ball Pendulum
USE .004" (.102mm) HOROLOVAR
Unit 26 (19 x 36)
See Appendix 67

Georg Wurthner c 1952

H. EULE

NO (0) JEWELS
UNADJUSTED
MADE IN
GERMANY
H. EULE

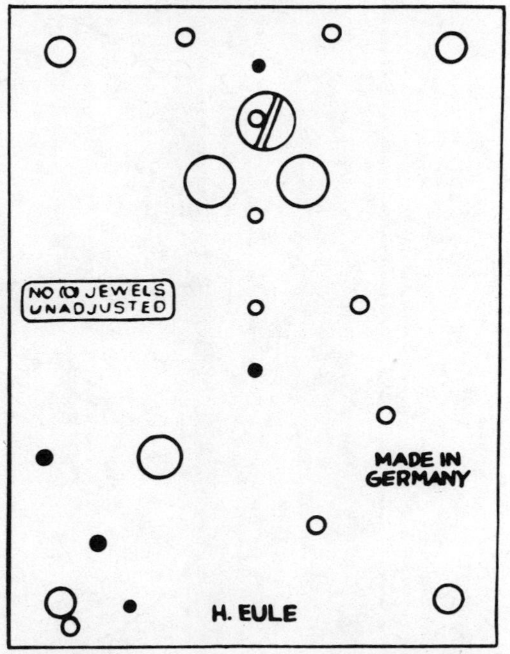

Plate 1245A 4-Ball Pendulum
USE .004" (.102mm) HOROLOVAR
Unit 26 (19 x 36)

Edgar Henn c 1952

H. EULE

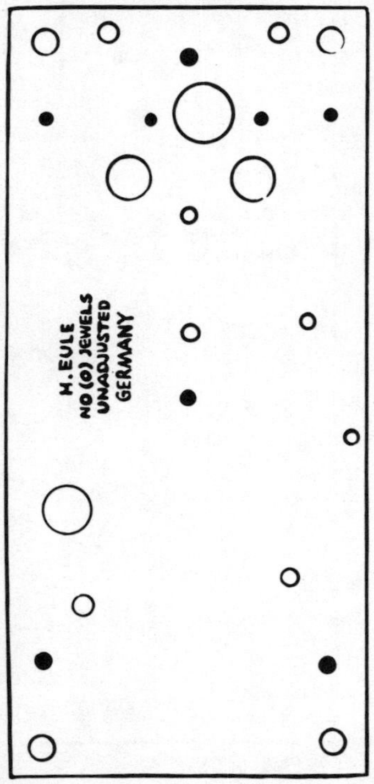

H. EULE
NO (0) JEWELS
UNADJUSTED
GERMANY

Plate 1246 4-Ball Pendulum
USE .0037" (.094mm) HOROLOVAR
Unit 32 (19 x 38)
See Appendix 39

Edgar Henn c 1951

H. EULE

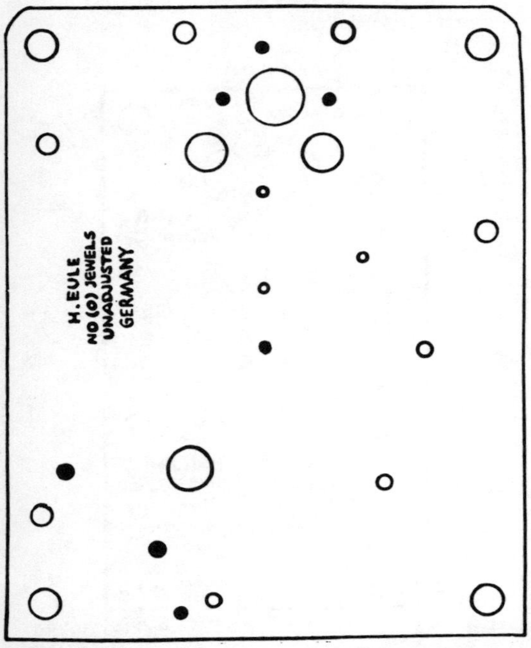

Plate 1246A 4-Ball Pendulum
USE .0035" (.089mm) HOROLOVAR
Unit 31 (19 x 38)
See Appendix 39

Uhrenfabrik Herr c 1952

H. EULE

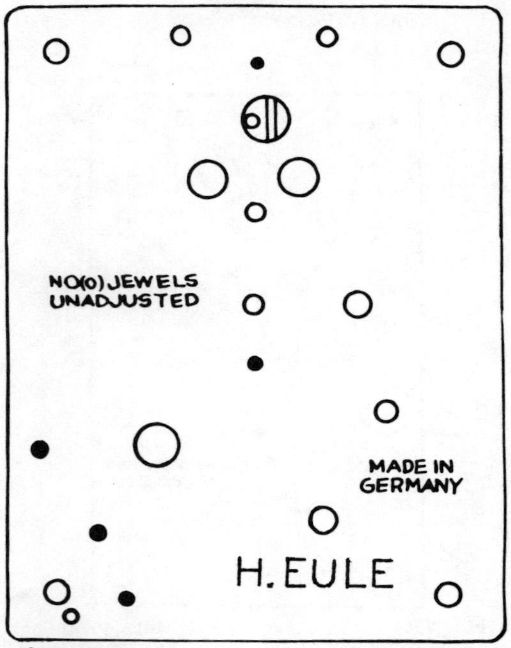

Plate 1246B 4-Ball Pendulum
USE .004" (.102mm) HOROLOVAR
Unit 27A (19 x 36)
See Appendix 130

Kienzle Clock Factories c 1912

HUBER UHREN

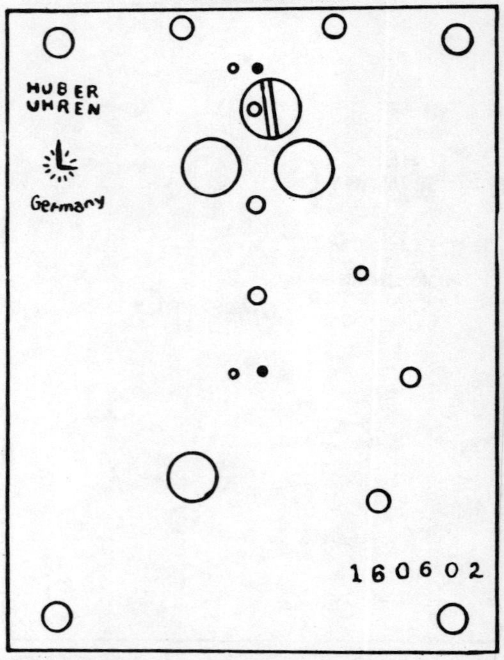

Plate 1247 4-Ball Pendulum
USE .0035" (.089mm) HOROLOVAR (19 x 38)
See Appendix 8

Badische Uhrenfabrik c 1890

HUBER UHREN

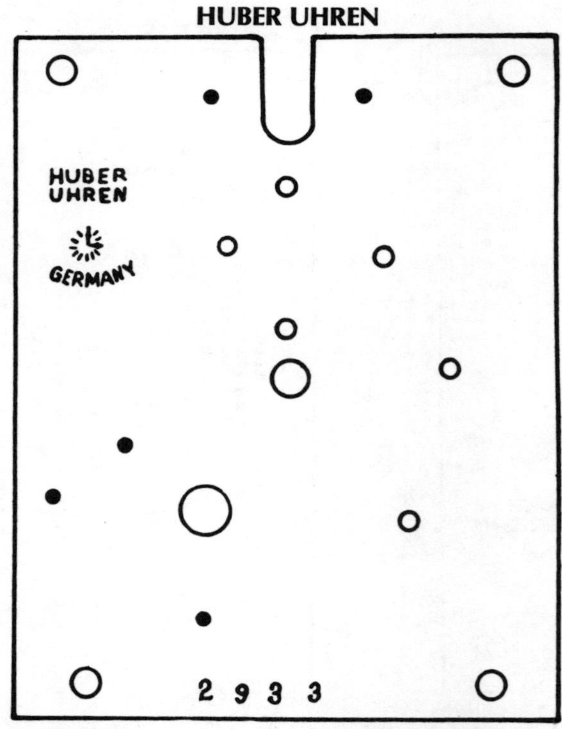

Plate 1251 Disc Pendulum
Pin Pallet Escapement
Lantern Pinions
USE .0037"* (.094mm*) HOROLOVAR
See Appendix 8, 12, 63 (20 x 38)

Manufacturer Not Known c 1906

HUBER UHREN

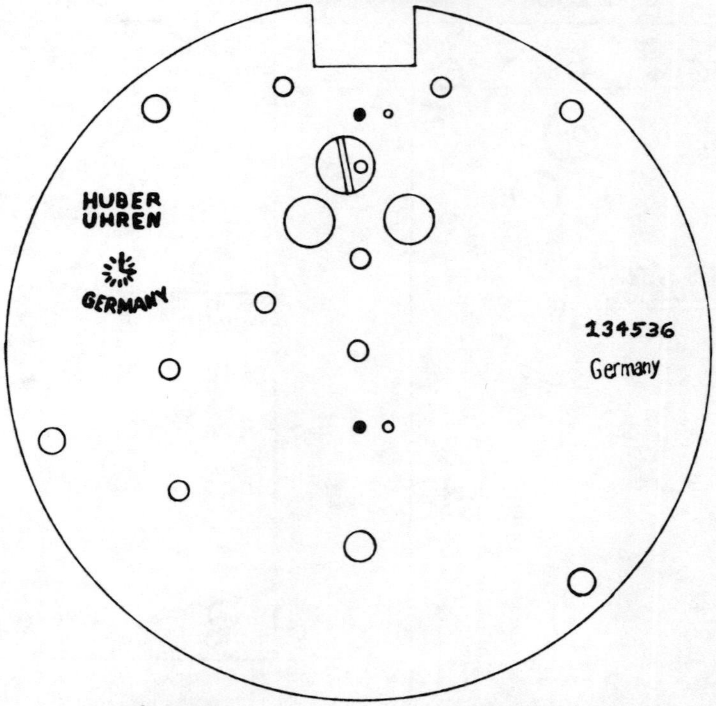

Plate 1251A Disc Pendulum
Front Wind
USE .0037"* (.094mm*) HOROLOVAR
See Appendix 73 (19 x 38)

Georg Wurthner c 1952

H.X.M.

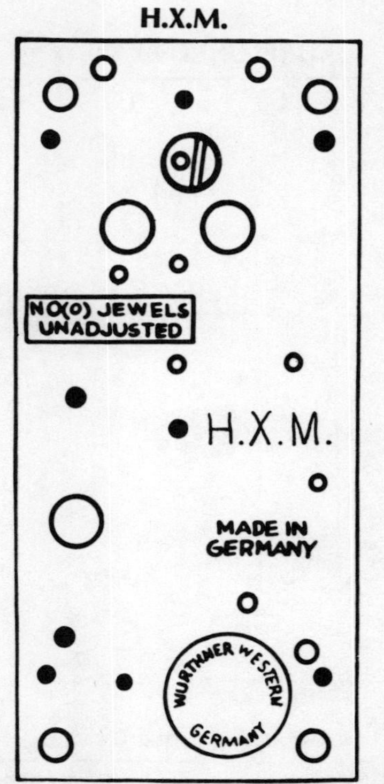

Plate 1252 4-Ball Pendulum
USE .004" (.102mm) HOROLOVAR
Unit 26 (19 x 36)
See Appendix 45, 67

Kieninger & Obergfell c 1930

H.Z.

H.Z.
MADE IN GERMANY

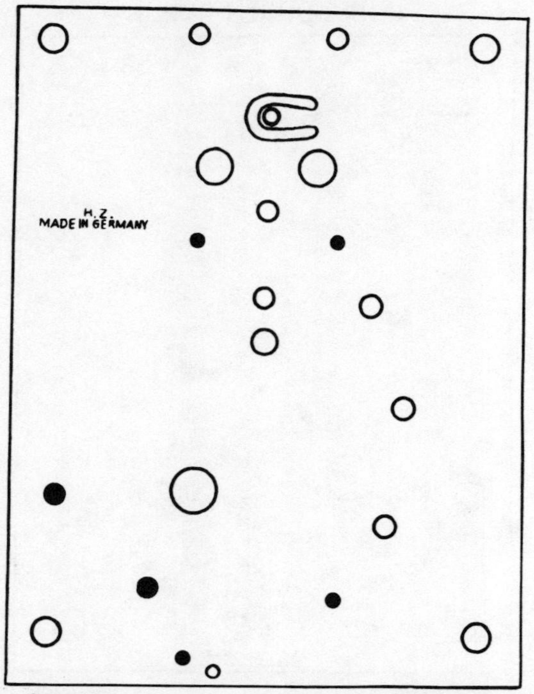

Plate 1253 4-Ball Pendulum
USE .0032" (.081mm) HOROLOVAR
Units 1, 3A (19 x 38)
See Appendix 76

Kieninger & Obergfell c 1930

H.Z.

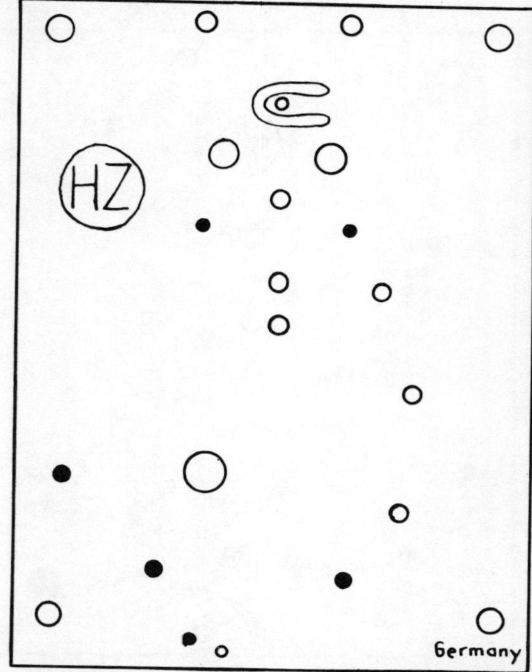

Plate 1254 4-Ball Pendulum
USE .0032" (.081mm) HOROLOVAR
Units 1, 3A (19 x 38)
See Appendix 76

Jahresuhrenfabrik c 1908

J

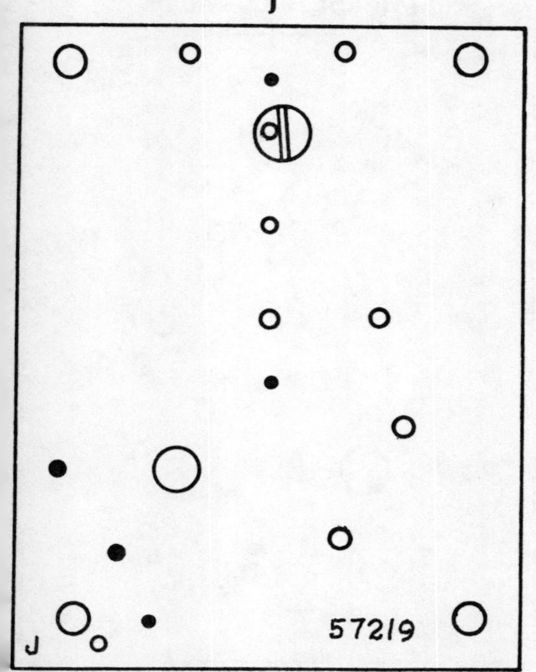

Plate 1255 Disc Pendulum
USE .004"* (.102mm*) HOROLOVAR
(19 x 36)

Jahresuhrenfabrik c 1910

JAHRESUHREN-FABRIK

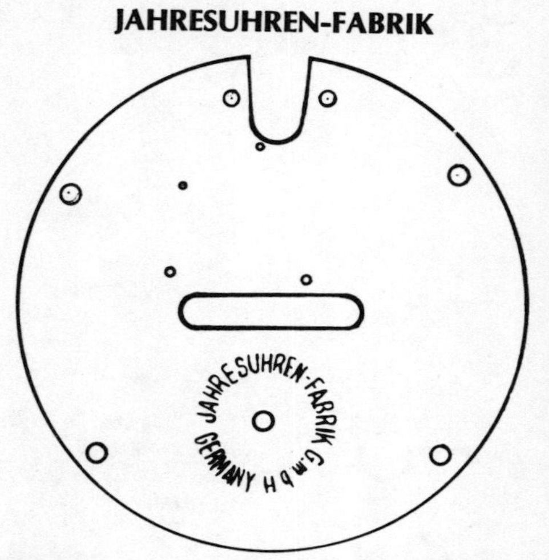

Plate 1257 2-Ball Pendulum
USE .0027" (.069mm) HOROLOVAR
See Appendix 63, 94

Jahresuhrenfabrik c 1910

JAHRESUHREN-FABRIK

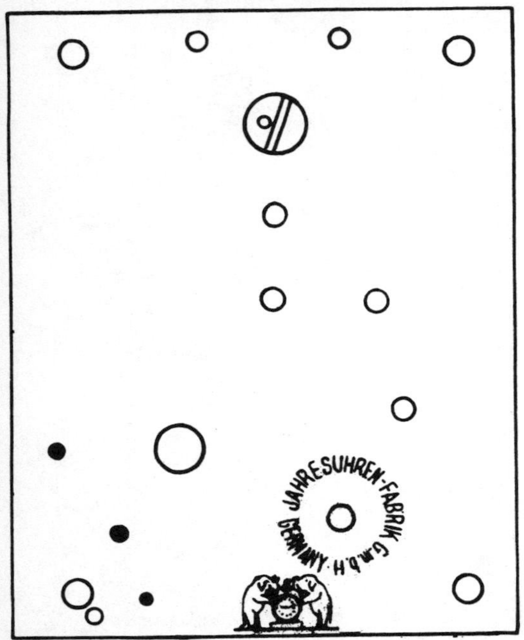

Plate 1259 Disc Pendulum
4-Ball Pendulum

USE .004" (.102mm) HOROLOVAR
Unit 6789A (19 x 36)
See Appendix 78

Jahresuhrenfabrik c 1930

JAHRESUHREN-FABRIK

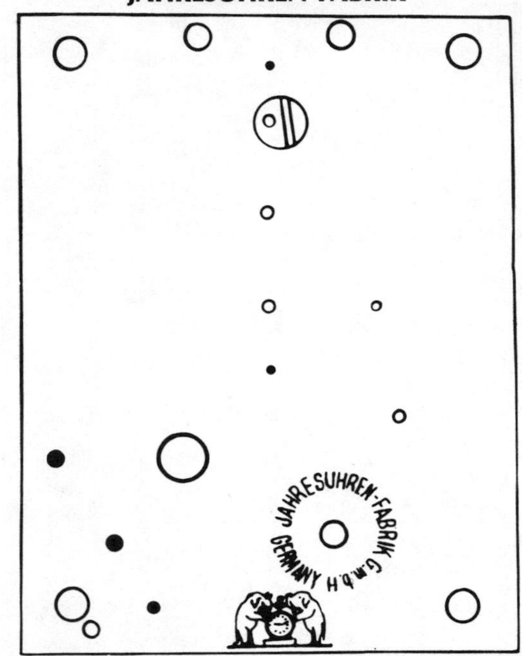

Plate 1260 Disc Pendulum
4-Ball Pendulum

USE .004" (.102mm) HOROLOVAR
Unit 6789A (19 x 36)
See Appendix 78

Jahresuhrenfabrik c 1930

JAHRESUHREN-FABRIK

JAHRESUHREN-FABRIK
GERMANY H.G.m.b.
459174

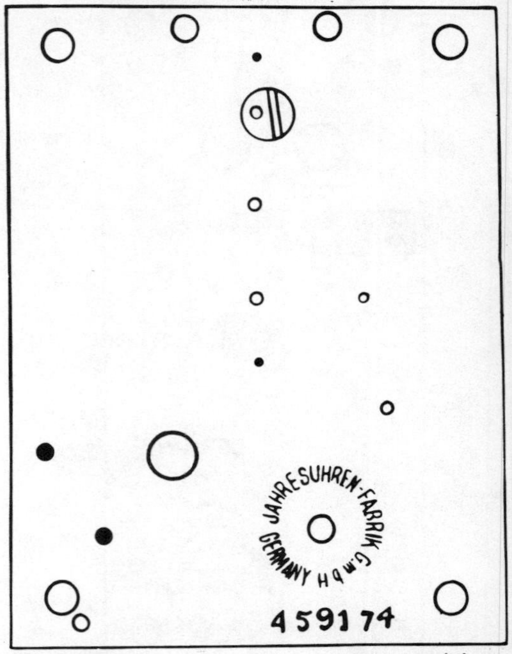

Plate 1263 Disc Pendulum
4-Ball Pendulum

USE .004" (.102mm) HOROLOVAR
Units 6, 7, 8, 9 (19 x 36)
See Appendix 78

Jahresuhrenfabrik c 1912

JAHRESUHREN-FABRIK

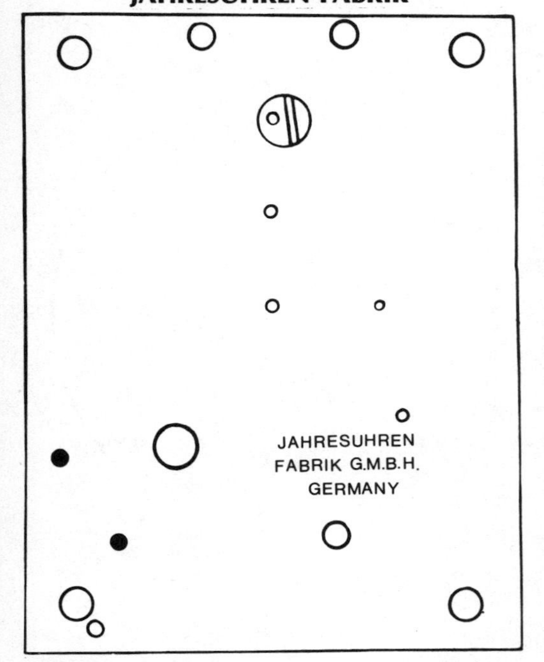

Plate 1264 Disc Pendulum
4-Ball Pendulum

USE .004" (.102mm) HOROLOVAR
Unit 6789A (19 x 36)
See Appendix 78

Jahresuhrenfabrik c 1910

JAHRESUHREN-FABRIK

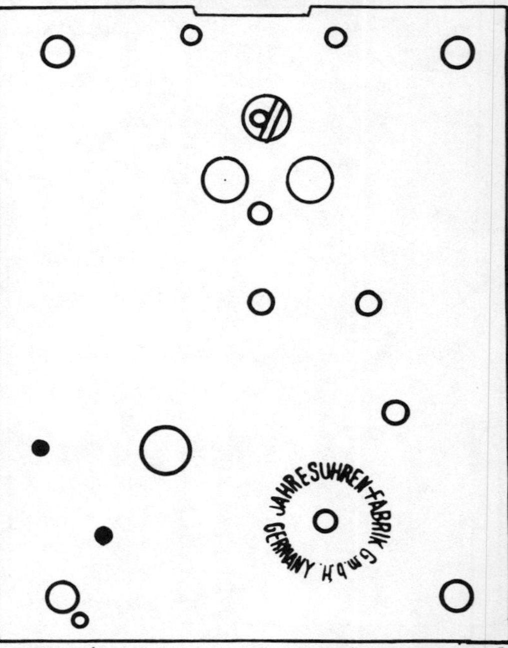

Plate 1265 Disc Pendulum
4-Ball Pendulum

USE .004" (.102mm) HOROLOVAR
Unit 6789A (19 x 36)
See Appendix 78

Jahresuhrenfabrik c 1910

JAHRESUHREN-FABRIK

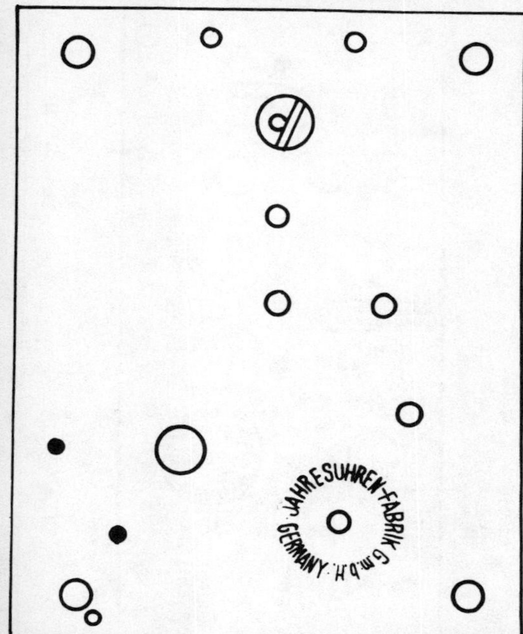

Plate 1267 4-Ball Pendulum
USE .004" (.102mm) HOROLOVAR
Unit 6789A (19 x 36)
See Appendix 78

Aug. Schatz & Sohne c 1949

JAHRESUHRENFABRIK

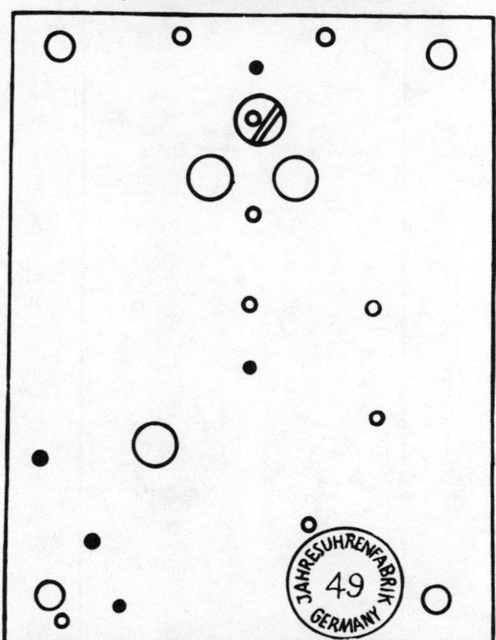

Plate 1271 4-Ball Pendulum
USE .004" (.102mm) HOROLOVAR
Units 6, 7, 8, 9 (19 x 36)
See Appendix 78

Aug. Schatz & Sohne c 1949

JAHRESUHRENFABRIK

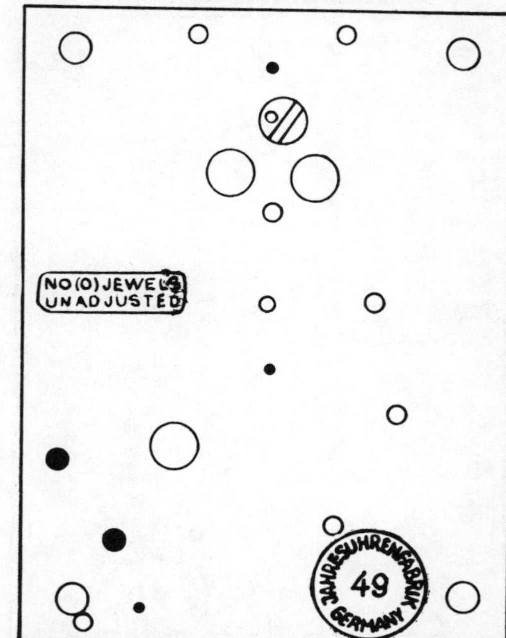

Plate 1278 4-Ball Pendulum
USE .004" (.102mm) HOROLOVAR
Units 6, 7, 8, 9 (19 x 36)
See Appendix 78

Aug. Schatz & Sohne c 1950

JAHRESUHRENFABRIK

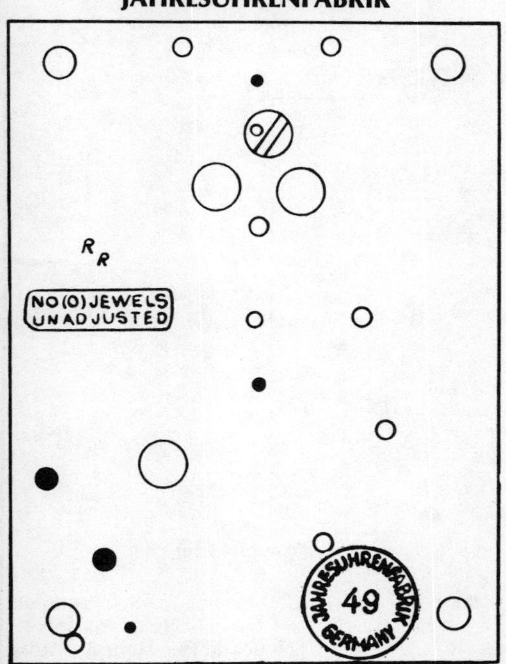

Plate 1279 4-Ball Pendulum
USE .004" (.102mm) HOROLOVAR
Units 6, 7, 8, 9 (19 x 36)
See Appendix 19, 78

Aug. Schatz & Sohne c 1949

JAHRESUHRENFABRIK

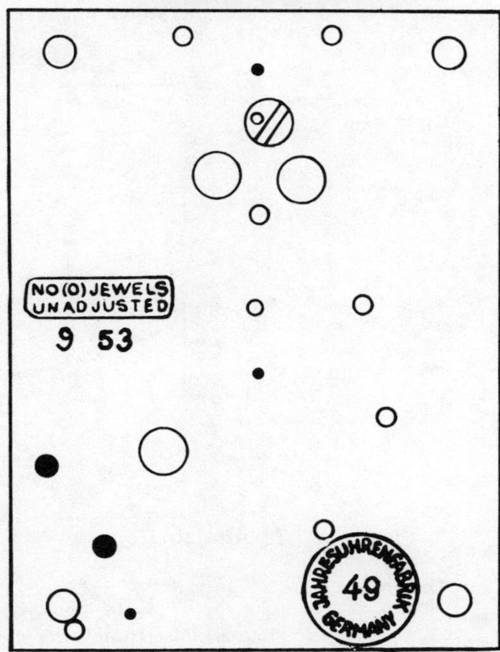

Plate 1281 4-Ball Pendulum
USE .004" (.102mm) HOROLOVAR
Units 6, 7, 8, 9 (19 x 36)
See Appendix 18, 78

Aug. Schatz & Sohne c 1949

JAHRESUHRENFABRIK

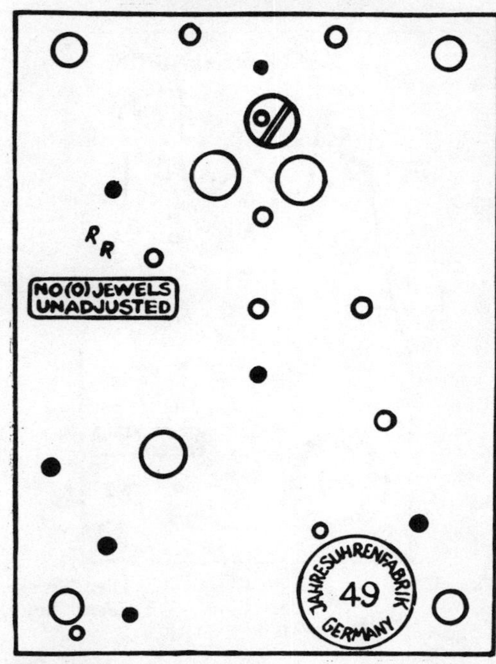

Plate 1283 4-Ball Pendulum
USE .004" (.102mm) HOROLOVAR
Units 6, 7, 8, 9 (19 x 36)
See Appendix 19, 78

Aug. Schatz & Sohne c 1954

JAHRESUHRENFABRIK

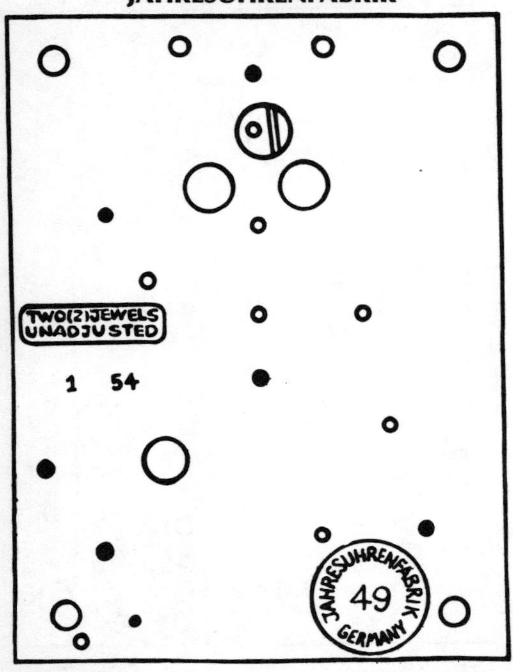

Plate 1287 4-Ball Pendulum
USE .004" (.102mm) HOROLOVAR
Units 6, 7, 8, 9 (19 x 36)
See Appendix 17, 18, 78

Aug. Schatz & Sohne c 1953

JAHRESUHRENFABRIK

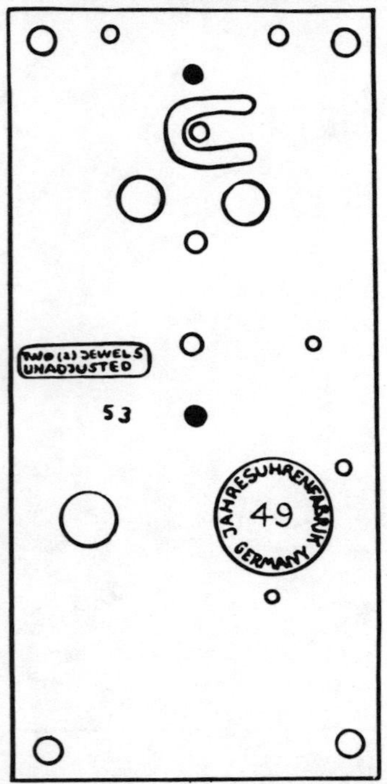

Plate 1291 4-Ball Pendulum
USE .004" (.102mm) HOROLOVAR
Units 6, 7, 8, 9 (19 x 36)
See Appendix 17, 18, 78

Aug. Schatz & Sohne c 1953

JAHRESUHRENFABRIK

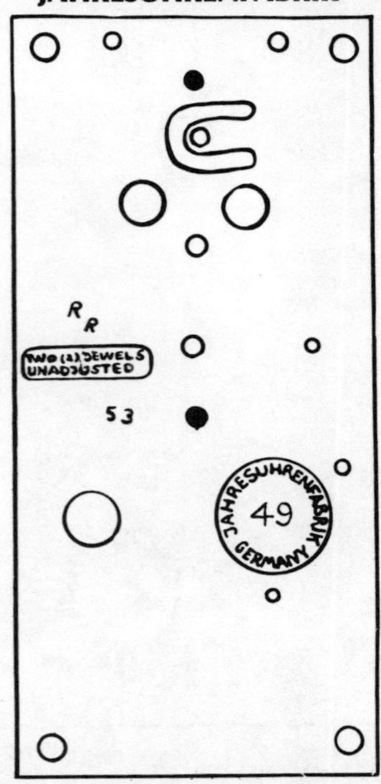

Plate 1291A 4-Ball Pendulum
USE .004" (.102mm) HOROLOVAR
Units 6, 7, 8, 9 (19 x 36)
See Appendix 17, 18, 19, 78

Aug. Schatz & Sohne c 1953

JAHRESUHRENFABRIK

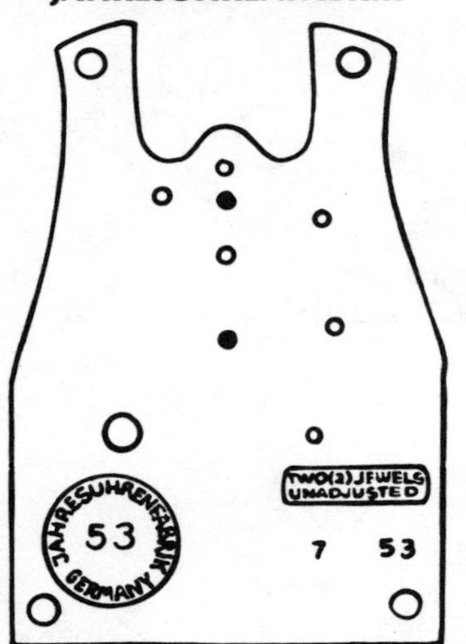

Plate 1299 Dome Clock—3-Ball Pendulum
Coach Clock—4-Ball Pendulum
USE .0023" (.058mm) HOROLOVAR
Unit 10A (13 x 30)
See Appendix 4, 5, 18, 63, 66, 79

R. Schneckenburger c 1898

JAHRESUHR SYLVESTER

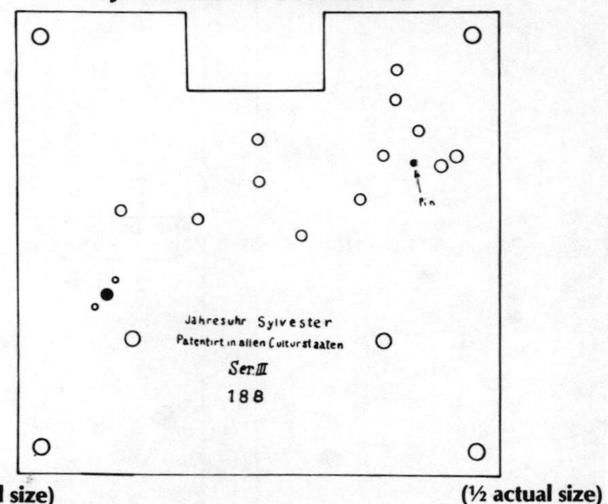

(½ actual size) **(½ actual size)**

Plate 1303 3" Ball Pendulum
Duplex Escapement
Hour and Half Hour Gong Strike
11" Suspension Spring
USE .006"— (.152mm—) HOROLOVAR
See Appendix 33

Plate 1307 3" Ball Pendulum
Duplex Escapement
Hour and Half Hour Bell Strike
4½" Suspension Spring
USE .0036" (.091mm) HOROLOVAR
See Appendix 33

Uhrenfabrik Herr c 1952

JENO BERGER

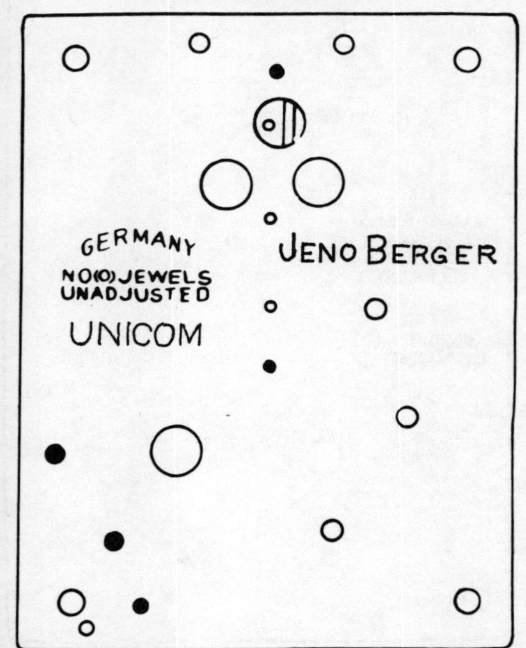

Plate 1307C 4-Ball Pendulum
USE .004″ (.102mm) HOROLOVAR
Unit 27A (19 x 36)
See Appendix 130

Jauch & Haller c 1951

J H

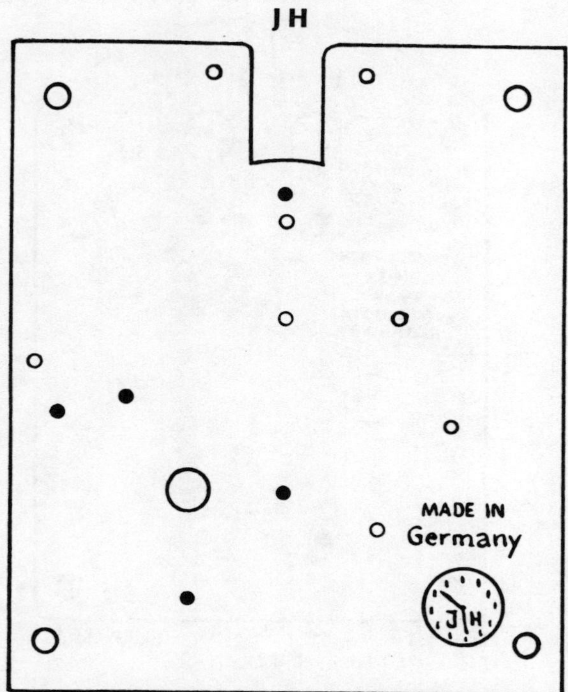

Plate 1308 4-Ball Pendulum
USE .0038″ (.097mm) HOROLOVAR
Unit 33 (19 x 38)
See Appendix 63, 64

Jauch & Haller c 1952

J H

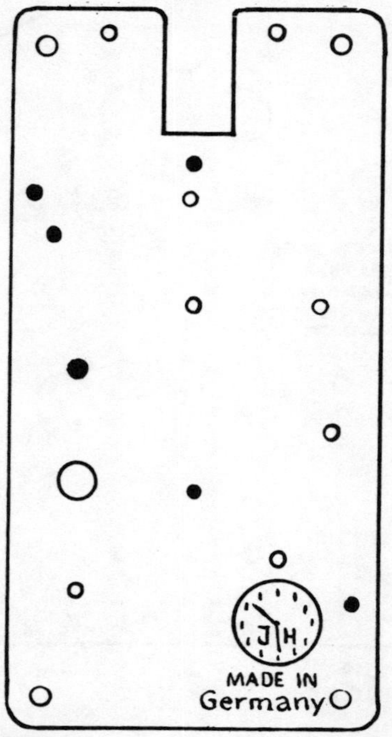

Plate 1308A 4-Ball Pendulum
USE .0038″ (.097mm) HOROLOVAR
Unit 33 (19 x 38)
See Appendix 63, 64

Uhrenfabrik J. Kaiser c 1954

J. KAISER G.M.B.H.

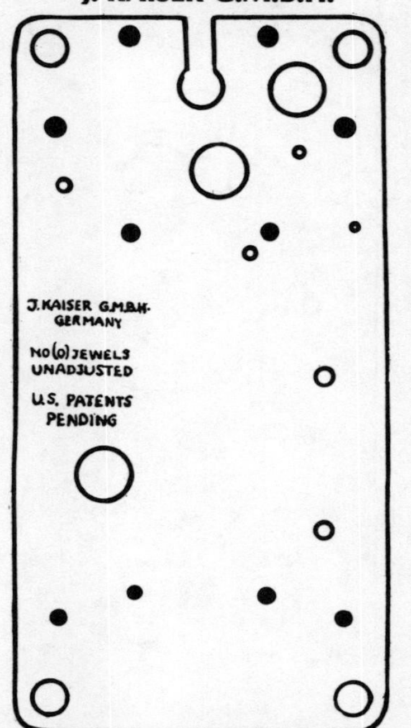

Plate 1309 Ball Pendulum
USE .003″ (.076mm) HOROLOVAR
Unit 34 (19 x 38)
See Appendix 21, 34, 63, 68

Uhrenfabrik J. Kaiser c 1954

J. KAISER G.M.B.H.

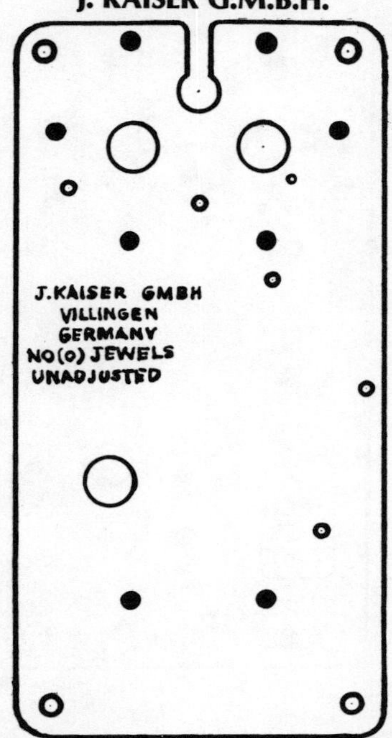

Plate 1309A Ball Pendulum
USE .003″ (.076mm) HOROLOVAR
Unit 34 (19 x 38)
See Appendix 20, 34, 63, 68

Uhrenfabrik J. Kaiser c 1954

J. KAISER G.M.B.H.

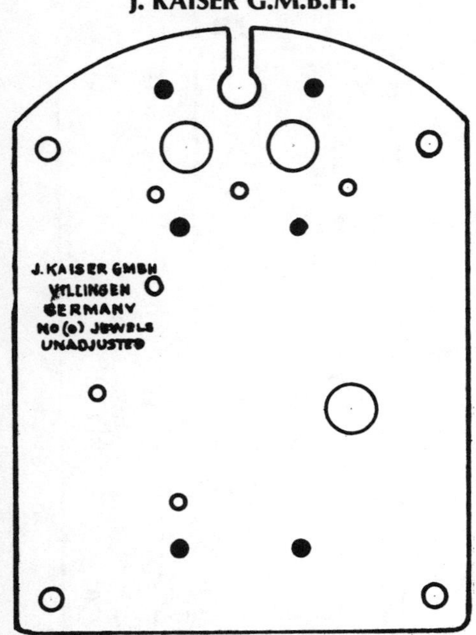

Plate 1310 Ball Pendulum
USE .003" (.076mm) HOROLOVAR
Unit 34 (19 x 38)
See Appendix 34, 63, 68

Uhrenfabrik J. Kaiser c 1954

J. KAISER G.M.B.H.

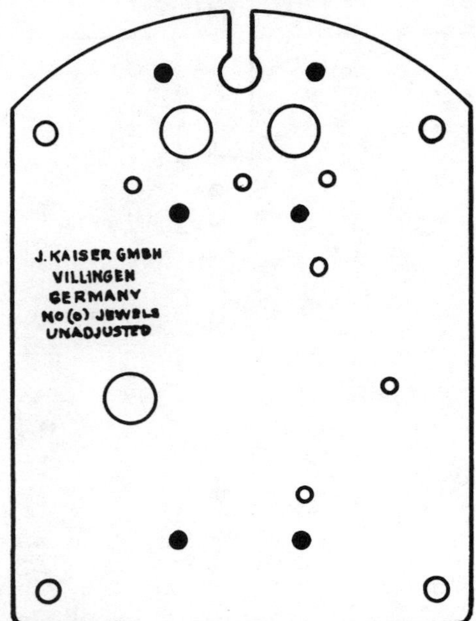

Plate 1311 Ball Pendulum
USE .003" (.076mm) HOROLOVAR
Unit 34 (19 x 38)
See Appendix 34, 63, 68

W. Petersen c 1954

J. KAISER G.M.B.H.

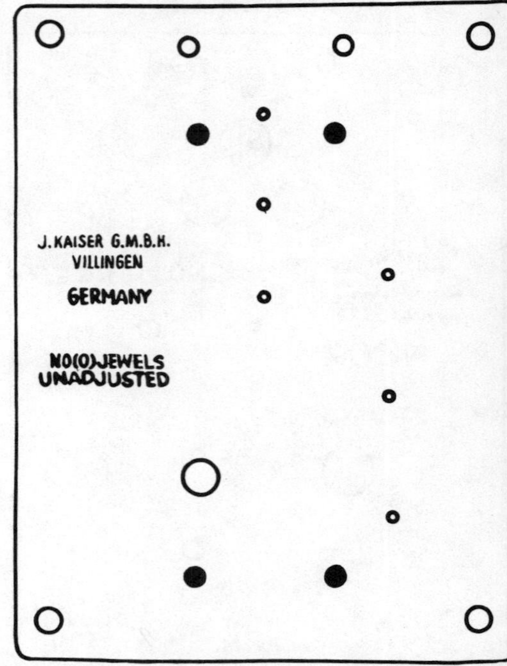

Plate 1312 4-Ball Pendulum
USE .0033" (.084mm) HOROLOVAR
Units 18B, 18C (19 x 38)
See Appendix 22, 59, 65

Konrad Mauch c 1954

JKG

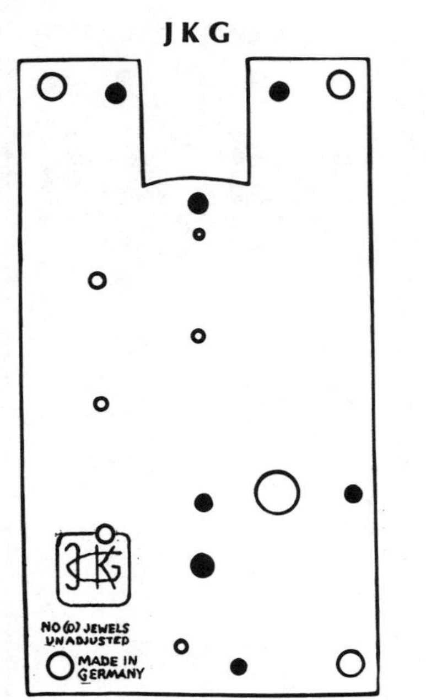

Plate 1313 4-Ball Pendulum
Miniature Movement
USE .0032" (.081mm) HOROLOVAR
Unit 14A
USE .003" (.076mm) HOROLOVAR
Unit 14B Original (13 x 32)
See Appendix 52, 64 Revised (15 x 32)

J. Link & Co. c 1952

J. LINK & CO.

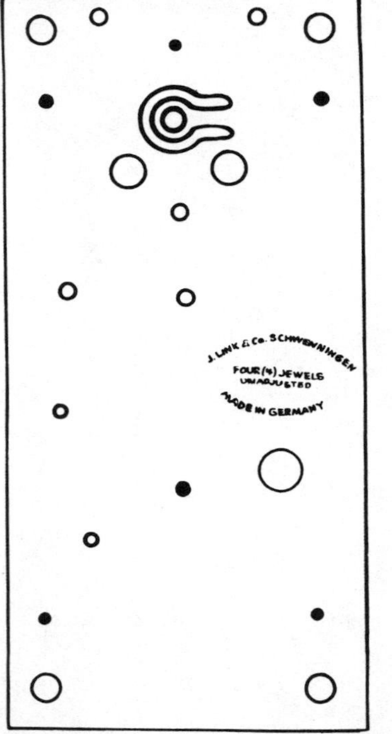

Plate 1314 4-Ball Pendulum
USE .0037" (.094mm) HOROLOVAR
Units 16, 17 (19 x 38)
See Appendix 38, 119

J. Link & Co. c 1952

J. LINK & CO.

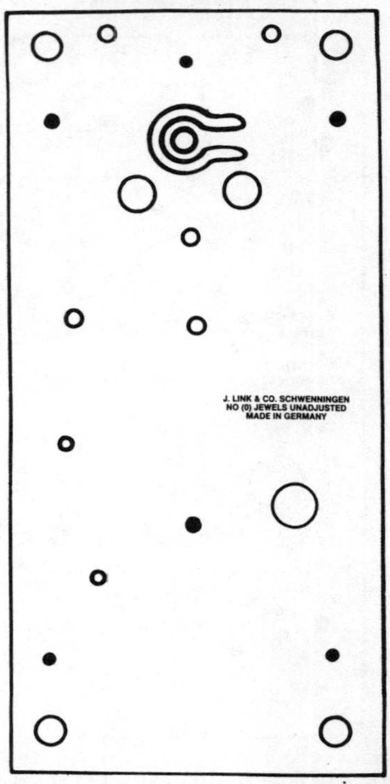

Plate 1315 4-Ball Pendulum
USE .0037" (.094mm) HOROLOVAR
Units 16, 17 (19 x 38)
See Appendix 38, 119

J. Link & Co. c 1956

J. LINK & CO.

J. LINK & CO. SCHWENNINGEN
NO (0) JEWELS UNADJUSTED
MADE IN GERMANY

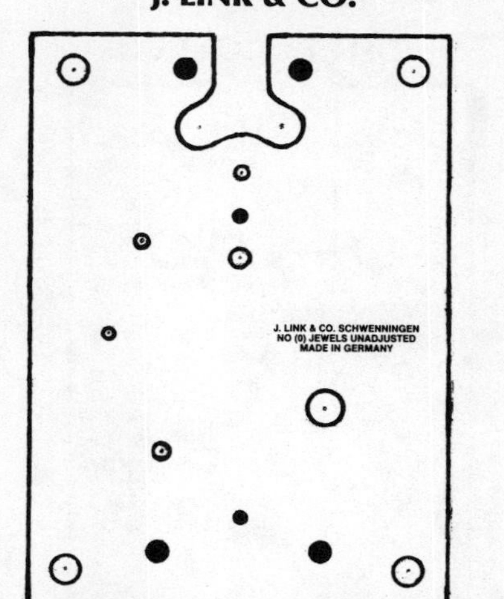

Plate 1316 4-Ball Pendulum
Miniature Clock
USE .0021″ (.053mm) HOROLOVAR
Unit 35 (14 x 28)
See Appendix 15, 63

Kieninger & Obergfell c 1911

J. MULLER & CO.

J. MULLER & Co
GERMANY
1484

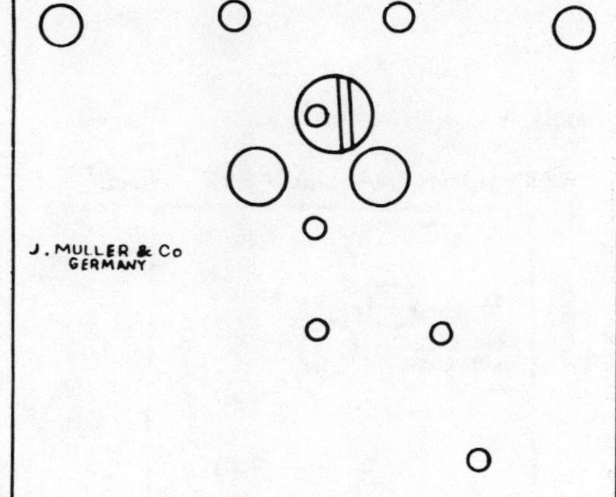

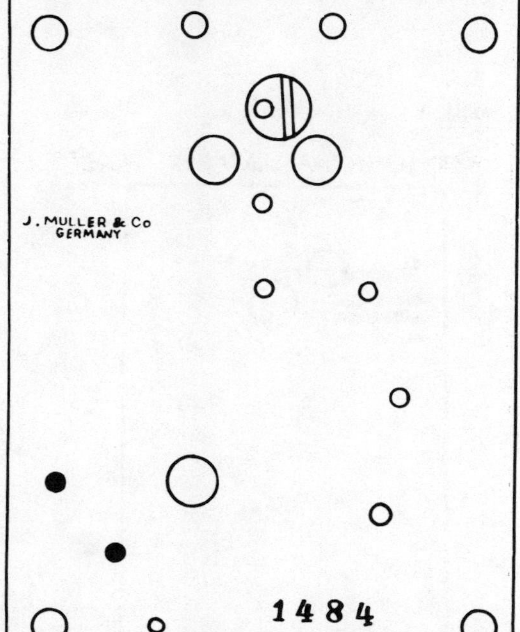

Plate 1317 4-Ball Pendulum
USE .0035″* (.089mm*) HOROLOVAR
See Appendix 76 (19 x 38)

Badische Uhrenfabrik c 1902

J. MULLER & CO.

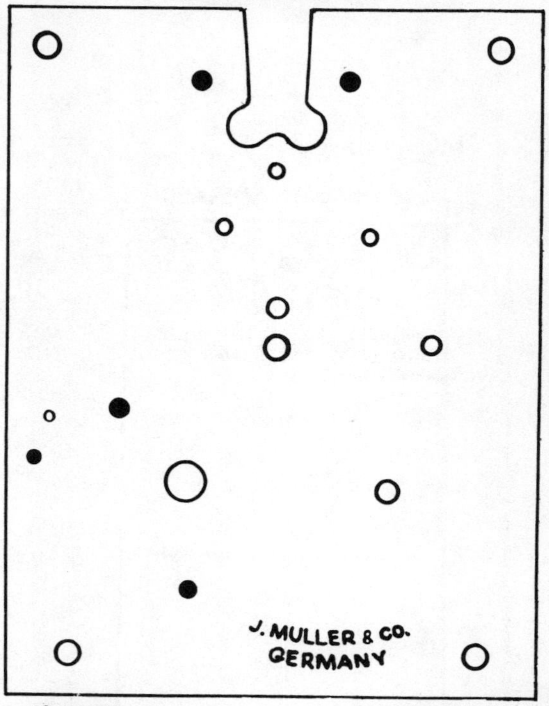

J. MULLER & CO.
GERMANY

Plate 1318 Disc Pendulum
3-Ball Pendulum
Pin Pallet Escapement
Lantern Pinions
USE .0035″* (.089mm*) HOROLOVAR
See Appendix 63 (20 x 38)

Uhrenfabrik Herr c 1951

JOHN WANAMAKER

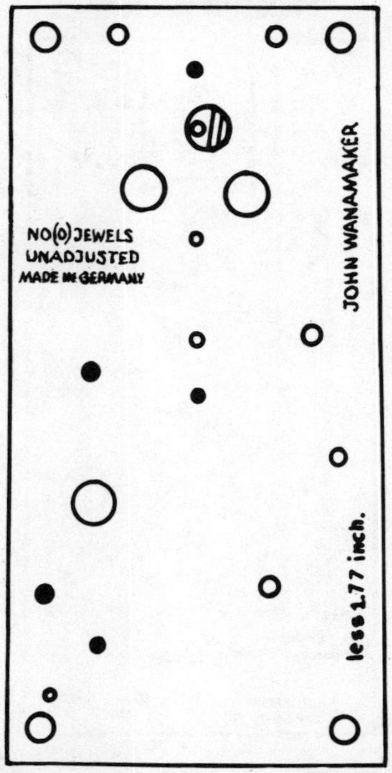

NO(0) JEWELS
UNADJUSTED
MADE IN GERMANY
JOHN WANAMAKER
less 1.77 inch.

Plate 1319 4-Ball Pendulum
USE .0035″ (.089mm) HOROLOVAR
Units 19, 20A, 21 (19 x 36)
See Appendix 109, 110

Uhrenfabrik Herr c 1952

JOHN WANAMAKER

NO(0) JEWELS
UNADJUSTED
MADE IN GERMANY
JOHN
WANAMAKER

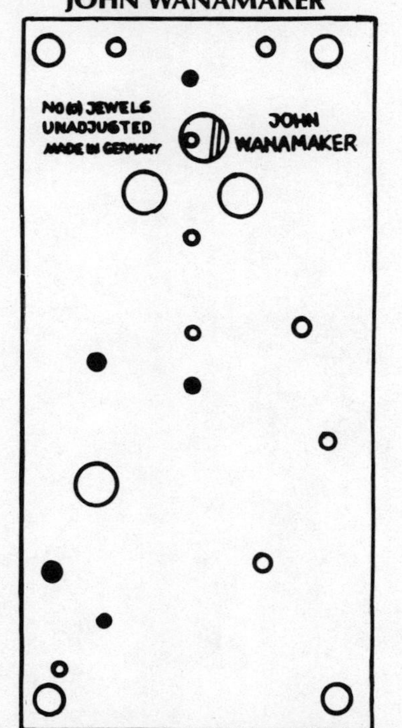

Plate 1320 4-Ball Pendulum
Dome Clock
USE .0035″ (.089mm) HOROLOVAR
Units 19, 20A, 21
Square Clock
USE .0033″ (.084mm) HOROLOVAR
Unit 20B (19 x 36)
See Appendix 109, 110

Uhrenfabrik Herr c 1952

JOHN WANAMAKER

NO(0) JEWELS
UNADJUSTED
MADE IN GERMANY
JOHN WANAMAKER

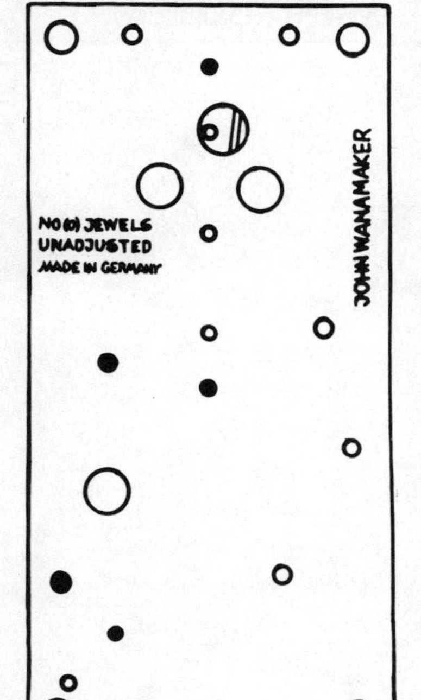

Plate 1323 4-Ball Pendulum
Dome Clock
USE .0035″ (.089mm) HOROLOVAR
Units 19, 20A, 21
Square Clock
USE .0033″ (.084mm) HOROLOVAR
Unit 20B (19 x 36)
See Appendix 109, 110

Uhrenfabrik M. Reiner c 1953

JOHN WANAMAKER

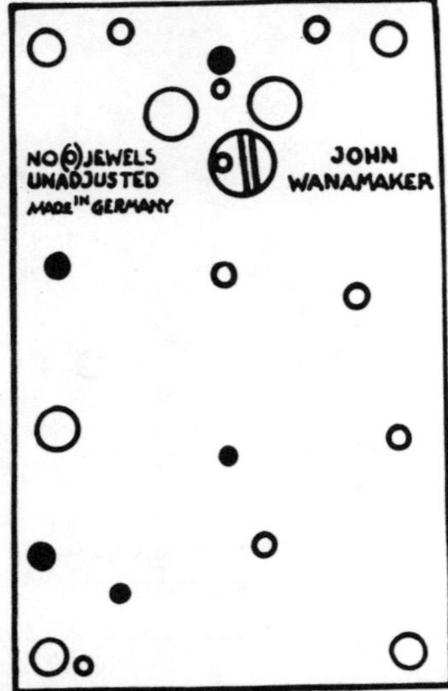

Plate 1325 Standard 4-Ball Pendulum
Miniature Movement
USE .003" (.076mm) HOROLOVAR
Units 22, 23A, 24 (16 x 36)
See Appendix 7, 25, 54, 109, 114, 115, 124

Uhrenfabrik M. Reiner c 1953

JOHN WANAMAKER

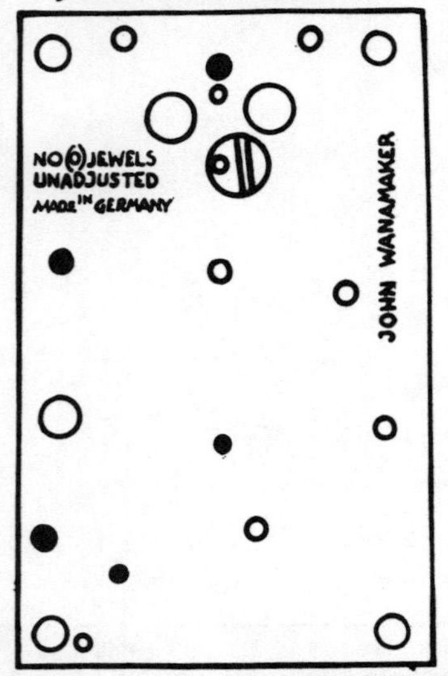

Plate 1326 Standard 4-Ball Pendulum
Miniature Movement
USE .003" (.076mm) HOROLOVAR
Units 22, 23A, 24 (16 x 36)
See Appendix 7, 25, 54, 109, 114, 115, 124

Sigfried Haller c 1951

JOSEF F. SCOTT CO.

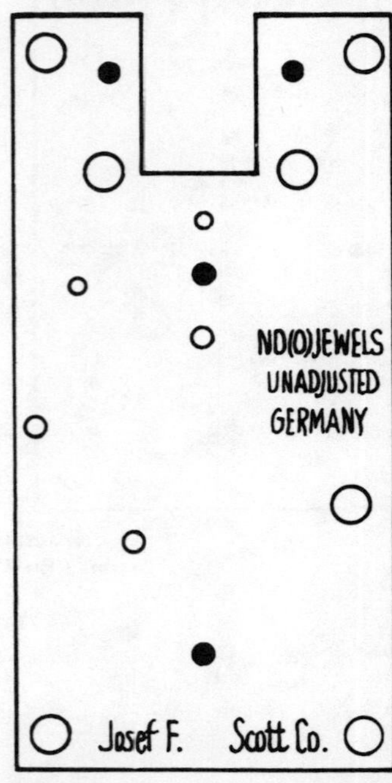

Plate 1332 4-Ball Pendulum
USE .0038" (.097mm) HOROLOVAR
Unit 33 (18 x 38)
See Appendix 63, 64

Kieninger & Obergfell c 1950

JOSEPH HORNE CO.

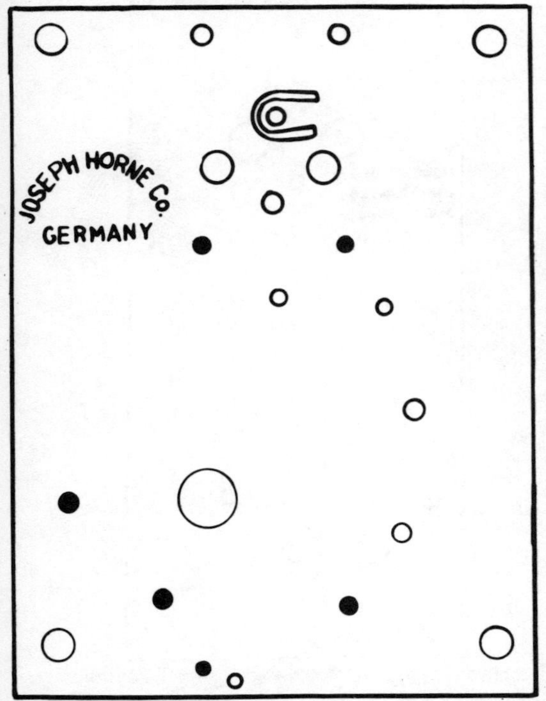

Plate 1338 4-Ball Pendulum
USE .0032" (.081mm) HOROLOVAR
Unit 1 (19 x 38)
See Appendix 76

Konrad Mauch c 1950

JOSEPH KAROL

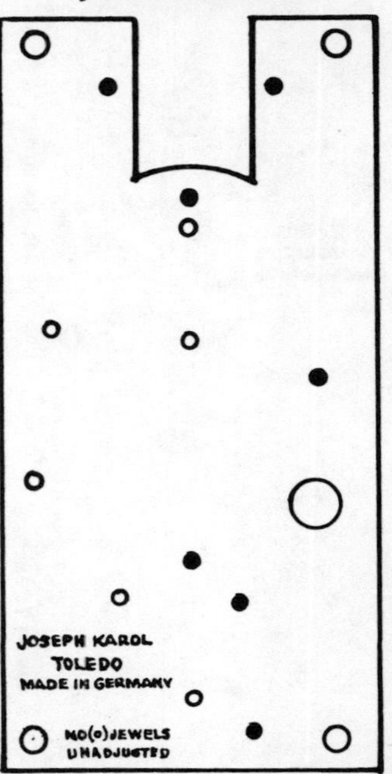

Plate 1340 4-Ball Pendulum
USE .0035" (.089mm) HOROLOVAR
Units 13A, 13B (20 x 38)
See Appendix 63, 64, 84

Kern & Sohne c 1961

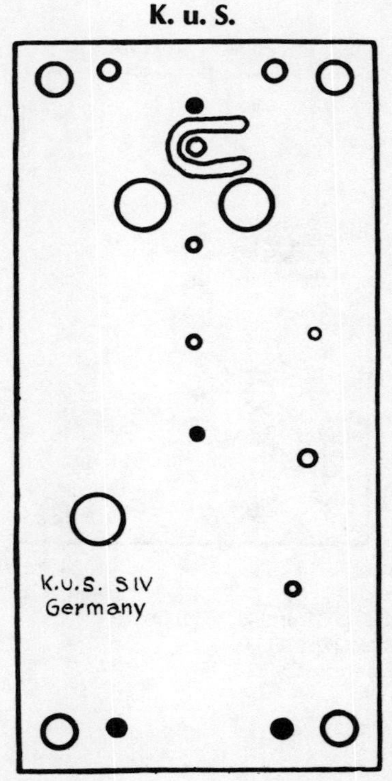

Plate 1340D 4-Ball Pendulum
USE .0036″ (.091mm) HOROLOVAR
Unit 11B (18 x 38)
See Appendix 81

Kern & Sohne c 1962

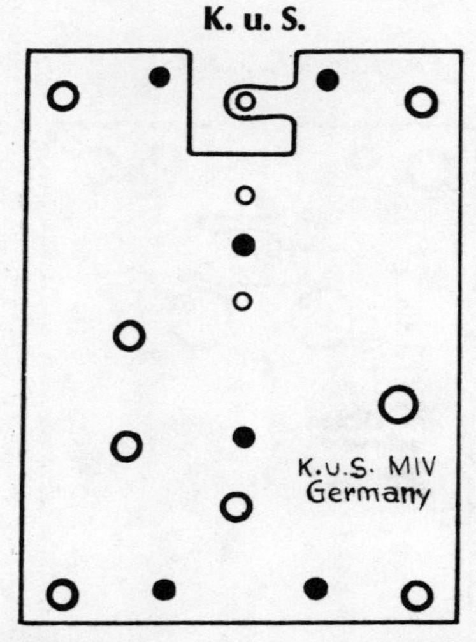

Plate 1340E 4-Ball Pendulum
Miniature Clock
USE .0023″ (.058mm) HOROLOVAR
Unit 12F (12 x 25)
See Appendix 82

Kern & Sohne c 1958

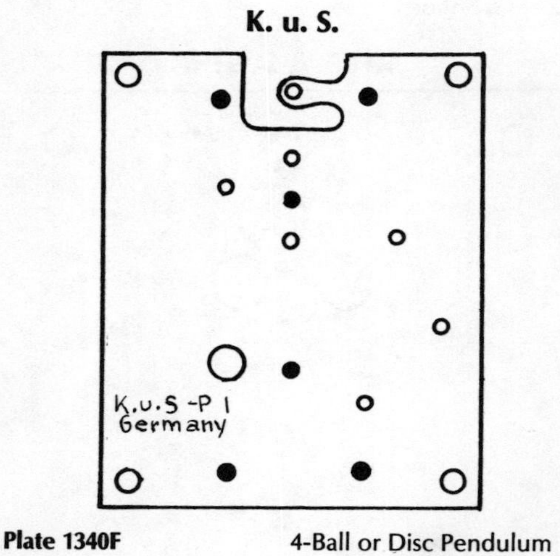

Plate 1340F 4-Ball or Disc Pendulum
Midget Clock
Pin Pallet Escapement
USE .0023″ (.058mm) HOROLOVAR
Unit 12E (14 x 25)
See Appendix 83

Kern & Link c 1929

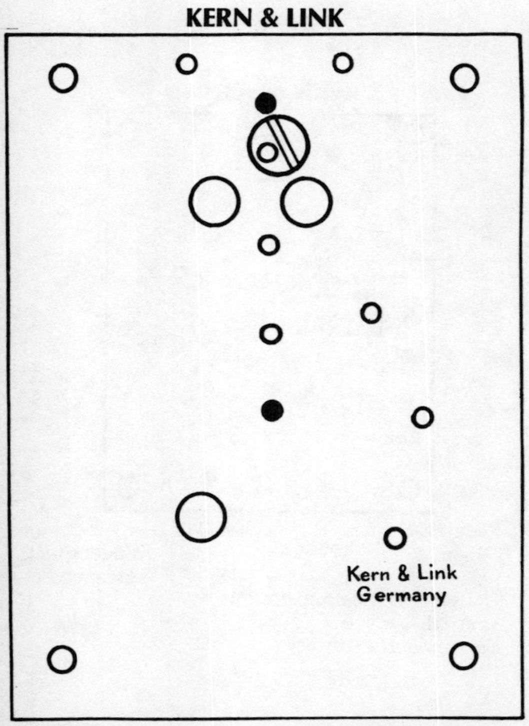

KERN & LINK

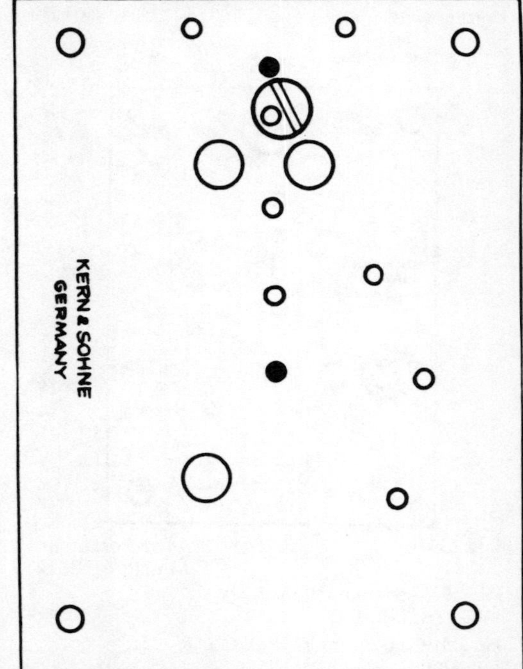

Kern & Link c 1929

KERN & LINK

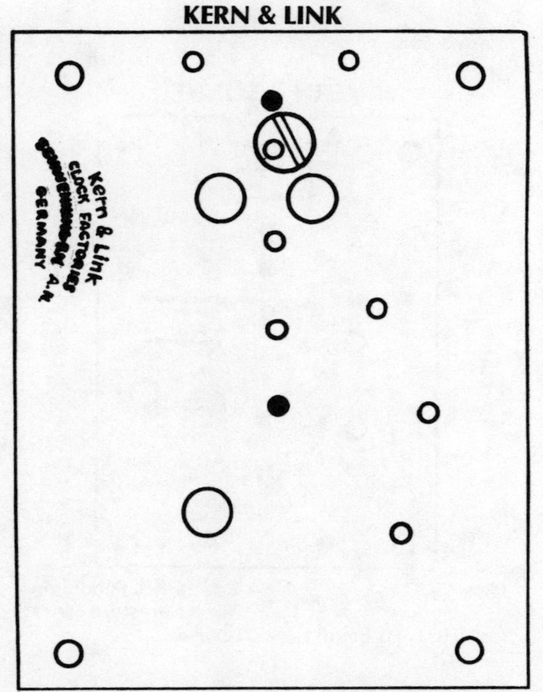

Plate 1342 4-Ball Pendulum
USE .0036″ (.091mm) HOROLOVAR
Unit 11A (18 x 38)

Kern & Sohne c 1949

KERN & SOHNE

Plate 1343 4-Ball Pendulum
USE .0036″ (.091mm) HOROLOVAR
Unit 11A (18 x 38)
See Appendix 81

Plate 1341 4-Ball Pendulum
USE .0036″ (.091mm) HOROLOVAR
Unit 11A (18 x 38)

Kern & Sohne c 1951

KERN & SOHNE

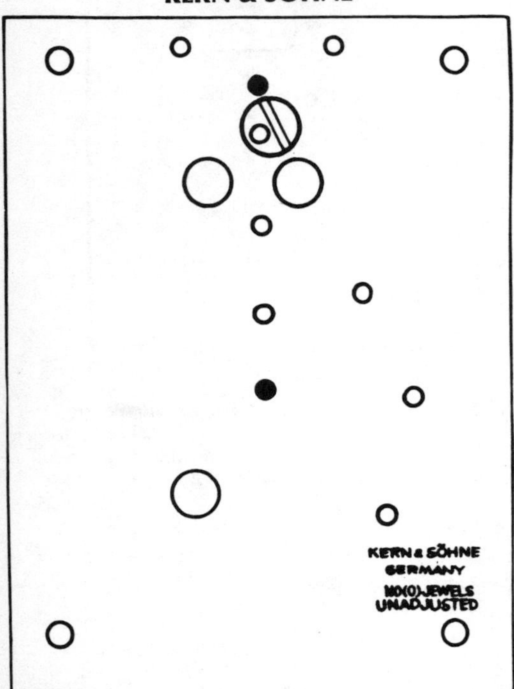

Plate 1343A 4-Ball Pendulum
USE .0036" (.091mm) HOROLOVAR
Unit 11A (18 x 38)
See Appendix 81

Kern & Sohne c 1961

KERN & SOHNE

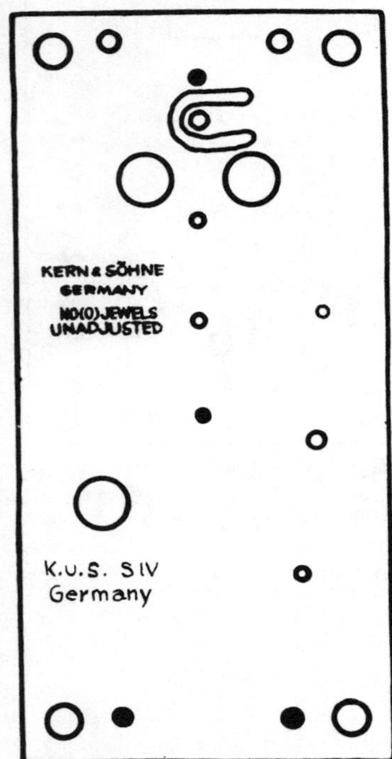

Plate 1345 4-Ball Pendulum
USE .0036" (.091mm) HOROLOVAR
Unit 11B (18 x 38)
See Appendix 81

Kern & Sohne c 1954

KERN & SOHNE

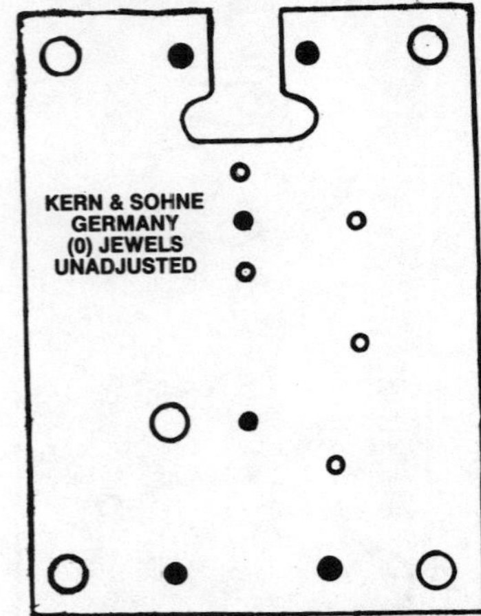

Plate 1348 4-Ball Pendulum Miniature Clock
USE .002" (.051mm) HOROLOVAR
Units 12A, 12B, 12C (12 x 25)
See Appendix 63, 82

Kern & Sohne c 1959

K S

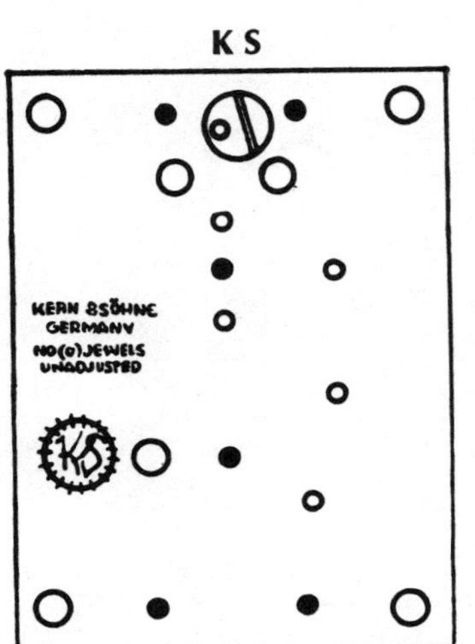

Plate 1348A 4-Ball Pendulum Miniature Clock
USE .002" (.051mm) HOROLOVAR
Units 12A, 12B, 12C
USE .0019" (.048mm) HOROLOVAR
Unit 12D (12 x 25)
See Appendix 82

Kern & Sohne c 1961

KERN & SOHNE

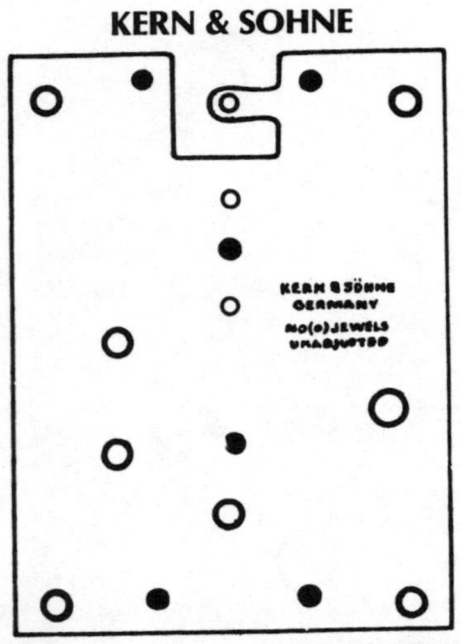

Plate 1349 4-Ball Pendulum Miniature Clock
USE .0023" (.058mm) HOROLOVAR
Unit 12F (14 x 30)
See Appendix 82

Kern & Sohne c 1958

KERN & SOHNE

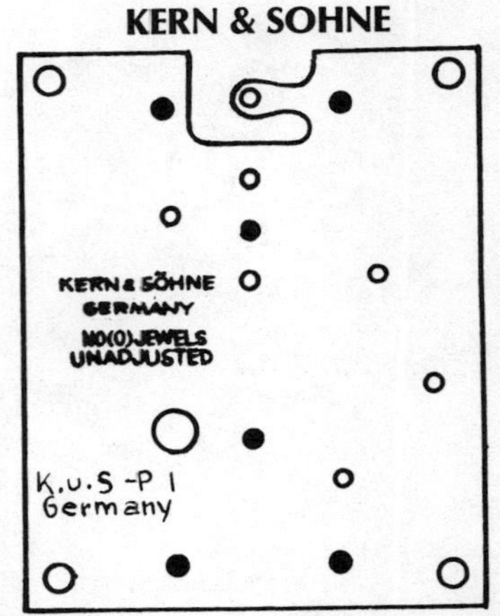

Plate 1350 4-Ball or Disc Pendulum Midget Clock Pin Pallet Escapement
USE .0023" (.058mm) HOROLOVAR
Unit 12E (14 x 25)
See Appendix 83

Kieninger & Obergfell c 1937

KIENINGER & OBERGFELL

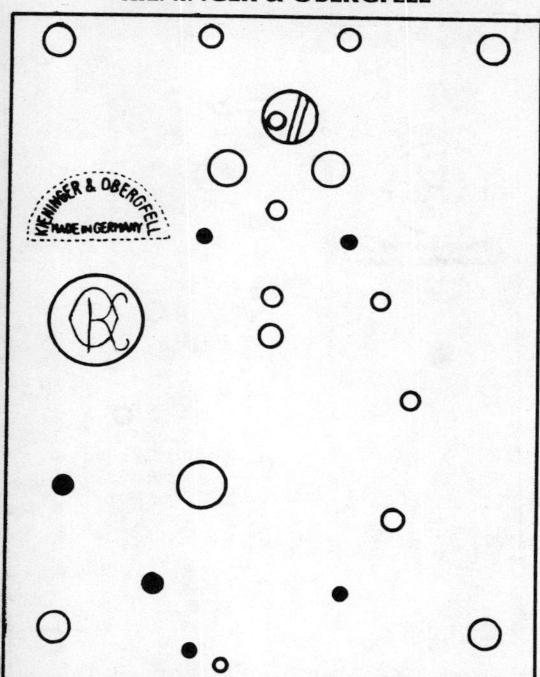

Plate 1351 4-Ball Pendulum
USE .0032″ (.081mm) HOROLOVAR
Unit 1 (19 x 38)
See Appendix 76

Kieninger & Obergfell c 1930

KIENINGER & OBERGFELL

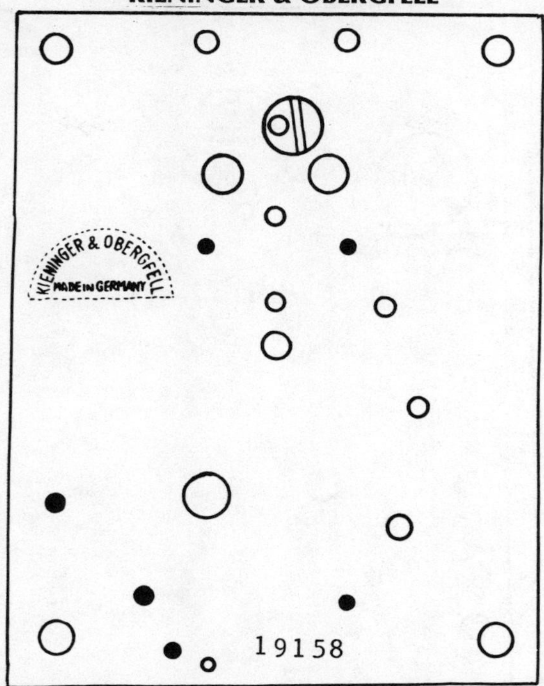

Plate 1355 4-Ball Pendulum
USE .0032″ (.081mm) HOROLOVAR
Unit 1 (19 x 38)
See Appendix 76

Kieninger & Obergfell c 1947

KIENINGER & OBERGFELL

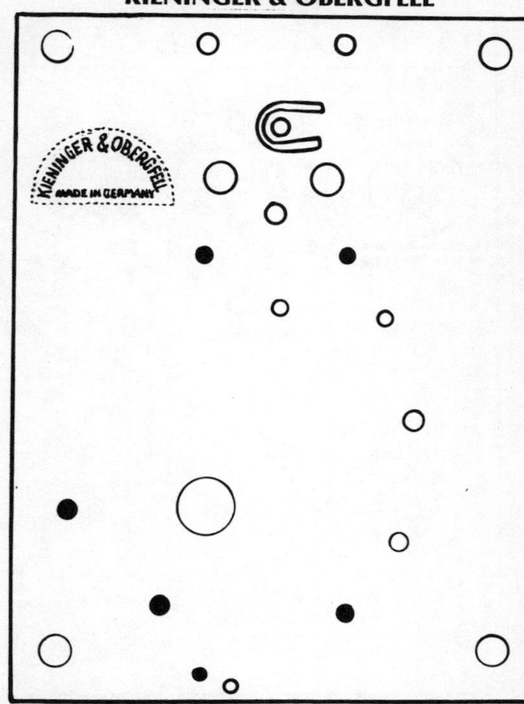

Plate 1359 4-Ball Pendulum
USE .0032″ (.081mm) HOROLOVAR
Unit 1 (19 x 38)
See Appendix 76

Kieninger & Obergfell c 1947

KIENINGER & OBERGFELL

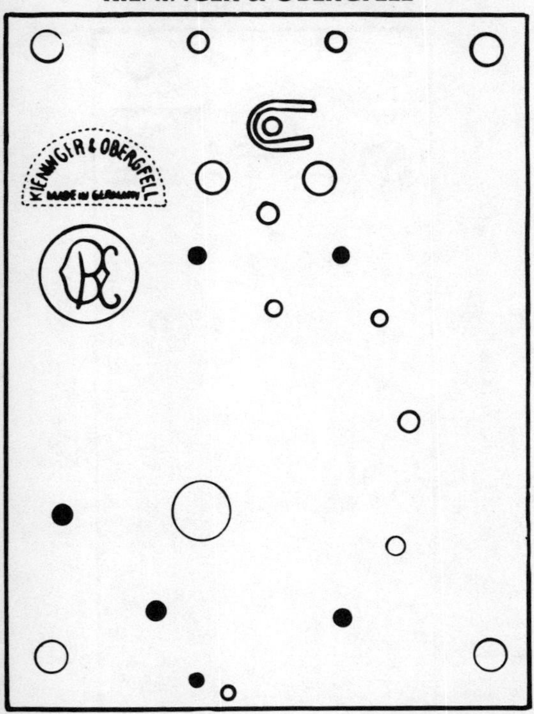

Plate 1363 4-Ball Pendulum
USE .0032″ (.081mm) HOROLOVAR
Unit 1 (19 x 38)
See Appendix 76

Kieninger & Obergfell c 1949

KIENINGER & OBERGFELL

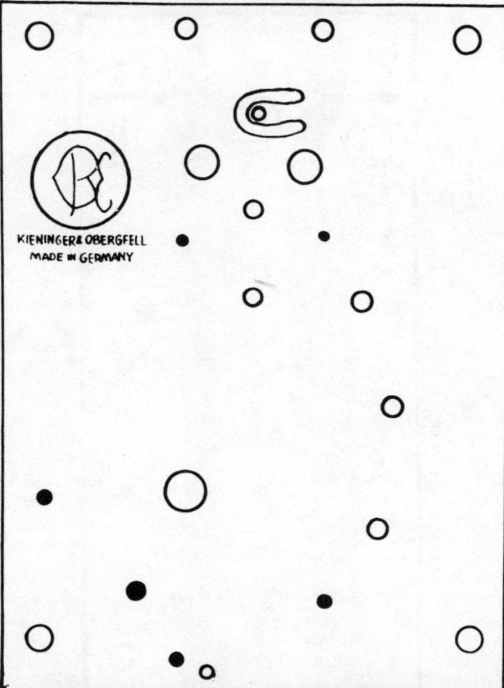

Plate 1367 4-Ball Pendulum
USE .0032″ (.081mm) HOROLOVAR
Unit 1, 3A (19 x 38)
See Appendix 76

Kieninger & Obergfell c 1950

KIENINGER & OBERGFELL

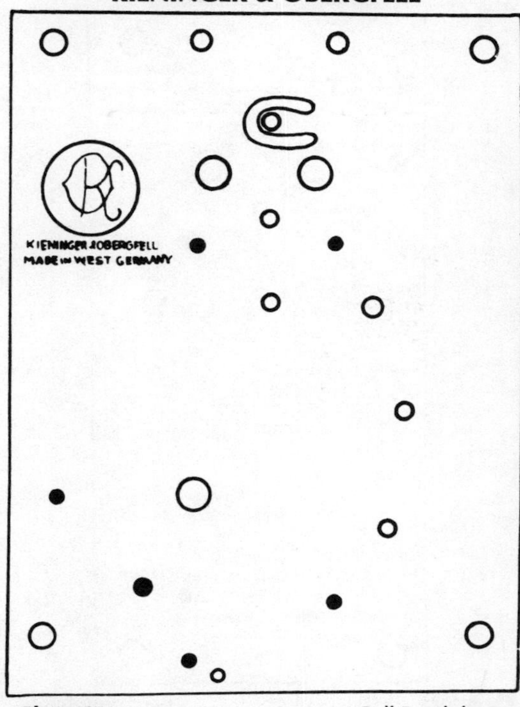

Plate 1371 4-Ball Pendulum
USE .0032″ (.081mm) HOROLOVAR
Units 1, 3A (19 x 38)
See Appendix 76

Kieninger & Obergfell c 1951

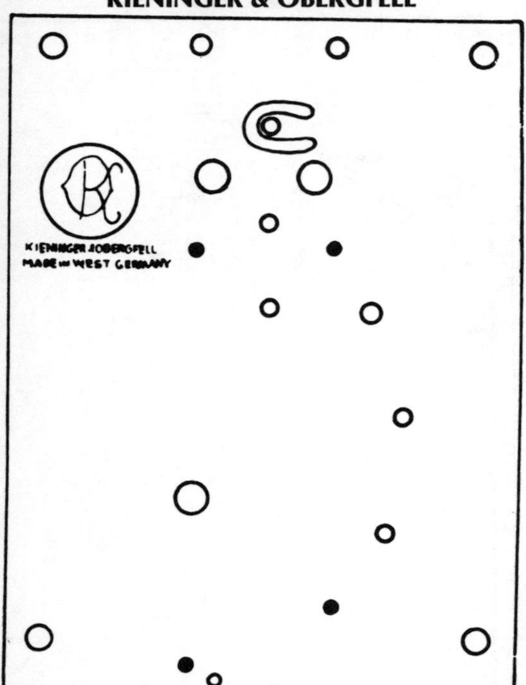

Plate 1371AAA 4-Ball Pendulum
USE .0032" (.081mm) HOROLOVAR
Units 1, 3A (19 x 38)
See Appendix 35, 76

Kieninger & Obergfell c 1951

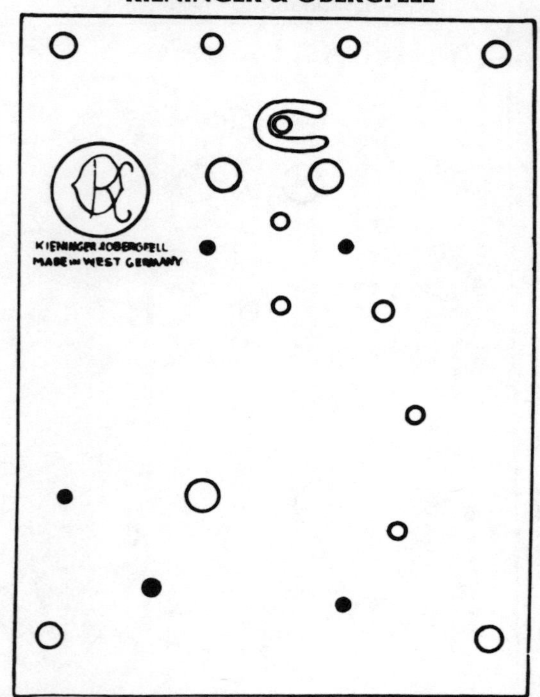

Plate 1371AA 4-Ball Pendulum
USE .0032" (.081mm) HOROLOVAR
Units 1, 3A (19 x 38)
See Appendix 35, 76

Kieninger & Obergfell c 1951

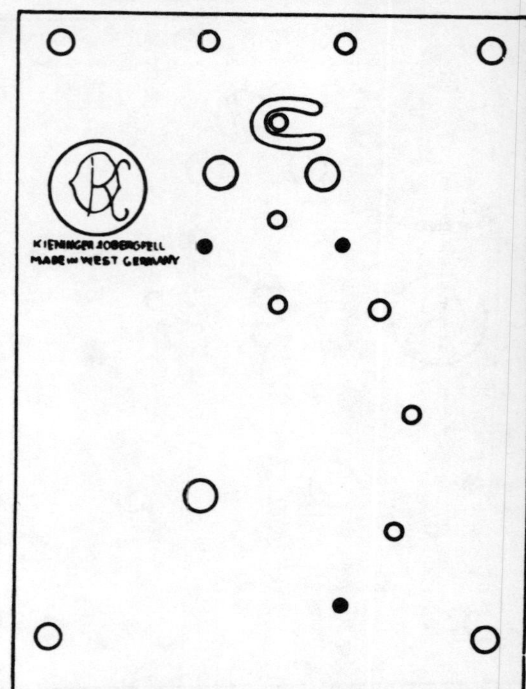

Plate 1371A 4-Ball Pendulum
USE .0032" (.081mm) HOROLOVAR
Units 1, 3A (19 x 38)
See Appendix 36, 76

Kieninger & Obergfell c 1951

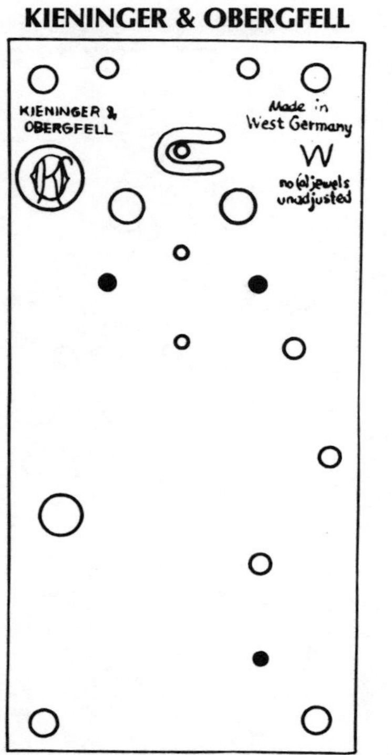

Plate 1375 4-Ball Pendulum
USE .0032" (.081mm) HOROLOVAR
Units 1, 3A, 3B, 3C (19 x 38)
See Appendix 76, 100

Kieninger & Obergfell c 1951

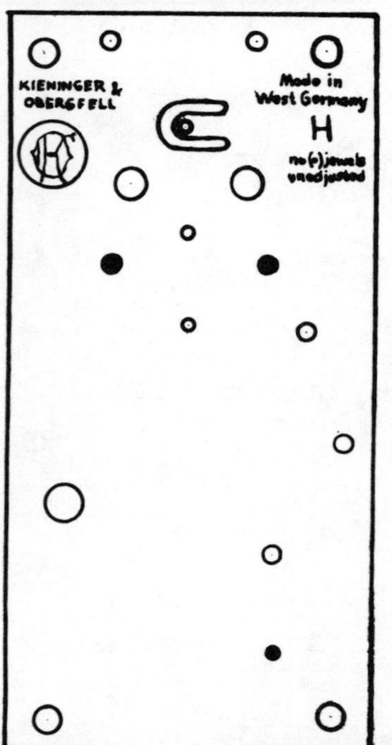

Plate 1376 4-Ball Pendulum
USE .0032" (.081mm) HOROLOVAR
Units 1, 3A, 3B, 3C (19 x 38)
See Appendix 76, 100

Kieninger & Obergfell c 1951

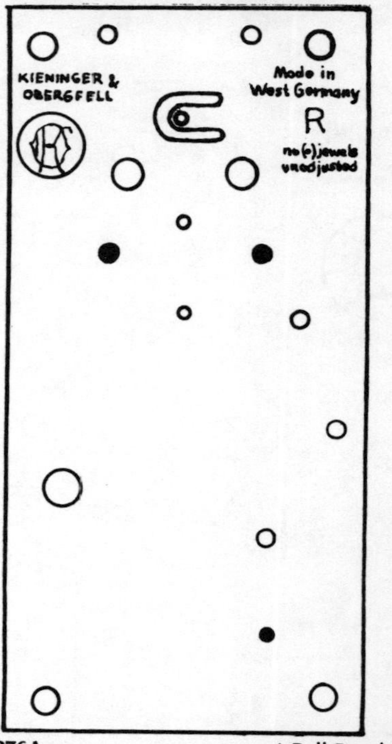

Plate 1376A 4-Ball Pendulum
USE .0032" (.081mm) HOROLOVAR
Units 1, 3A, 3B, 3C (19 x 38)
See Appendix 76, 100

Kieninger & Obergfell c 1951

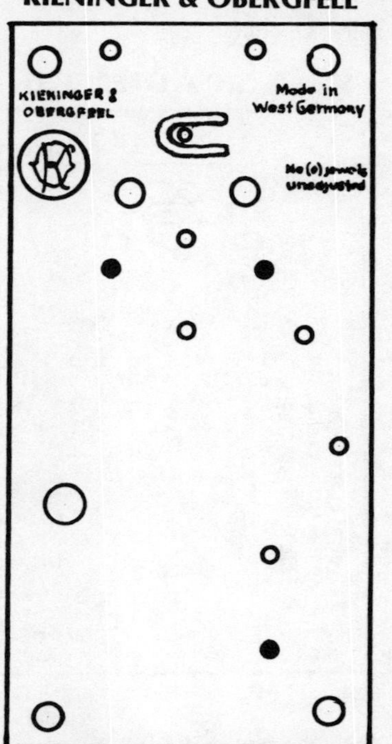

Plate 1377 4-Ball Pendulum
USE .0032" (.081mm) HOROLOVAR
Units 1, 3A, 3B, 3C (19 x 38)
See Appendix 76

Kieninger & Obergfell c 1951

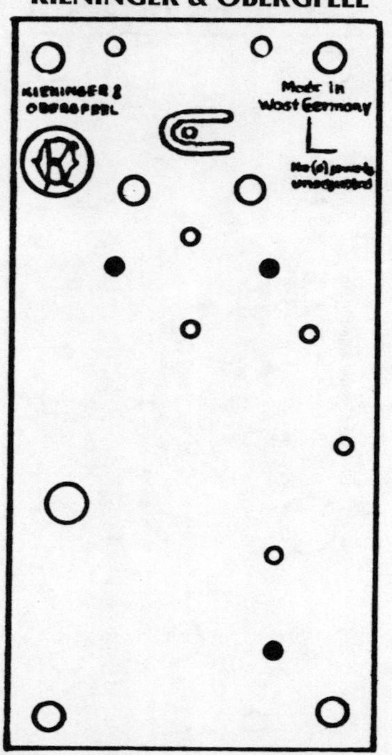

Plate 1378 4-Ball Pendulum
USE .0032" (.081mm) HOROLOVAR
Units 1, 3A, 3B, 3C (19 x 38)
See Appendix 76

Kieninger & Obergfell c 1952

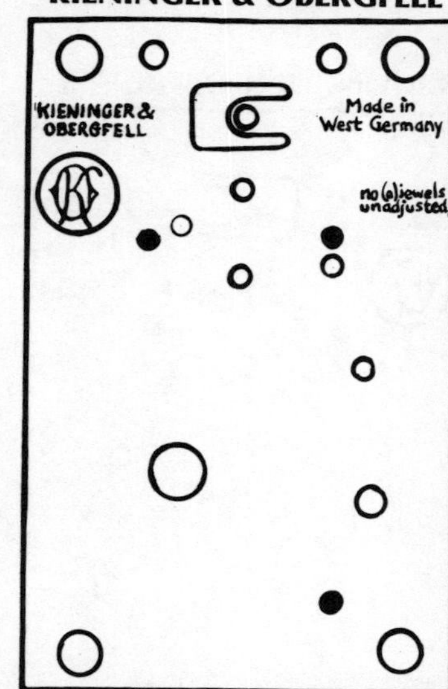

Plate 1379 4-Ball Pendulum Miniature Clock
USE .0023" (.058mm) HOROLOVAR
Units 5A, 5B, 5C, 5D, 5E (14 x 30)
See Appendix 77, 111

Kieninger & Obergfell c 1974

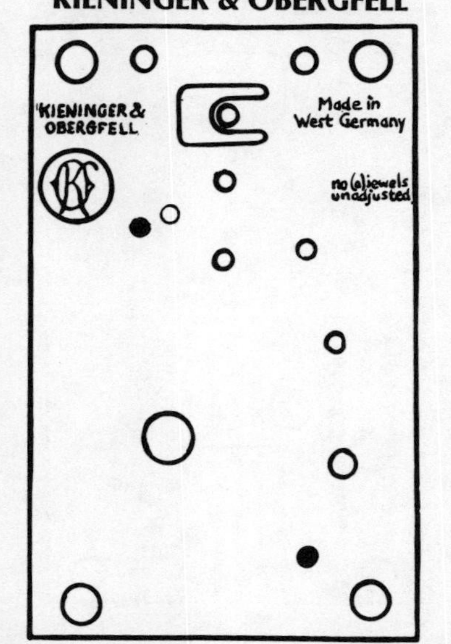

Plate 1379A 4-Ball Pendulum Miniature Clock
USE .0023" (.058mm) HOROLOVAR
Units 5C, 5D, 5E (14 x 30)
See Appendix 77, 111

Kieninger & Obergfell c 1955

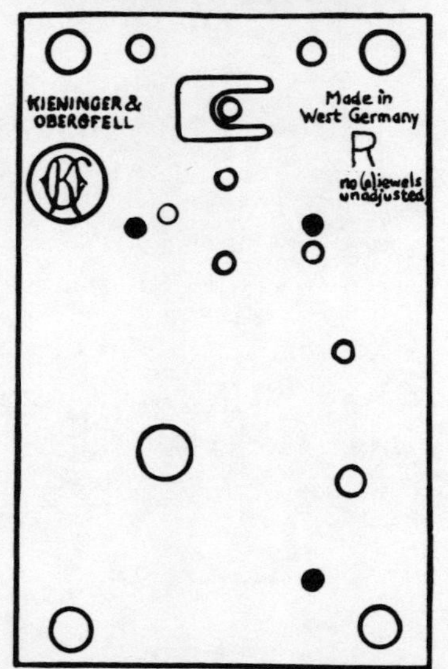

Plate 1379B 4-Ball Pendulum Miniature Clock
USE .0023" (.058mm) HOROLOVAR
Units 5A, 5B, 5C, 5D, 5E (14 x 30)
See Appendix 77, 100

Kieninger & Obergfell c 1955

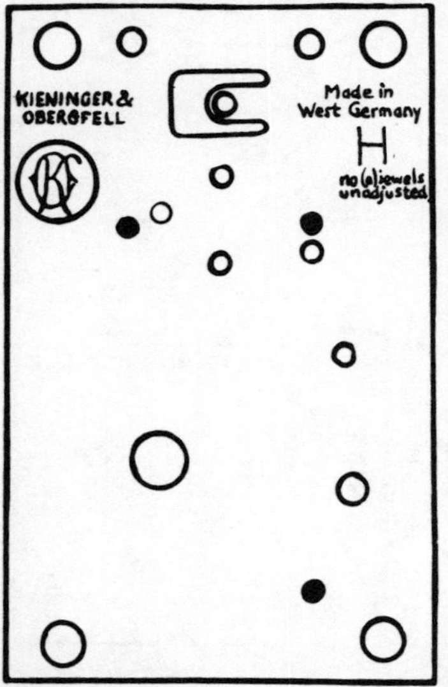

Plate 1379C 4-Ball Pendulum Miniature Clock
USE .0023" (.058mm) HOROLOVAR
Units 5A, 5B, 5C, 5D, 5E (14 x 30)
See Appendix 77, 100

Kieninger & Obergfell c 1955

KIENINGER & OBERGFELL

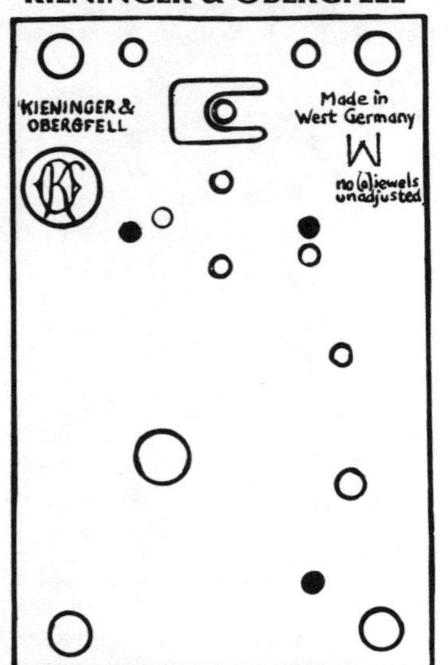

Plate 1379D 4-Ball Pendulum Miniature Clock
USE .0023" (.058mm) HOROLOVAR
Units 5A, 5B, 5C, 5D, 5E (14 x 30)
See Appendix 77, 100

Kieninger & Obergfell c 1976

KIENINGER & OBERGFELL

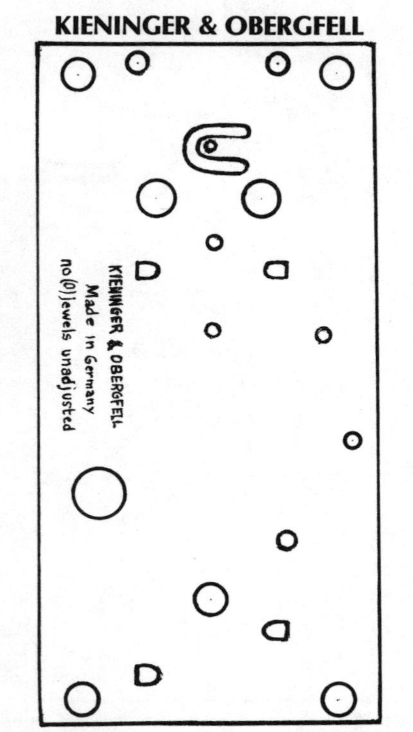

Plate 1380 4-Ball Pendulum
USE .0032" (.081mm) HOROLOVAR
Unit 3C (19 x 38)
See Appendix 76, 101

Kieninger & Obergfell c 1976

KIENINGER & OBERGFELL

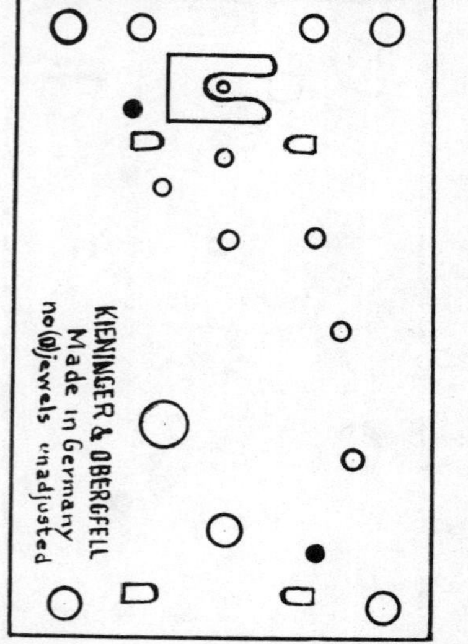

Plate 1381 4-Ball Pendulum Miniature Clock
USE .0023" (.058mm) HOROLOVAR
Unit 5E (14 x 30)
See Appendix 77, 101

Kieninger & Obergfell c 1965

KIENINGER & OBERGFELL

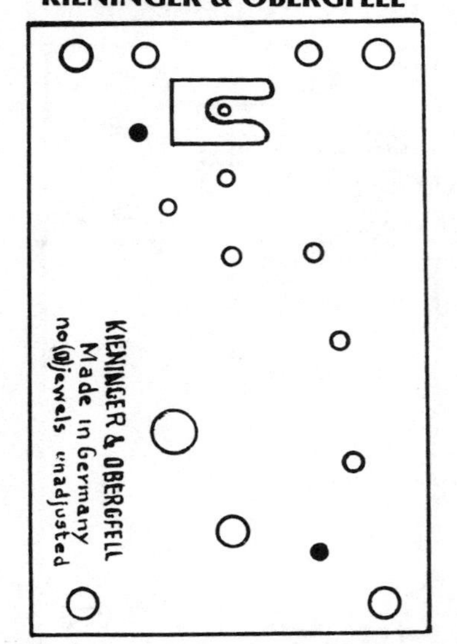

Plate 1381A 4-Ball Pendulum Miniature Clock
USE .0023" (.058mm) HOROLOVAR
Unit 5E (14 x 30)
See Appendix 77

Kieninger & Obergfell c 1960

KIENINGER & OBERGFELL

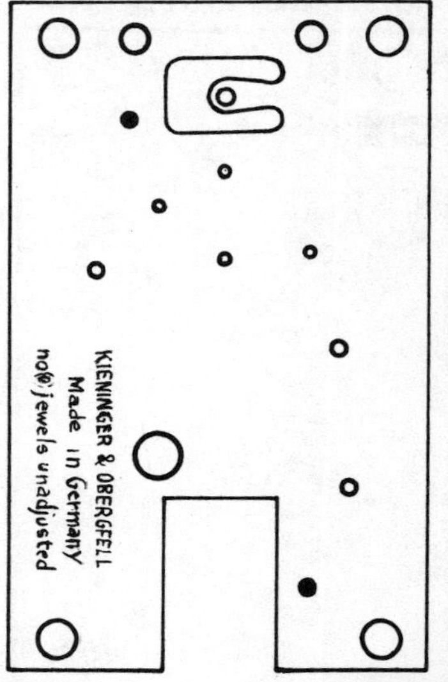

Plate 1382 4-Ball Pendulum Midget Clock
USE .0022"— (.056mm—) HOROLOVAR
Unit 5F (14 x 30)
See Appendix 16, 77

Kieninger & Obergfell c 1976

KIENINGER & OBERGFELL

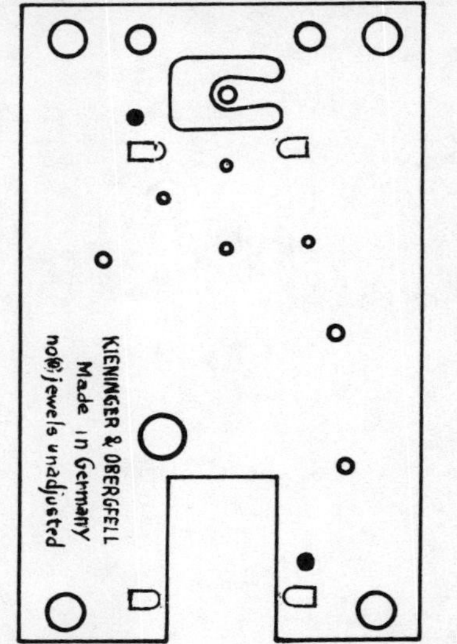

Plate 1382A 4-Ball Pendulum
Midget Clock

USE .0022″— (.056mm—) HOROLOVAR
Unit 5F (14 x 30)
See Appendix 16, 77, 101

Kienzle Clock Factories c 1908

KIENZLE CLOCK FACTORIES

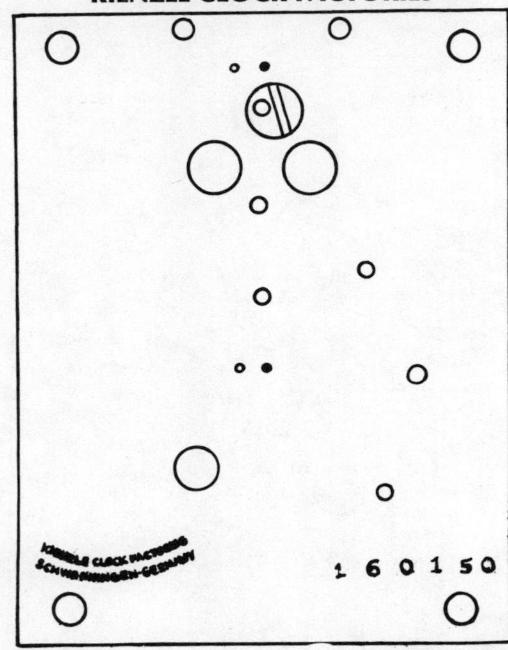

Plate 1383 4-Ball Pendulum
USE .0036″* (.091mm*) HOROLOVAR
(18 x 38)

Badische Uhrenfabrik c 1898

KIENZLE CLOCK FACTORIES

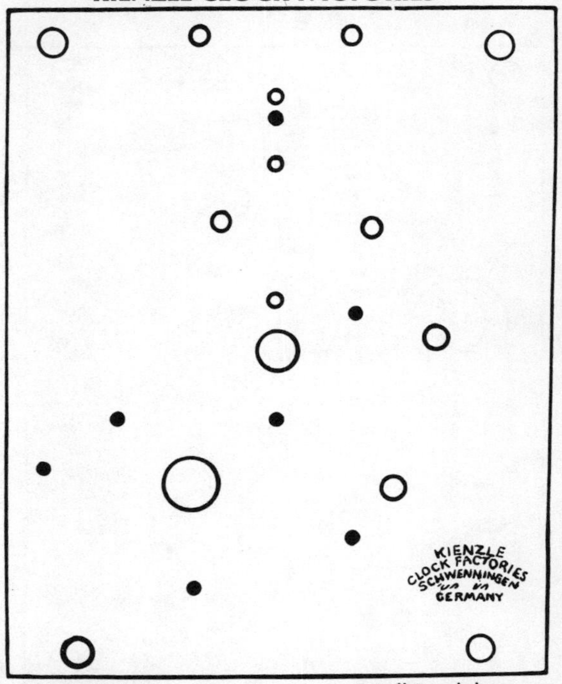

Plate 1384 4-Ball Pendulum
Pin Pallet Escapement
Lantern Pinions

USE .0035″* (.089mm*) HOROLOVAR
See Appendix 8, 10 (20 x 38)

Badische Uhrenfabrik c 1908

KIENZLE CLOCK FACTORIES

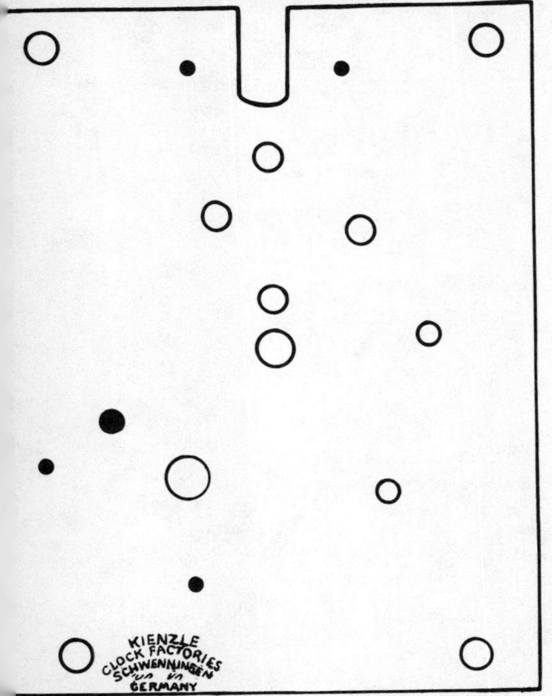

Plate 1385 4-Ball Pendulum
Pin Pallet Escapement
Lantern Pinions

USE .0035″* (.089mm*) HOROLOVAR
See Appendix 8, 63 (20 x 38)

Badische Uhrenfabrik c 1908

KIENZLE CLOCK FACTORIES

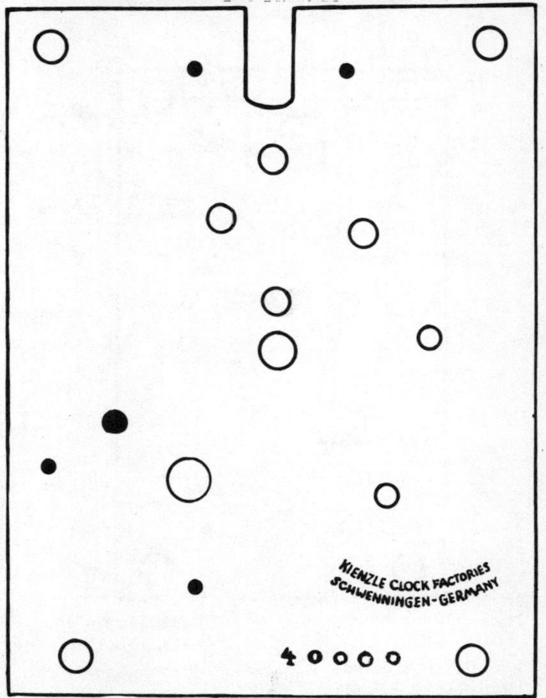

Plate 1387 4-Ball Pendulum
Pin Pallet Escapement
Lantern Pinions

USE .0035″* (.089mm*) HOROLOVAR
See Appendix 8, 63, 96 (20 x 38)

Kieninger & Obergfell c 1930

KO

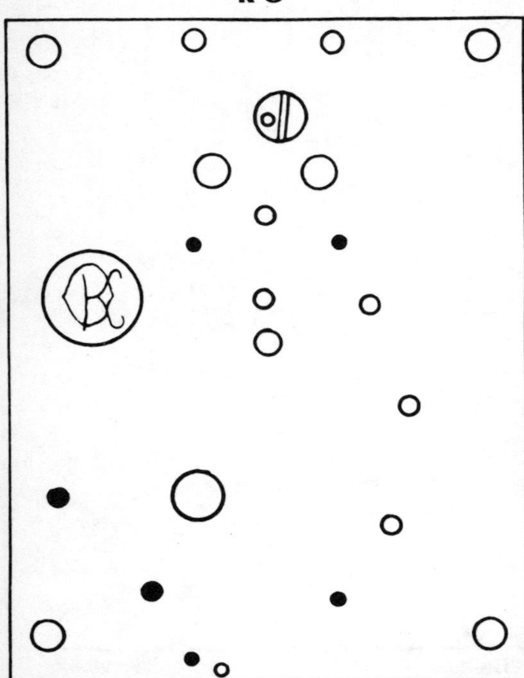

Plate 1388 4-Ball Pendulum
USE .0032″ (.081mm) HOROLOVAR
Unit 1 (19 x 38)
See Appendix 76

Kieninger & Obergfell c 1935

KO

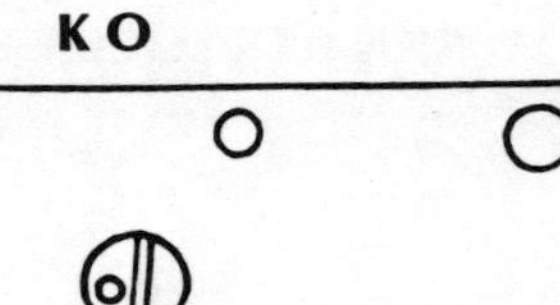

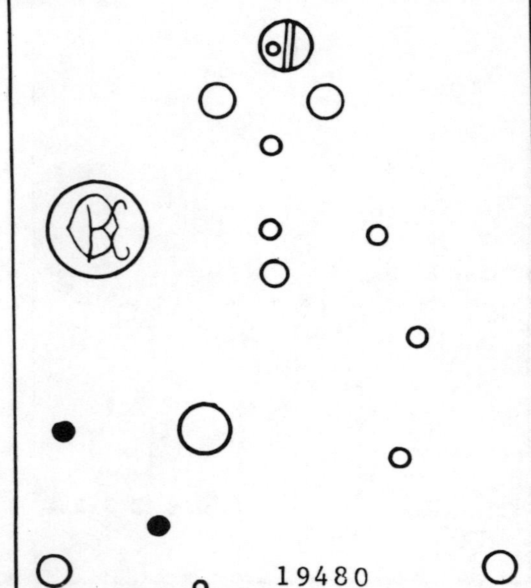

Plate 1388A 4-Ball Pendulum
USE .0032″ (.081mm) HOROLOVAR
Unit 1 (19 x 38)
See Appendix 76

Kieninger & Obergfell c 1950

KO

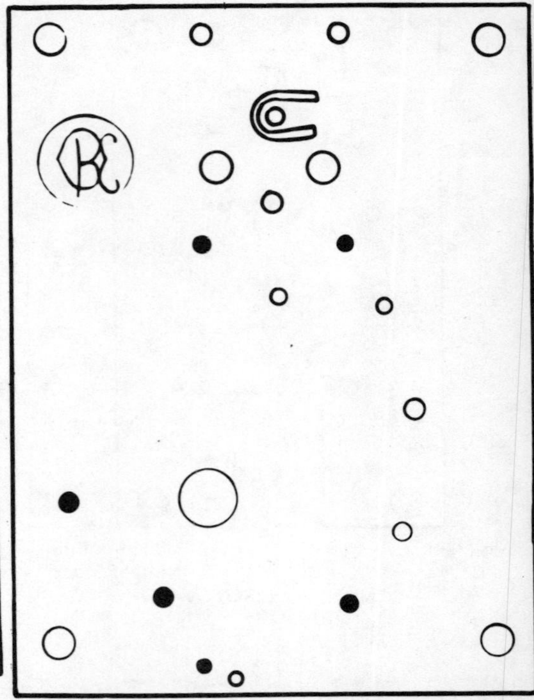

Plate 1389 4-Ball Pendulum
USE .0032″ (.081mm) HOROLOVAR
Units 1, 3A (19 x 38)
See Appendix 76

Kieninger & Obergfell c 1952

KO

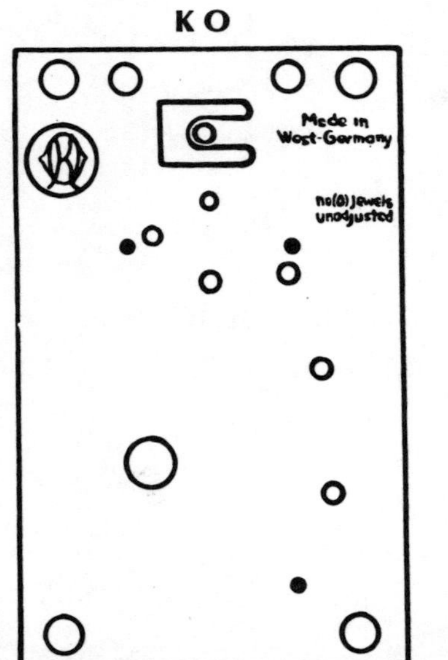

Plate 1390 4-Ball Pendulum
Miniature Clock
USE .0023″ (.058mm) HOROLOVAR
Units 5A, 5B, 5C, 5D, 5E (14 x 30)
See Appendix 77

Konrad Mauch c 1954

KOMA

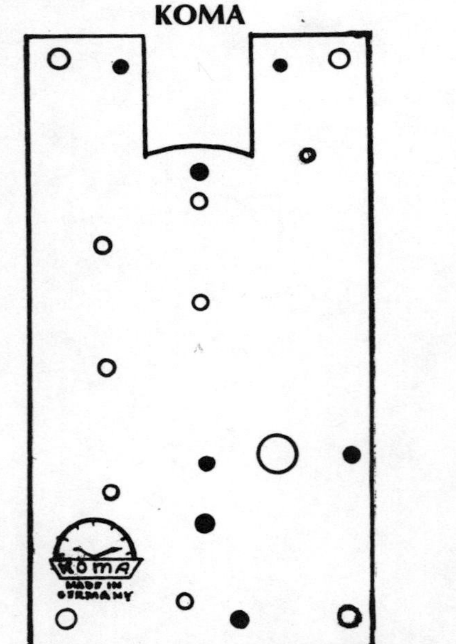

Plate 1392 4-Ball Pendulum
Miniature Clock
USE .0032″ (.081mm) HOROLOVAR
Unit 14A
USE .003″ (.076mm) HOROLOVAR
Unit 14B Original (13 x 32)
See Appendix 52, 63, 64, 85 Revised (15 x 32)

Konrad Mauch c 1950

KOMA

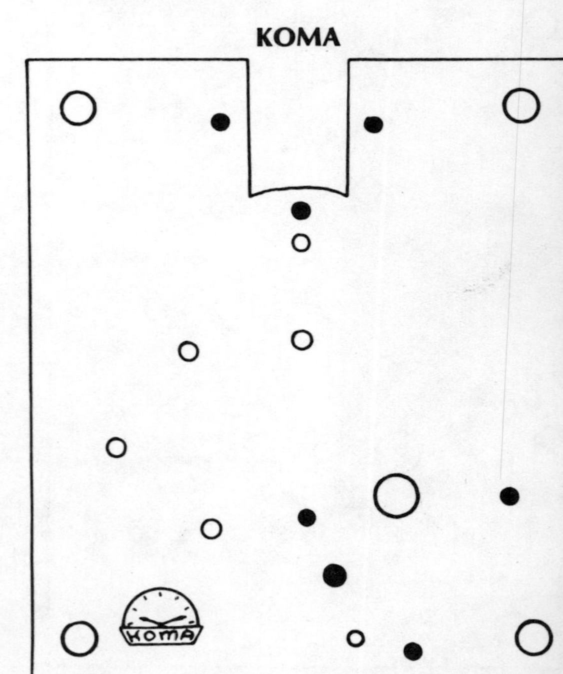

Plate 1393 4-Ball Pendulum
USE .0035″ (.089mm) HOROLOVAR
Units 13A, 13B (20 x 38)
See Appendix 63, 64, 84

Konrad Mauch c 1951

KOMA

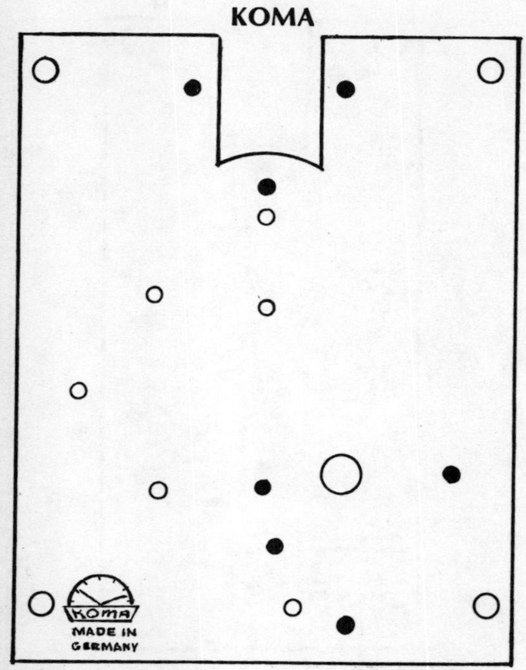

Plate 1393A 4-Ball Pendulum
USE .0035" (.089mm) HOROLOVAR
Units 13A, 13B (20 x 38)
See Appendix 63, 64, 84

Konrad Mauch c 1957

KOMA

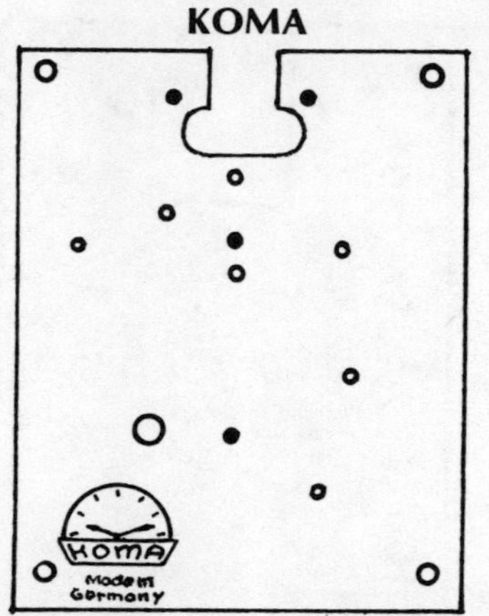

Plate 1393B 3-Ball Pendulum
Midget Clock
Pin Pallet Escapement
USE .0022" (.056mm) HOROLOVAR
Units 38, 38A Original (12 x 25)
See Appendix 53, 63, 64, 86 Revised (14 x 25)

Konrad Mauch c 1959

KOMA

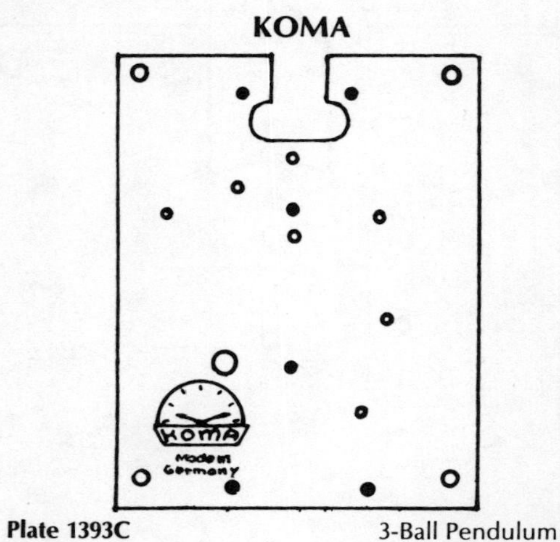

Plate 1393C 3-Ball Pendulum
Midget Clock
Pin Pallet Escapement
USE .0022" (.056mm) HOROLOVAR
Units 38, 38A Original (12 x 25)
See Appendix 53, 63, 64, 86 Revised (14 x 25)

Konrad Mauch c 1960

KOMA

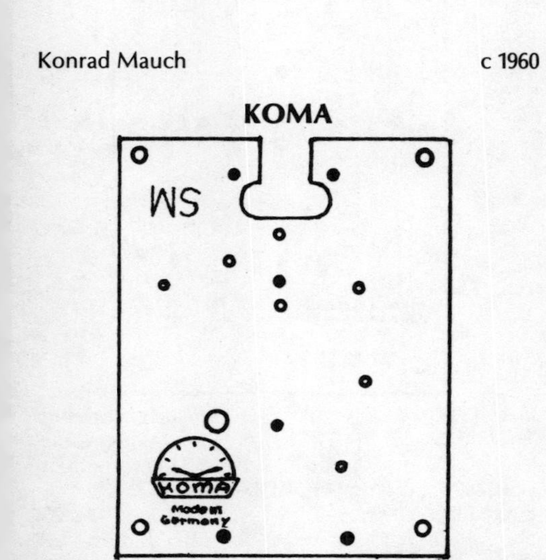

Plate 1393D 3-Ball Pendulum
Midget Clock
Pin Pallet Escapement
USE .0022" (.056mm) HOROLOVAR
Units 38, 38A Original (12 x 25)
See Appendix 53, 63, 64, 86 Revised (14 x 25)

Konrad Mauch c 1957

KOMA

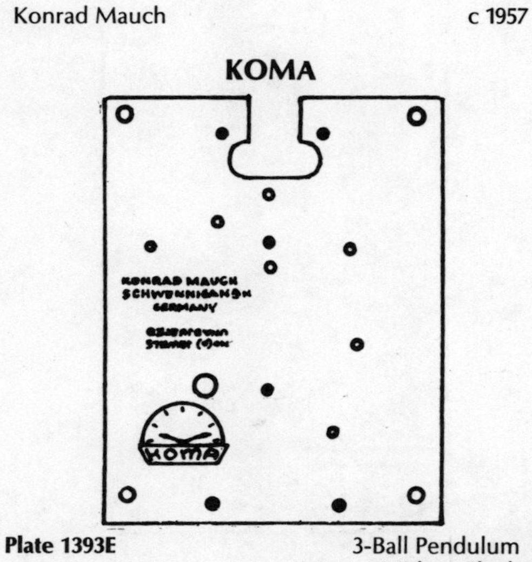

Plate 1393E 3-Ball Pendulum
Midget Clock
Pin Pallet Escapement
USE .0022" (.056mm) HOROLOVAR
Units 38, 38A Original (12 x 25)
See Appendix 53, 63, 64, 86 Revised (14 x 25)

Konrad Mauch c 1950

KONRAD MAUCH

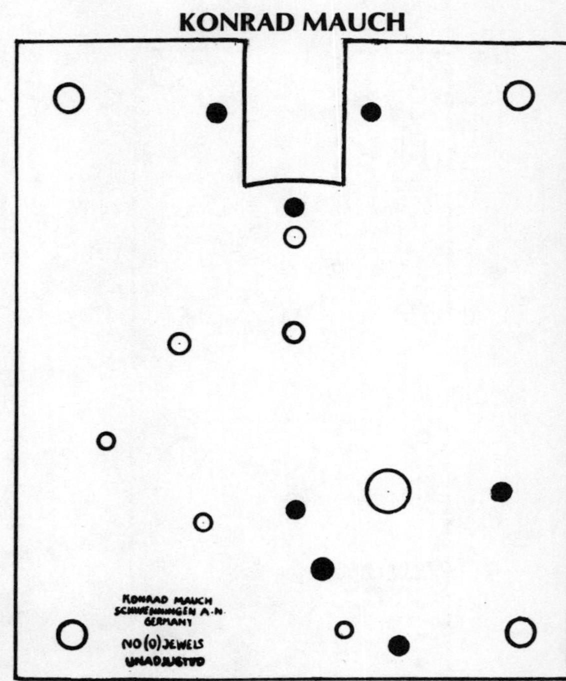

Plate 1394 4-Ball Pendulum
USE .0035" (.089mm) HOROLOVAR
Units 13A, 13B (20 x 38)
See Appendix 63, 64, 84

Konrad Mauch c 1951

KONRAD MAUCH

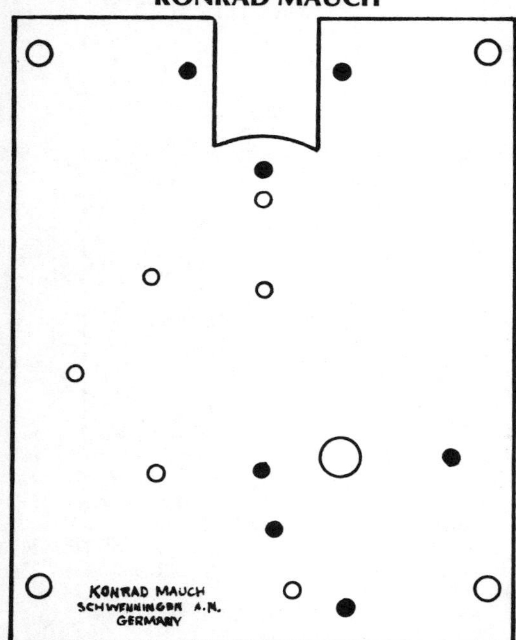

Plate 1394A 4-Ball Pendulum
USE .0035" (.089mm) HOROLOVAR
Units 13A, 13B (20 x 38)
See Appendix 63, 64, 84

Konrad Mauch c 1951

KONRAD MAUCH

KONRAD MAUCH
SCHWENNINGEN A.N.
GERMANY
NO(0) JEWELS
UNADJUSTED

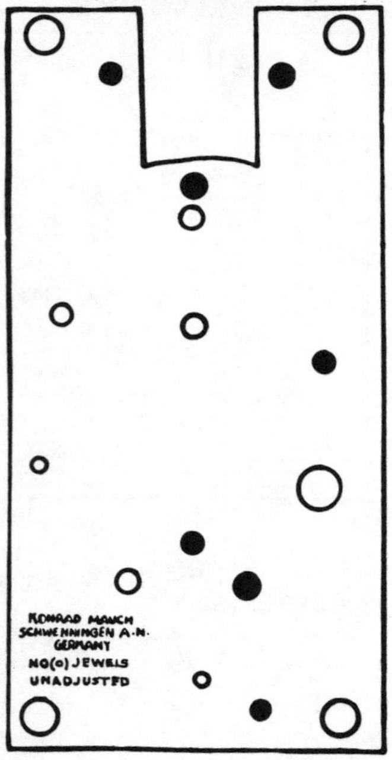

Plate 1395 4-Ball Pendulum
USE .0035" (.089mm) HOROLOVAR
Units 13A, 13B, 13C (20 x 38)
See Appendix 63, 64, 84

Konrad Mauch c 1951

KONRAD MAUCH

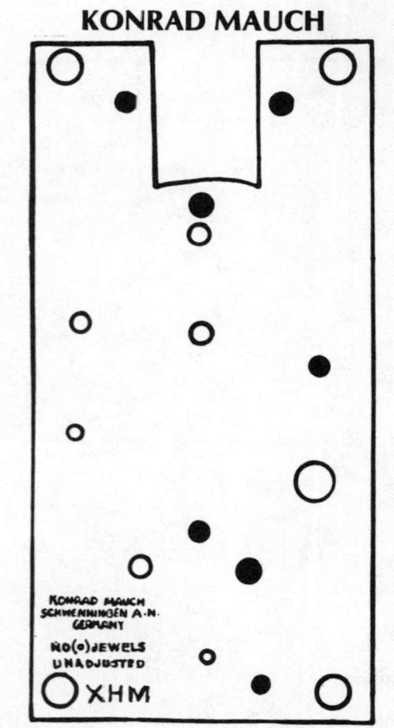

Plate 1395A 4-Ball Pendulum
USE .0035" (.089mm) HOROLOVAR
Units 13A, 13B, 13C (20 x 38)
See Appendix 63, 64, 84

Konrad Mauch c 1952

KONRAD MAUCH

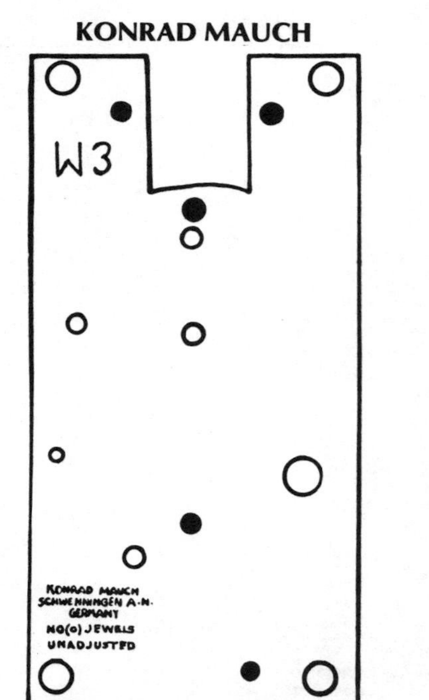

Plate 1395B 4-Ball Pendulum
USE .0035" (.089mm) HOROLOVAR
Units 13A, 13B, 13C (20 x 38)
See Appendix 63, 64, 84

Konrad Mauch c 1953

KONRAD MAUCH

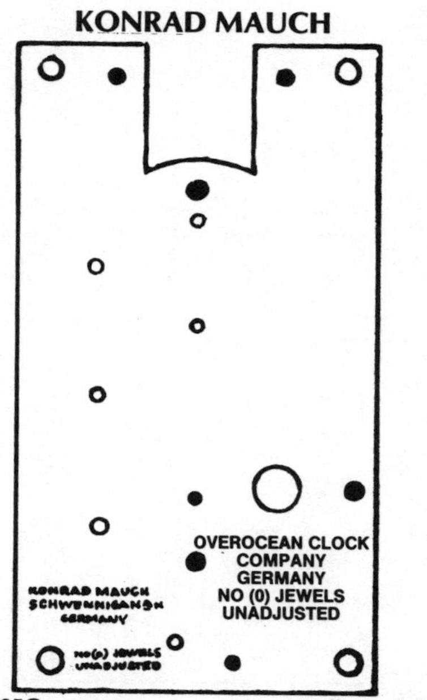

Plate 1395C 4-Ball Pendulum
Miniature Clock
USE .0032" (.081mm) HOROLOVAR
Unit 14A
USE .003" (.076mm) HOROLOVAR
Unit 14B Original (13 x 32)
See Appendix 52, 63, 64, 85 Revised (15 x 32)

Konrad Mauch c 1957

KONRAD MAUCH

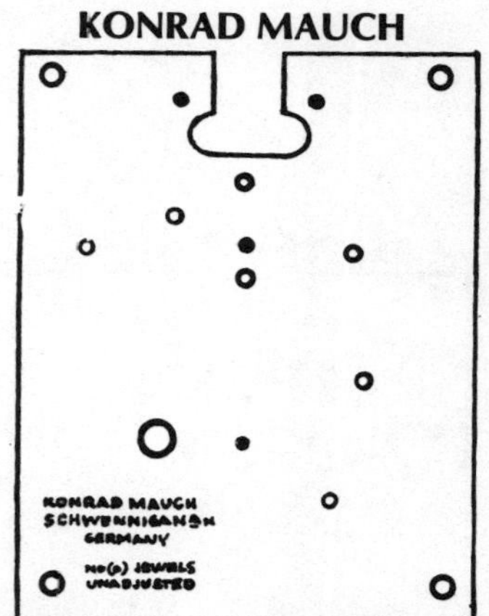

Plate 1396 3-Ball Pendulum
Midget Clock
Pin Pallet Escapement
USE .0022" (.056mm) HOROLOVAR
Units 38, 38A Original (12 x 25)
See Appendix 53, 63, 64, 86 Revised (14 x 25)

Konrad Mauch c 1959

KONRAD MAUCH

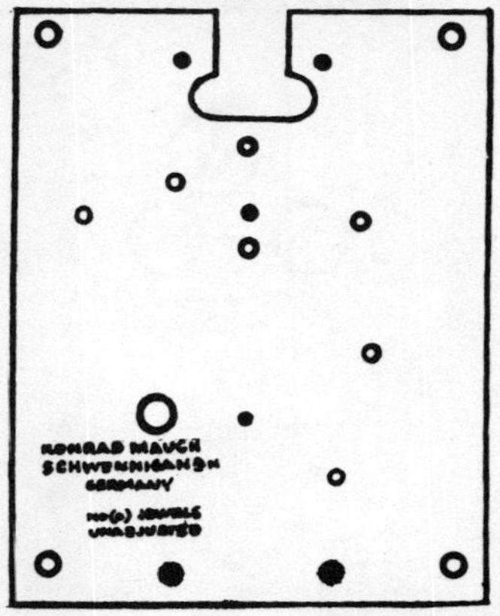

Plate 1396A 3-Ball Pendulum Midget Clock Pin Pallet Escapement

USE .0022" (.056mm) HOROLOVAR
Units 38, 38A Original (12 x 25)
See Appendix 53, 63, 64, 86 Revised (14 x 25)

Konrad Mauch c 1954

KONRAD MAUCH

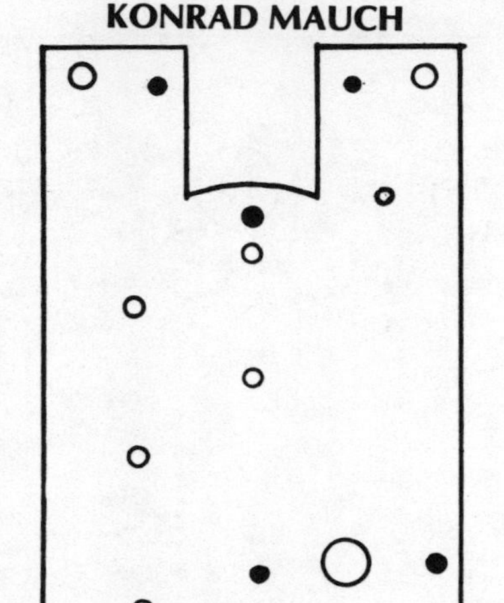

Plate 1398 4-Ball Pendulum Miniature Clock

USE .0032" (.081mm) HOROLOVAR
Unit 14A
USE .003" (.076mm) HOROLOVAR
Unit 14B Original (13 x 32)
See Appendix 52, 63, 64, 85 Revised (15 x 32)

Konrad Mauch c 1956

KONRAD MAUCH

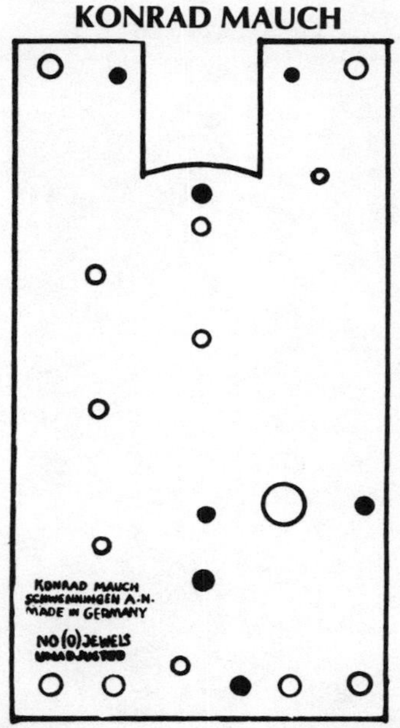

Plate 1398A 4-Ball Pendulum Miniature Clock

USE .0032" (.081mm) HOROLOVAR
Unit 14A
USE .003" (.076mm) HOROLOVAR
Unit 14B Original (13 x 32)
See Appendix 52, 63, 64, 85 Revised (15 x 32)

Konrad Mauch c 1954

KONRAD MAUCH

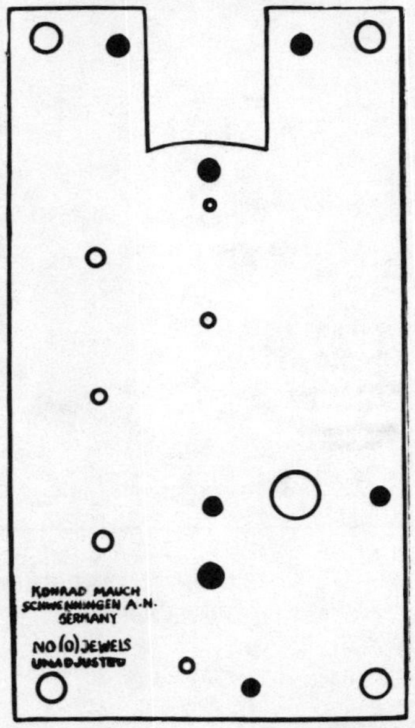

Plate 1399 4-Ball Pendulum Miniature Clock

USE .0032" (.081mm) HOROLOVAR
Unit 14A
USE .003" (.076mm) HOROLOVAR
Unit 14B Original (13 x 32)
See Appendix 52, 63, 64, 85 Revised (15 x 32)

Konrad Mauch c 1954

KONRAD MAUCH

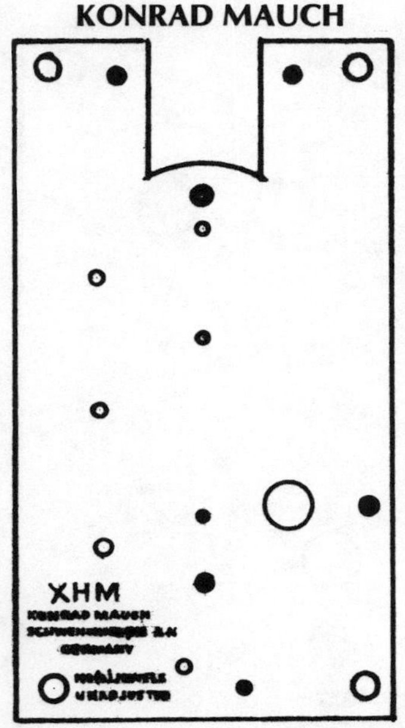

Plate 1399A 4-Ball Pendulum Miniature Clock

USE .0032" (.081mm) HOROLOVAR
Unit 14A
USE .003" (.076mm) HOROLOVAR
Unit 14B Original (13 x 32)
See Appendix 52, 63, 64, 85 Revised (15 x 32)

Kern & Sohne c 1949

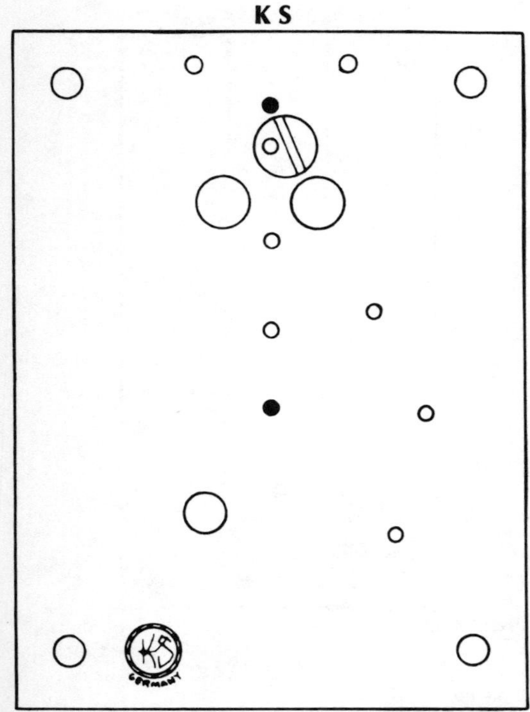

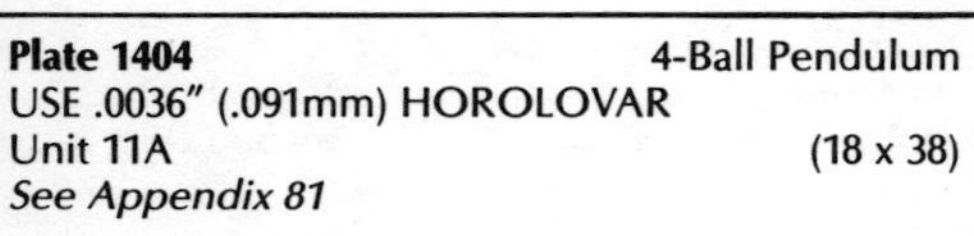

Plate 1404 4-Ball Pendulum
USE .0036″ (.091mm) HOROLOVAR
Unit 11A (18 x 38)
See Appendix 81

Kern & Sohne c 1952

K S

GERMANY

Plate 1404A 4-Ball Pendulum
USE .0036″ (.091mm) HOROLOVAR
Unit 11B (18 x 38)
See Appendix 81

Kern & Sohne c 1953

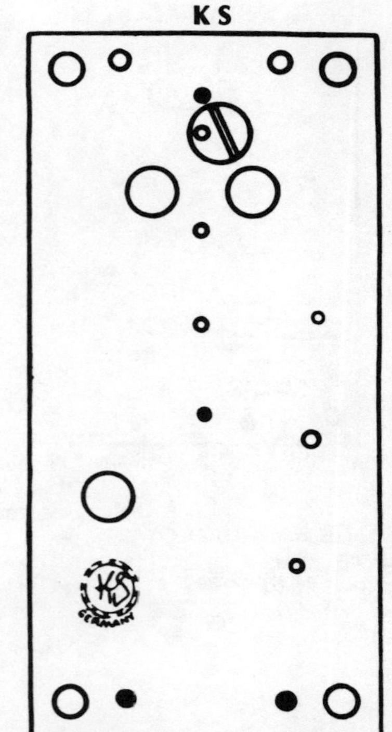

Plate 1405 4-Ball Pendulum
USE .0036″ (.091mm) HOROLOVAR
Unit 11B (18 x 38)
See Appendix 81

Kern & Sohne c 1961

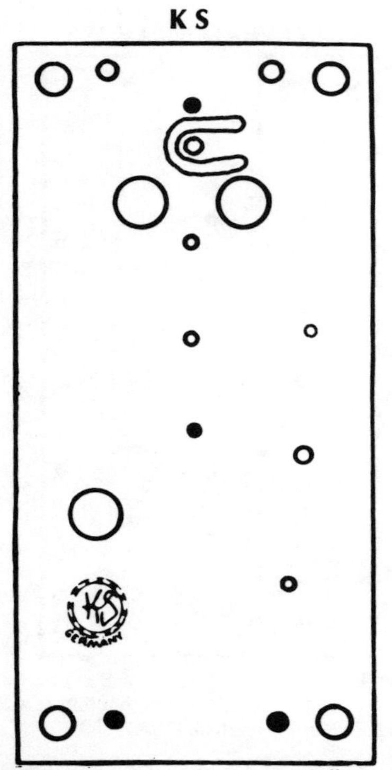

Plate 1405A 4-Ball Pendulum
USE .0036″ (.091mm) HOROLOVAR
Unit 11B (18 x 38)
See Appendix 81

Kern & Sohne c 1954

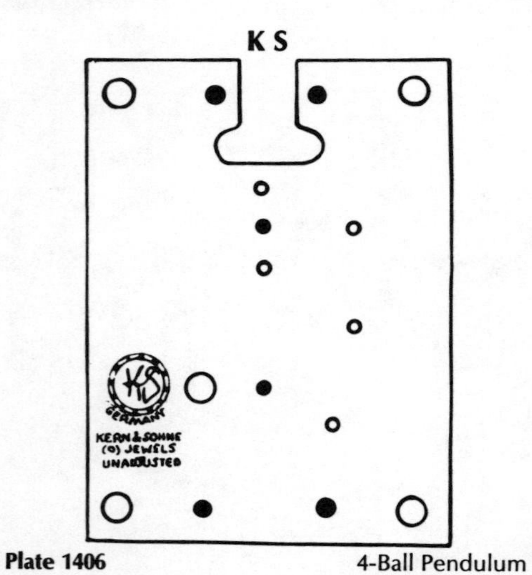

Plate 1406 4-Ball Pendulum Miniature Clock
USE .002″ (.051mm) HOROLOVAR
Units 12A, 12B, 12C
USE .0019″ (.048mm) HOROLOVAR
Unit 12D (12 x 25)
See Appendix 63, 82

Kern & Sohne c 1959

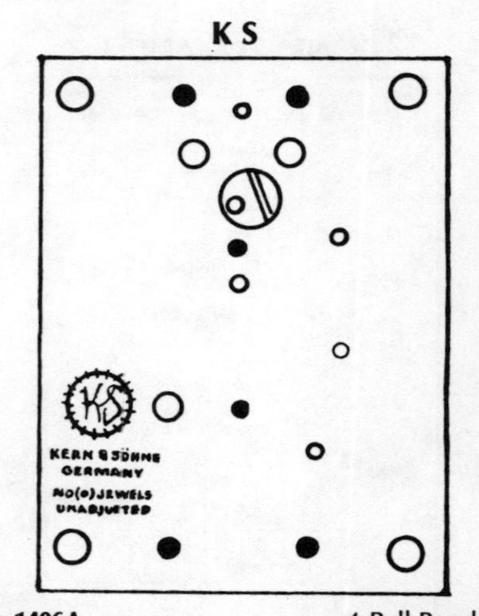

Plate 1406A 4-Ball Pendulum Miniature Clock
USE .002″ (.051mm) HOROLOVAR
Units 12A, 12B, 12C
USE .0019″ (.048mm) HOROLOVAR
Unit 12D (12 x 25)
See Appendix 25, 82

Kern & Sohne c 1960

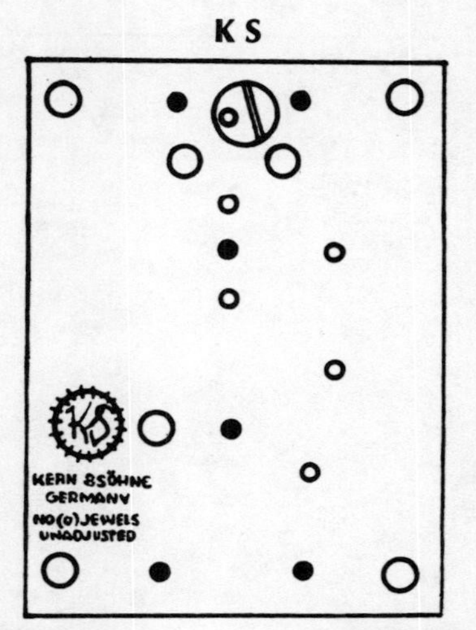

Plate 1406B 4-Ball Pendulum Miniature Clock

USE .002" (.051mm) HOROLOVAR
Units 12A, 12B, 12C
USE .0019" (.048mm) HOROLOVAR
Unit 12D (12 x 25)
See Appendix 82

Kern & Sohne c 1958

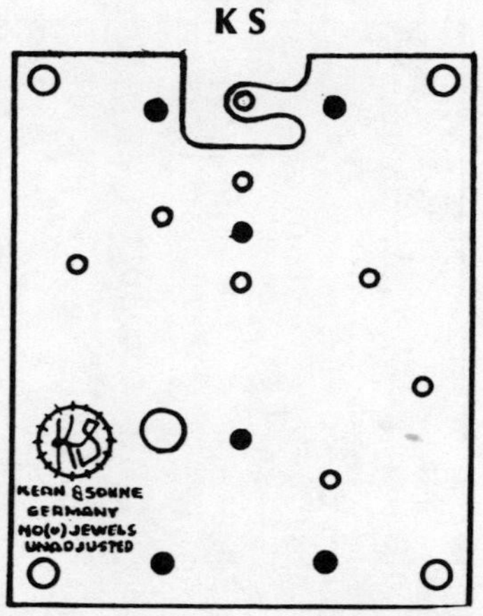

Plate 1406C 4-Ball Pendulum Midget Clock Pin Pallet Escapement

USE .0023" (.058mm) HOROLOVAR
Unit 12E (14 x 25)
See Appendix 83

Kern & Sohne c 1960

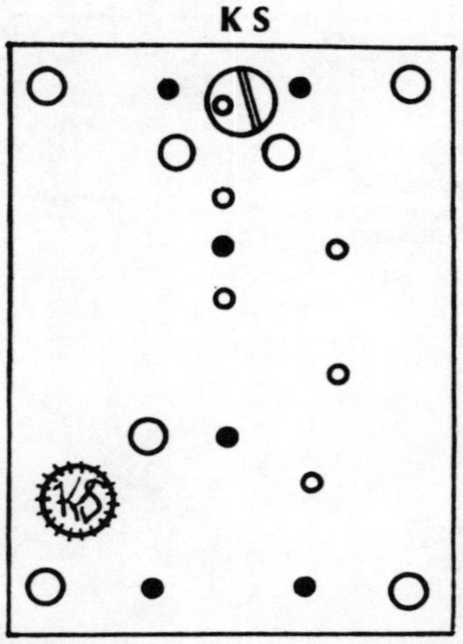

Plate 1406D 4-Ball Pendulum Miniature Clock

USE .002" (.051mm) HOROLOVAR
Units 12A, 12B, 12C
USE .0019" (.048mm) HOROLOVAR
Unit 12D (12 x 25)
See Appendix 82

Kern & Sohne c 1958

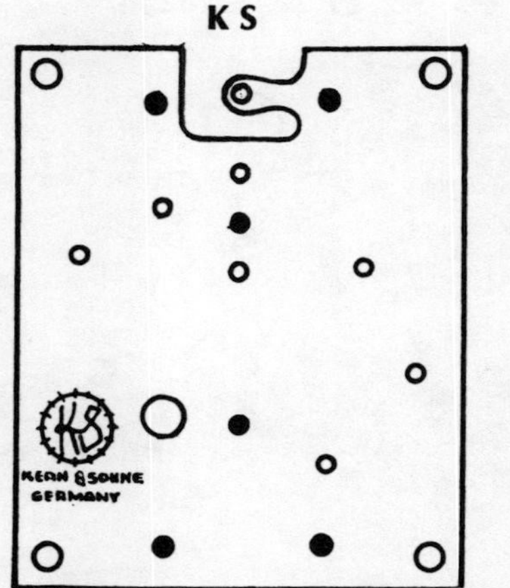

Plate 1406E 4-Ball Pendulum Midget Clock Pin Pallet Escapement

USE .0023" (.058mm) HOROLOVAR
Unit 12E (14 x 25)
See Appendix 83

Kern & Sohne c 1961

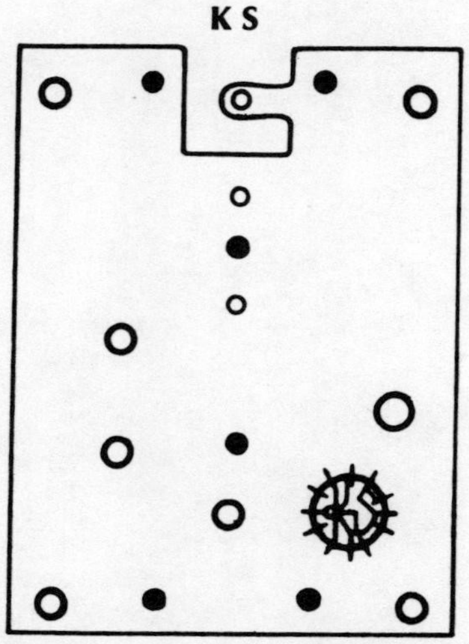

Plate 1406F 4-Ball Pendulum Miniature Clock

USE .0023" (.058mm) HOROLOVAR
Unit 12F (14 x 30)
See Appendix 82

Kern & Sohne c 1960

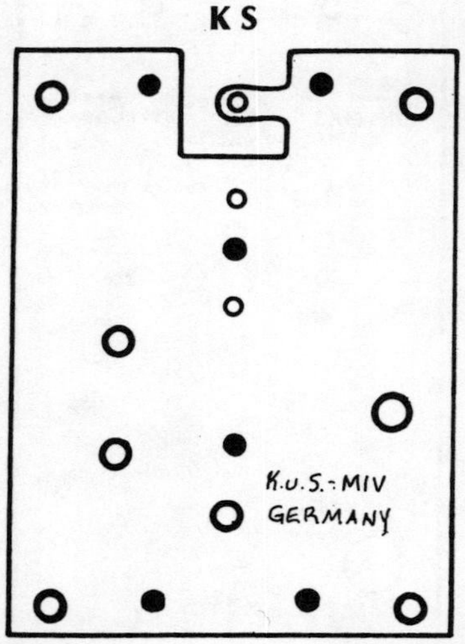

Plate 1406G 4-Ball Pendulum Miniature Clock

USE .0023" (.058mm) HOROLOVAR
Unit 12F (14 x 30)
See Appendix 82

Kieninger & Obergfell c 1970

KUNDO

Made in
West Germany
Kundo
KIENINGER &
OBERGFELL
no(o)jewels
unadjusted

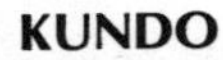

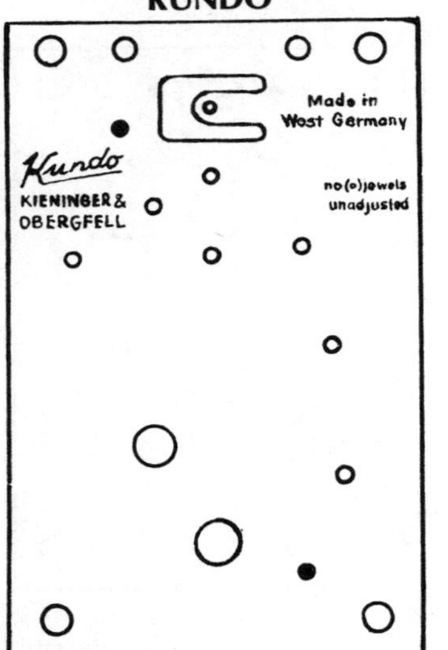

Plate 1406H 4-Ball Pendulum Miniature Clock

USE .0023" (.058mm) HOROLOVAR

Unit 5E (14 x 30)

See Appendix 77

Kieninger & Obergfell c 1970

KUNDO

KIENINGER & OBERGFELL
Made in Germany
no(o) jewels unadjusted

Plate 1406J 4-Ball Pendulum Miniature Clock

USE .0023" (.058mm) HOROLOVAR

Unit 5E (14 x 30)

See Appendix 77

Kieninger & Obergfell c 1965

KUNDO

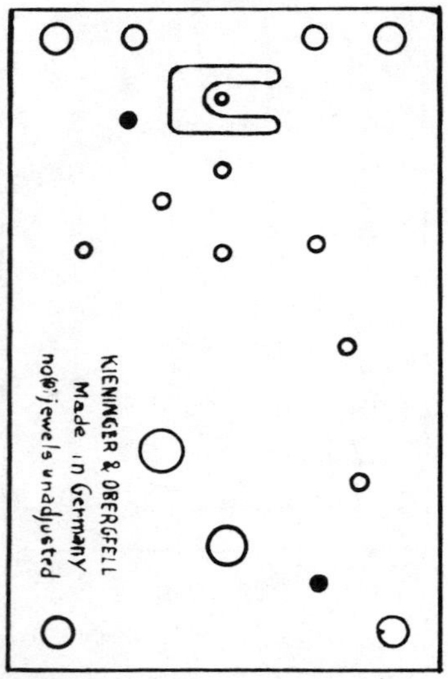

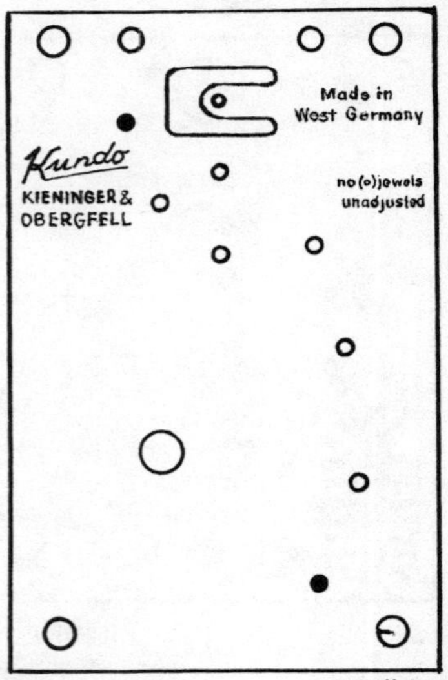

Made in
West Germany
Kundo
KIENINGER &
OBERGFELL
no(o)jewels
unadjusted

Plate 1407A 4-Ball Pendulum Miniature/Midget Clock

USE .0023" (.058mm) HOROLOVAR

Units 5E, 5F (14 x 30)

See Appendix 77

Kieninger & Obergfell c 1951

KUNDO

Kundo
KIENINGER &
OBERGFELL
Made in
West-Germany
no(o) jewels
unadjusted

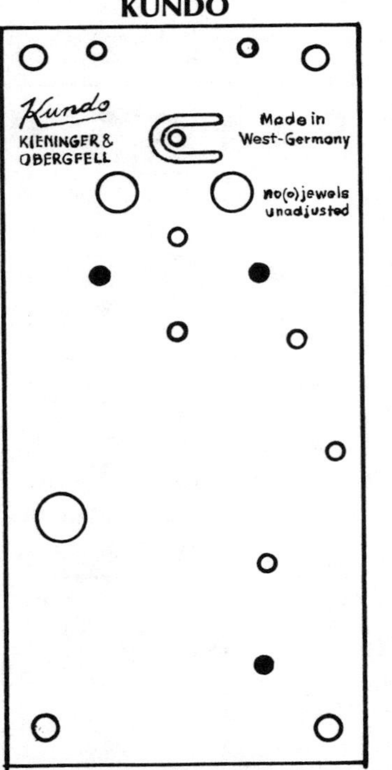

Plate 1407B 4-Ball Pendulum

USE .0032" (.081mm) HOROLOVAR

Units 3A, 3B, 3C (19 x 38)

See Appendix 76

Kieninger & Obergfell c 1952

KUNDO

Made in
West-Germany
Kundo
KIENINGER &
OBERGFELL
no(o)jewels
unadjusted

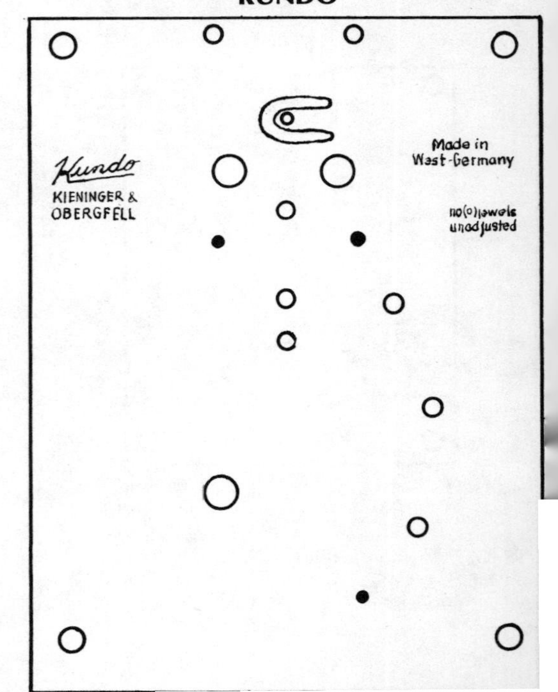

Plate 1407D 4-Ball Pendulum

USE .0032" (.081mm) HOROLOVAR

Unit 3C (19 x 38)

See Appendix 36, 76

Kieninger & Obergfell c 1965

KUNDO

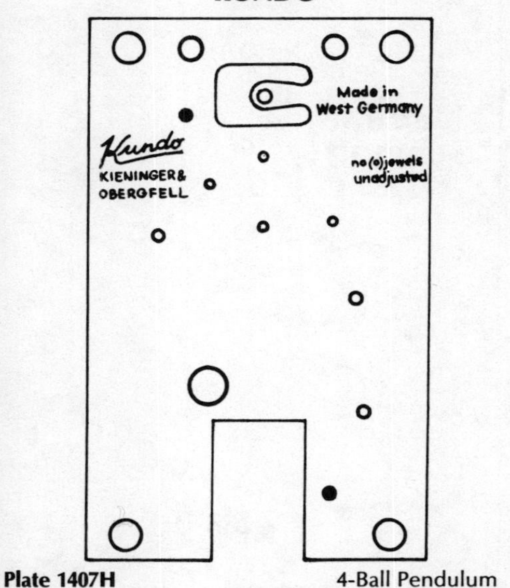

Plate 1407H 4-Ball Pendulum
Midget Clock
USE .0022"— (.056mm—) HOROLOVAR
Unit 5F (14 x 30)
See Appendix 16, 77

Uhrenfabrik Herr **c 1952**

L. HERR SOHNE

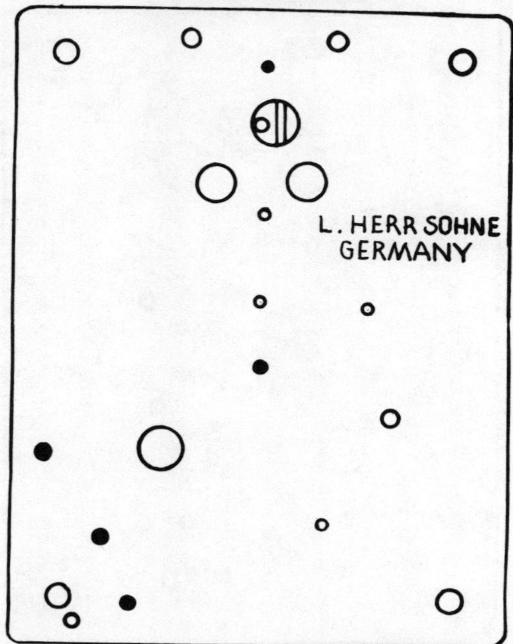

Plate 1408 4-Ball Pendulum
USE .004" (.102mm) HOROLOVAR
Unit 27A (19 x 36)
See Appendix 130

Kieninger & Obergfell c 1930

L. LURIA & SON

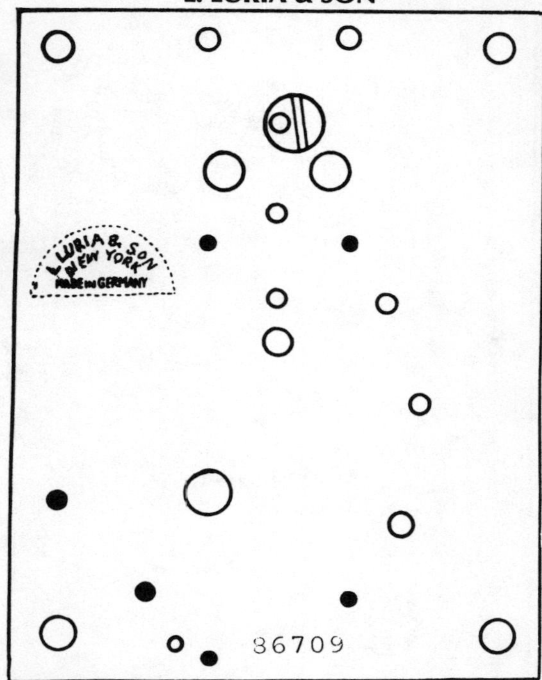

Plate 1408D 4-Ball Pendulum
USE .0032" (.081mm) HOROLOVAR
Unit 1 (19 x 38)
See Appendix 76

Jahresuhrenfabrik c 1910

LUDW. HAKENJOS

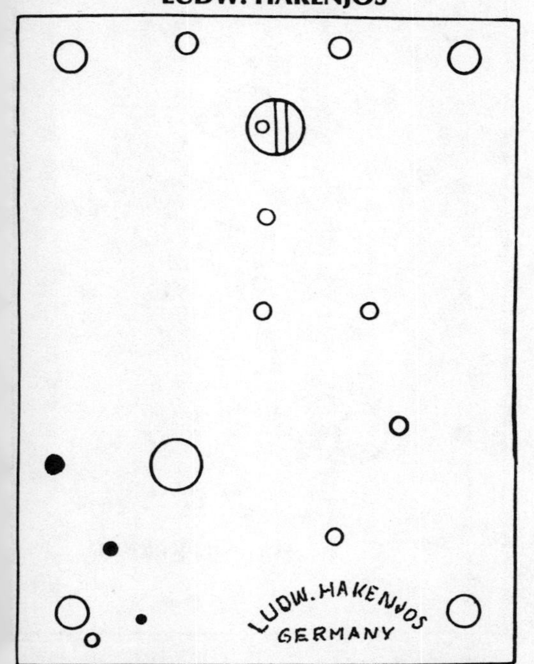

Plate 1408F 4-Ball Pendulum
USE .004" (.102mm) HOROLOVAR
(19 x 36)

Konrad Mauch c 1950

L. von der Burg G.m.b.H.

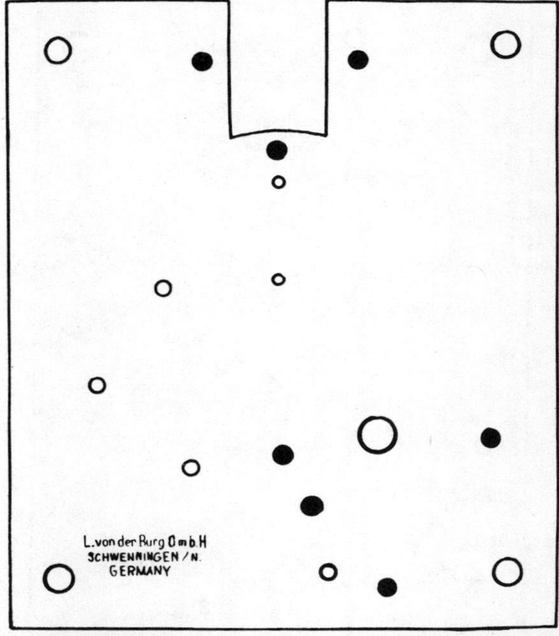

Plate 1409 4-Ball Pendulum
USE .0035" (.089mm) HOROLOVAR
Units 13A, 13B (20 x 38)
See Appendix 63, 64, 84

Kern & Sohne c 1949

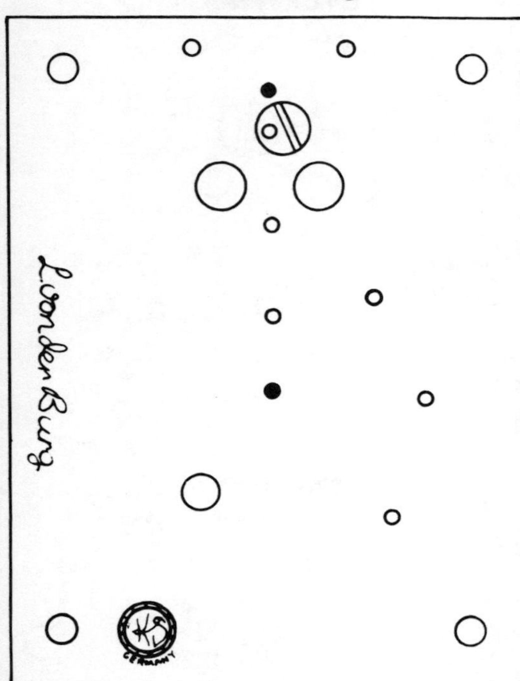

Plate 1411 4-Ball Pendulum
USE .0036" (.091mm) HOROLOVAR
Unit 11A (18 x 38)
See Appendix 81

Phillipp Haas c 1904

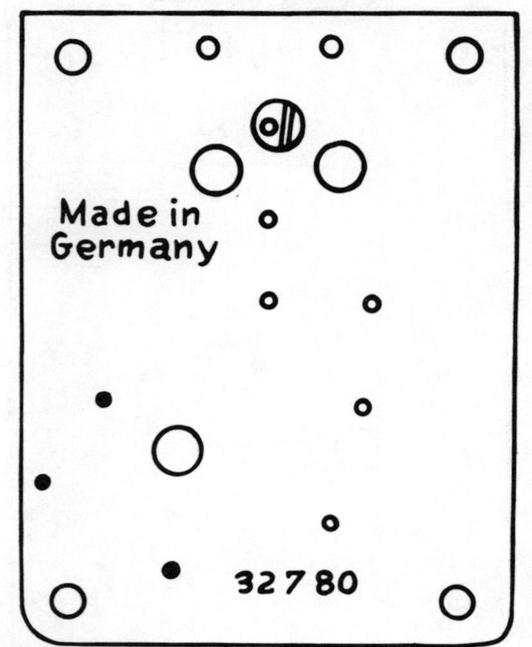

Plate 1415 Chronometer Balance
"Temperature Compensating" Pendulum
USE .0045"—* (.114mm—*) HOROLOVAR
See Appendix 89, 118 (19 x 36)

Phillipp Haas c 1905

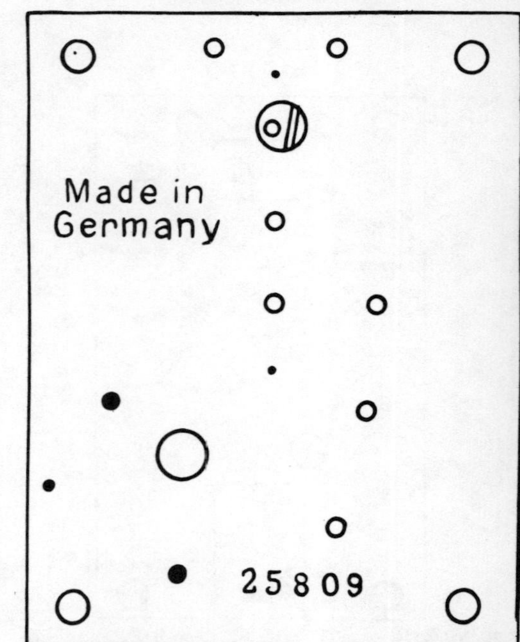

Plate 1419 Disc Pendulum
4-Ball Pendulum
USE .0045"* (.114mm*) HOROLOVAR
See Appendix 89, 118 (19 x 36)

Kienzle Clock Factories c 1905

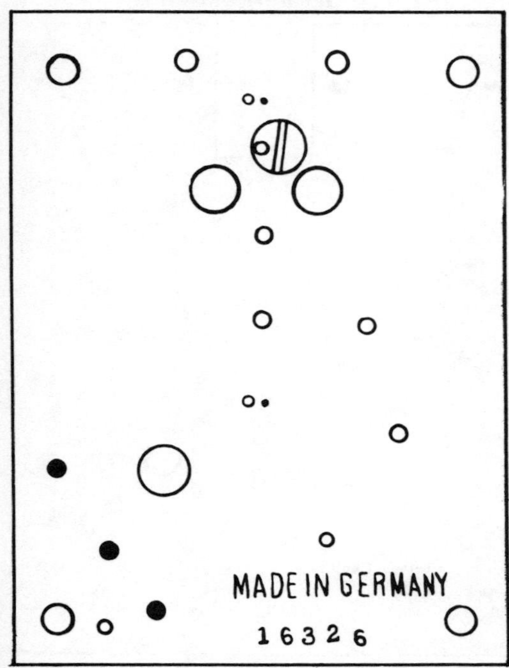

Plate 1423 Disc Pendulum
USE .0038" (.097mm) HOROLOVAR
(19 x 36)

Phillipp Haas c 1907

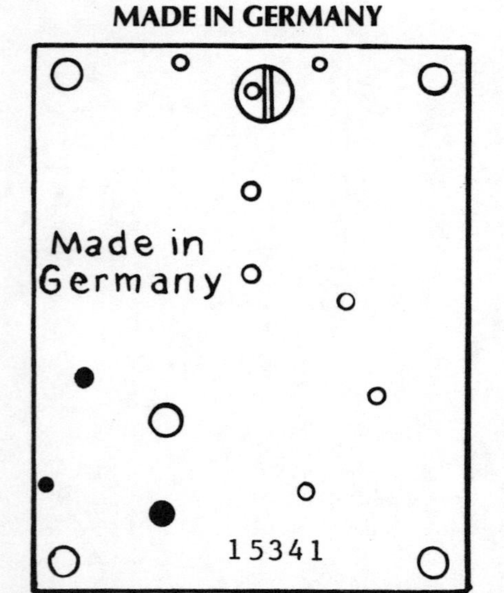

Plate 1425 Standard Disc Pendulum
Miniature Movement
Miniature Pendulum
USE .0038"* (.097mm*) HOROLOVAR
See Appendix 118

Kienzle Clock Factories c 1905

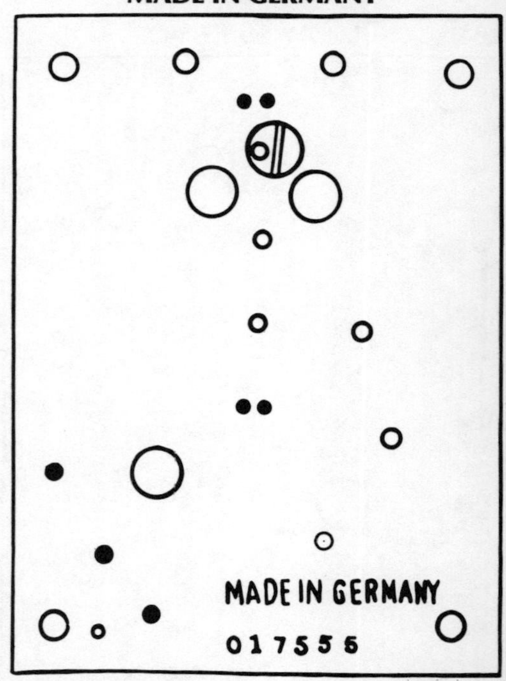

Plate 1427 Disc Pendulum
4-Ball Pendulum
USE .0038" (.097mm) HOROLOVAR
(19 x 36)

Kienzle Clock Factories c 1908

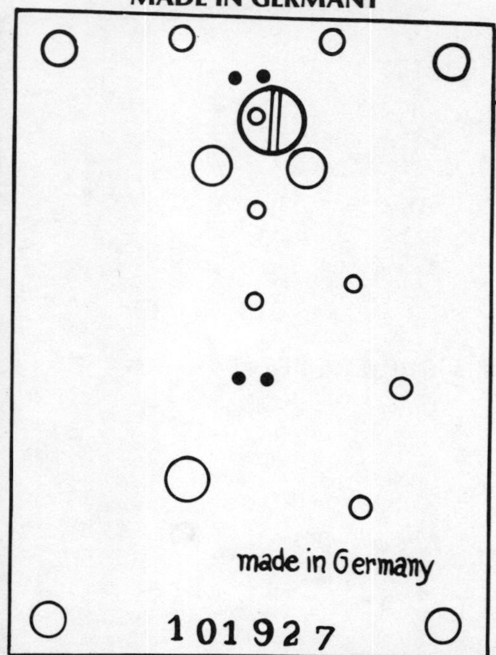

Plate 1431 Disc Pendulum
4-Ball Pendulum
USE .0038"* (.097mm*) HOROLOVAR
See Appendix 126 (18 x 38)

Kienzle Clock Factories c 1908

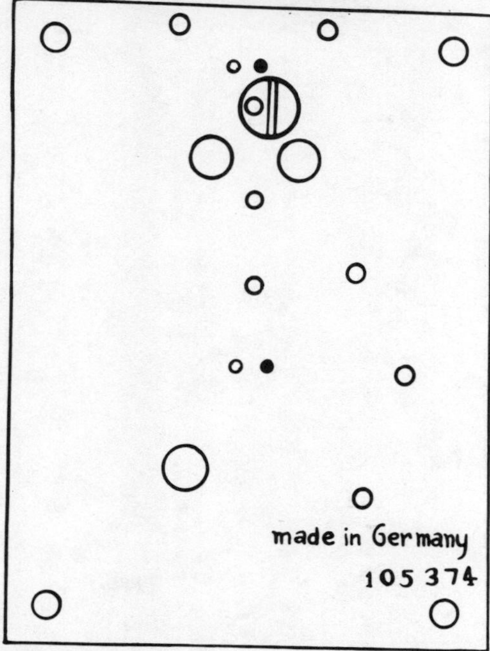

Plate 1435 Disc Pendulum
4-Ball Pendulum
USE .0038"* (.097mm*) HOROLOVAR
See Appendix 41, 126 (18 x 38)

Kienzle Clock Factories c 1909

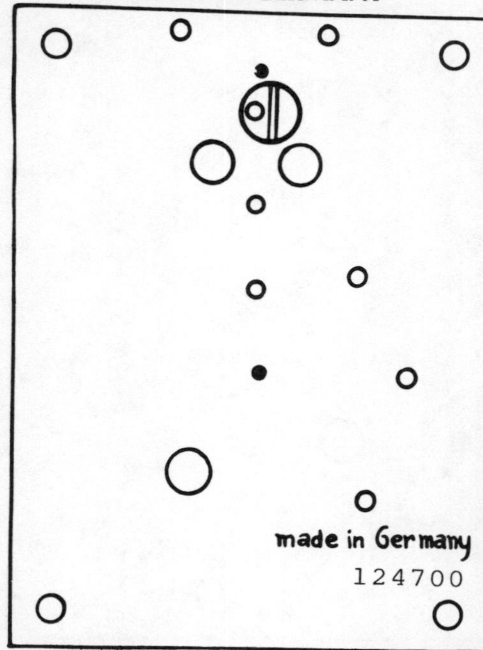

Plate 1436 Disc Pendulum
4-Ball Pendulum
USE .0038"* (.097mm*) HOROLOVAR
See Appendix 126 (18 x 38)

Manufacturer Not Known c 1900

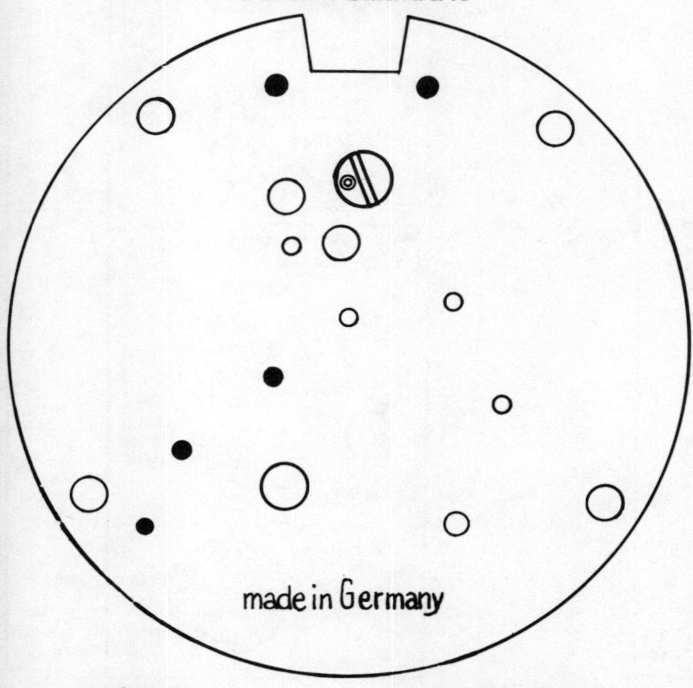

Plate 1437 Disc Pendulum
Offset Anchor
USE .004"* (.102mm*) HOROLOVAR
See Appendix 73 (19 x 36)

Manufacturer Not Known c 1900

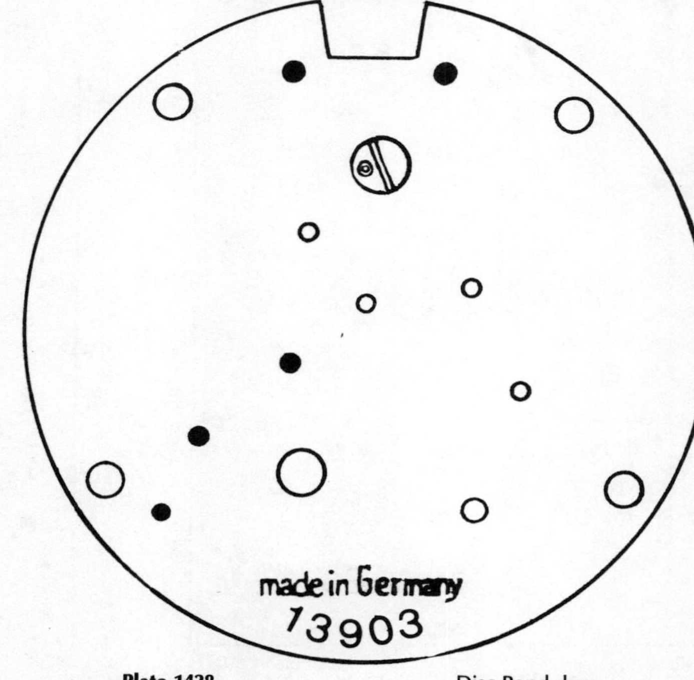

Plate 1438 Disc Pendulum
Offset Anchor
USE .004"* (.102mm*) HOROLOVAR
See Appendix 73 (19 x 36)

Jahresuhrenfabrik c 1912

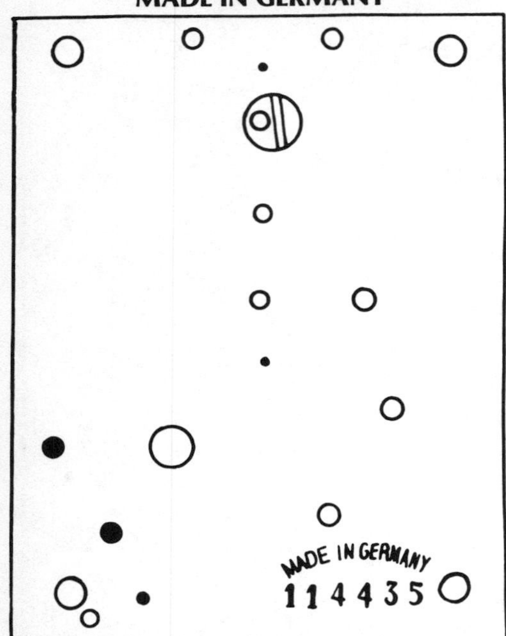

Plate 1439 Disc Pendulum
4-Ball Pendulum
USE .004"* (.102mm*) HOROLOVAR
See Appendix 118 (19 x 36)

Jahresuhrenfabrik c 1904

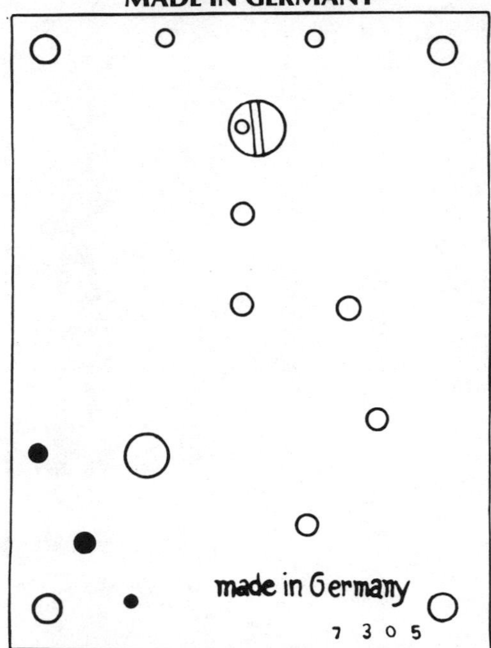

Plate 1440 Disc Pendulum
4-Ball Pendulum
USE .004"* (.102mm*) HOROLOVAR
(19 x 36)

'Jahresuhrenfabrik? c 1902

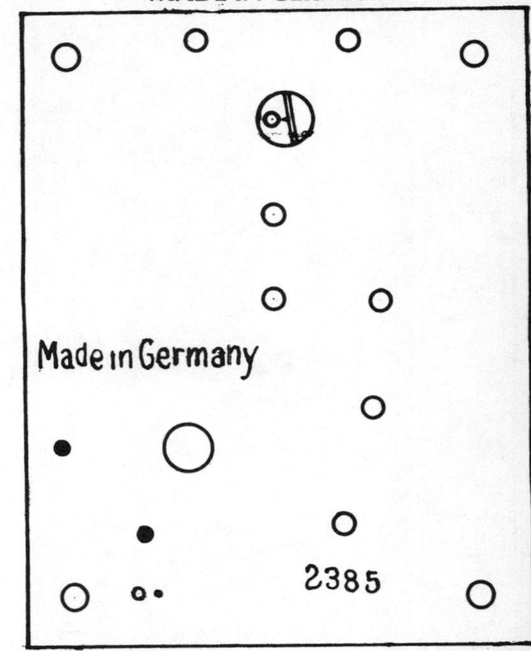

Plate 1441 Disc Pendulum
4-Ball Pendulum
USE .004"* (.102mm*) HOROLOVAR
(19 x 38)

Badische Uhrenfabrik c 1900

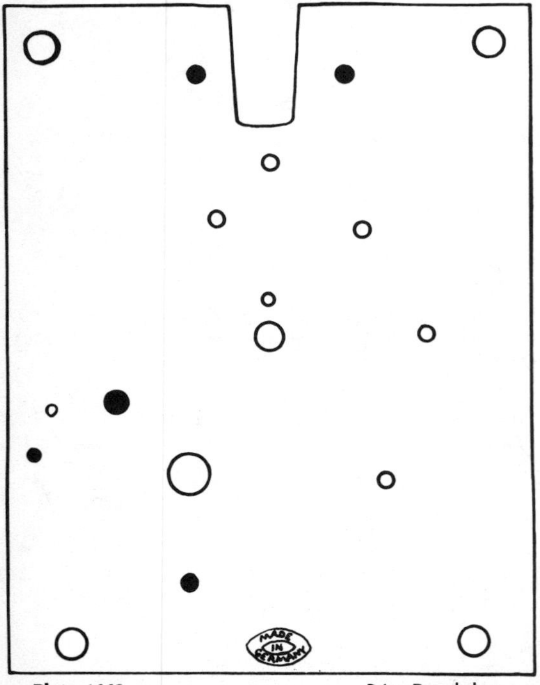

Plate 1443 Disc Pendulum
4-Ball Pendulum
Lantern Pinions
USE .004"* (.102mm*) HOROLOVAR
See Appendix 63 (20 x 38)

Georg Wurthner c 1952

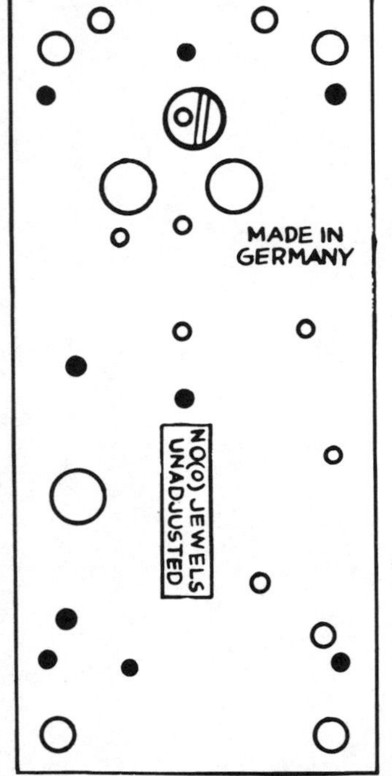

Plate 1447 4-Ball Pendulum
USE .004" (.102mm) HOROLOVAR
Unit 26 (19 x 36)
See Appendix 67

Georg Wurthner c 1953

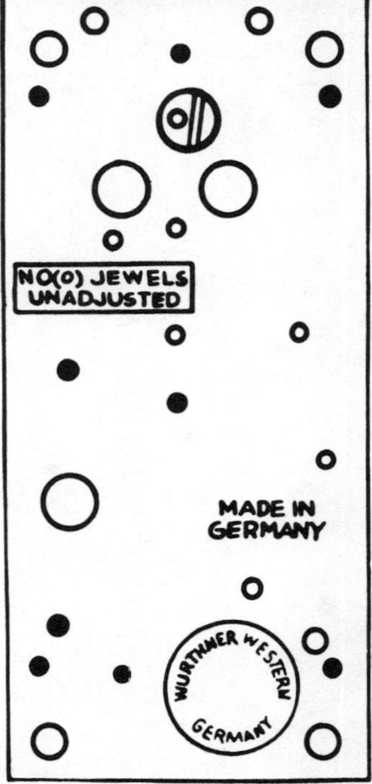

Plate 1448 4-Ball Pendulum
USE .004" (.102mm) HOROLOVAR
Unit 26 (19 x 36)
See Appendix 67, 105

Georg Wurthner c 1957

MADE IN GERMANY

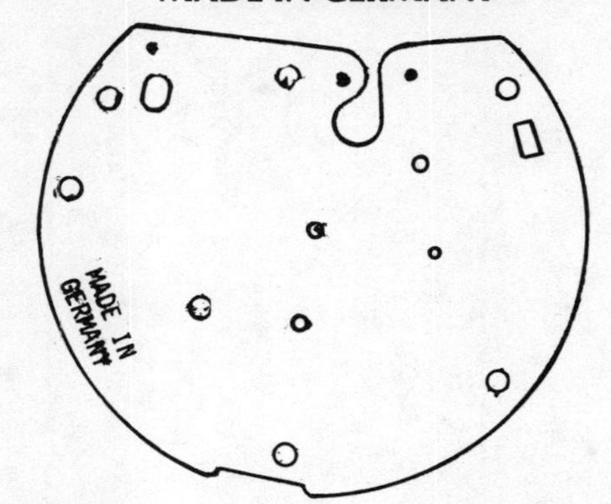

Plate 1449

4-Ball Pendulum
Sub Midget
14-Day Clock

See Appendix 104

Manufacturer Not Known c 1904

MADE IN GERMANY

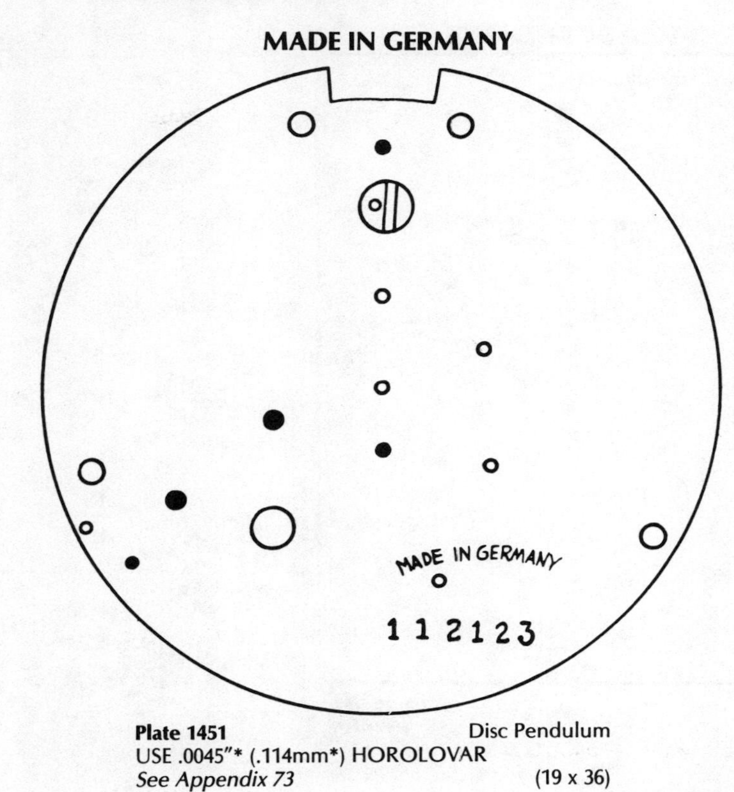

Plate 1451 Disc Pendulum
USE .0045"* (.114mm*) HOROLOVAR
See Appendix 73 (19 x 36)

Manufacturer Not Known c 1906

MADE IN GERMANY

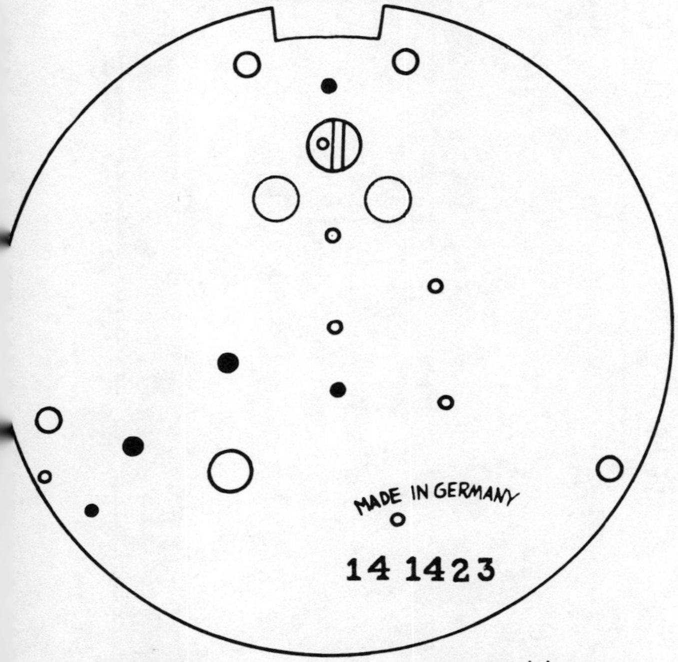

MADE IN GERMANY
14 1423

Plate 1451A Disc Pendulum
USE .0045"* (.114mm*) HOROLOVAR
See Appendix 73 (19 x 36)

Uhrenfabrik Herr c 1951

MADE IN GERMANY

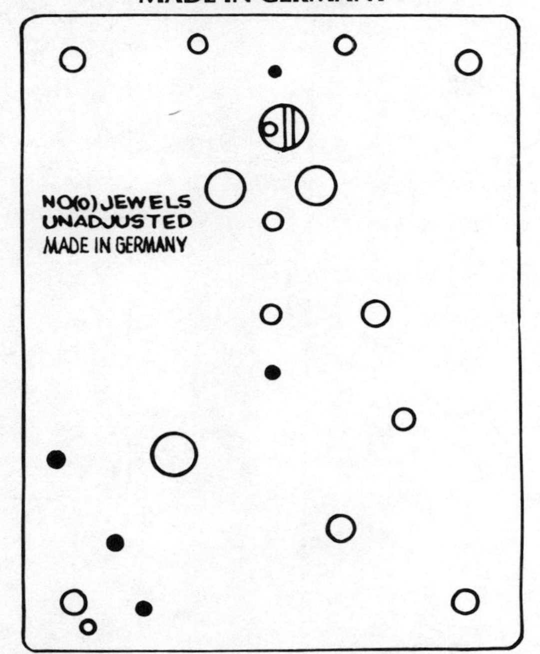

Plate 1453 4-Ball Pendulum
USE .004" (.102mm) HOROLOVAR
Unit 27A (19 x 36)
See Appendix 130

Uhrenfabrik Herr c 1952

MADE IN GERMANY

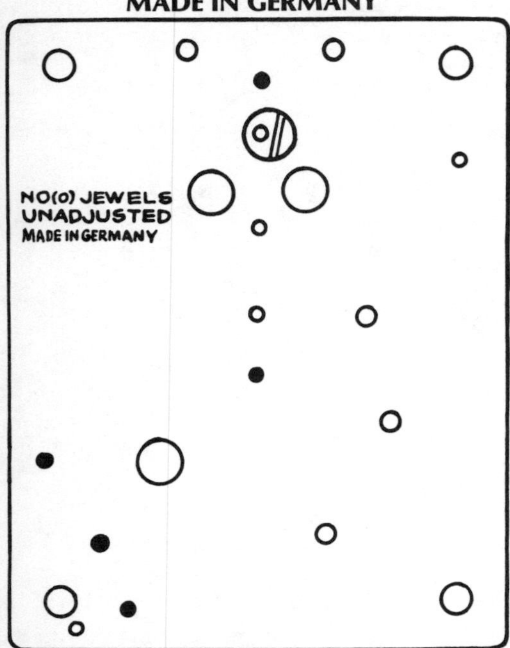

Plate 1453A 4-Ball Pendulum Hour Striker

USE .004" (.102mm) HOROLOVAR
Unit 27A (19 x 36)
See Appendix 56

Uhrenfabrik M. Reiner c 1951

MADE IN GERMANY

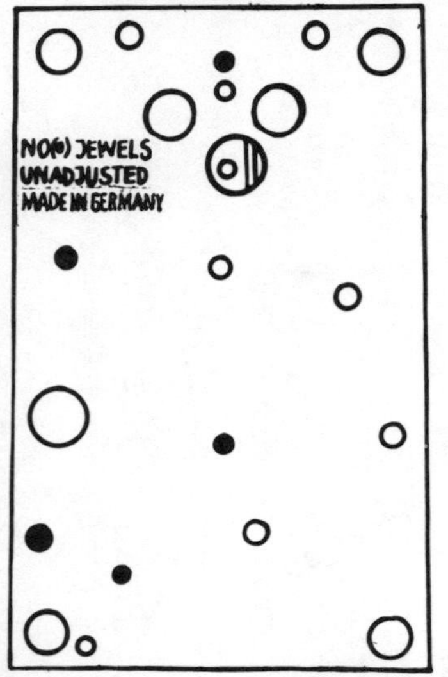

Plate 1456 4-Ball Pendulum Miniature Clock

USE .0025" (.064mm) HOROLOVAR
Units 28A, 28B (16 x 36)
See Appendix 25, 54, 109

Uhrenfabrik M. Reiner c 1954

MADE IN GERMANY

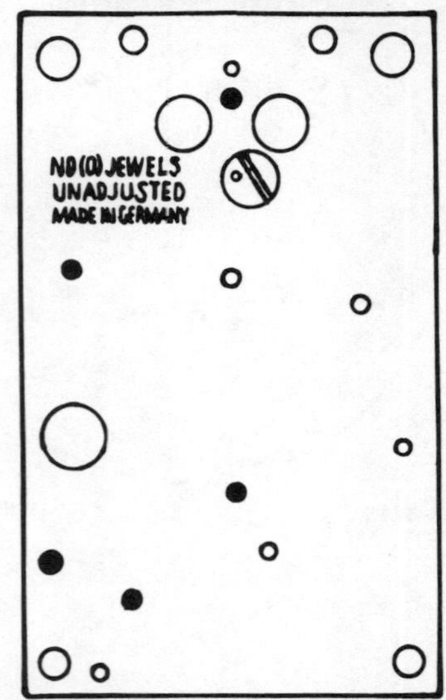

Plate 1457 4-Ball Pendulum Miniature Clock

USE .0028" (.071mm) HOROLOVAR
Unit 28C (16 x 36)
See Appendix 25, 55, 109

Uhrenfabrik Herr c 1951

MADE IN GERMANY

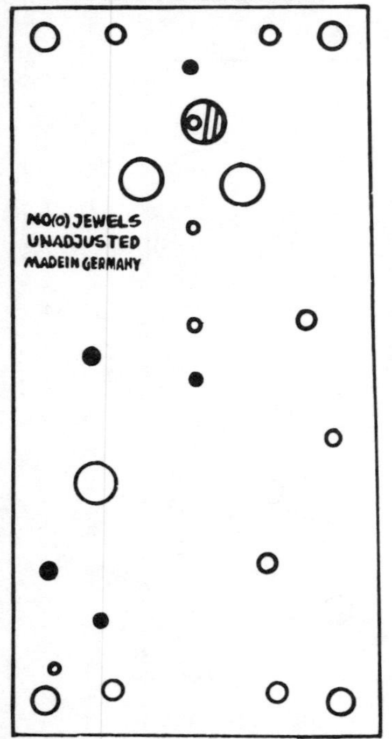

Plate 1458 4-Ball Pendulum

Dome Clock
USE .0035" (.089mm) HOROLOVAR
Units 19, 20A, 21

Square Clock
USE .0033" (.084mm) HOROLOVAR
Units 20B (19 x 36)
See Appendix 109, 110

Phillipp Haas c 1907

MADE IN GERMANY

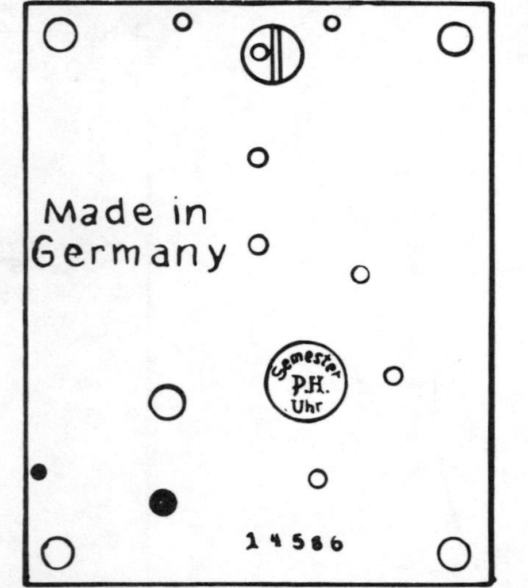

Plate 1459 Standard Disc Pendulum Miniature Movement

USE .004" (.102mm) HOROLOVAR
See Appendix 61, 118

Uhrenfabrik Herr c 1953

MADE IN GERMANY

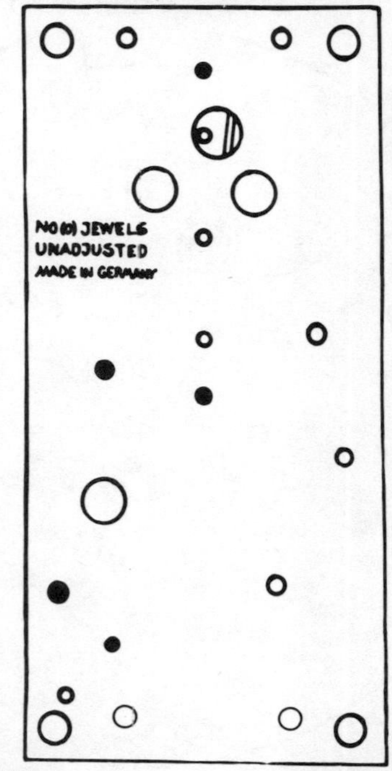

Plate 1460 4-Ball Pendulum

USE .0035" (.089mm) HOROLOVAR
Units 19, 20A, 21

USE .0038" (.097mm) HOROLOVAR
Unit 27B (19 x 36)
See Appendix 106, 109, 110

Manufacturer Not Known c 1912

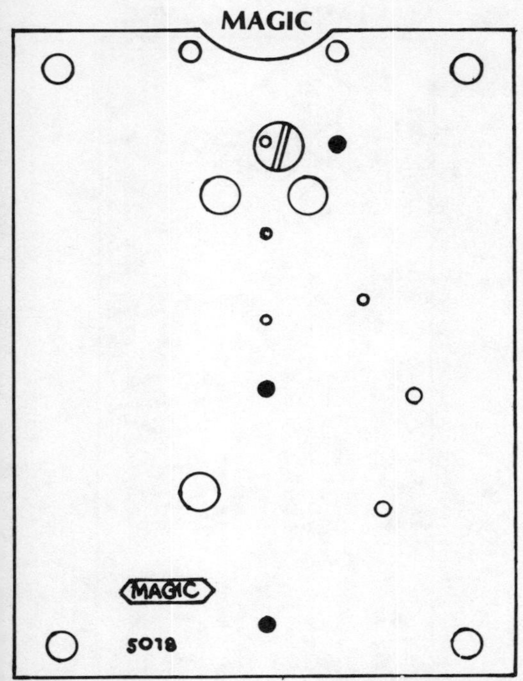

Plate 1461 4-Ball Pendulum
USE .004"* (.102mm*) HOROLOVAR
See Appendix 73 (19 x 38)

Uhrenfabrik Herr c 1951

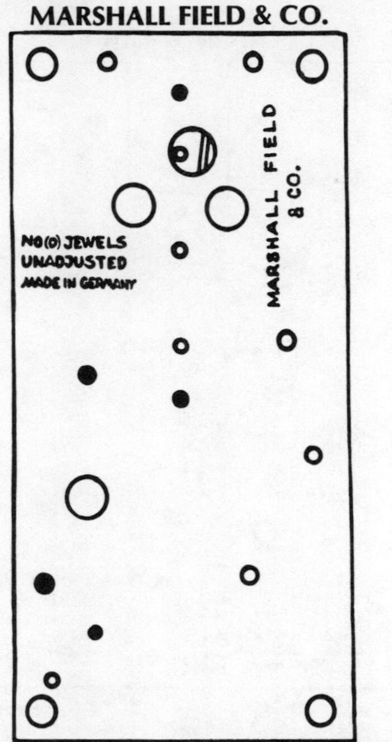

Plate 1461B 4-Ball Pendulum
USE .0035" (.089mm) HOROLOVAR
Units 19, 20A, 21
USE .0038" (.097mm) HOROLOVAR
Unit 27B (19 x 36)
See Appendix 106, 109, 110

Nisshindo Watch Co. c 1955

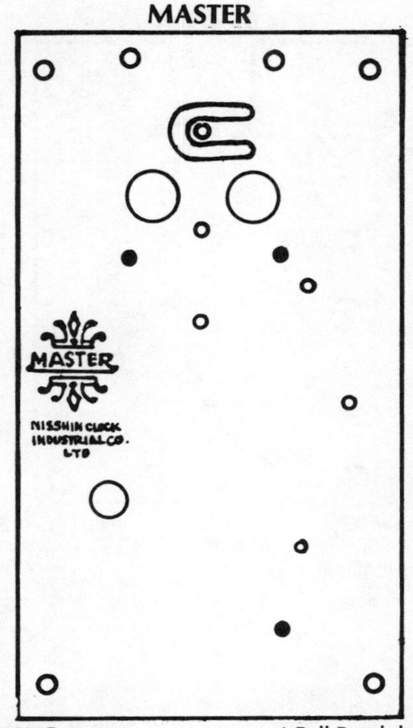

Plate 1461C 4-Ball Pendulum
USE .0035" (.089mm) HOROLOVAR
Units 41A, 41B (19 x 38)
See Appendix 57

Nisshindo Watch Co. c 1957

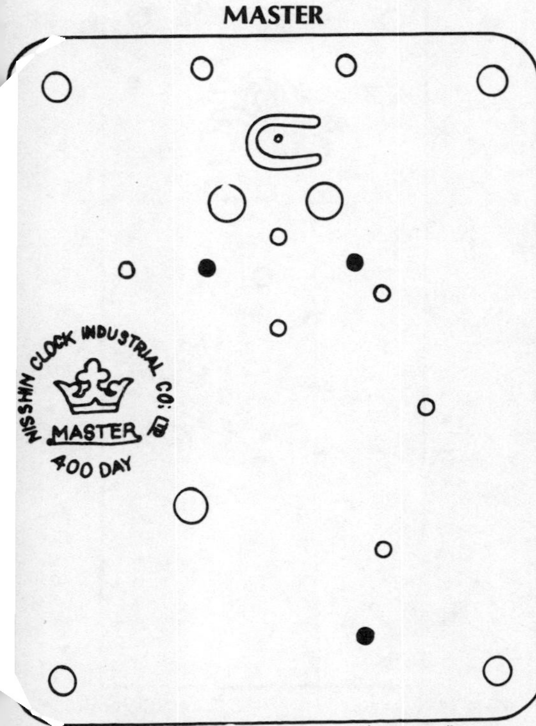

Plate 1461D 4-Ball Pendulum
USE .0034" (.086mm) HOROLOVAR
Units 45, 45A (19 x 38)

Nisshindo Watch Co. c 1957

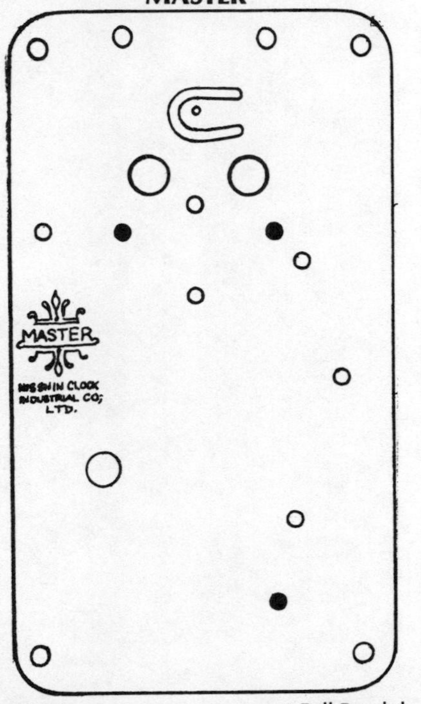

Plate 1461E 4-Ball Pendulum
USE .0034" (.086mm) HOROLOVAR
Units 45, 45A (19 x 38)
See Appendix 58

Nisshindo Watch Co. c 1957

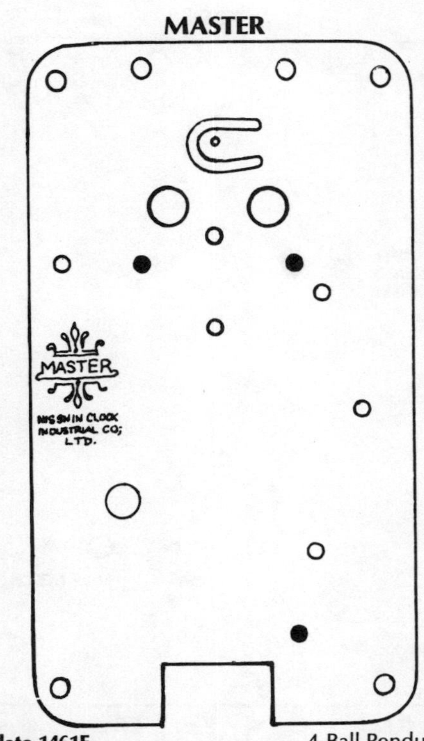

Plate 1461F 4-Ball Pendulum
USE .0034" (.086mm) HOROLOVAR
Units 45, 45A (19 x 38)
See Appendix 58

Nisshindo Watch Co. c 1968

MASTER

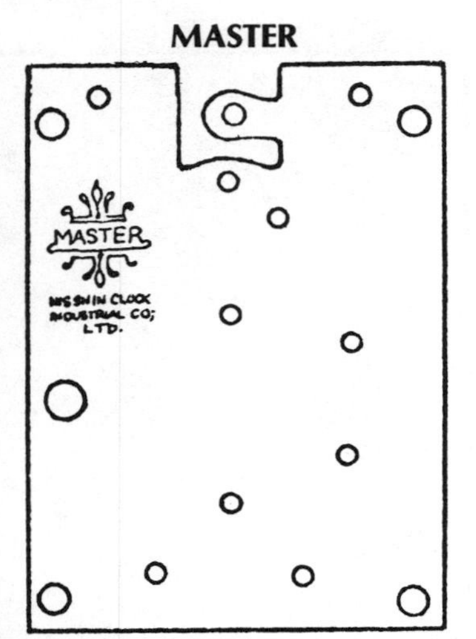

Plate 1461G 4-Ball Pendulum
USE .0028" (.071mm) HOROLOVAR
Units 50A, 50B (14 x 32)

Konrad Mauch c 1950

MAUCH-FRICK

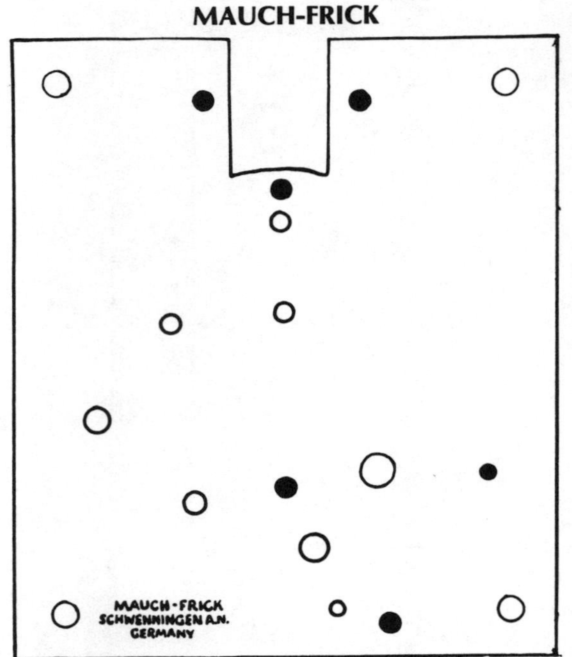

Plate 1462 4-Ball Pendulum
USE .0035" (.089mm) HOROLOVAR
Units 13A, 13B (20 x 38)
See Appendix 63, 64, 84

J. Link & Co. c 1953

MAX LANDAU & CO. INC.

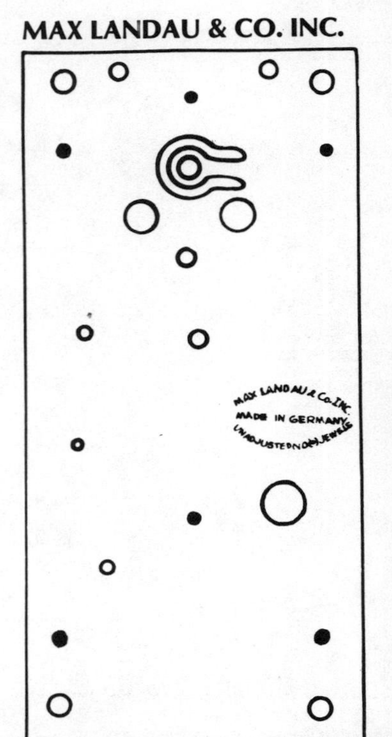

Plate 1463 4-Ball Pendulum
USE .0037" (.094mm) HOROLOVAR
Units 16, 17 (19 x 38)
See Appendix 38, 119

Kieninger & Obergfell c 1930

MAYER

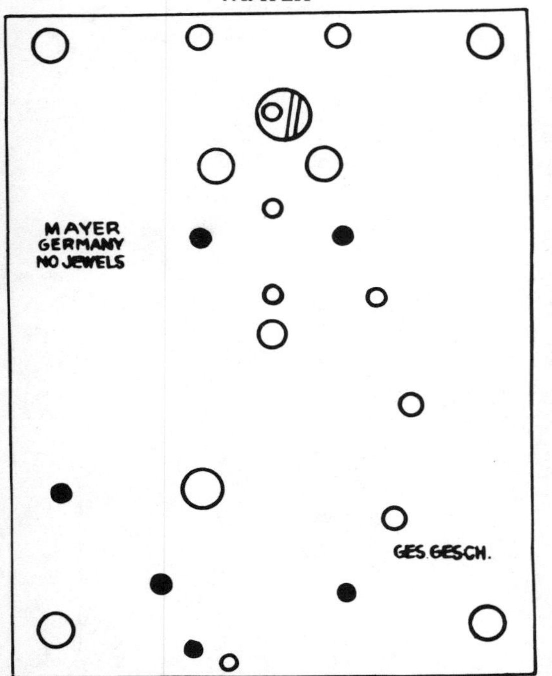

Plate 1467 4-Ball Pendulum
USE .0032" (.081mm) HOROLOVAR
Unit 1 (19 x 38)
See Appendix 76

J. Link & Co. c 1953

METASCO INC.

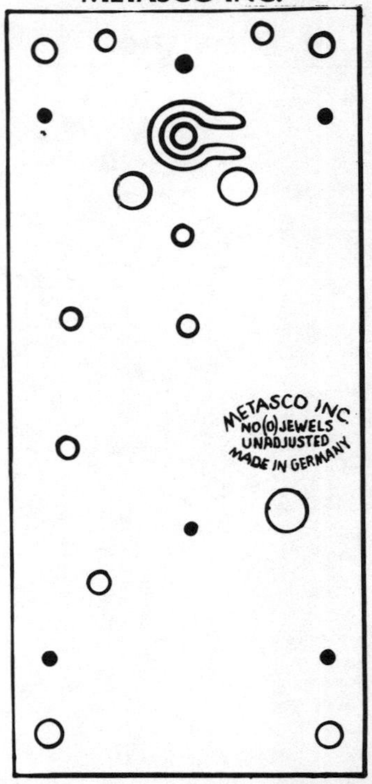

Plate 1468 4-Ball Pendulum
USE .0037" (.094mm) HOROLOVAR
Units 16, 17 (19 x 38)
See Appendix 38, 119

Uhrenfabrik Herr c 1952

MOHERTUS TRADING CO.

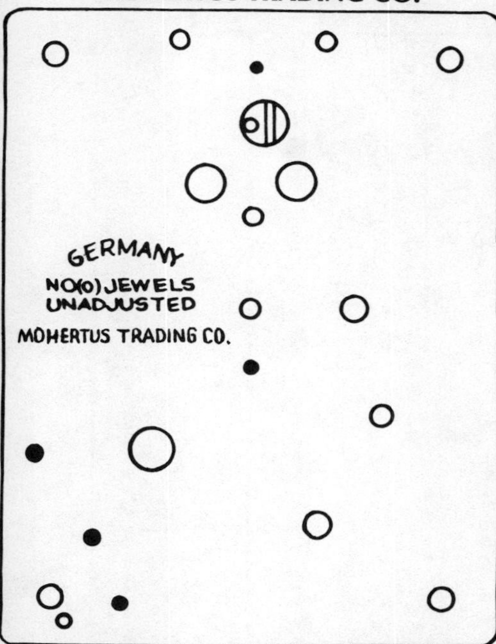

Plate 1469 4-Ball Pendulum
USE .004" (.102mm) HOROLOVAR
Unit 27A (19 x 36)
See Appendix 130

Nisshindo Watch Co. c 1968

MONTGOMERY WARD & CO.

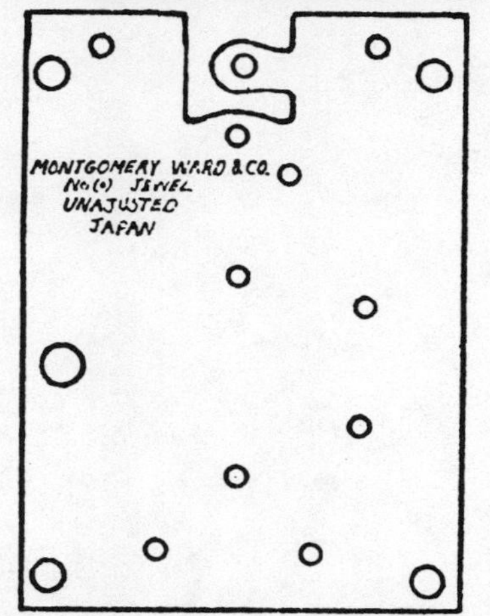

Plate 1469A 4-Ball Pendulum
USE .0028" (.071mm) HOROLOVAR
Units 50A, 50B (14 x 32)

Jahresuhrenfabrik c 1910

NATIONAL SILVER CO.

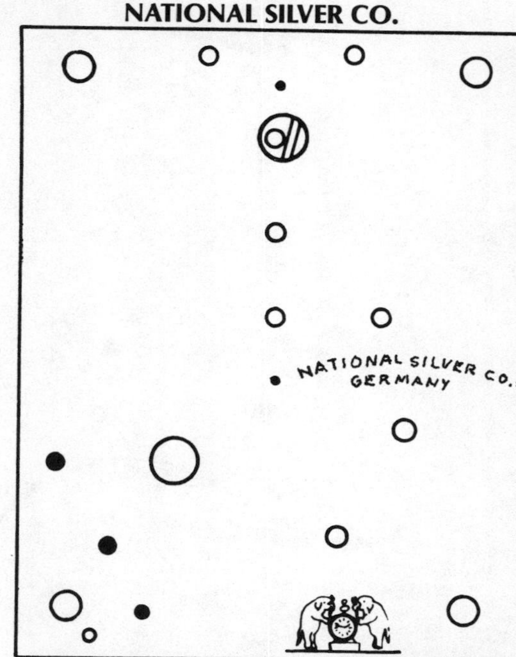

Plate 1469B 4-Ball Pendulum
USE .004" (.102mm) HOROLOVAR
Unit 6789A (19 x 36)

Uhrenfabrik Neueck c 1956

NEUECK

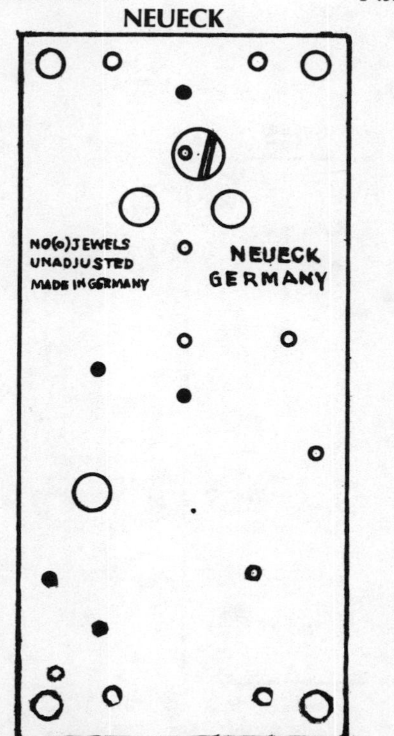

Plate 1469C 4-Ball Pendulum
USE .0036" (.091mm) HOROLOVAR
Unit 27C (19 x 36)
See Appendix 44, 110

Nisshindo Watch Co. c 1970

NISSHIN

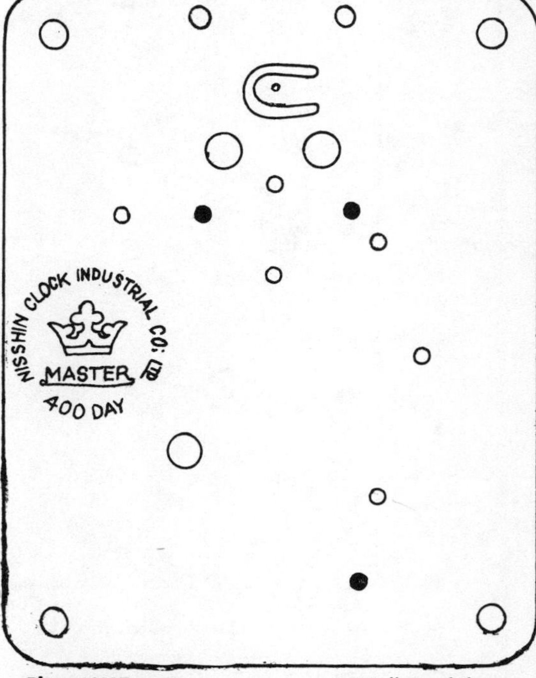

Plate 1469D 4-Ball Pendulum
USE .0034" (.086mm) HOROLOVAR
Units 45, 45A (19 x 38)

Nisshindo Watch Co. c 1957

NISSHIN

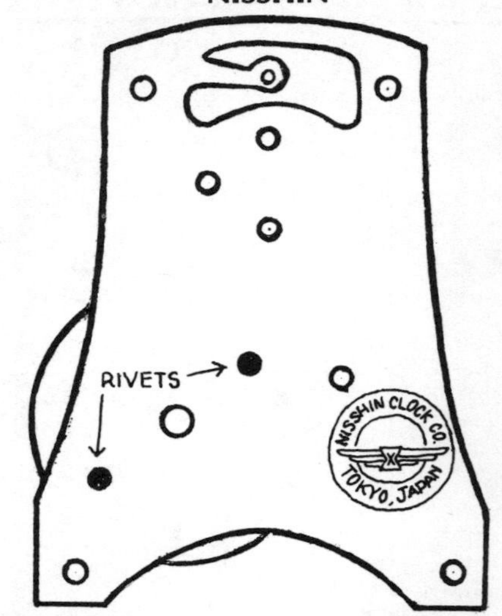

Plate 1469E 3 or 4-Ball Pendulum
100-Day Clock
USE .0028" (.071mm) HOROLOVAR
Units 46, 46A (12 x 32)
See Appendix 121

Uhrenfabrik Herr c 1952

NO (0) JEWELS

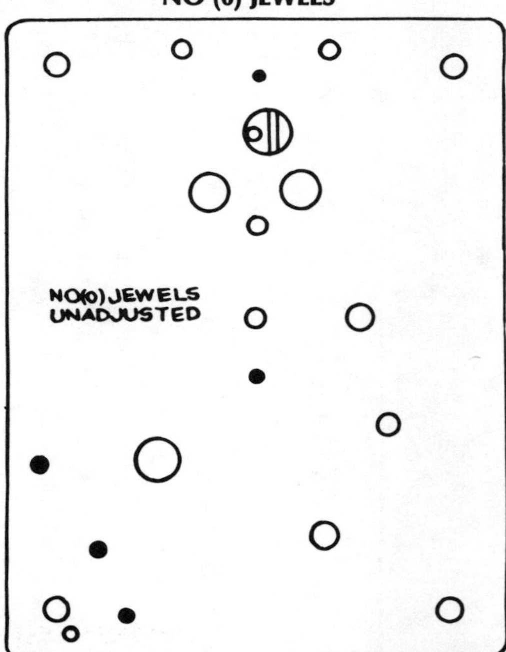

Plate 1469G 4-Ball Pendulum
USE .004" (.102mm) HOROLOVAR
Unit 27A (19 x 36)
See Appendix 130

Uhrenfabrik Neueck c 1956

NO (0) JEWELS
(Plate marking error)

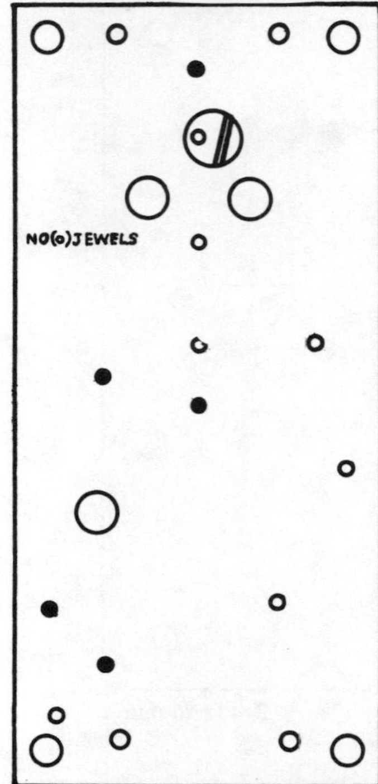

Plate 1469F 4-Ball Pendulum
USE .0036" (.091mm) HOROLOVAR
Unit 27C (19 x 36)
See Appendix 44, 110

Uhrenfabrik M. Reiner c 1954

NO (0) JEWELS

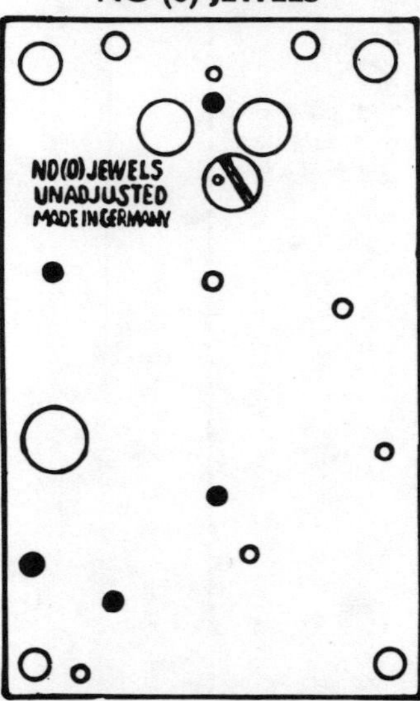

Plate 1469H Standard 4-Ball Pendulum
Miniature Movement
USE .0032" (.081mm) HOROLOVAR
Unit 23B (16 x 36)
See Appendix 25, 55, 109

Kieninger & Obergfell c 1930

OSCAR HEYMANN

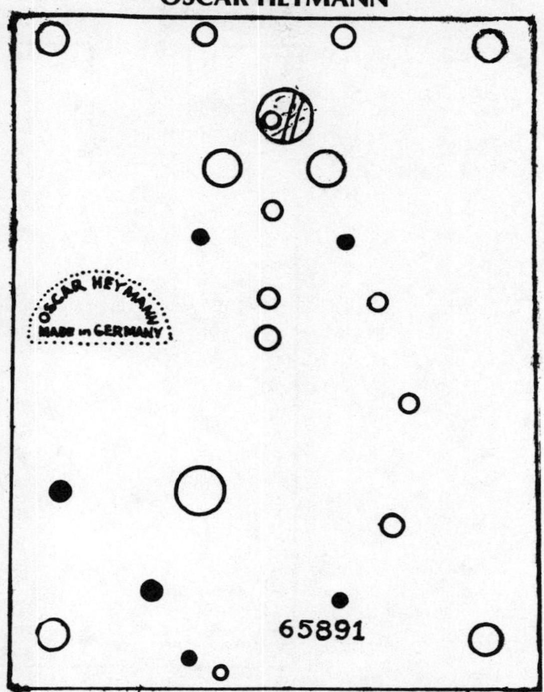

Plate 1470 4-Ball Pendulum
USE .0032" (.081mm) HOROLOVAR
Unit 1 (19 x 38)
See Appendix 76

Konrad Mauch c 1954

OVEROCEAN CLOCK COMPANY

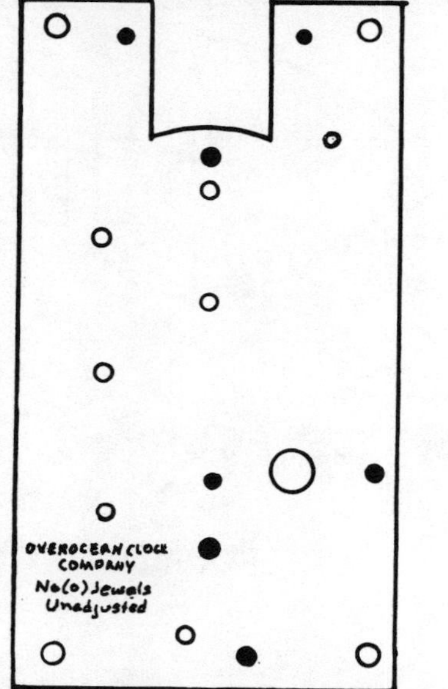

Plate 1470B 4-Ball Pendulum
Miniature Clock
USE .0032" (.081mm) HOROLOVAR
Unit 14A
USE .003" (.076mm) HOROLOVAR
Unit 14B Original (13 x 32)
See Appendix 63, 64, 85 Revised (15 x 32)

Manufacturer Not Known c 1900

PATENT

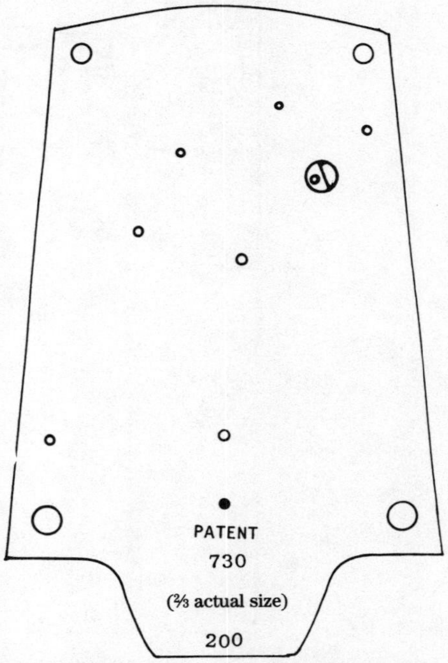

PATENT
730
(⅔ actual size)
200

Plate 1470E Special Pendulum
Constant Force Escapement
See Appendix 73, 92 Front Wind

Jahresuhrenfabrik c 1902

PATENT ANGEMELDET

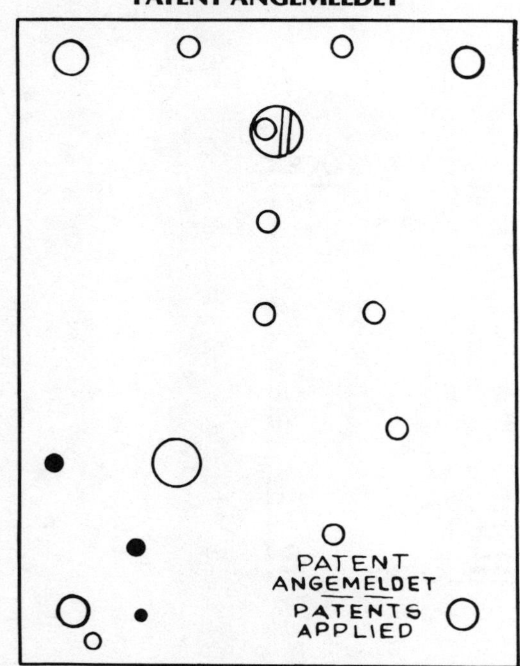

Plate 1471 Disc Pendulum
USE .004" (.102mm) HOROLOVAR
See Appendix 120 (19 x 36)

Grivolas c 1910

PENDULE 400 JOURS

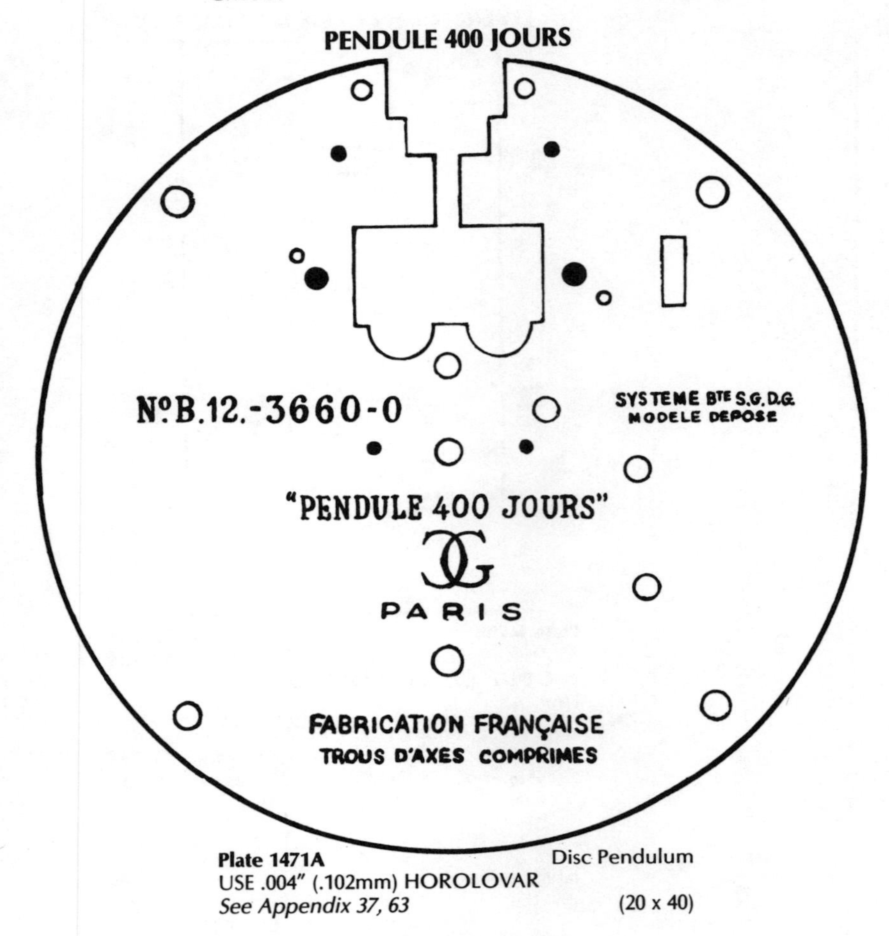

Plate 1471A Disc Pendulum
USE .004" (.102mm) HOROLOVAR
See Appendix 37, 63 (20 x 40)

Konrad Mauch c 1957

PERFECTA WATCH & CLOCK CO.

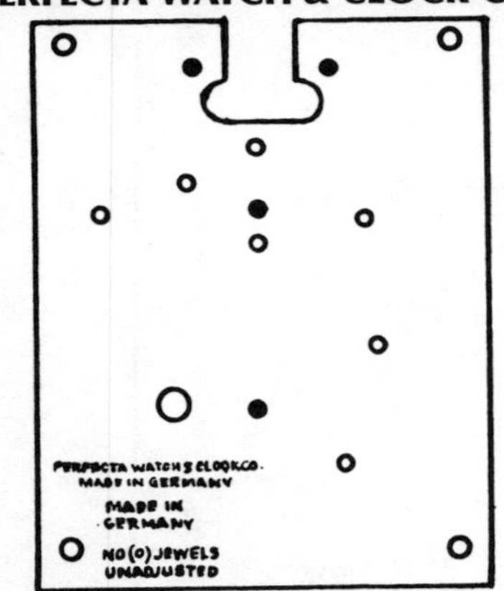

Plate 1471C 3-Ball Pendulum
Midget Clock
Pin Pallet Escapement
USE .0022" (.056mm) HOROLOVAR
Units 38, 38A Original (12 x 25)
See Appendix 53, 64, 86 Revised (14 x 25)

Konrad Mauch c 1952

PERFECTA WATCH & CLOCK CO.

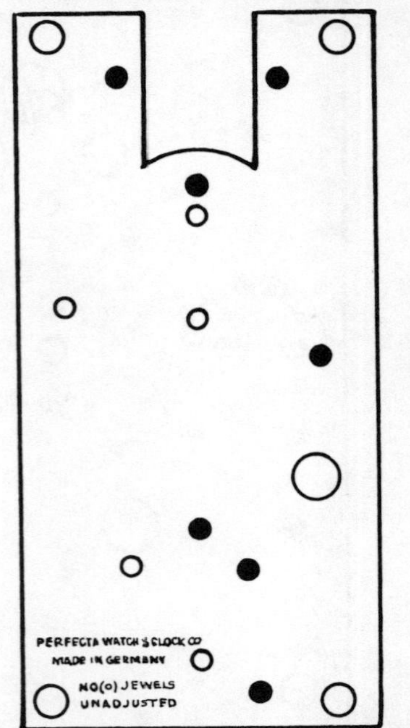

Plate 1471B 4-Ball Pendulum
USE .0035" (.089mm) HOROLOVAR
Units 13A, 13B, 13C (20 x 38)
See Appendix 63, 64, 84

Kieninger & Obergfell c 1951

PERFECTA WATCH & CLOCK CO.

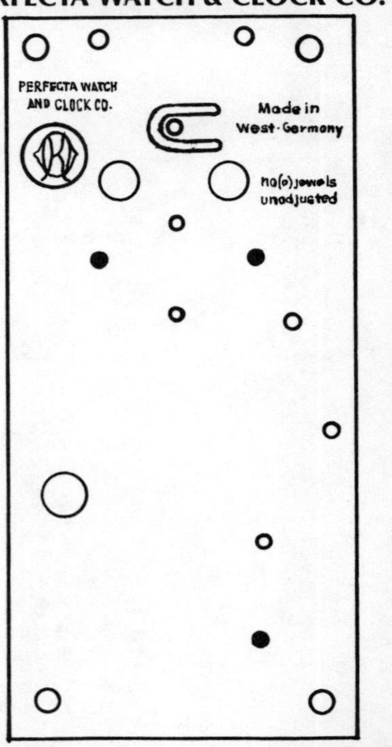

Plate 1471D 4-Ball Pendulum
USE .0032" (.081mm) HOROLOVAR
Units 3A, 3B, 3C (19 x 38)
See Appendix 76

Kern & Sohne c 1960

PERFECTA WATCH & CLOCK CO.

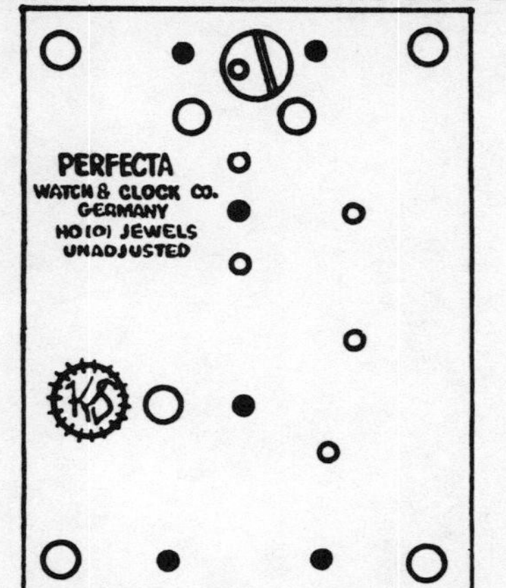

Plate 1471E 4-Ball Pendulum Miniature Clock

USE .002" (.051mm)
Units 12A, 12B, 12C
USE .0019" (.048mm) HOROLOVAR
Unit 12D (12 x 25)

Konrad Mauch c 1954

PERFECTA WATCH & CLOCK CO.

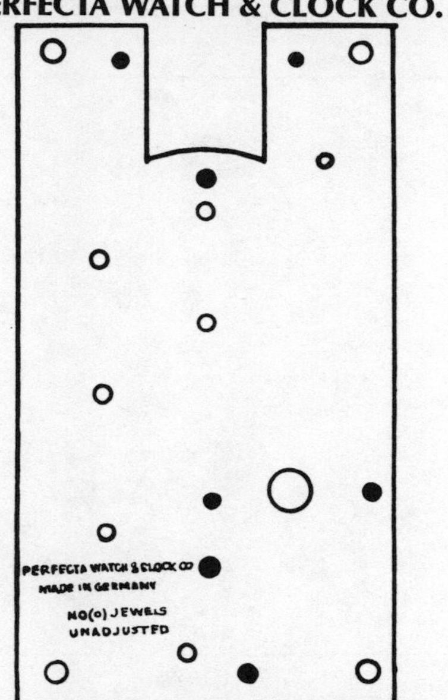

Plate 1471F 4-Ball Pendulum Miniature Clock

USE .0032" (.081mm) HOROLOVAR
Unit 14A
USE .003" (.076mm) HOROLOVAR
Unit 14B Original (13 x 32)
See Appendix 52, 63, 64, 85 Revised (15 x 32)

W. Petersen c 1954

PETERSEN

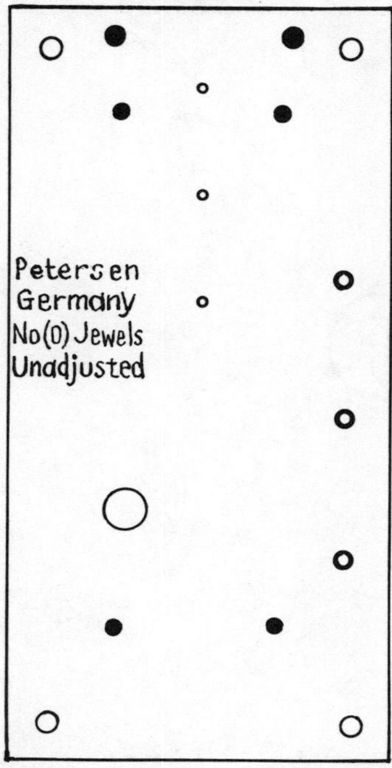

Plate 1472 4-Ball Pendulum *or* 4-Figurine Pendulum

USE .0033" (.084mm) HOROLOVAR
Units 18B, 18C (19 x 38)
See Appendix 59, 65

Kieninger & Obergfell c 1951

P. R. MYERS & CO.

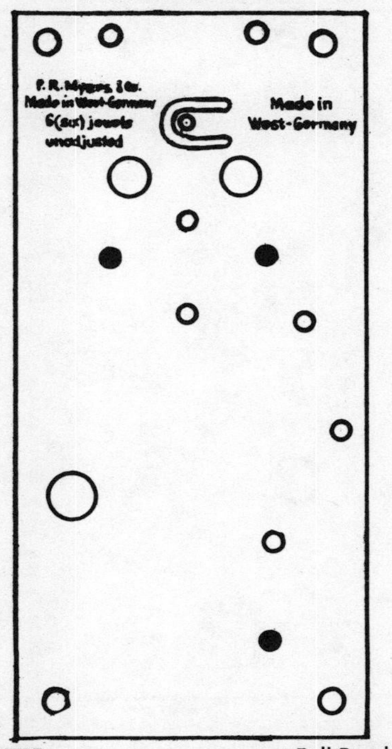

Plate 1472E 4-Ball Pendulum

USE .0032" (.081mm) HOROLOVAR
Unit 3A (19 x 38)
See Appendix 76

Kieninger & Obergfell c 1965

P. R. MYERS & CO.

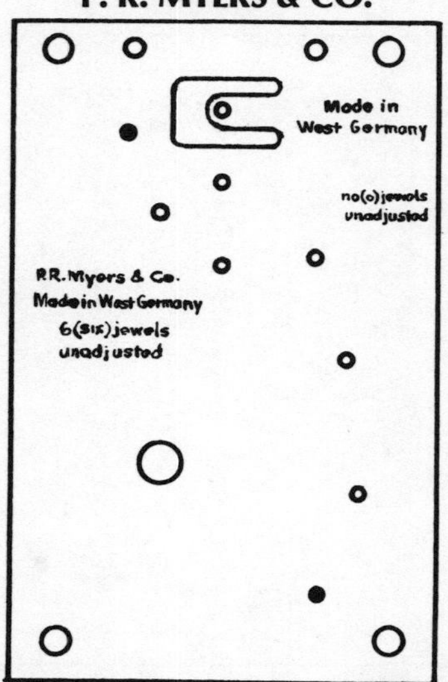

Plate 1472F 4-Ball Pendulum Miniature Clock

USE .0023" (.058mm) HOROLOVAR
Unit 5E (14 x 30)
See Appendix 77

Kieninger & Obergfell c 1912

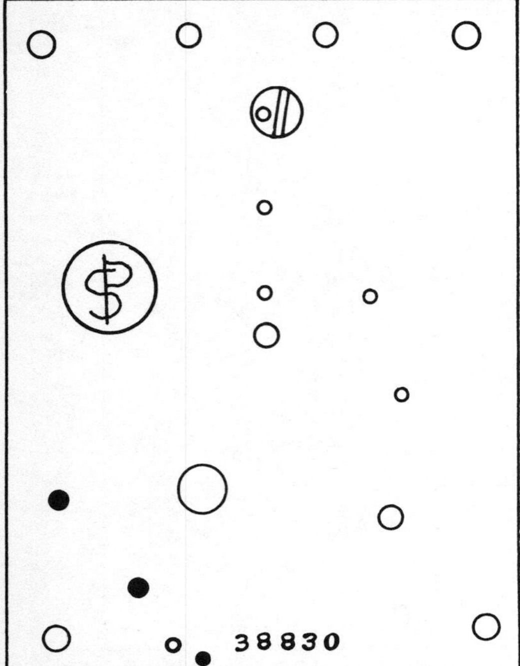

Plate 1472H 4-Ball Pendulum
USE .0032" (.081mm) HOROLOVAR
Units 1, 3A (19 x 38)

Manufacturer Not Known c 1885

R

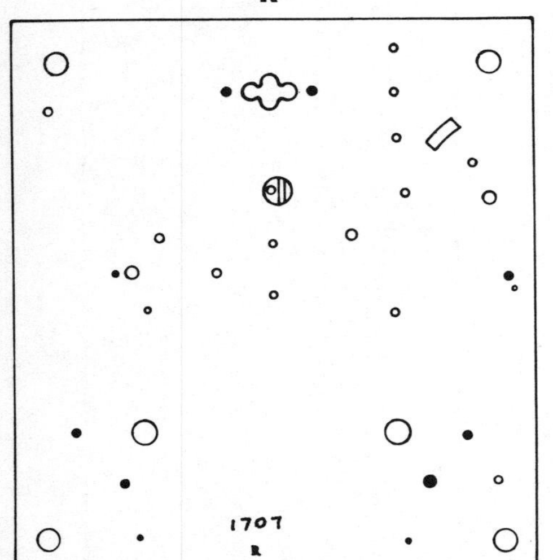

(½ actual size)

Plate 1473B Disc Pendulum
Hour and Half Hour Gong Strike
USE .0045"— (.114mm—) HOROLOVAR
See Appendix 63, 73

Jahresuhrenfabrik c 1885

R

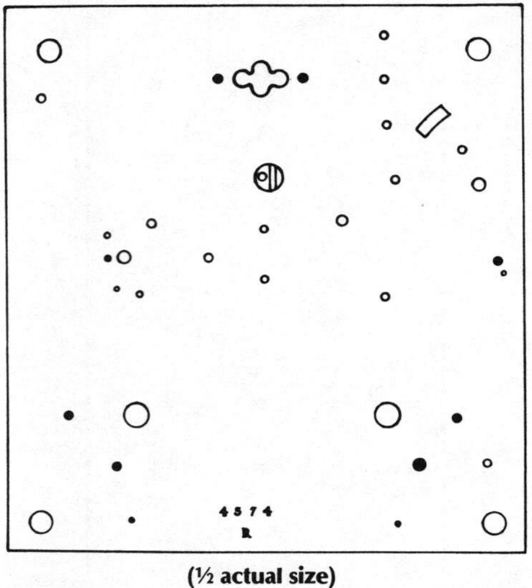

(½ actual size)

Plate 1473 Disc Pendulum
Hour and Half Hour Gong Strike
USE .0045"— (.114mm—) HOROLOVAR
See Appendix 63

Jahresuhrenfabrik c 1885

R

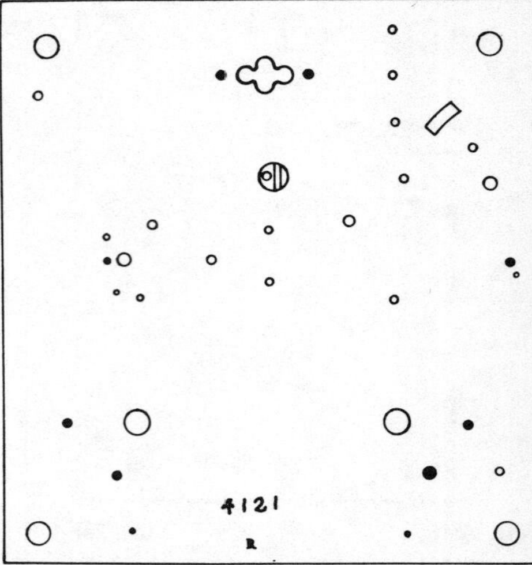

(½ actual size)

Plate 1473A Disc Pendulum
Hour and Half Hour Bell Strike
USE .005"— (.127mm—) HOROLOVAR
See Appendix 63

Jahresuhrenfabrik c 1885

R

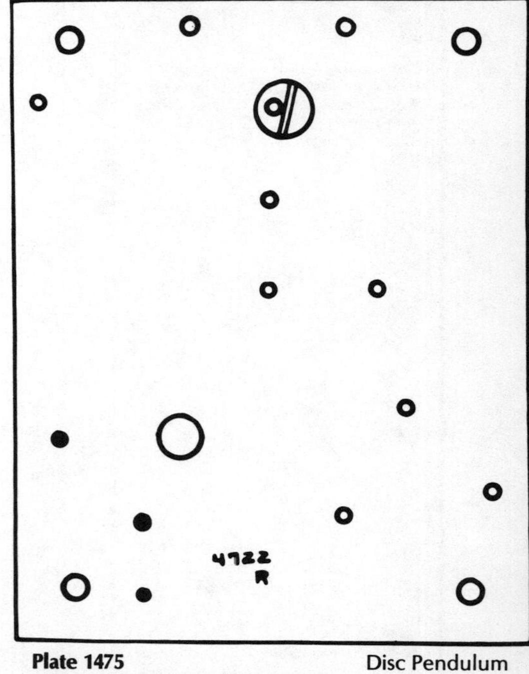

Plate 1475 Disc Pendulum
USE .004" (.102mm) HOROLOVAR
See Appendix 23 (19 x 36)

Jahresuhrenfabrik c 1908

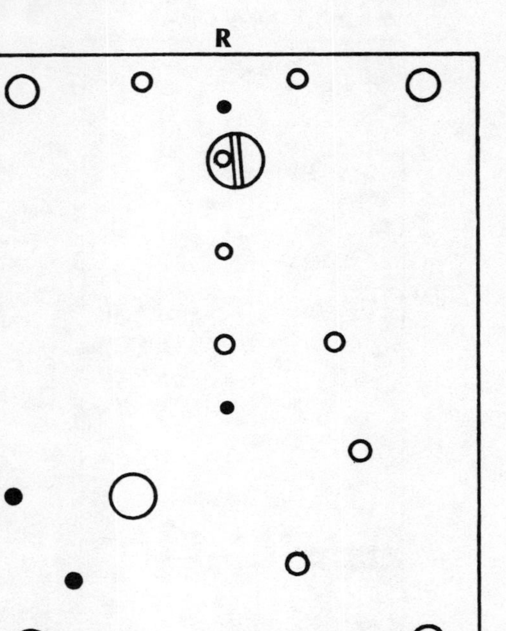

Plate 1476 Disc Pendulum
USE .004"* (.102mm*) HOROLOVAR
(19 x 36)

J. Link & Co. c 1953

RALPH HERMAN CLOCK HOUSE

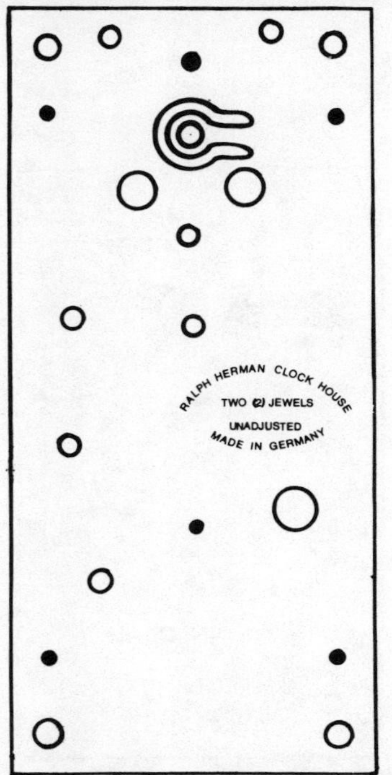

Plate 1477 4-Ball Pendulum
USE .0037" (.094mm) HOROLOVAR
Units 16, 17 (19 x 38)
See Appendix 38, 119

Kieninger & Obergfell c 1951

RENSIE WATCH CO. INC.

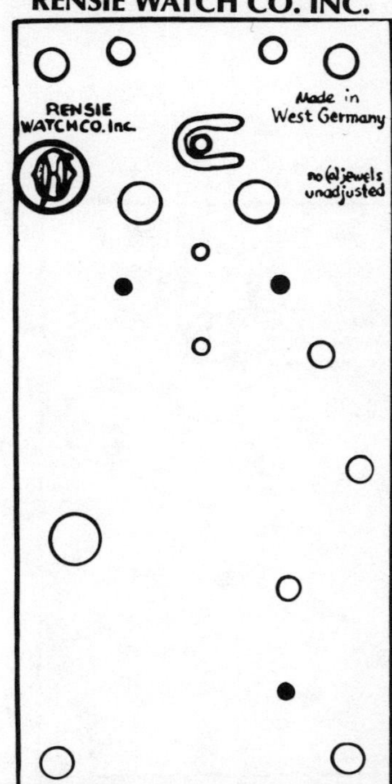

Plate 1479 4-Ball Pendulum
USE .0032" (.081mm) HOROLOVAR
Units 1, 3A, 3B, 3C (19 x 38)
See Appendix 76

Kieninger & Obergfell c 1954

RENSIE WATCH CO. INC.

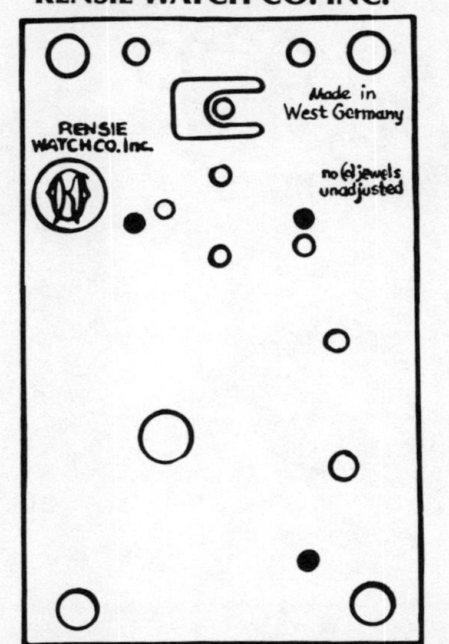

Plate 1483 4-Ball Pendulum
Miniature Clock
USE .0023" (.058mm) HOROLOVAR
Units 5A, 5B, 5C, 5D, 5E (14 x 30)
See Appendix 77, 111

Uhrenfabrik Herr c 1951

REX

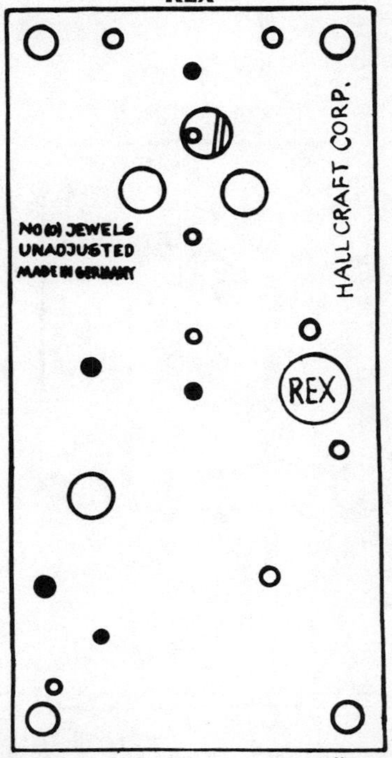

Plate 1484 4-Ball Pendulum
USE .0035" (.089mm) HOROLOVAR
Units 19, 20A, 21
USE .0038" (.097mm) HOROLOVAR
Unit 27B (19 x 36)
See Appendix 106, 109, 110

Uhrenfabrik M. Reiner c 1952

REX

NO(O)JEWELS
UNADJUSTED
MADE IN GERMANY
HALL CRAFT CORP.
REX

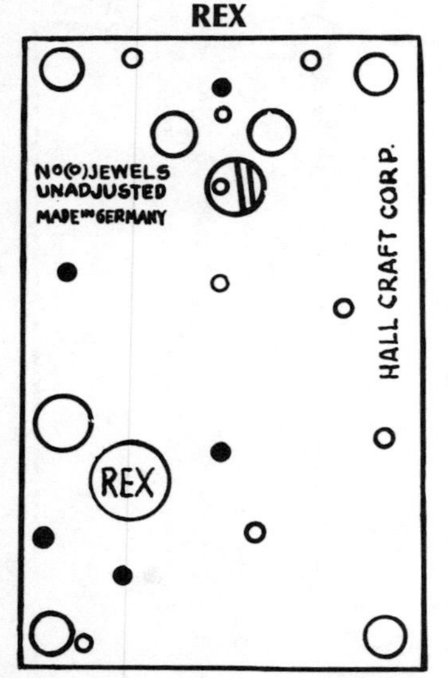

Plate 1484A 4-Ball Pendulum Miniature Clock
USE .0025" (.064mm) HOROLOVAR
Units 28A, 28B (16 x 36)
See Appendix 25, 54, 109

Uhrenfabrik M. Reiner c 1954

REX

NO(O)JEWELS
UNADJUSTED
MADE IN GERMANY
HALL CRAFT CORP.
REX

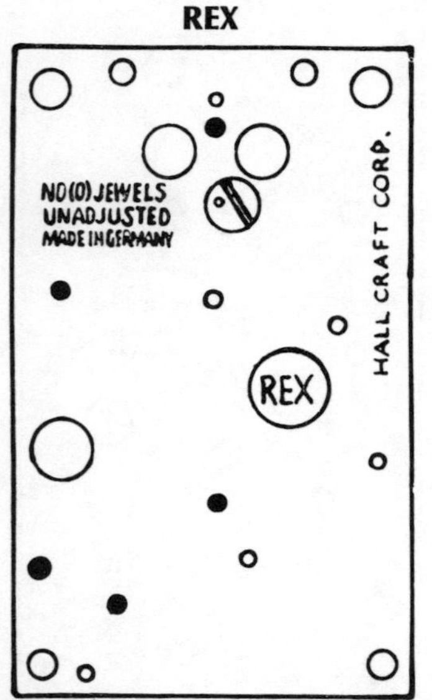

Plate 1484B 4-Ball Pendulum Miniature Clock
USE .0028" (.071mm) HOROLOVAR
Unit 28C (16 x 36)
See Appendix 25, 55, 109

Sigfried Haller c 1951

ROBANNE CORP.

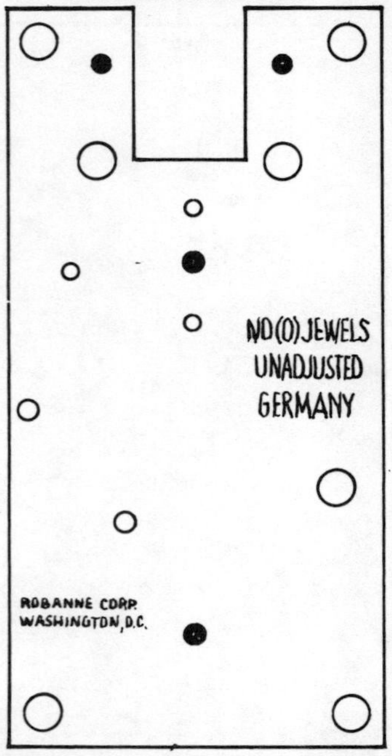

NO(O)JEWELS
UNADJUSTED
GERMANY
ROBANNE CORP.
WASHINGTON, D.C.

Plate 1485 4-Ball Pendulum
USE .0038" (.097mm) HOROLOVAR
Unit 33 (18 x 38)
See Appendix 63, 64

Uhrenfabrik M. Reiner c 1952

ROKI CORP.

NO(O)JEWELS
UNADJUSTED
MADE IN GERMANY
ROKI CORP.
Made in Germany

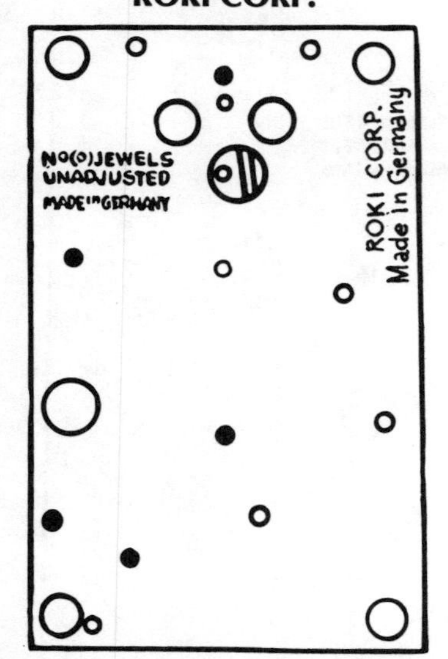

Plate 1486 Standard 4-Ball Pendulum Miniature Movement
USE .003" (.076mm) HOROLOVAR
Units 22, 23A, 24 (16 x 36)
See Appendix 25, 54, 109

Edgar Henn c 1951

ROTH BROS.

ROTH BROS.
MADE IN GERMANY

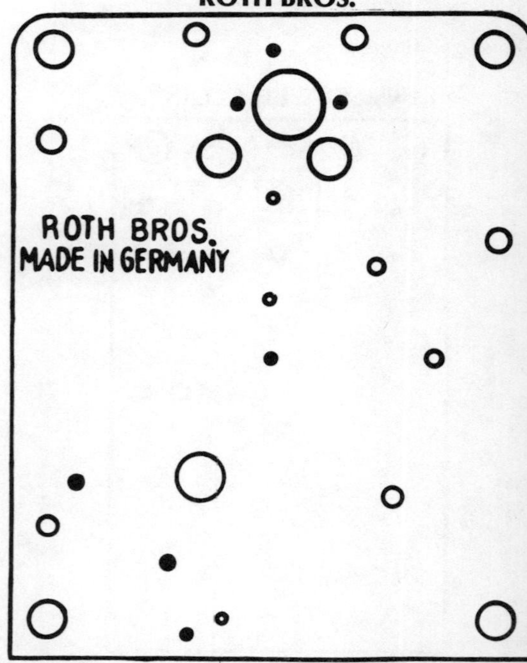

Plate 1487 4-Ball Pendulum
USE .0035" (.089mm) HOROLOVAR
Unit 31 (19 x 38)
See Appendix 39

Kieninger & Obergfell c 1951

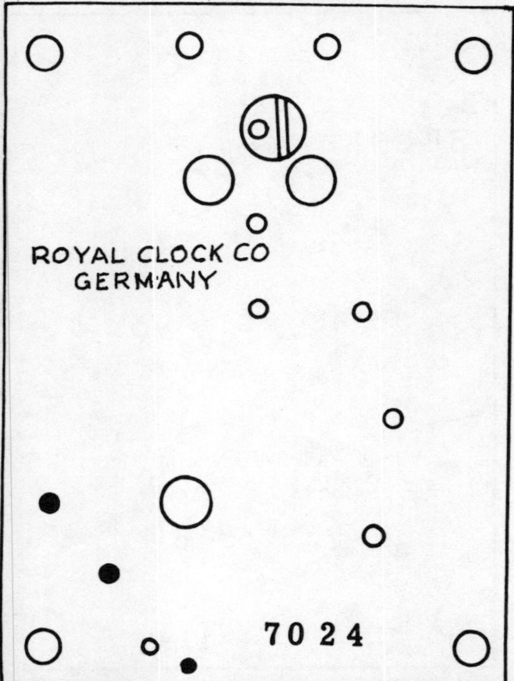

Plate 1490 4-Ball Pendulum
USE .0035"* (.089mm*) HOROLOVAR
See Appendix 76 (19 x 38)

Kieninger & Obergfell c 1930

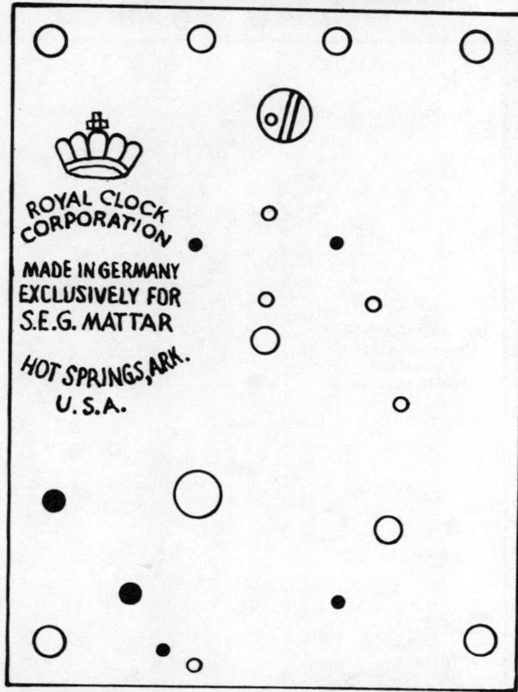

Plate 1491 4-Ball Pendulum
USE .0032" (.081mm) HOROLOVAR
Unit 1 (19 x 38)
See Appendix 76

Kieninger & Obergfell c 1950

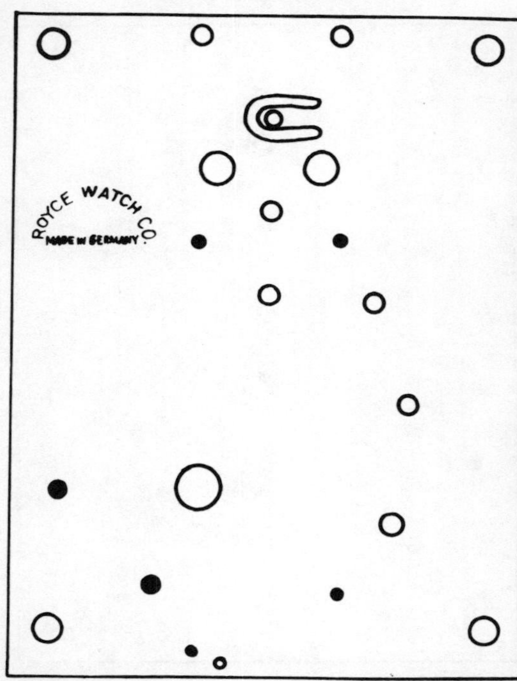

Plate 1495 4-Ball Pendulum
USE .0032" (.081mm) HOROLOVAR
Unit 1 (19 x 38)
See Appendix 76

Kieninger & Obergfell c 1951

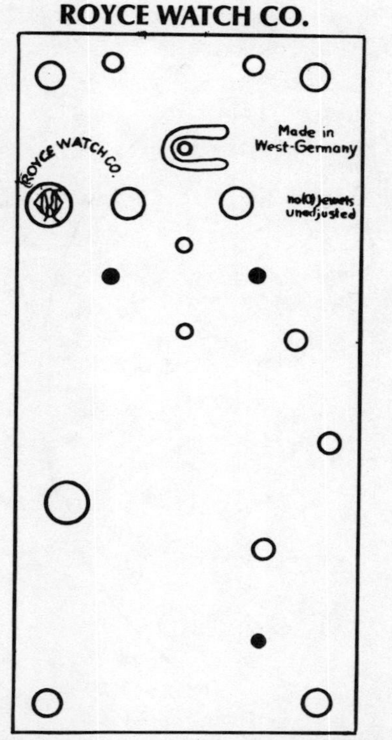

Plate 1499 4-Ball Pendulum
USE .0032" (.081mm) HOROLOVAR
Units 1, 3A, 3B, 3C (19 x 38)
See Appendix 76

Kieninger & Obergfell c 1953

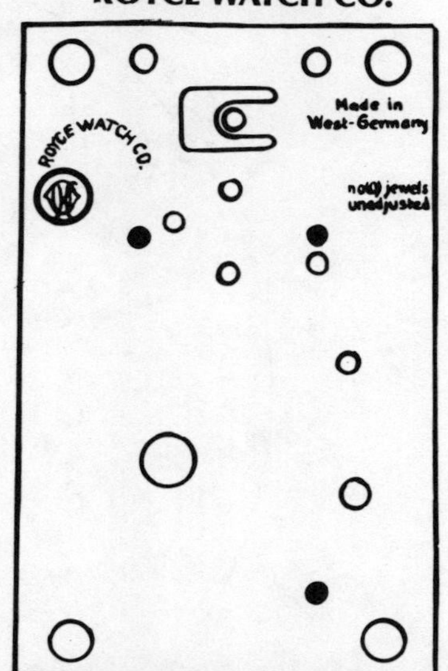

Plate 1503 4-Ball Pendulum
Miniature clock
USE .0023" (.058mm) HOROLOVAR
Units 5A, 5B, 5C, 5D, 5E (14 x 30)
See Appendix 77, 111

R. Schneckenburger c 1894

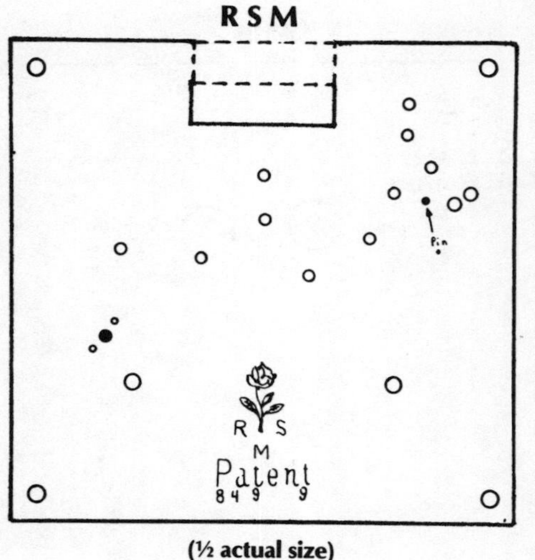

(½ actual size)

Plate 1504 3-Ball Pendulum
Duplex or Detent Escapement
Hour and Half Hour Bell Strike
4½" to 9" Suspension Spring
See Appendix 32

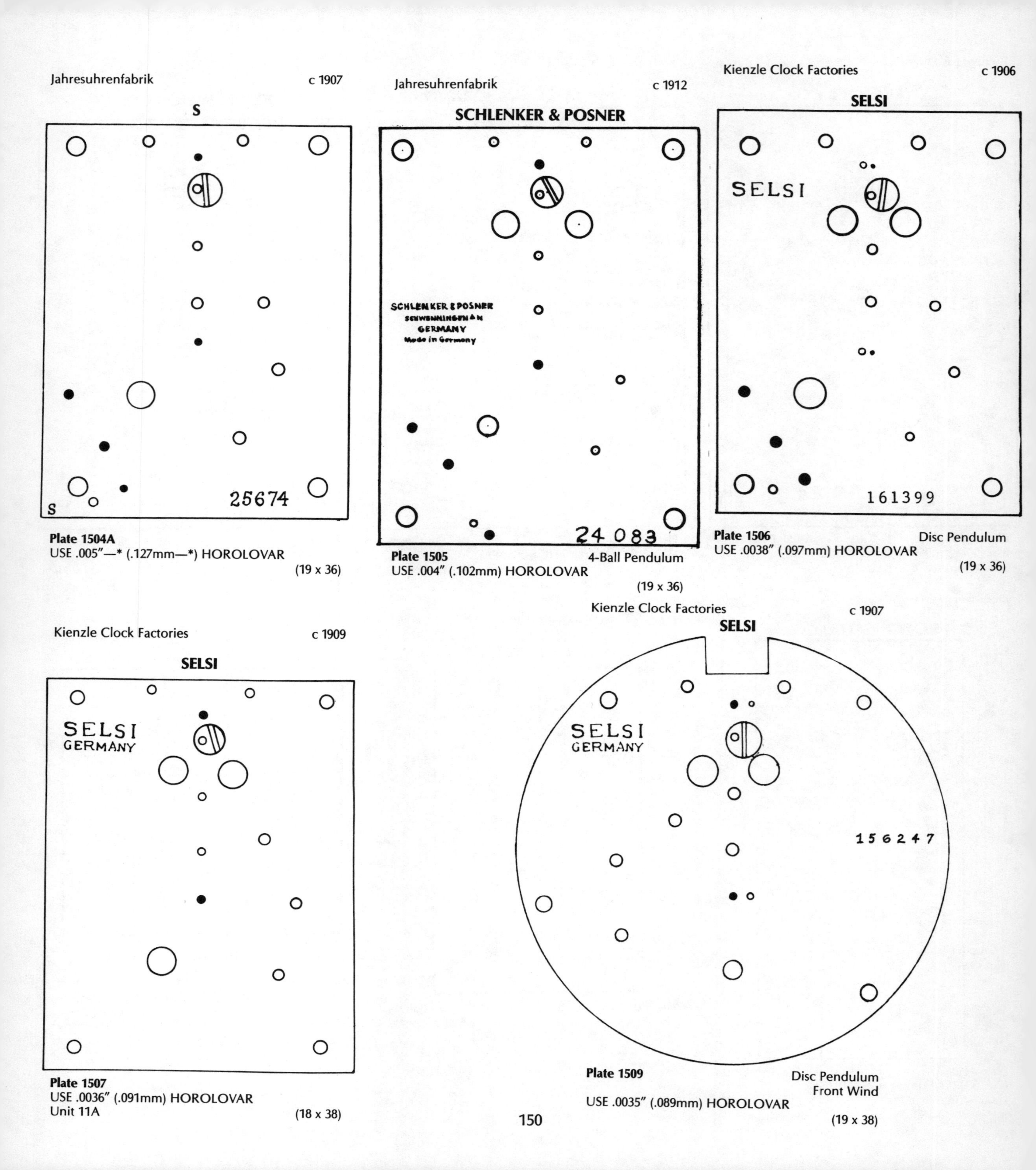

Plate 1504A
USE .005"—* (.127mm—*) HOROLOVAR
(19 x 36)

Plate 1505 4-Ball Pendulum
USE .004" (.102mm) HOROLOVAR
(19 x 36)

Plate 1506 Disc Pendulum
USE .0038" (.097mm) HOROLOVAR
(19 x 36)

Plate 1507
USE .0036" (.091mm) HOROLOVAR
Unit 11A
(18 x 38)

Plate 1509 Disc Pendulum
Front Wind
USE .0035" (.089mm) HOROLOVAR
(19 x 38)

Kienzle Clock Factories c 1912

SELSI

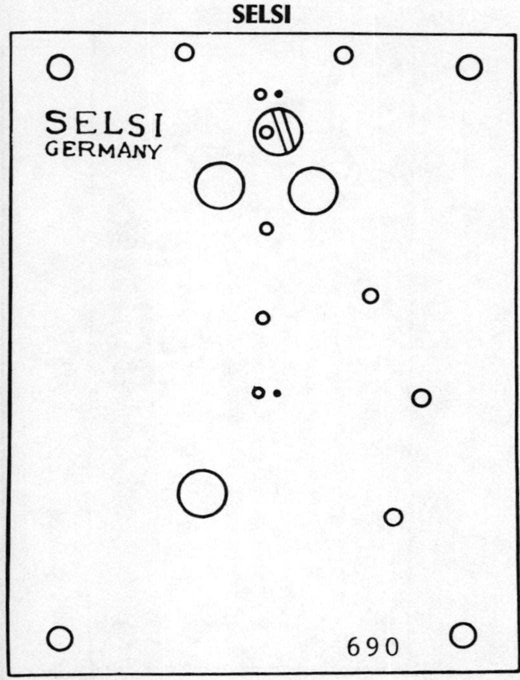

Plate 1511
USE .0035"* (.089mm*) HOROLOVAR
(18 x 38)

Badische Uhrenfabrik c 1904

SELSI

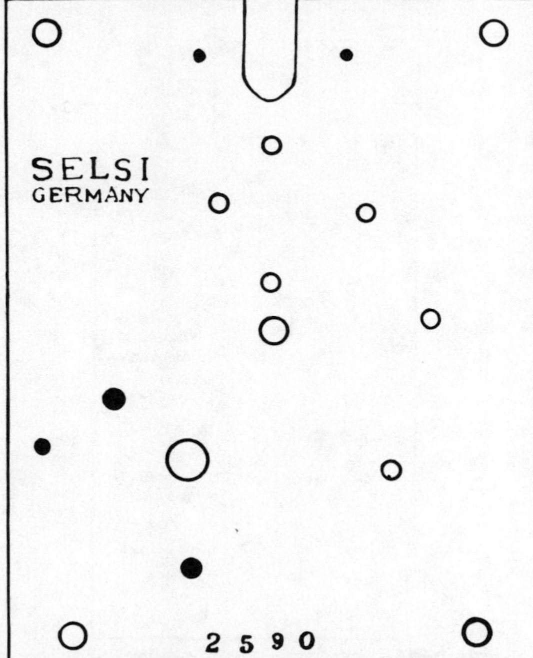

Plate 1515 4-Ball Pendulum
Lantern Pinions
USE .0035"* (.089mm*) HOROLOVAR
See Appendix 63 (20 x 38)

Sigfried Haller c 1960

SEMCA

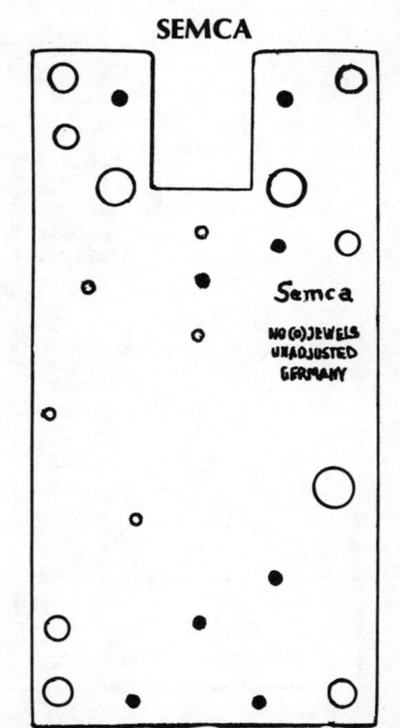

Plate 1517 4-Ball Pendulum
USE .003" (.076mm) HOROLOVAR
Units 42, 42A (18 x 38)
See Appendix 63, 64, 108

Phillipp Haas c 1907

SEMESTER UHR.

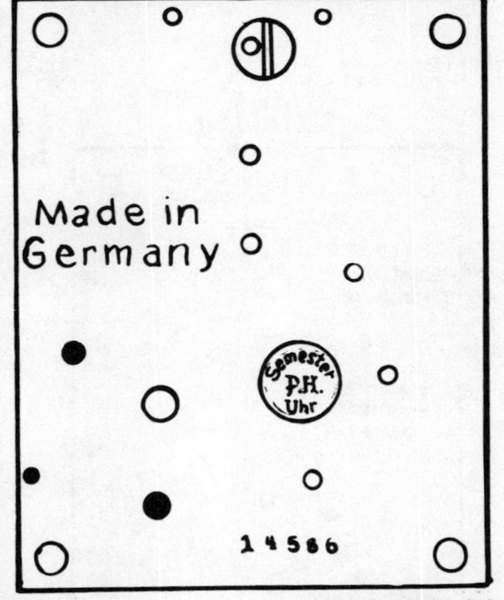

Plate 1519 Standard Disc Pendulum
Miniature Movement
Miniature Pendulum
USE .0038"* (.097mm*) HOROLOVAR
See Appendix 61, 118

Phillipp Haas c 1907

SEMESTER UHR.

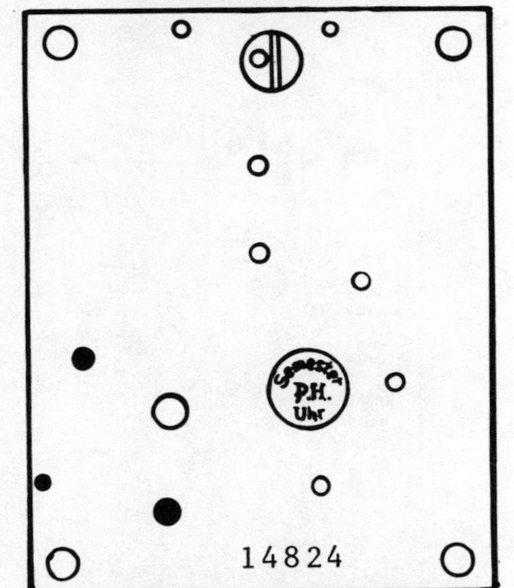

Plate 1519A Standard Disc Pendulum
Miniature Movement
Miniature Pendulum
USE .0038"* (.097mm*) HOROLOVAR
See Appendix 61, 118

Sigfried Haller c 1960

S. HALLER

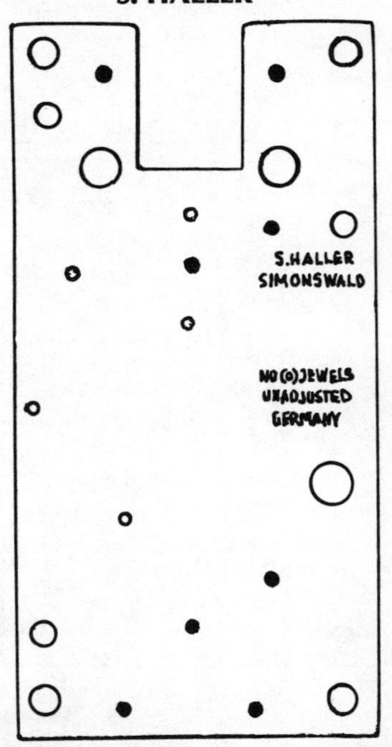

Plate 1520 4-Disc Pendulum
USE .003" (.076mm) HOROLOVAR
Units 42, 42A (18 x 38)
See Appendix 63, 64, 108

Sigfried Haller c 1960

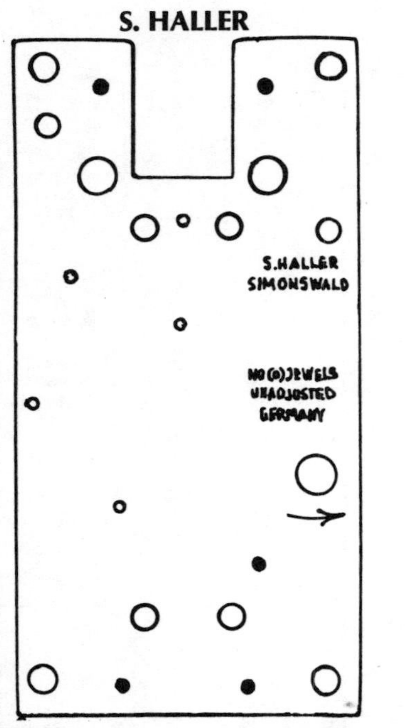

Plate 1520A 4-Ball Pendulum
USE .003" (.076mm) HOROLOVAR
Units 42, 42A (18 x 38)
See Appendix 63, 64, 108

Sigfried Haller c 1960

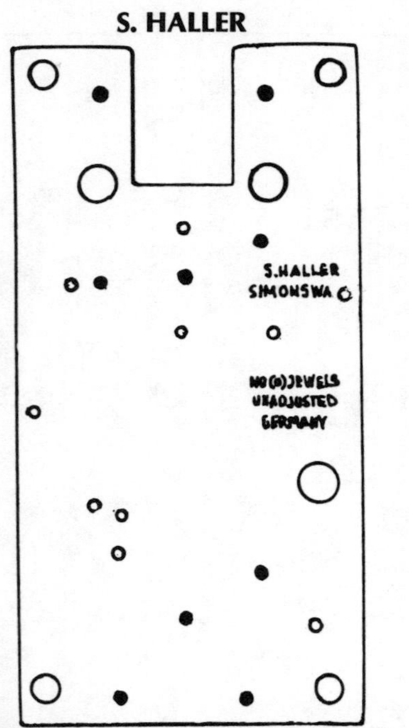

Plate 1520AA 4-Ball Pendulum
USE .003" (.076mm) HOROLOVAR
Units 42, 42A (18 x 38)
See Appendix 63, 64, 108

Sigfried Haller c 1960

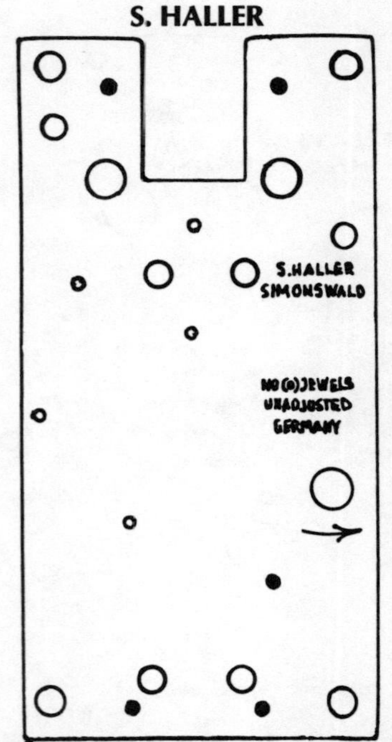

Plate 1520AAA 4-Ball Pendulum
USE .003" (.076mm) HOROLOVAR
Units 42, 42A (18 x 38)
See Appendix 63, 64, 108

Sigfried Haller c 1970

S. HALLER

Plate 1521 4-Ball Pendulum
Miniature Clock
USE .0022" (.056mm) HOROLOVAR
Units 43, 43A (12 x 24)
See Appendix 63, 64, 108

Sigfried Haller c 1974

S. HALLER

Plate 1521A 4-Ball Pendulum
Miniature Clock
Pin Pallet Escapement
USE .0022" (.056mm) HOROLOVAR
Units 43, 43A (12 x 24)
See Appendix 63, 64, 108

Sigfried Haller c 1970

S. HALLER

Plate 1521AA 4-Ball Pendulum
Miniature Clock
Pin Pallet Escapement
USE .0022" (.056mm) HOROLOVAR
Units 43, 43A (12 x 24)
See Appendix 63, 64, 108

Sigfried Haller c 1972

S. HALLER

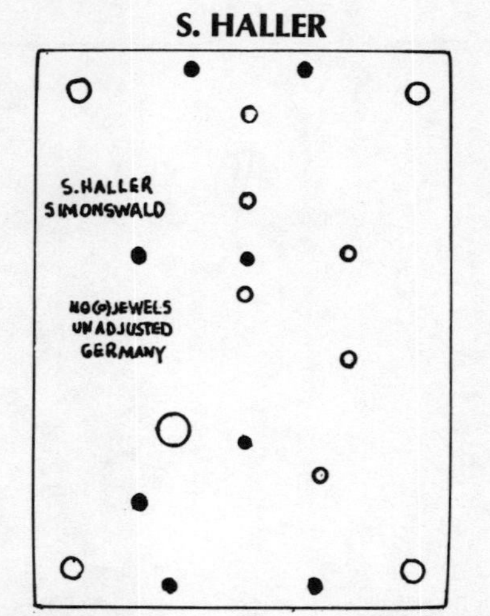

Plate 1522 4-Ball Pendulum
Midget Clock
Pin Pallet Escapement
USE .0018" (.046mm) HOROLOVAR
Units 43, 43A, 44, 44A (12 x 24)
See Appendix 108

Sigfried Haller c 1974

S. HALLER

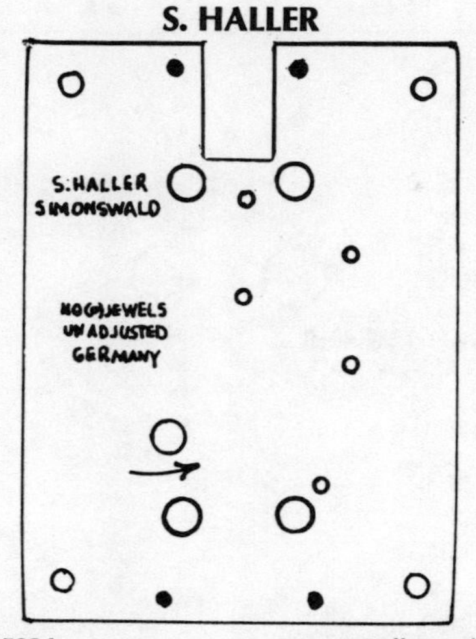

Plate 1522A 4-Ball Pendulum
Midget Clock
Pin Pallet Escapement
USE .0018" (.046mm) HOROLOVAR
Units 44, 44A (12 x 24)
See Appendix 63, 64, 108

Sigfried Haller c 1986

S. HALLER

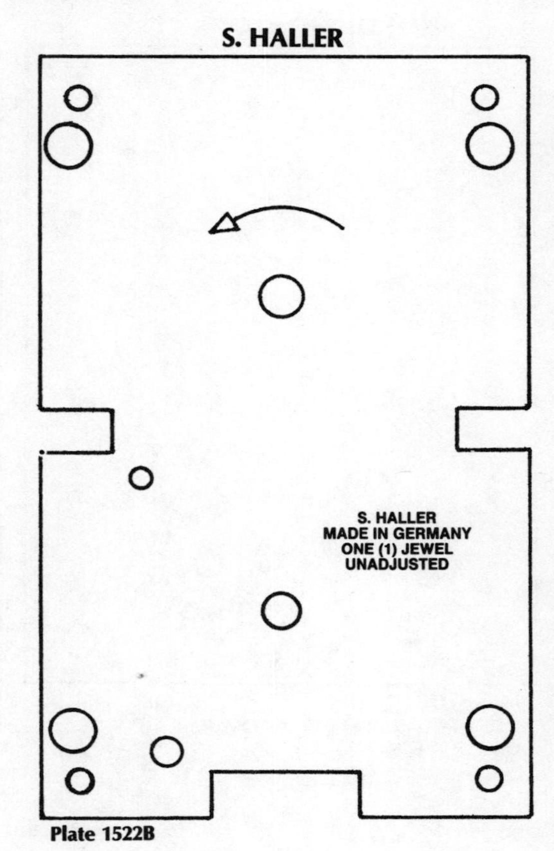

Plate 1522B

Jahresuhrenfabrik c 1900

S. THOMA

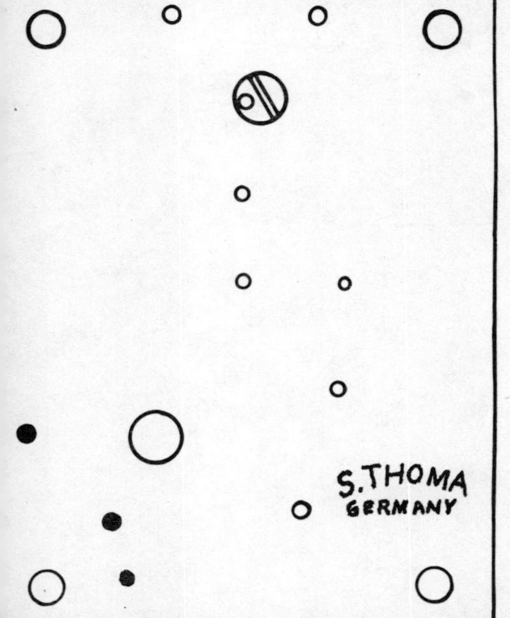

Plate 1522D Disc Pendulum
SE .004" (.102mm) HOROLOVAR
(19 x 36)

Kieninger & Obergfell c 1930

SOKOL MONTAG & CO.

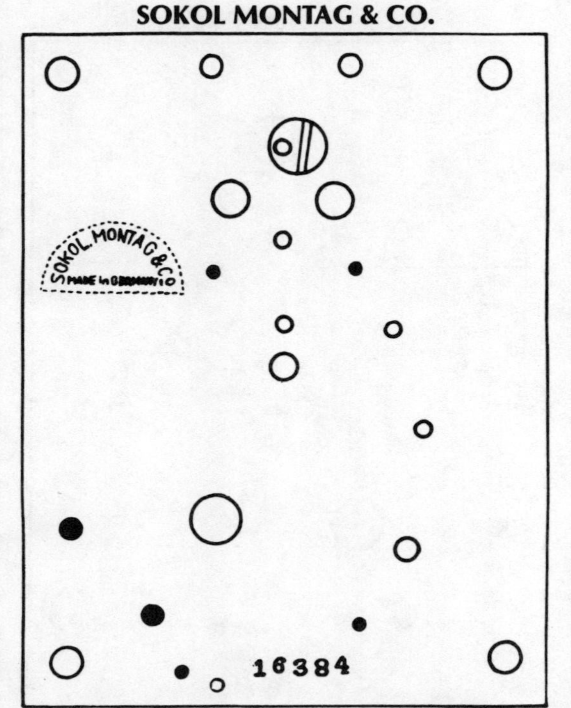

Plate 1523 4-Ball Pendulum
USE .0032" (.081mm) HOROLOVAR
Unit 1 (19 x 38)
See Appendix 76

Kieninger & Obergfell c 1930

SOKOL MONTAG & CO.

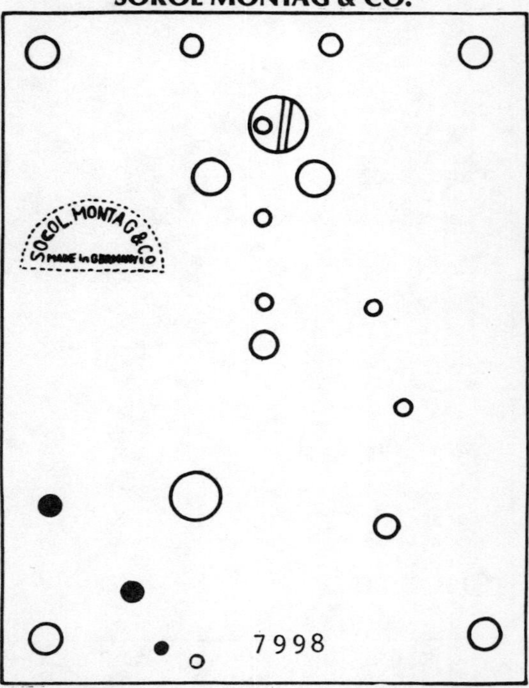

Plate 1524 4-Ball Pendulum
USE .0032" (.081mm) HOROLOVAR
Unit 1 (19 x 38)
See Appendix 76

Jahresuhrenfabrik c 1912

SOKOL MONTAG & CO.

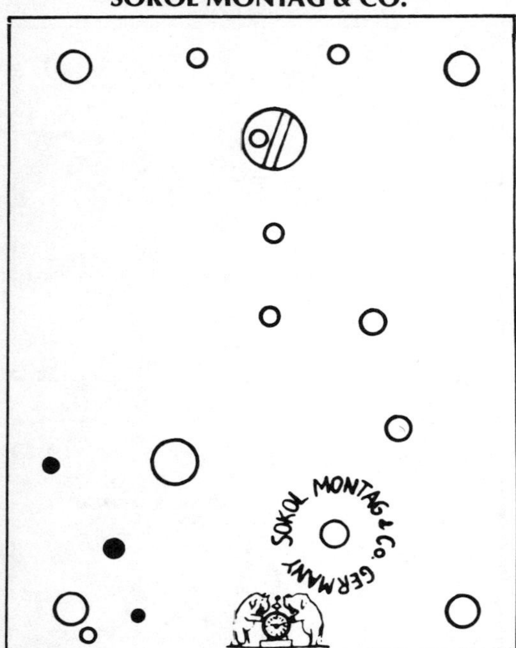

Plate 1527 4-Ball Pendulum
USE .004" (.102mm) HOROLOVAR
Units 6, 7, 8, 9 (19 x 36)

Kieninger & Obergfell c 1912

SPECIALTY TRADING CO.

Plate 1529 4-Ball Pendulum
USE .0032" (.081mm) HOROLOVAR
Unit 1 (19 x 36)
See Appendix 76

Jahresuhrenfabrik c 1900

THE CARP

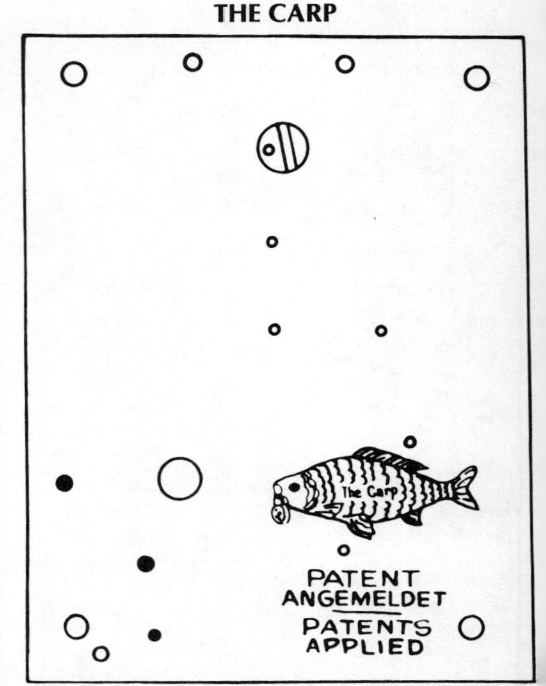

Plate 1531 Twin Loop "Temperature Compensating" Pendulum
USE .004" (.102mm) HOROLOVAR
See Appendix 42 (19 x 36)

Kieninger & Obergfell c 1950

THE GERSON CO.

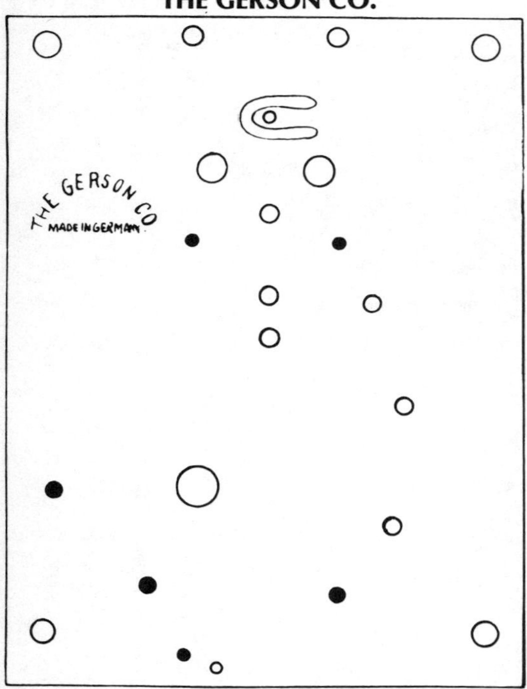

Plate 1533 4-Ball Pendulum
USE .0032" (.081mm) HOROLOVAR
Units 1, 3A (19 x 38)
See Appendix 76

Kieninger & Obergfell c 1930

THE J. L. HUDSON CO.

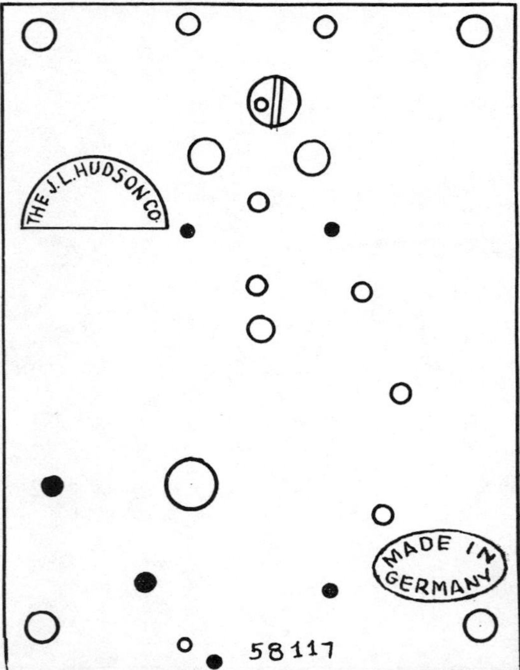

Plate 1534 4-Ball Pendulum
USE .0032" (.081mm) HOROLOVAR
Unit 1 (19 x 38)
See Appendix 76

Kieninger & Obergfell c 1950

THE J. L. HUDSON CO.

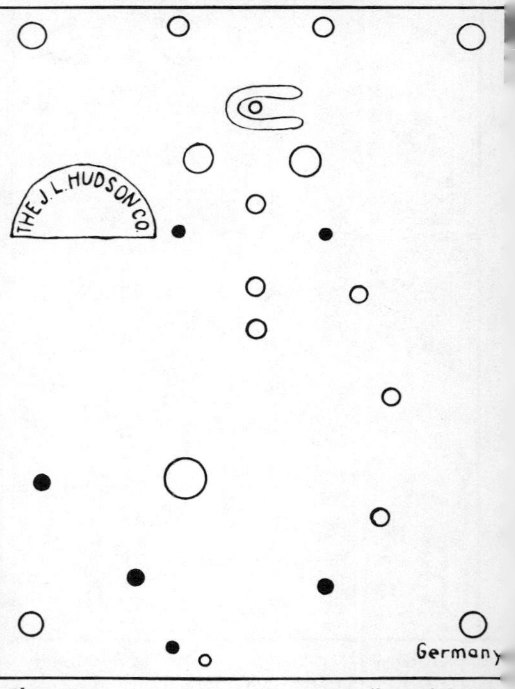

Plate 1535 4-Ball Pendulum
USE .0032" (.081mm) HOROLOVAR
Units 1, 3A (19 x 38)
See Appendix 76

Kieninger & Obergfell c 1949

THE J. L. HUDSON CO.

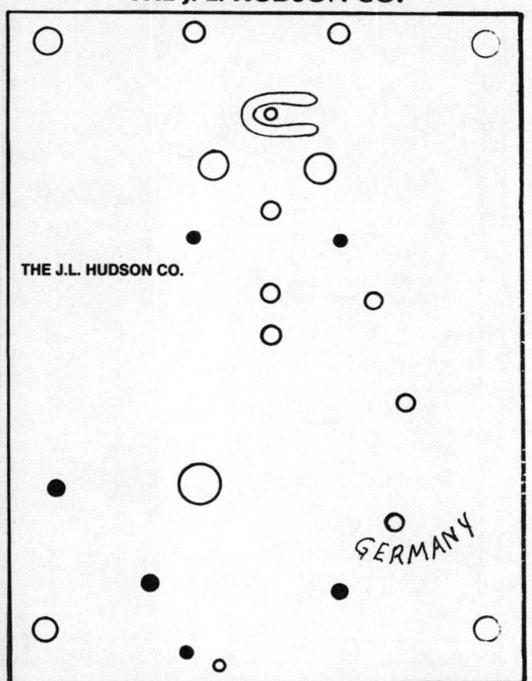

Plate 1536 4-Ball Pendulum
USE .0032" (.081mm) HOROLOVAR
Units 1, 3A (19 x 38)
See Appendix 76

Kieninger & Obergfell c 1951

THE J. L. HUDSON CO.

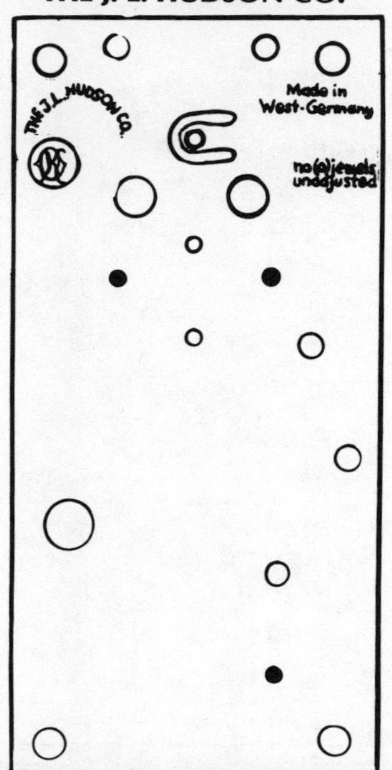

Plate 1539 4-Ball Pendulum
USE .0032" (.081mm) HOROLOVAR
Units 1, 3A, 3B, 3C (19 x 38)
See Appendix 76

Kieninger & Obergfell c 1954

THE J. L. HUDSON CO.

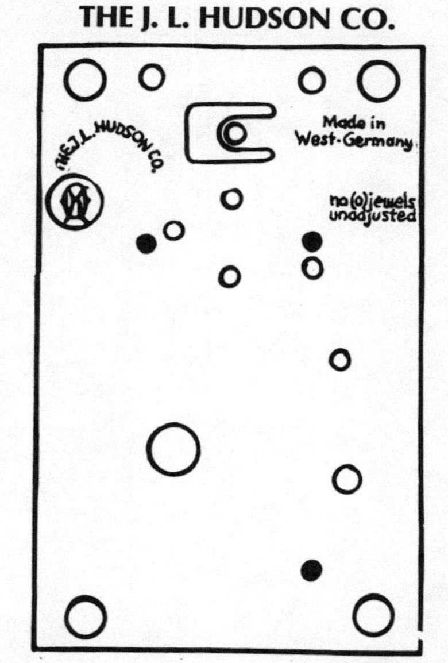

Plate 1543 4-Ball Pendulum Miniature Clock
USE .0023" (.058mm) HOROLOVAR
Units 5A, 5B, 5C, 5D, 5E (14 x 30)
See Appendix 77

Jahresuhrenfabrik c 1912

THE NATIONAL SILVER CO.

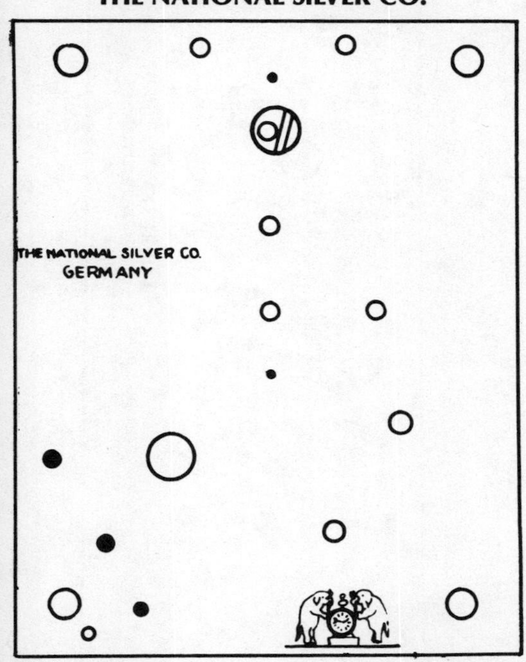

Plate 1547 4-Ball Pendulum
USE .004" (.102mm) HOROLOVAR
Units 6, 7, 8, 9 (19 x 36)

Jahresuhrenfabrik c 1912

THE NATIONAL SILVER CO.

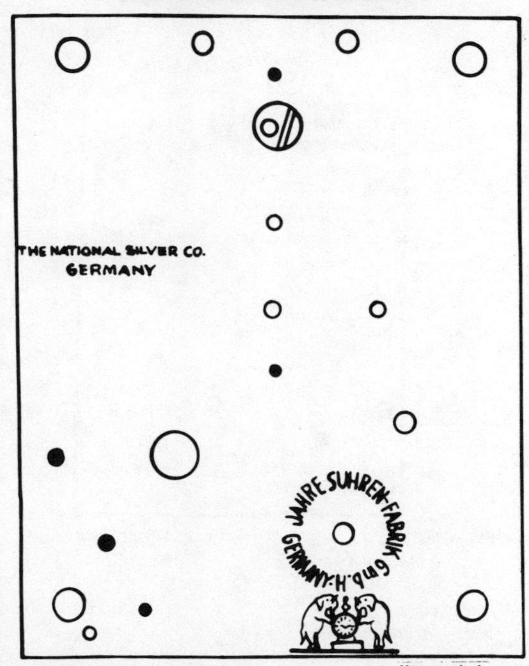

Plate 1551 4-Ball Pendulum
USE .004" (.102mm) HOROLOVAR
Units 6, 7, 8, 9 (19 x 36)

Kieninger & Obergfell c 1930

THE NATIONAL SILVER CO.

THE NATIONAL SILVER CO.
MADE IN GERMANY
58916

Plate 1555 4-Ball Pendulum
USE .0032" (.081mm) HOROLOVAR
Unit 1 (19 x 38)
See Appendix 76

Kieninger & Obergfell c 1912

THE NATIONAL SILVER CO.

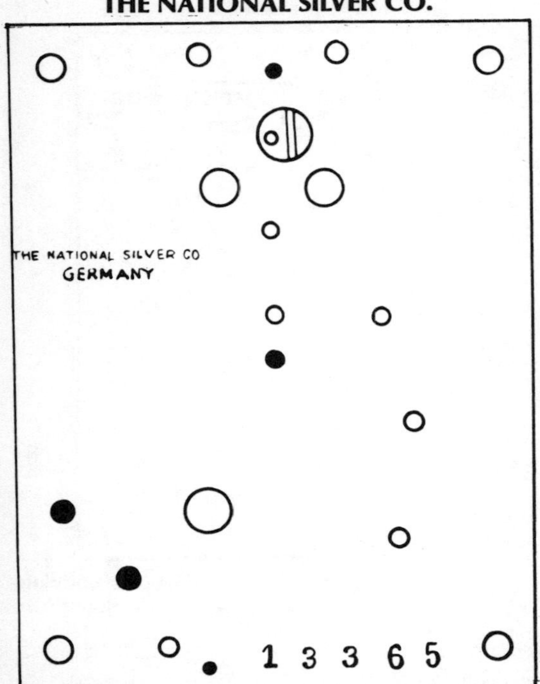

Plate 1559 4-Ball Pendulum
USE .0032" (.081mm) HOROLOVAR
See Appendix 76 (19 x 38)

Kieninger & Obergfell c 1930

TRANSATLANTIC CLOCK CO.

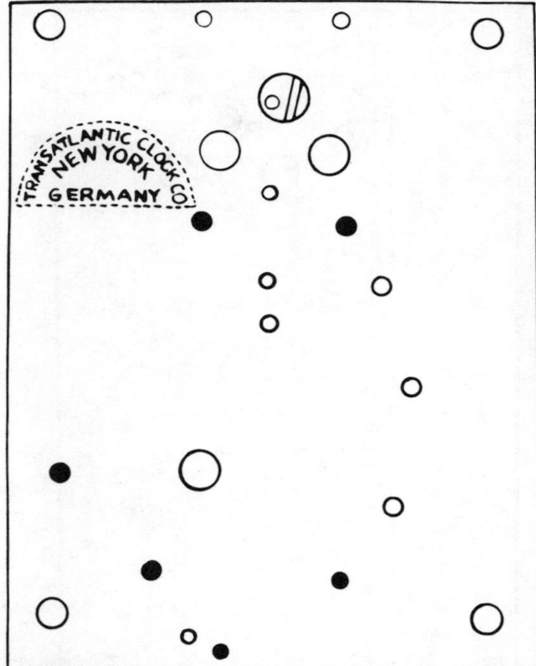

Plate 1563 4-Ball Pendulum
USE .0032" (.081mm) HOROLOVAR
Unit 1 (19 x 38)
See Appendix 76

Uhrenfabrik Herr c 1951

UHRENFABRIK HERR

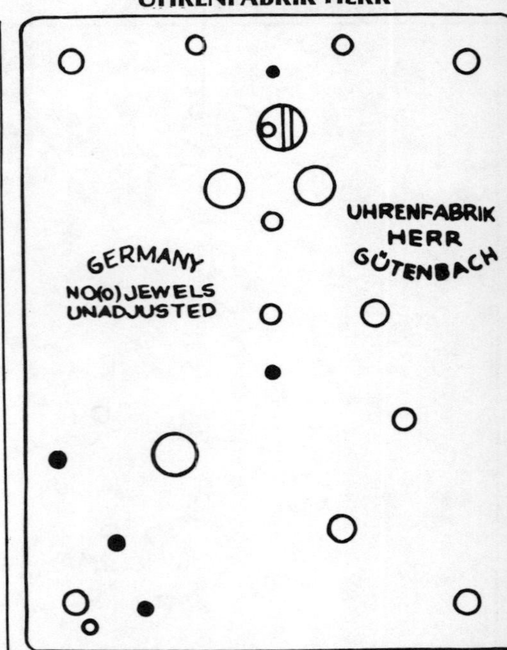

Plate 1567 4-Ball Pendulum
USE .004" (.102mm) HOROLOVAR
Unit 27A (19 x 36)
See Appendix 130

Uhrenfabrik Herr c 1951

UHRENFABRIK HERR

Plate 1571 4-Ball Pendulum
USE .004" (.102mm) HOROLOVAR
Unit 27A (19 x 36)
See Appendix 130

Uhrenfabrik M. Reiner c 1951

UHRENFABRIK HERR

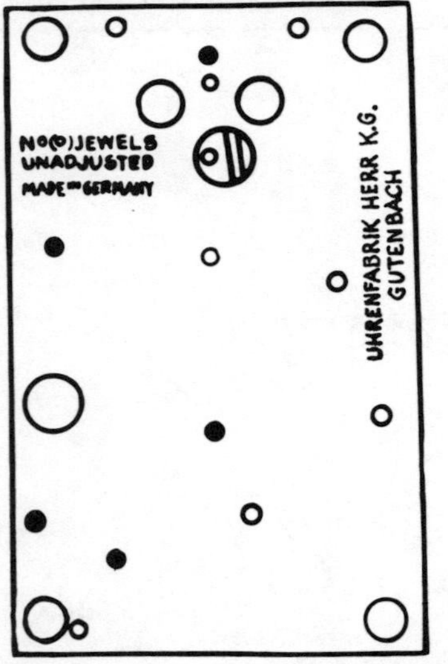

Plate 1572 4-Ball Pendulum
Miniature Clock
USE .0025" (.064mm) HOROLOVAR
Units 28A, 28B (16 x 36)
See Appendix 25, 43, 54, 109

Uhrenfabrik Herr c 1951

UHRENFABRIK HERR

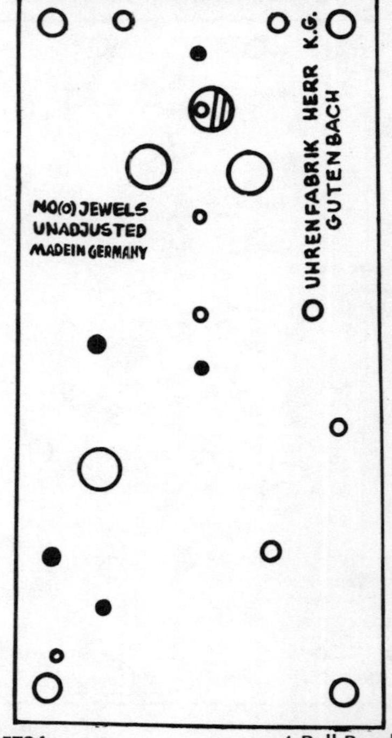

Plate 1572A 4-Ball Pendulum
USE .0038" (.097mm) HOROLOVAR
Unit 27B (19 x 36)
See Appendix 25, 43, 106, 109, 110

Uhrenfabrik M. Reiner c 1951

UHRENFABRIK HERR

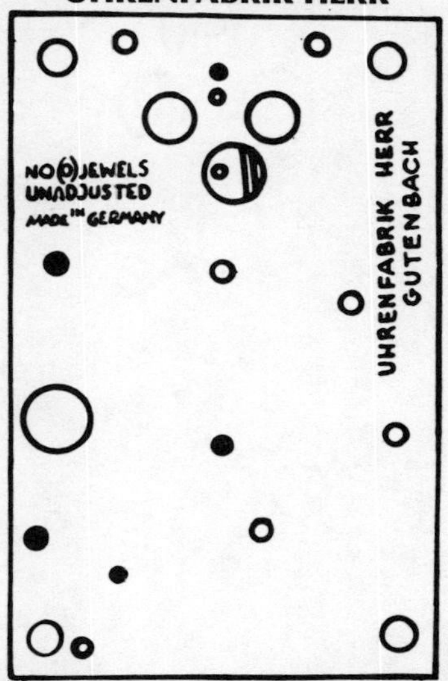

Plate 1572B Standard 4-Ball Pendulum Miniature Movement
USE .003" (.076mm) HOROLOVAR
Units 22, 23A, 24 (16 x 36)
See Appendix 25, 43, 54, 109

Uhrenfabrik M. Reiner c 1951

UHRENFABRIK K.G.

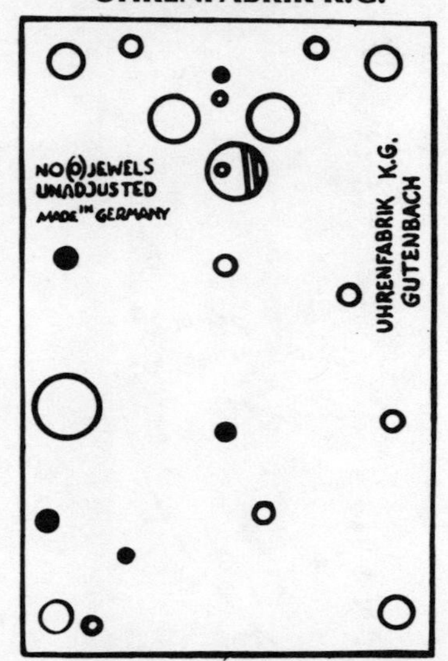

Plate 1573 Standard 4-Ball Pendulum Miniature Movement
USE .003" (.076mm) HOROLOVAR
Units 22, 23A, 24 (16 x 36)
See Appendix 25, 43, 54, 109

Uhrenfabrik M. Reiner c 1954

UHRENFABRIK M. REINER & CO.

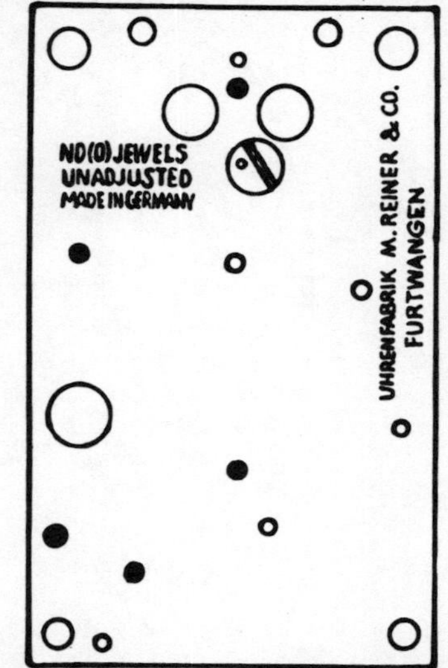

Plate 1574 Standard 4-Ball Pendulum Miniature Movement
USE .0032" (.081mm) HOROLOVAR
Unit 23B (16 x 36)
See Appendix 25, 43, 55, 109

Uhrenfabrik M. Reiner c 1954

UHRENFABRIK M. REINER & CO.

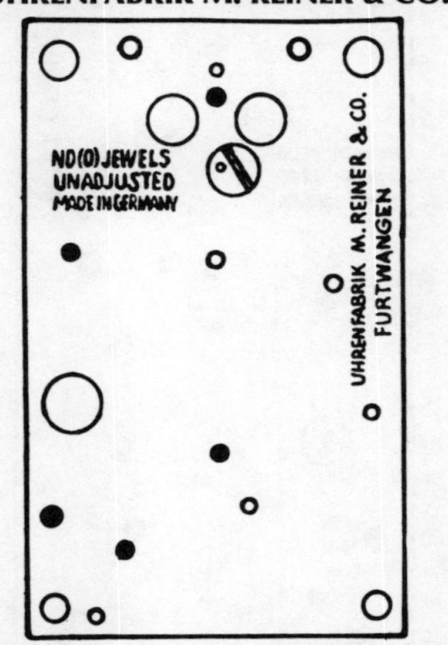

Plate 1574A 4-Ball Pendulum Miniature Clock
USE .0028" (.071mm) HOROLOVAR
Unit 28C
See Appendix 25, 43, 55, 109 (16 x 36)

Uhrenfabrik Herr c 1951

UHRENFABRIK M. REINER & CO.

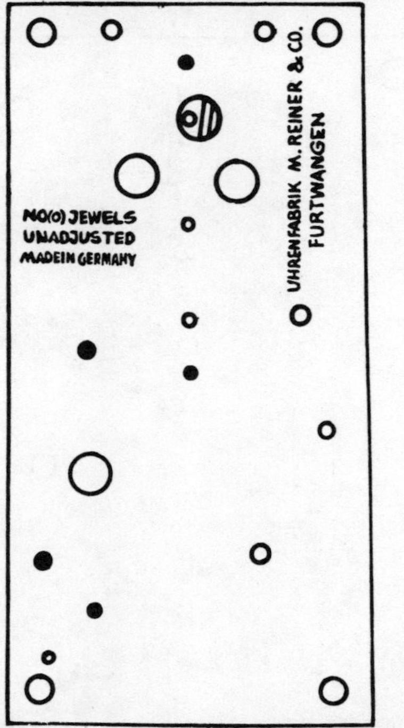

Plate 1574AA 4-Ball Pendulum
USE .0038" (.097mm) HOROLOVAR
Unit 27B (19 x 36)
See Appendix 43, 106, 109, 110

Uhrenfabrik Neueck c 1956

UHRENFABRIK NEUECK

NO(0) JEWELS
UNADJUSTED
MADE IN GERMANY
UHRENFABRIK
NEUECK GmbH

Plate 1574B 4-Ball Pendulum
USE .0036" (.091mm) HOROLOVAR
Unit 27C (19 x 36)
See Appendix 44, 110

Kienzle Clock Factories c 1905

URANIA

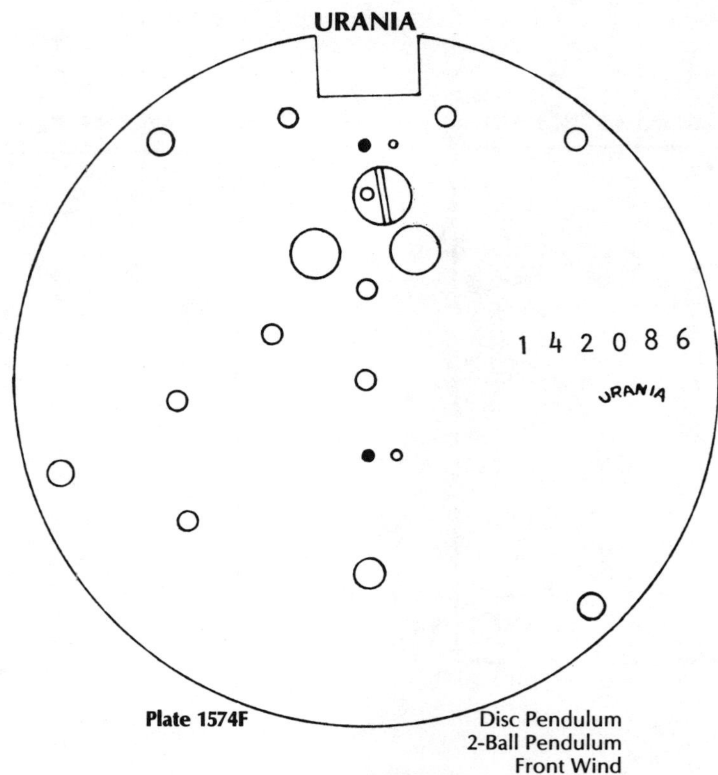

Plate 1574F Disc Pendulum
2-Ball Pendulum
Front Wind

USE .0035" (.089mm) HOROLOVAR
See Appendix 8 (19 x 38)

Kienzle Clock Factories c 1910

URANIA

Plate 1575 Disc Pendulum
4-Ball Pendulum

USE .0038"* (.097mm*) HOROLOVAR
See Appendix 41 (18 x 38)

Kienzle Clock Factories c 1910

URANIA

Plate 1579 Disc Pendulum
4-Ball Pendulum

USE .0038"* (.097mm*) HOROLOVAR
See Appendix 41 (18 x 38)

Wintermantel Uhrenfabrik c 1905

W

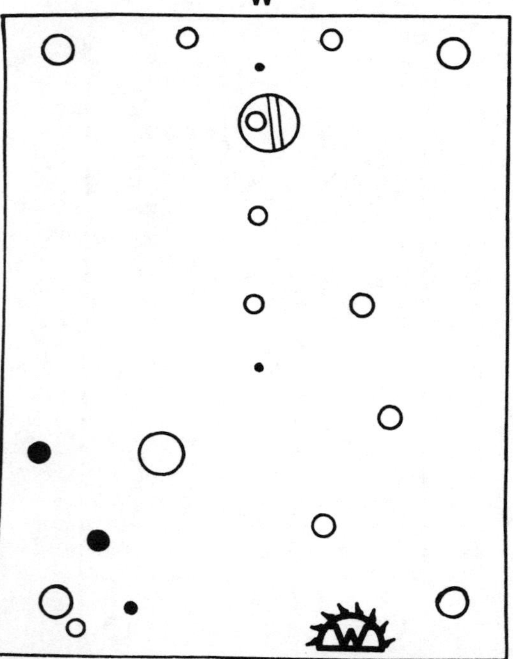

Plate 1579AA 4-Ball Pendulum

USE .004" (.102mm) HOROLOVAR
See Appendix 107 (19 x 36)

Uhrenfabrik Herr c 1952

WELBY CORP.

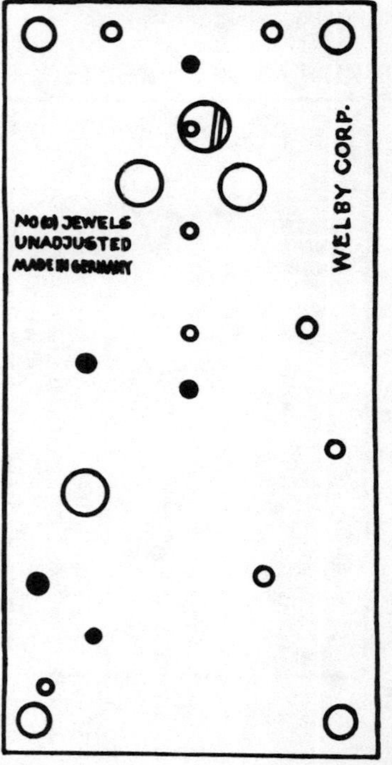

Plate 1579A 4-Ball Pendulum

USE .0035" (.089mm) HOROLOVAR
Units 19, 20A, 21

USE .0038" (.097mm) HOROLOVAR
Unit 27B (19 x 36)
See Appendix 106, 109, 110

Kieninger & Obergfell c 1974

WELBY CORPORATION

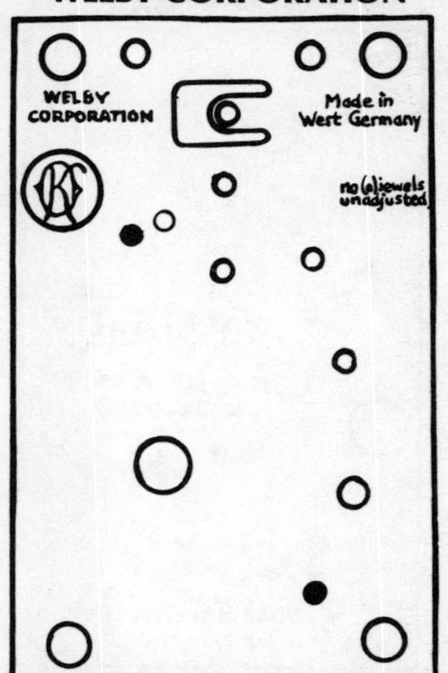

Plate 1579B 4-Ball Pendulum Miniature Clock

USE .0023" (.058mm) HOROLOVAR

Units 5C, 5D, 5E (14 x 30)

See Appendix 77

Franz Hermle c 1954

WELBY CORPORATION

WELBY CORPORATION GERMANY NO (O) JEWELS UNADJUSTED

Plate 1581 4-Ball Pendulum Miniature Clock

USE .0022" (.056mm) HOROLOVAR

Unit 25B (13 x 30)

See Appendix 63

Kieninger & Obergfell c 1951

WELBY CORPORATION

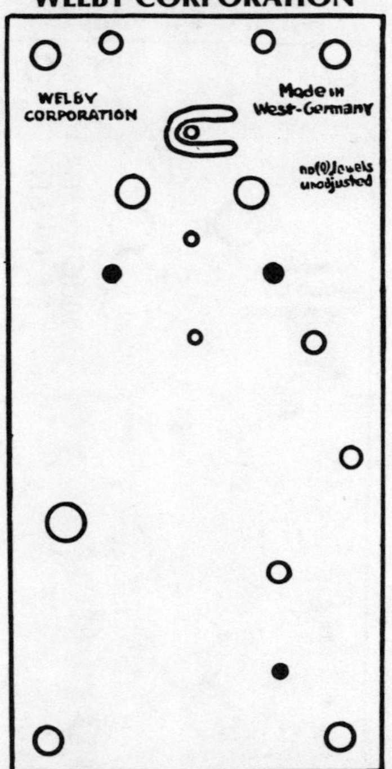

Plate 1580 4-Ball Pendulum

USE .0032" (.081mm) HOROLOVAR

Units 1, 3A, 3B, 3C (19 x 38)

See Appendix 76

Kieninger & Obergfell c 1965

WELBY CORPORATION

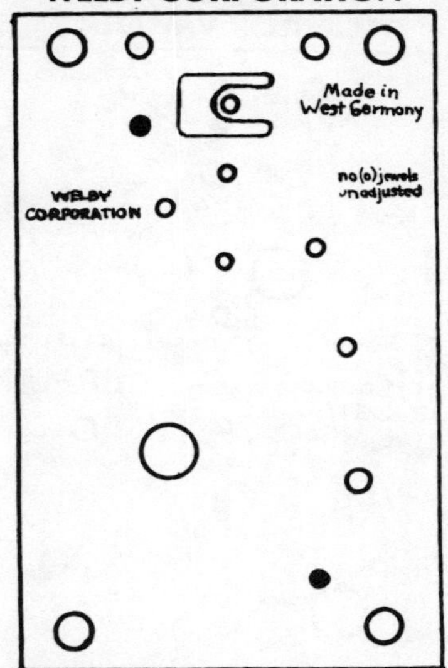

Plate 1580A 4-Ball Pendulum Miniature Clock

USE .0023" (.058mm) HOROLOVAR

Units 5E, 5F (14 x 30)

See Appendix 77

Franz Hermle c 1954

WELBY CORPORATION

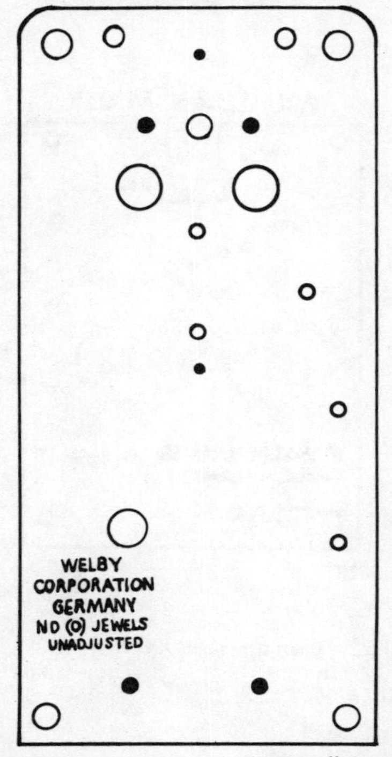

Plate 1581A 4-Ball Pendulum

USE .0033" (.084mm) HOROLOVAR

Unit 25A (18 x 38)

See Appendix 71

Uhrenfabrik Herr c 1952

WILHELM V. MAIER

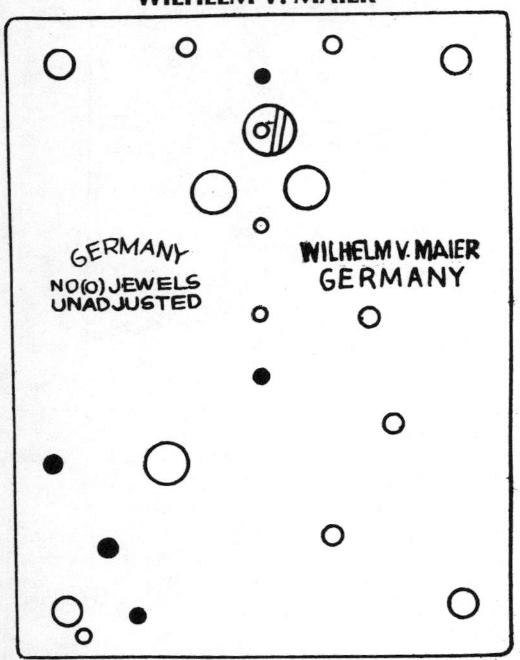

Plate 1582 4-Ball Pendulum
USE .004" (.102mm) HOROLOVAR
Unit 27A (19 x 36)
See Appendix 130

Konrad Mauch c 1957

WILHELM V. MAIER

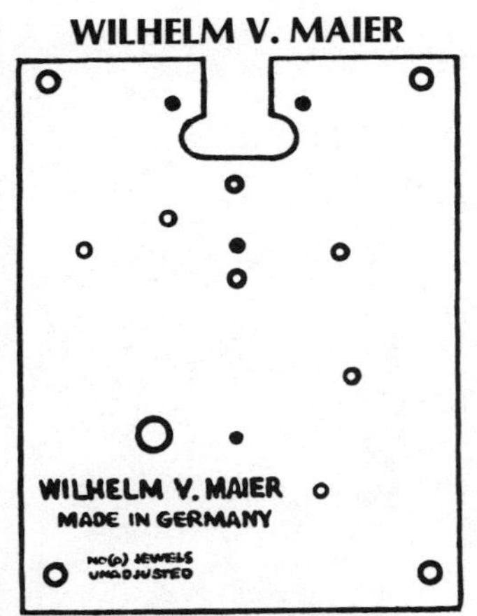

Plate 1582B 3-Ball Pendulum
Midget Clock
Pin Pallet Escapement
USE .0022" (.056mm) HOROLOVAR
Units 38, 38A Original (12 x 25)
See Appendix 63, 64, 86 Revised (14 x 25)

Uhrenfabrik Herr c 1951

WILHELM V. MAIER

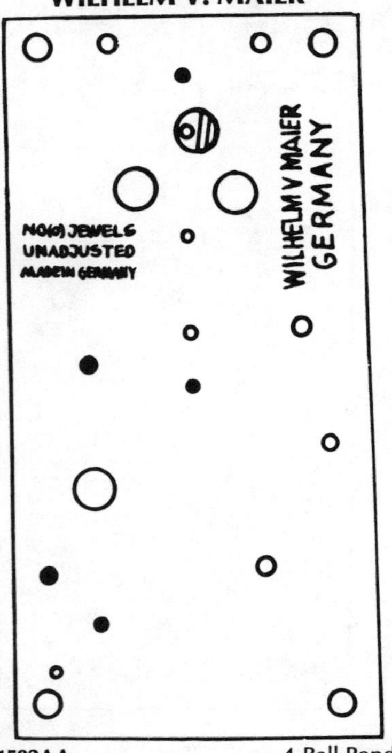

Plate 1582AA 4-Ball Pendulum
USE .0035" (.089mm) HOROLOVAR
Units 19, 20A, 21
USE .0038" (.097mm) HOROLOVAR
Unit 27B (19 x 36)
See Appendix 106, 109, 110

Uhrenfabrik M. Reiner c 1951

WILHELM V. MAIER

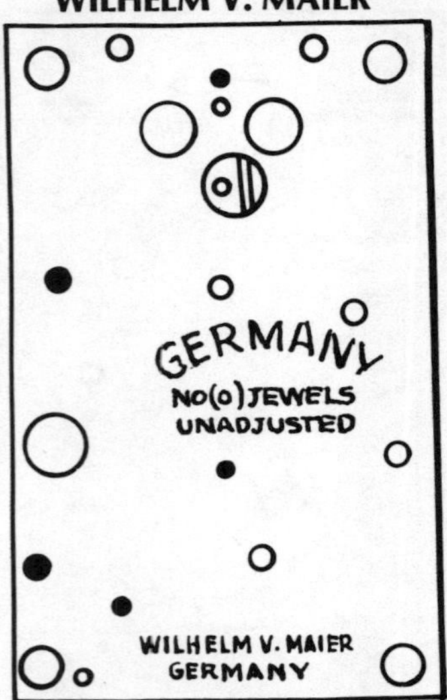

Plate 1582A Standard 4-Ball Pendulum
Miniature Movement
USE .003" (.076mm) HOROLOVAR
Units 22, 23A, 24 (16 x 36)
See Appendix 25, 54, 109

Kern & Sohne c 1954

WILMAC

NO(O) JEWELS
UNADJUSTED
Made in Germany
WILMAC
GERMANY

Plate 1582C 4-Ball Pendulum
USE .0036" (.091mm) HOROLOVAR
Unit 11B (18 x 38)
See Appendix 81

Uhrenfabrik Neueck c 1956

WILMAC

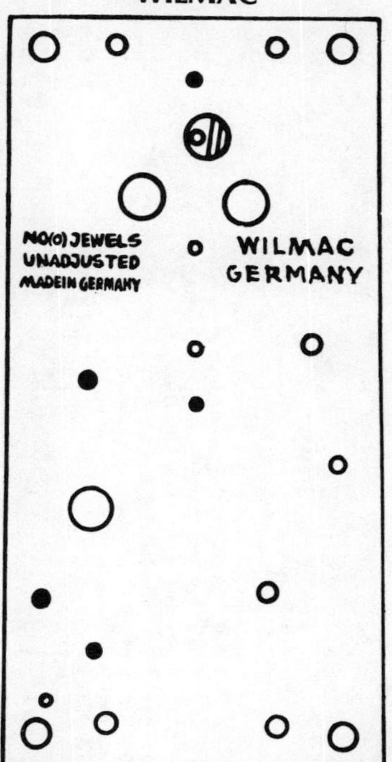

Plate 1582D 4-Ball Pendulum
USE .0036" (.091mm) HOROLOVAR
Unit 27C (19 x 36)
See Appendix 43

W. Petersen c 1952

W. PETERSEN

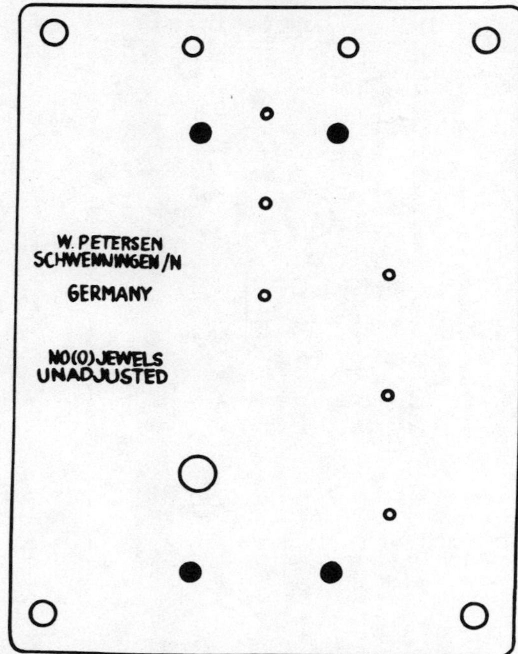

Plate 1583 4-Ball Pendulum
USE .0033" (.084mm) HOROLOVAR
Units 18B, 18C (19 x 38)
See Appendix 59, 65

Georg Wurthner c 1951

WURTHNER WESTERN

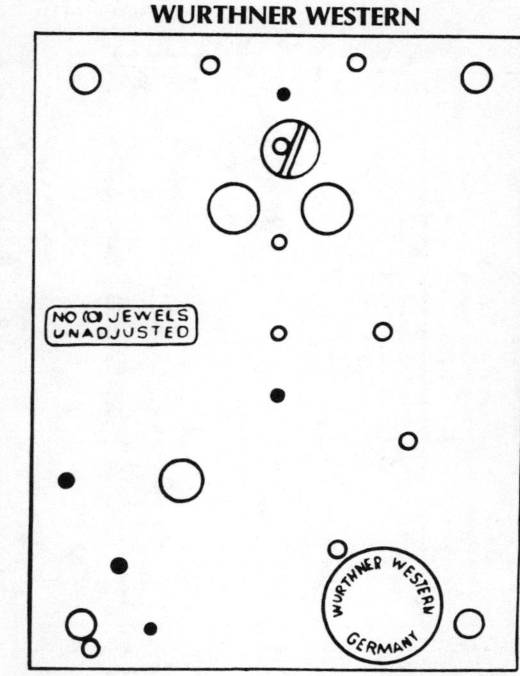

Plate 1587 4-Ball Pendulum
USE .004" (.102mm) HOROLOVAR
Unit 26 (19 x 36)

Georg Wurthner c 1952

WURTHNER WESTERN

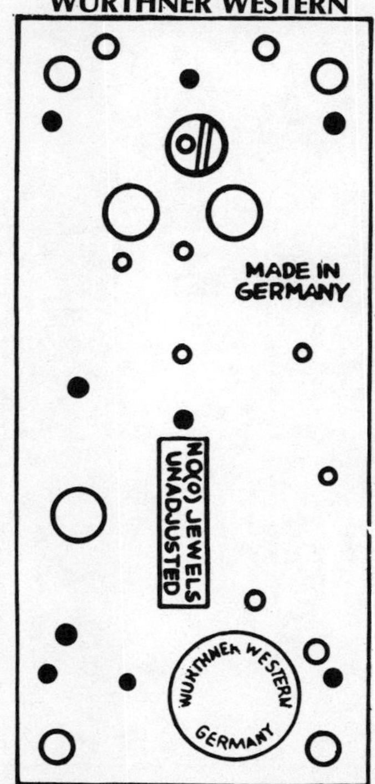

Plate 1591
USE .004" (.102mm) HOROLOVAR
Unit 26 (19 x 36)
See Appendix 67, 105

Georg Wurthner c 1953

WURTHNER WESTERN

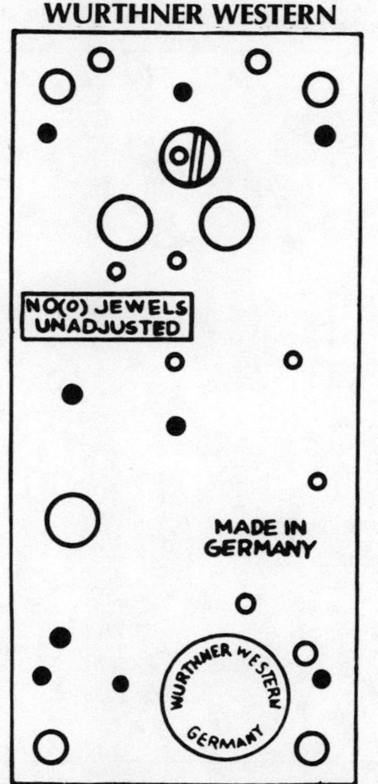

Plate 1592 4-Ball Pendulum
USE .004" (.102mm) HOROLOVAR
Unit 26 (19 x 36)
See Appendix 67, 105

Georg Wurthner c 1957

WURTHNER WESTERN

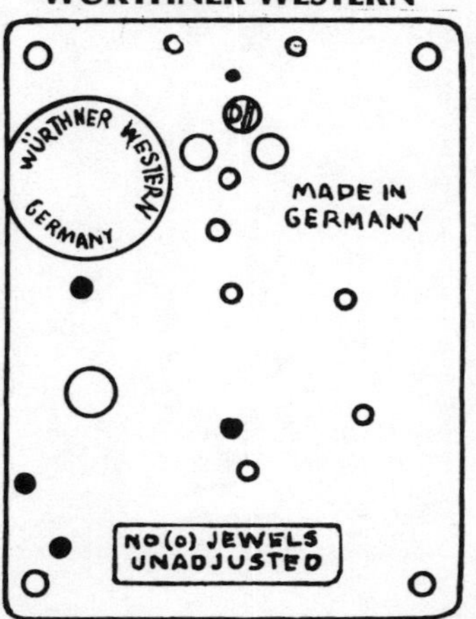

Plate 1593 4-Ball Pendulum
Midget Clock
USE .002" (.051mm) HOROLOVAR
Unit 37 (9.5 x 27)
See Appendix 27

Konrad Mauch c 1954

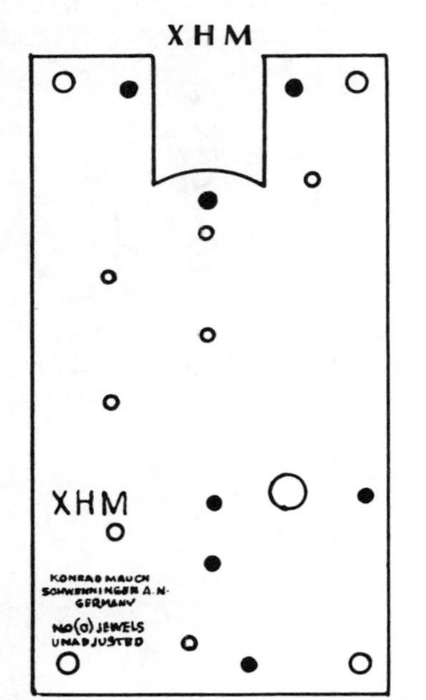

Plate 1594 4-Ball Pendulum
Miniature Clock
USE .0032" (.081mm) HOROLOVAR
Unit 14A
USE .003" (.076mm) HOROLOVAR
Unit 14B Original (13 x 32)
See Appendix 52, 63, 64, 85 Revised (15 x 32)

Georg Wurthner c 1952

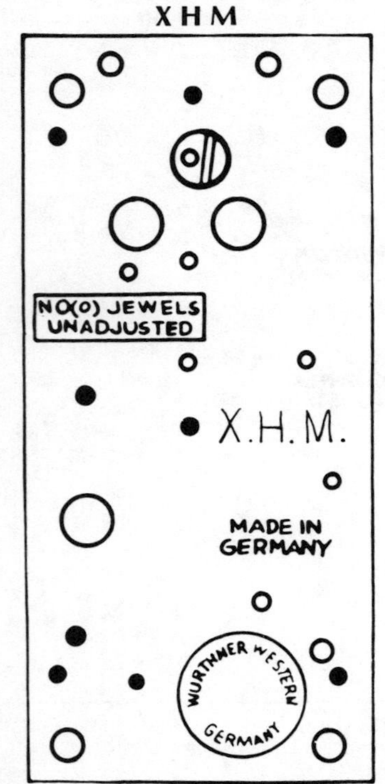

Plate 1594A 4-Ball Pendulum
USE .004" (.102mm) HOROLOVAR
Unit 26 (19 x 36)
See Appendix 67

Jahresuhrenfabrik c 1912

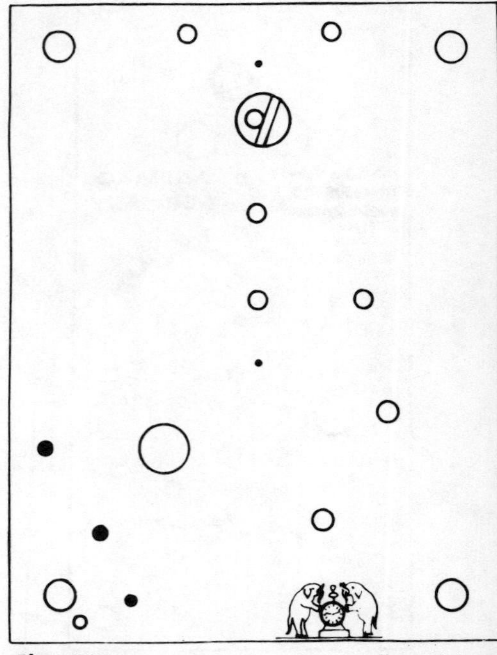

Plate 1595 4-Ball Pendulum
USE .004" (.102mm) HOROLOVAR
Units 6789A (19 x 36)

Jahresuhrenfabrik c 1911

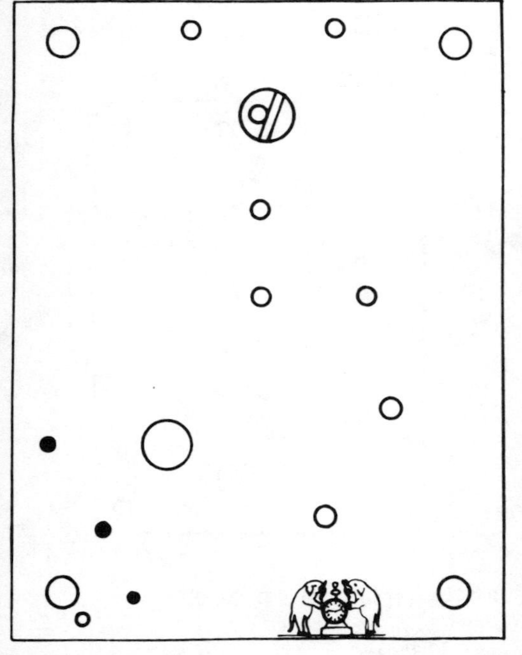

Plate 1595A 4-Ball Pendulum
USE .004" (.102mm) HOROLOVAR
Units 6789A (19 x 36)

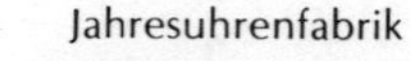

Jahresuhrenfabrik c 1903

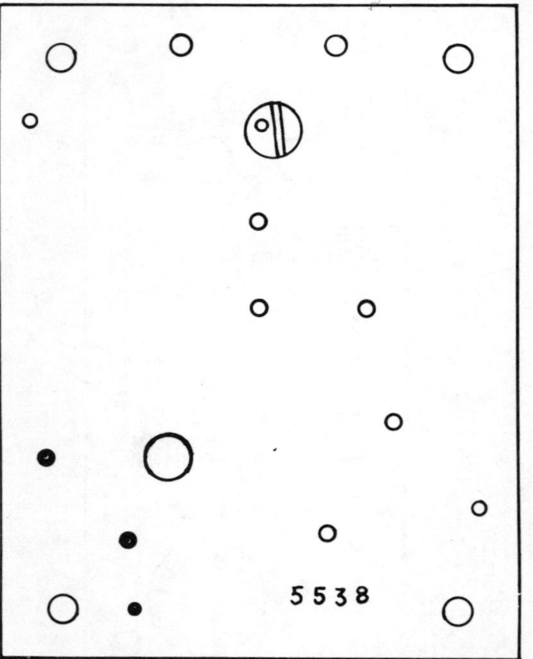

Plate 1597 Disc Pendulum
4-Ball Pendulum
USE .0035"* (.089mm*) HOROLOVAR
(19 x 36)

Manufacturer Not Known c 1902

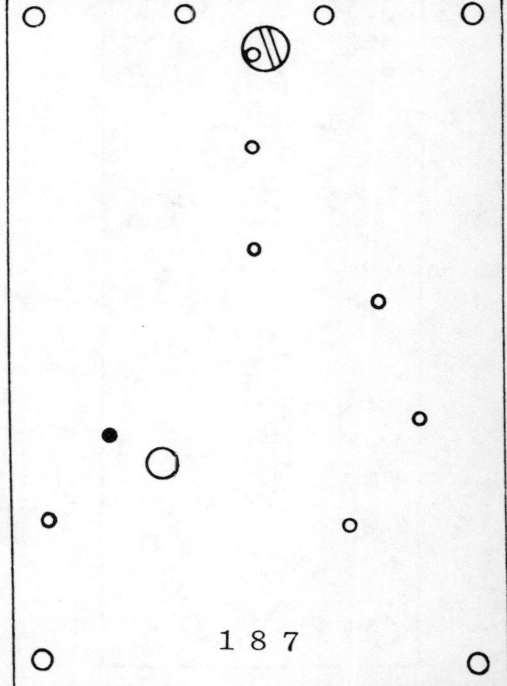

Plate 1599A Disc Pendulum
See Appendix 73 (19 x 36)

Kienzle Clock Factories c 1909

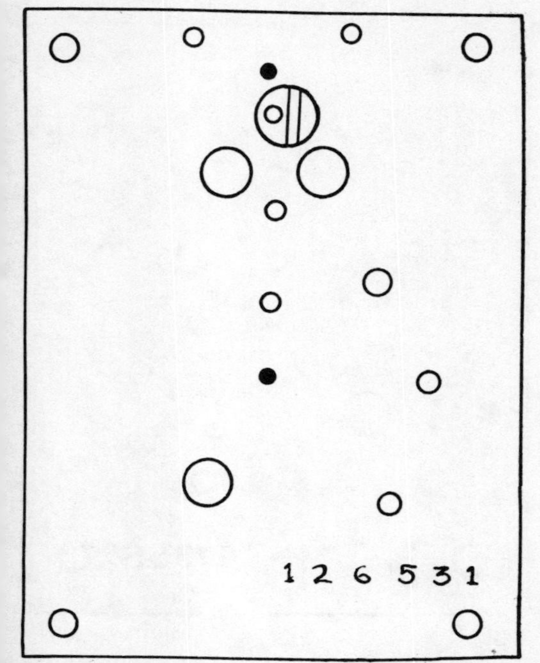

Plate 1601 Disc Pendulum
4-Ball Pendulum
USE .0036" (.091mm) HOROLOVAR
Unit 11A (18 x 38)
See Appendix 41

Kienzle Clock Factories c 1908

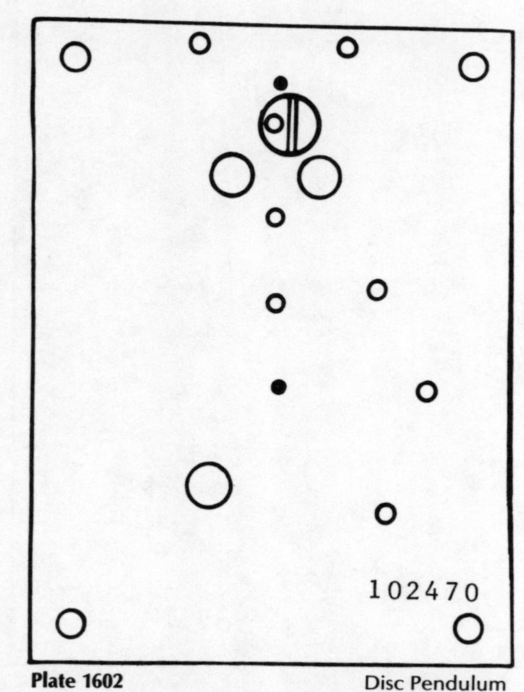

Plate 1602 Disc Pendulum
4-Ball Pendulum
USE .0038"* (.097mm*) HOROLOVAR
See Appendix 41 (18 x 38)

Jahresuhrenfabrik c 1905

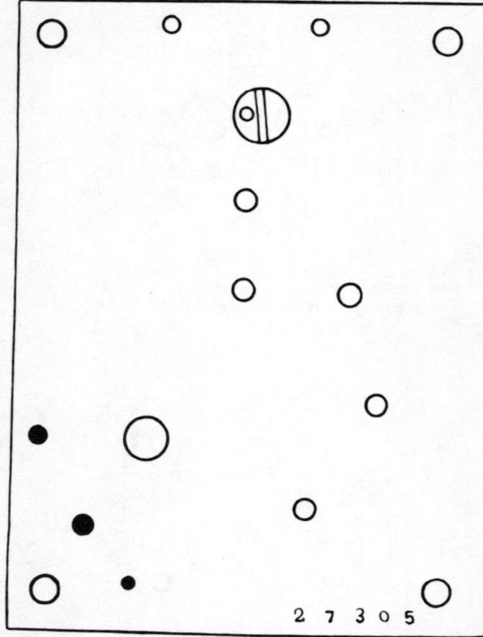

Plate 1603 Disc Pendulum
4-Ball Pendulum
USE .004"* (.102mm*) HOROLOVAR
(19 x 36)

Phillipp Haas c 1904

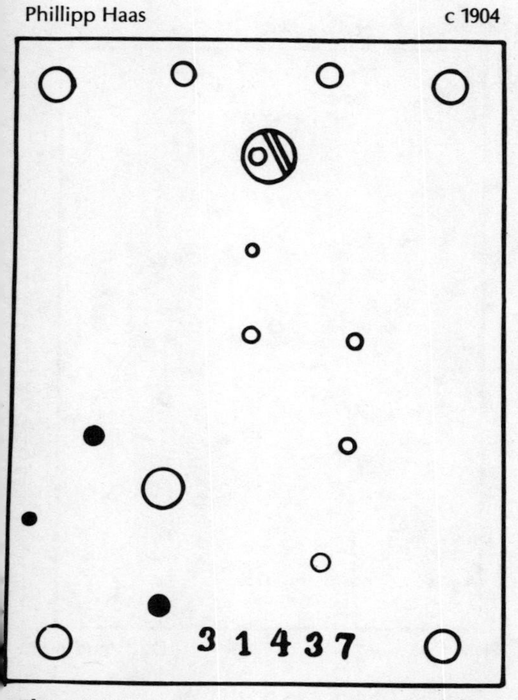

Plate 1607 Disc Pendulum
4-Ball Pendulum
USE .0038"* (.097mm*) HOROLOVAR
See Appendix 89, 118 (19 x 36)

Jahresuhrenfabrik c 1905

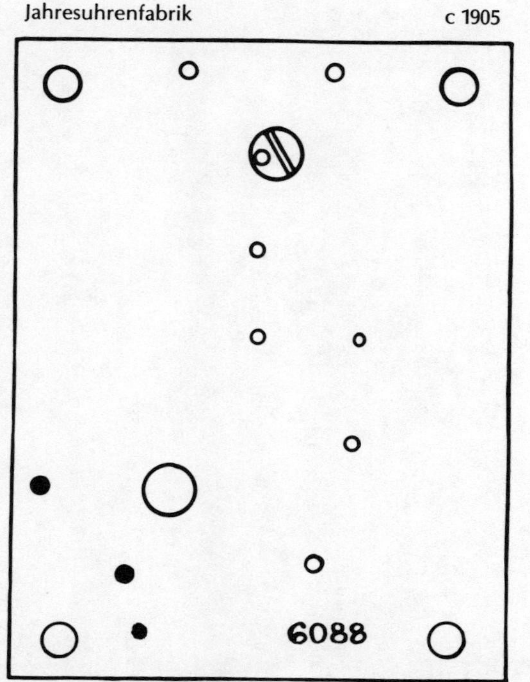

Plate 1610 Disc Pendulum
USE .004"* (.102mm*) HOROLOVAR
(19 x 36)

Jahresuhrenfabrik c 1906

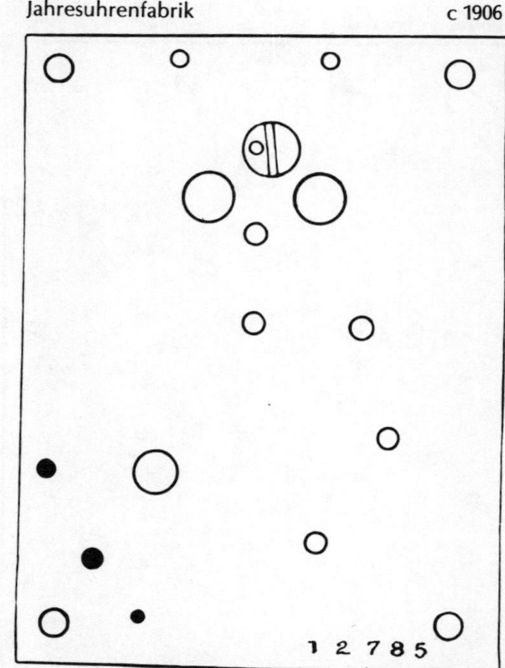

Plate 1613 Disc Pendulum
4-Ball Pendulum
USE .004"* (.102mm*) HOROLOVAR
(19 x 36)

Aug. Schatz & Sohne c 1981

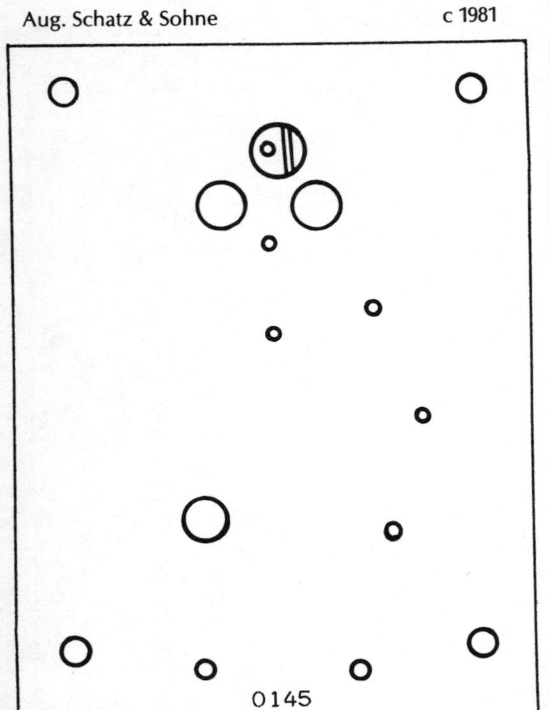

Plate 1614 Disc Pendulum
Schatz Jubilee Clock
USE .0036" (.091mm) HOROLOVAR
Unit 10D (19 x 38)

Kieninger & Obergfell c 1930

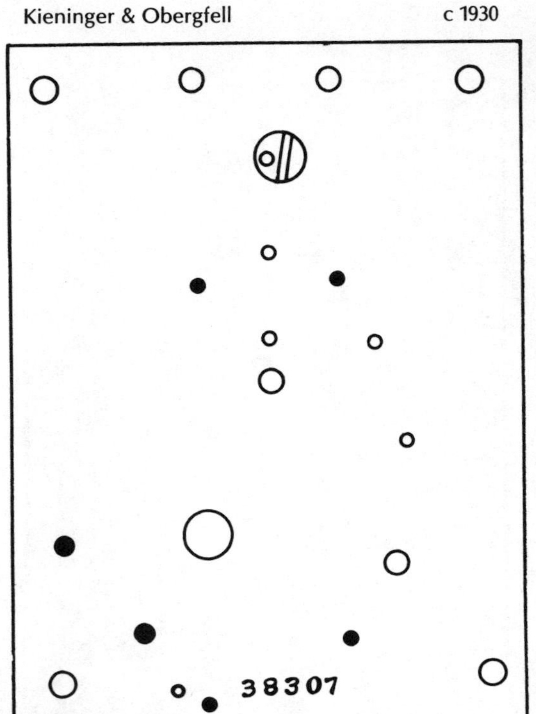

Plate 1615 4-Ball Pendulum
USE .0032" (.081mm) HOROLOVAR
Unit 1 (19 x 38)
See Appendix 76

Jahresuhrenfabrik c 1908

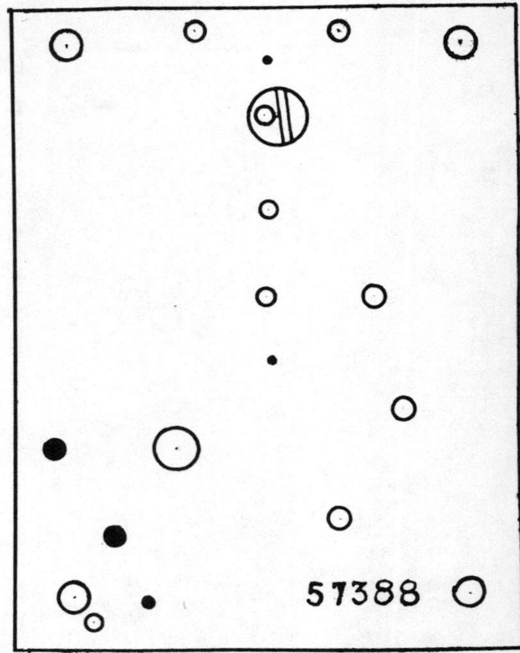

Plate 1617 Disc Pendulum
4-Ball Pendulum
USE .004"* (.102mm*) HOROLOVAR
(19 x 36)

Jahresuhrenfabrik c 1905

165771

Plate 1618 Disc Pendulum
4-Ball Pendulum
USE .0035"* (.089mm*) HOROLOVAR
(19 x 36)

Manufacturer Not Known c 1908

151644

Plate 1619 Disc Pendulum
Miniature Clock
USE .0025" (.064mm) HOROLOVAR
See Appendix 73

Manufacturer Not Known c 1910

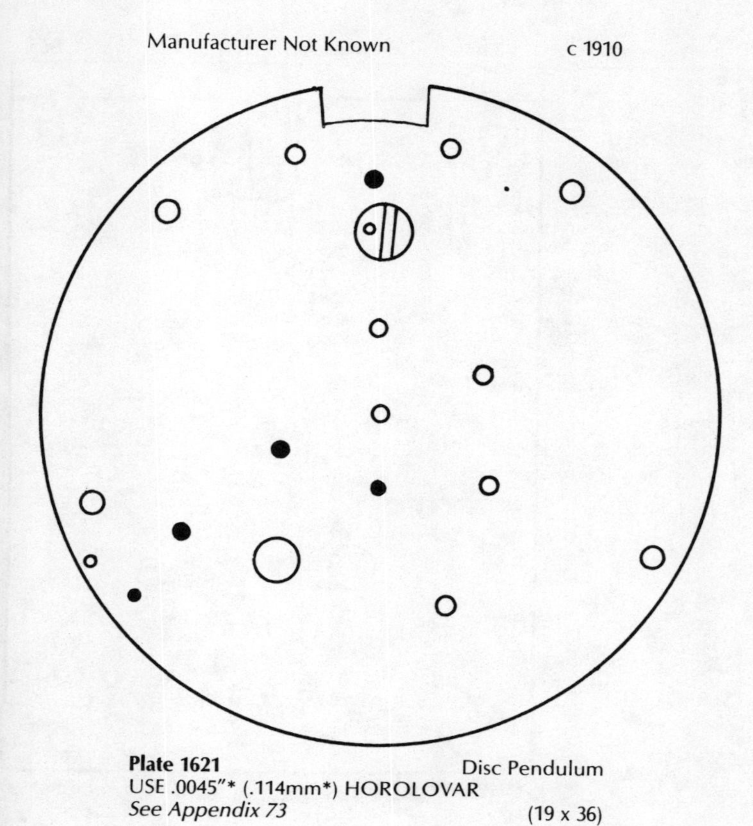

Plate 1621 Disc Pendulum
USE .0045"* (.114mm*) HOROLOVAR
See Appendix 73 (19 x 36)

Manufacturer Not Known c 1910

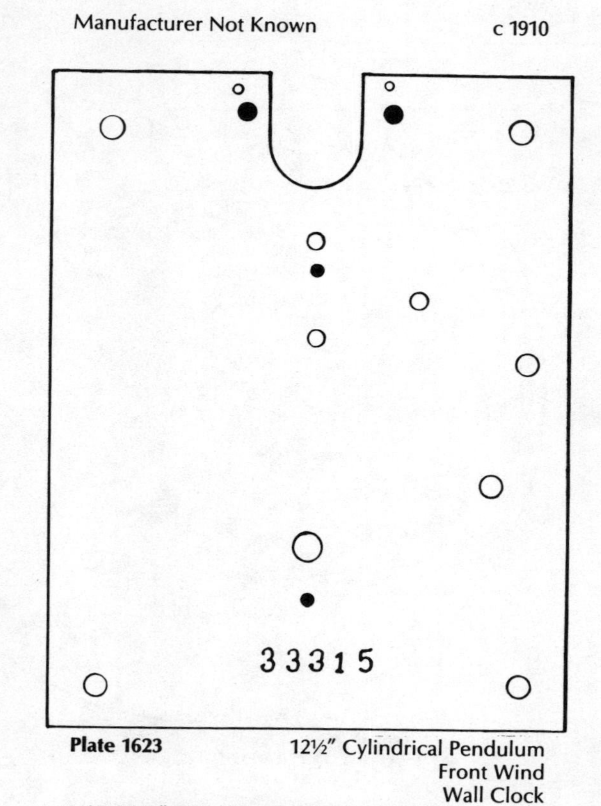

Plate 1623 12½" Cylindrical Pendulum
Front Wind
Wall Clock
USE .0045"—* (.114mm—*) HOROLOVAR
See Appendix 63, 73 (19 x 38)

Kieninger & Obergfell c 1910

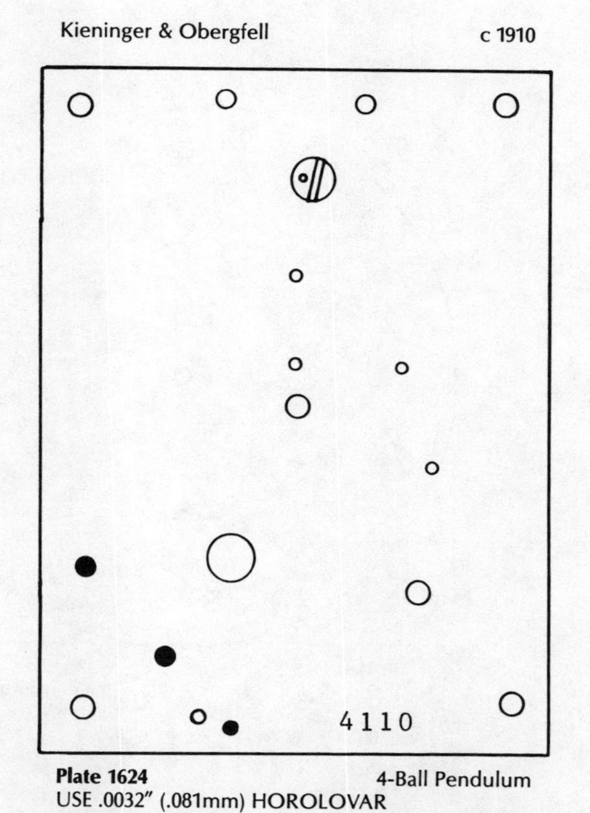

Plate 1624 4-Ball Pendulum
USE .0032" (.081mm) HOROLOVAR
Unit 1 (19 x 38)
See Appendix 76

Gebr. Junghans c 1910

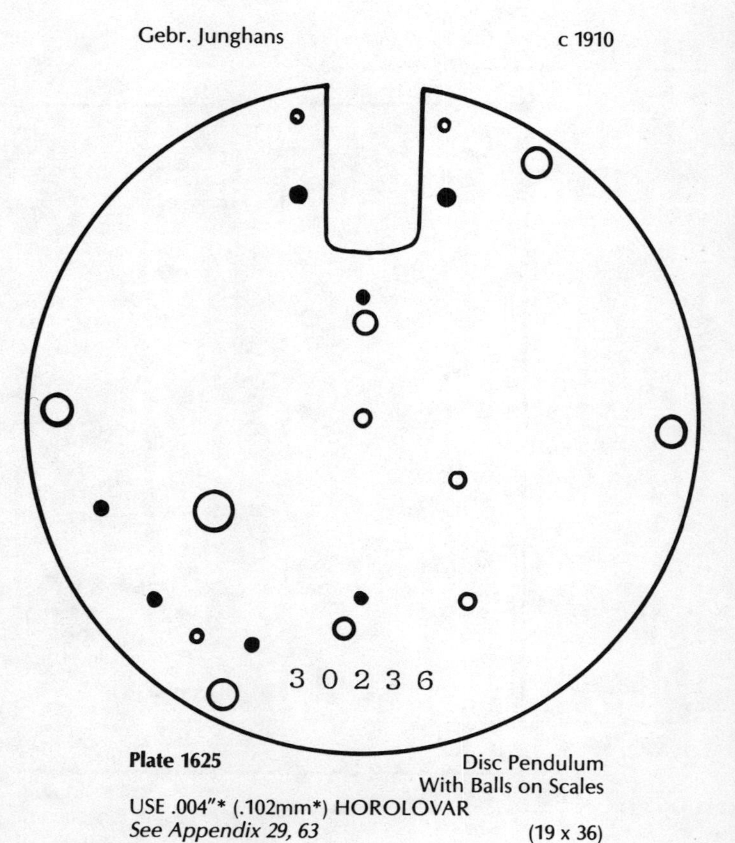

Plate 1625 Disc Pendulum
With Balls on Scales
USE .004"* (.102mm*) HOROLOVAR
See Appendix 29, 63 (19 x 36)

Manufacturer Not Known c 1904

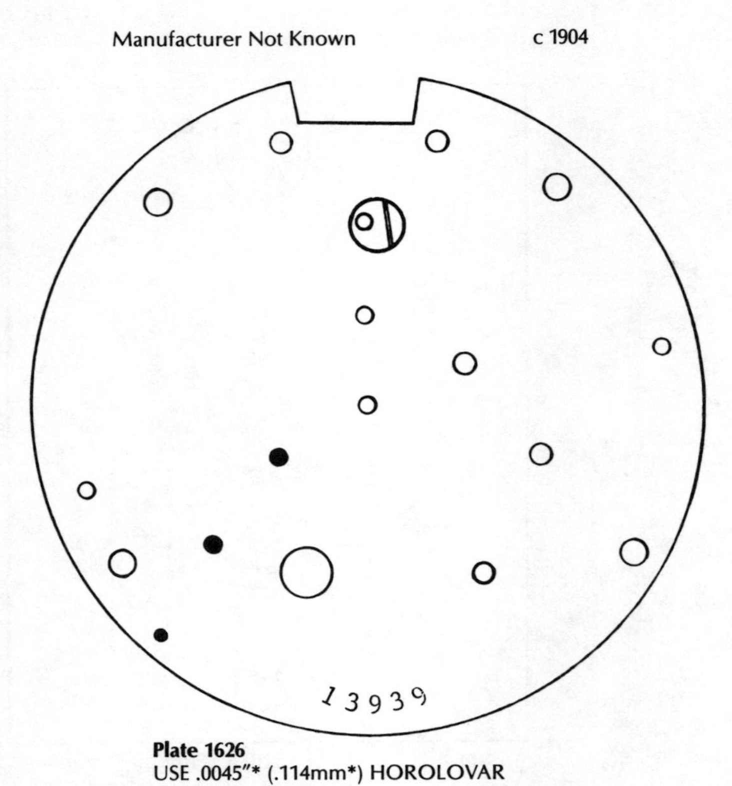

Plate 1626
USE .0045"* (.114mm*) HOROLOVAR
See Appendix 73

Badische Uhrenfabrik c 1905

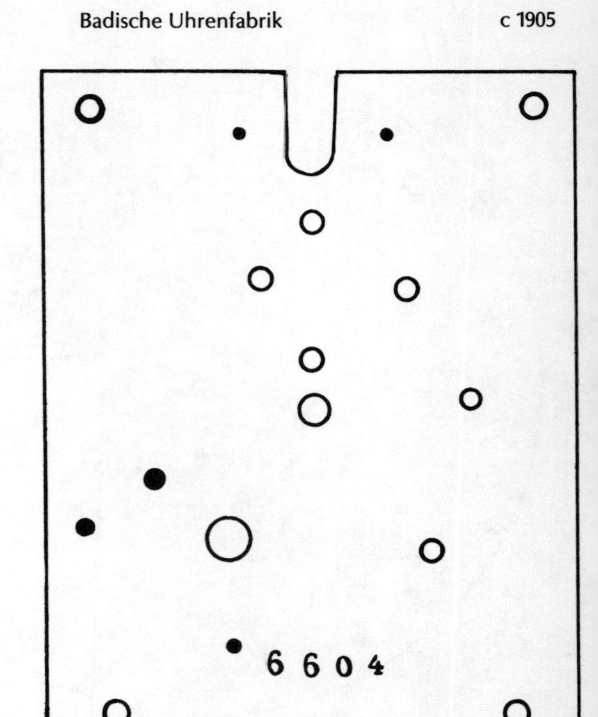

Plate 1627
4-Ball Pendulum
Pin Pallet Escapement
Lantern Pinions
USE .0035"* (.089mm*) HOROLOVAR
See Appendix 63, 96 (19 x 38)

Badische Uhrenfabrik c 1910

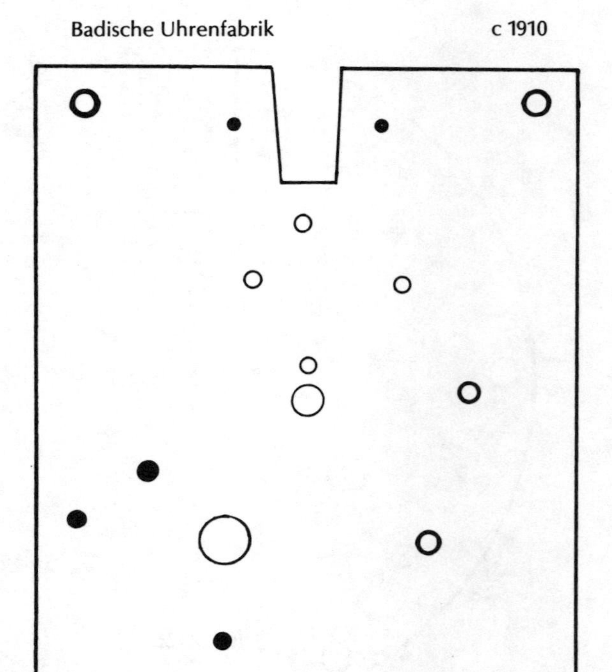

Plate 1628
4-Ball Pendulum
Pin Pallet Escapement
Solid Pallet Escapement
Lantern Pinions
USE .0035"* (.089mm*) HOROLOVAR
See Appendix 63 (20 x 38)

Jahresuhrenfabrik c 1905

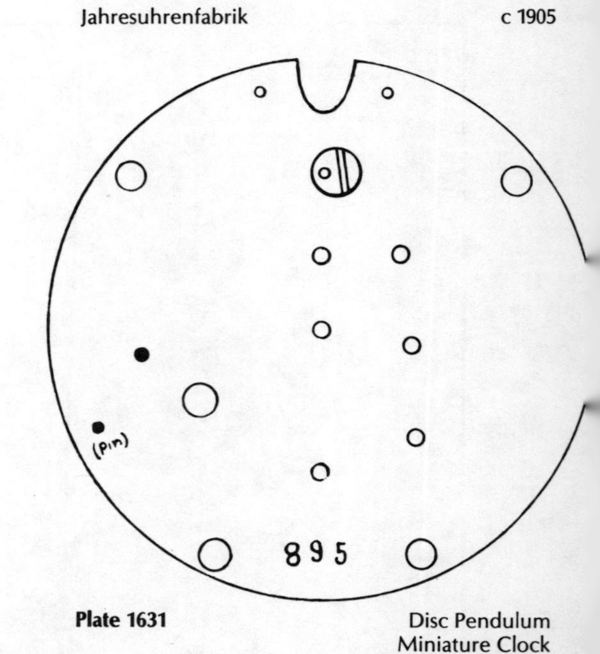

Plate 1631
Disc Pendulum
Miniature Clock
8" High 4-Glass Case
USE .0025"* (.064mm*) HOROLOVAR

Phillipp Haas c 1908

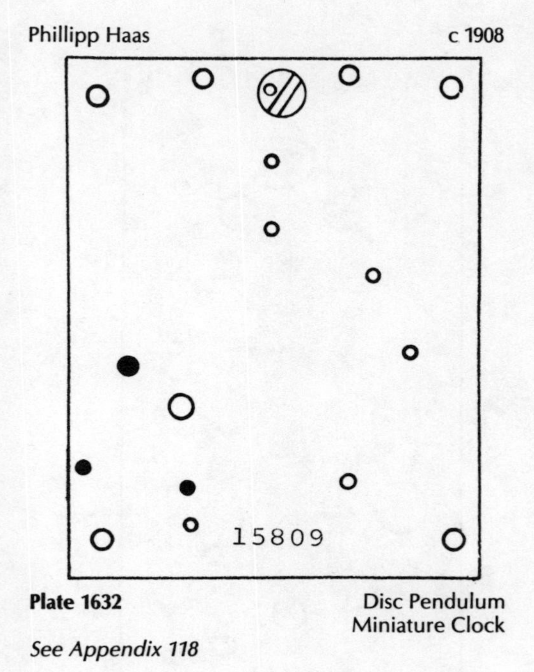

Plate 1632 Disc Pendulum
Miniature Clock

See Appendix 118

A. Willmann Co.? c 1879

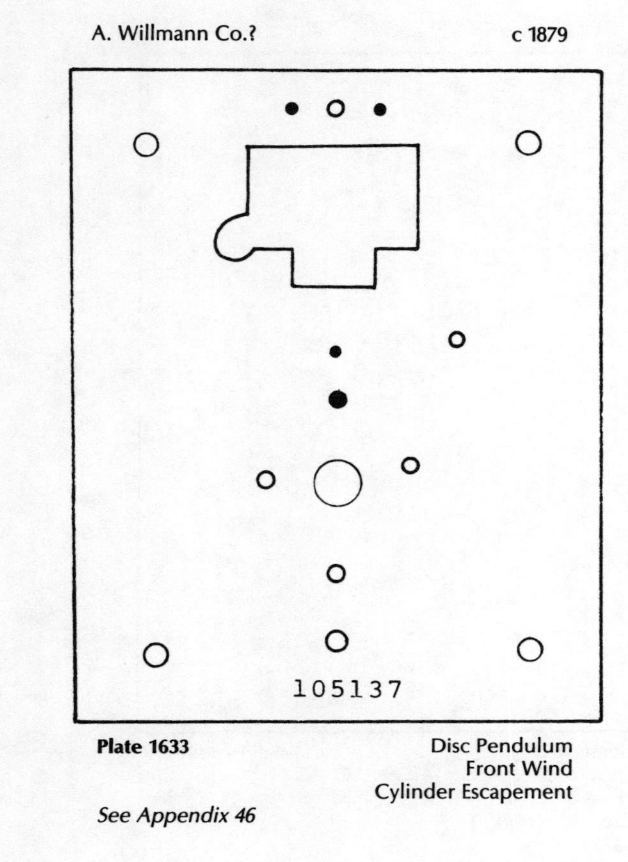

Plate 1633 Disc Pendulum
Front Wind
Cylinder Escapement

See Appendix 46

Gustav Becker c 1907

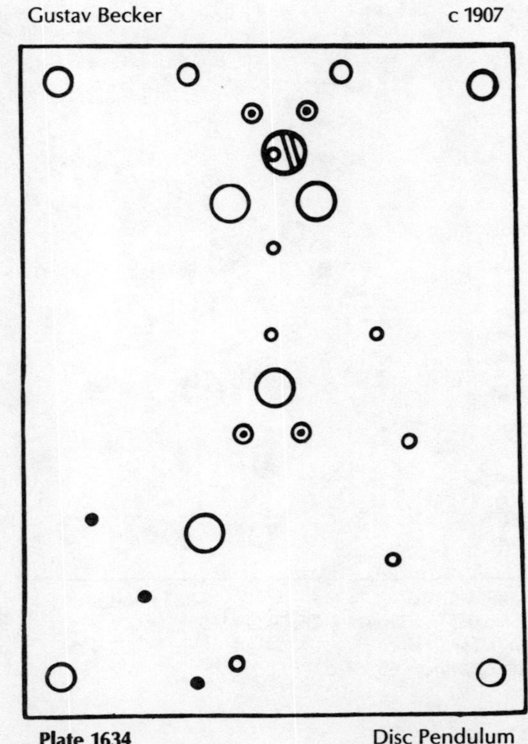

Plate 1634 Disc Pendulum
4-Ball Pendulum

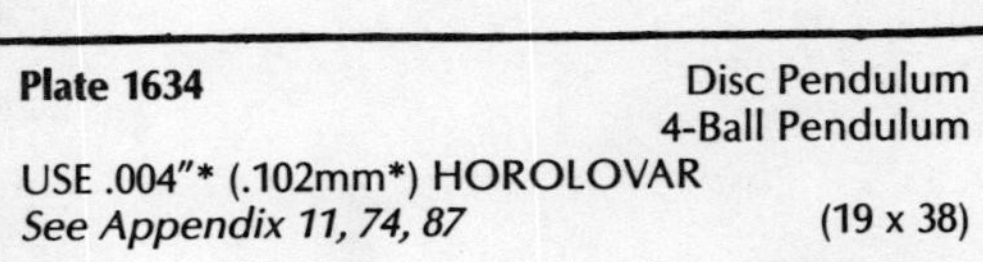

USE .004"* (.102mm*) HOROLOVAR
See Appendix 11, 74, 87 (19 x 38)

Manufacturer Not Known c 1902

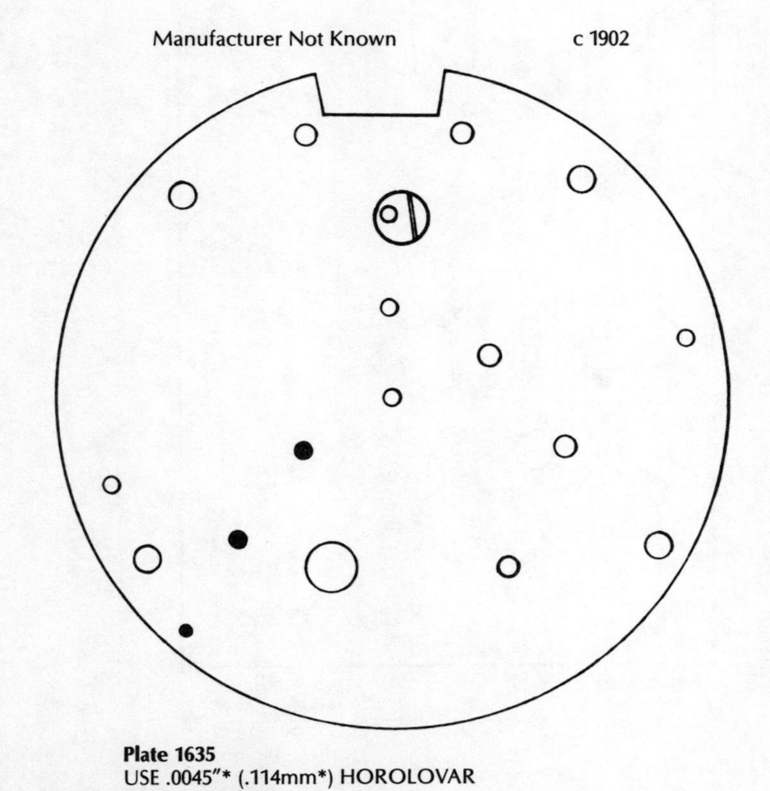

Plate 1635
USE .0045"* (.114mm*) HOROLOVAR
See Appendix 73

Kieninger & Obergfell c 1930

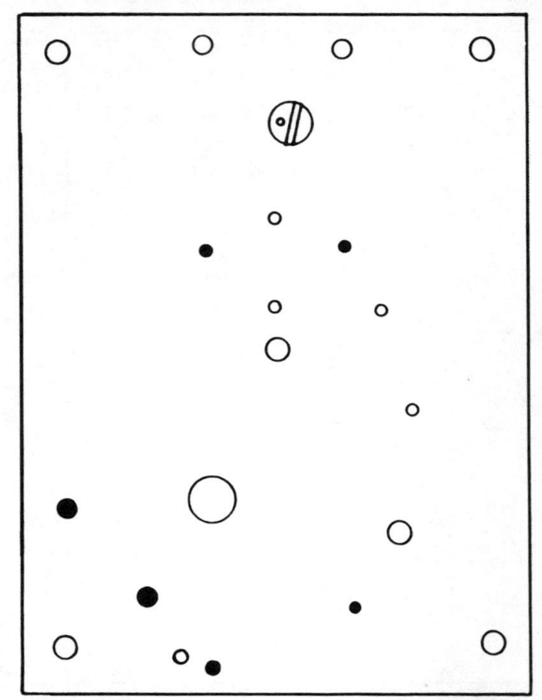

Plate 1639 4-Ball Pendulum
USE .0032" (.081mm) HOROLOVAR
Unit 1 (19 x 38)
See Appendix 76

Kieninger & Obergfell c 1976

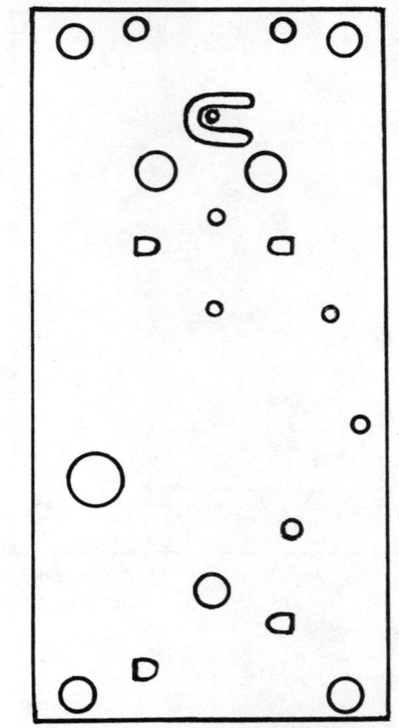

Plate 1641 4-Ball Pendulum
USE .0032" (.081mm) HOROLOVAR
Unit 3C (19 x 38)
See Appendix 76, 101

Badische Uhrenfabrik c 1902

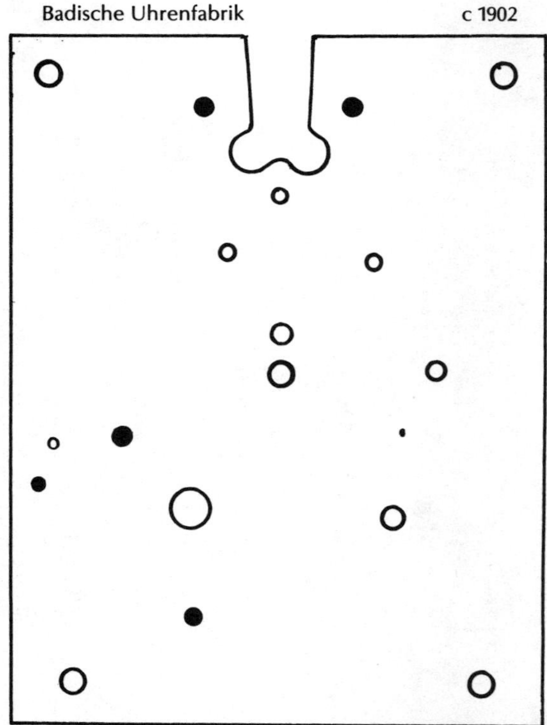

Plate 1643 Disc Pendulum
3-Ball Pendulum
Pin Pallet Escapement
Lantern Pinions
USE .0035"* (.089mm*) HOROLOVAR
See Appendix 63 (20 x 38)

Konrad Mauch c 1950

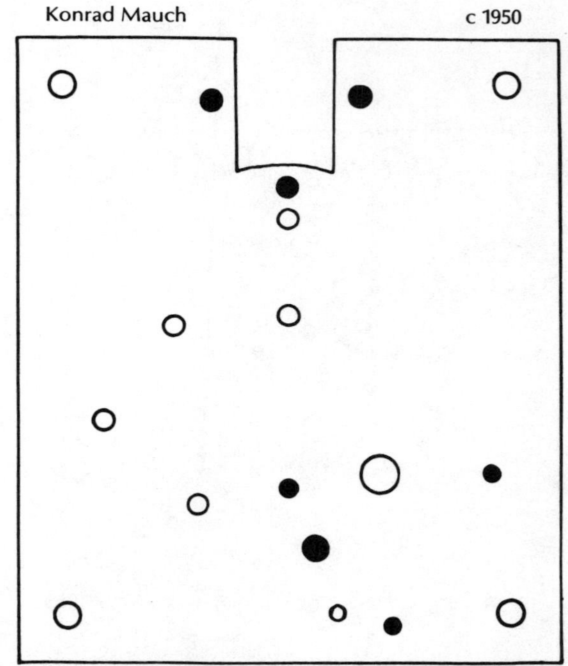

Plate 1647 4-Ball Pendulum
USE .0035" (.089mm) HOROLOVAR
Units 13A, 13B (20 x 38)
See Appendix 63, 64, 84

Uhrenfabrik Herr c 1951

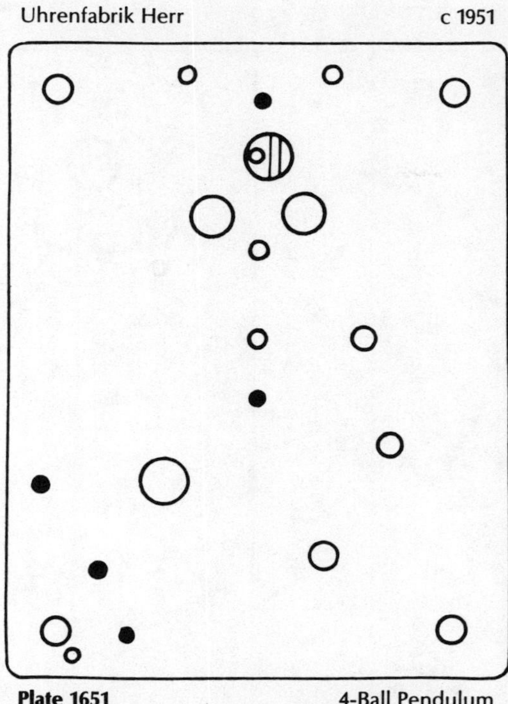
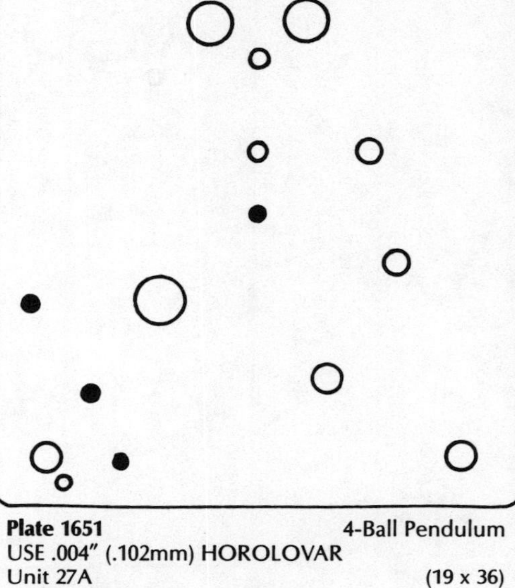

Plate 1651 4-Ball Pendulum
USE .004″ (.102mm) HOROLOVAR
Unit 27A (19 x 36)
See Appendix 130

Jahresuhrenfabrik c 1905

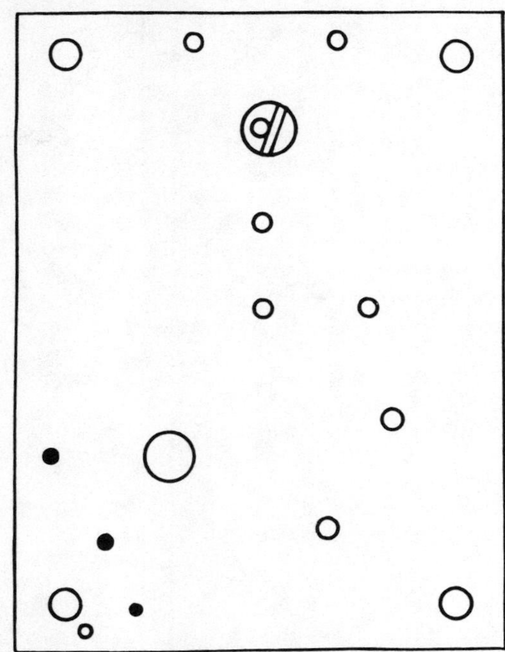

Plate 1651A 4-Ball Pendulum
USE .004″ (.102mm) HOROLOVAR
Unit 6789A (19 x 36)

Uhrenfabrik Herr c 1952

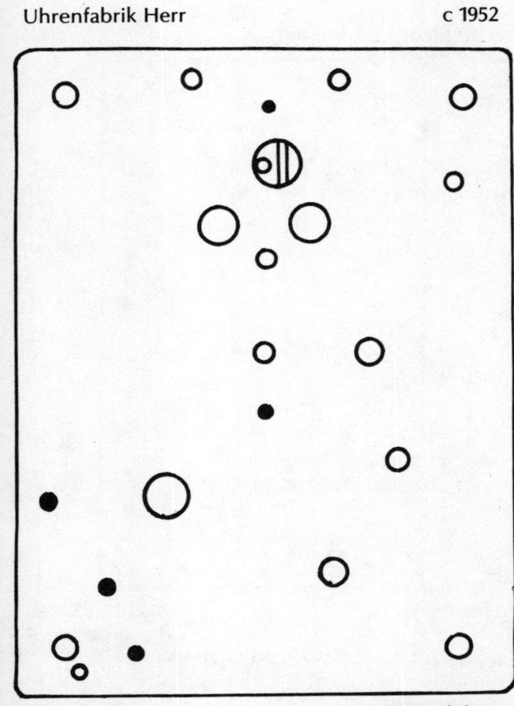

Plate 1652 4-Ball Pendulum
Hour Striker
USE .004″ (.102mm) HOROLOVAR
Unit 27A (19 x 36)
See Appendix 56

Edgar Henn c 1951

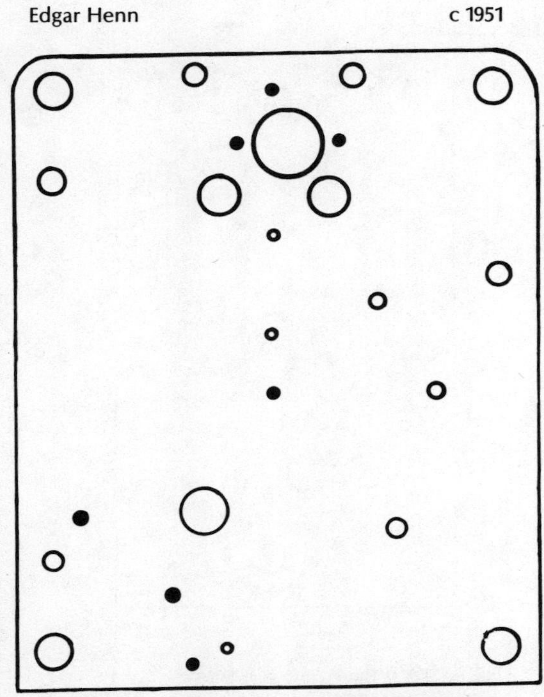

Plate 1659 4-Ball Pendulum
USE .0035″ (.089mm) HOROLOVAR
Unit 31 (19 x 38)
See Appendix 39

Badische Uhrenfabrik c 1898

Plate 1663 Disc Pendulum
Pin Pallet Escapement
Lantern Pinions
USE .0037″* (.094mm*) HOROLOVAR
See Appendix 10, 96 (20 x 38)

Kern & Link c 1932

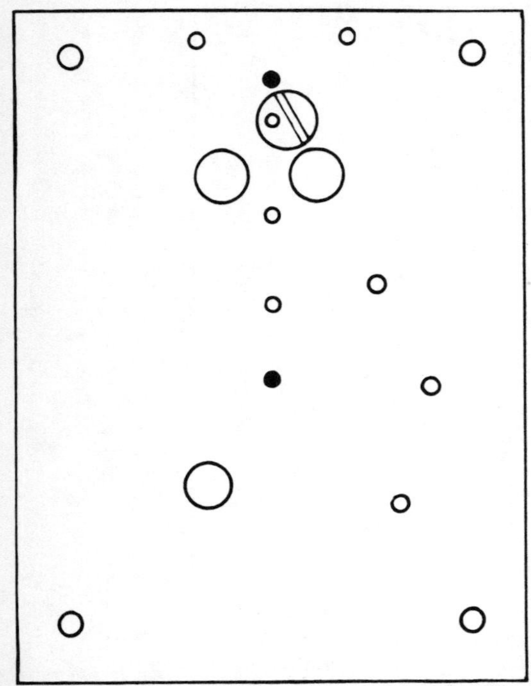

Plate 1667 4-Ball Pendulum
USE .0036" (.091mm) HOROLOVAR
Unit 11A (18 x 38)
See Appendix 81

Kern & Sohne c 1954

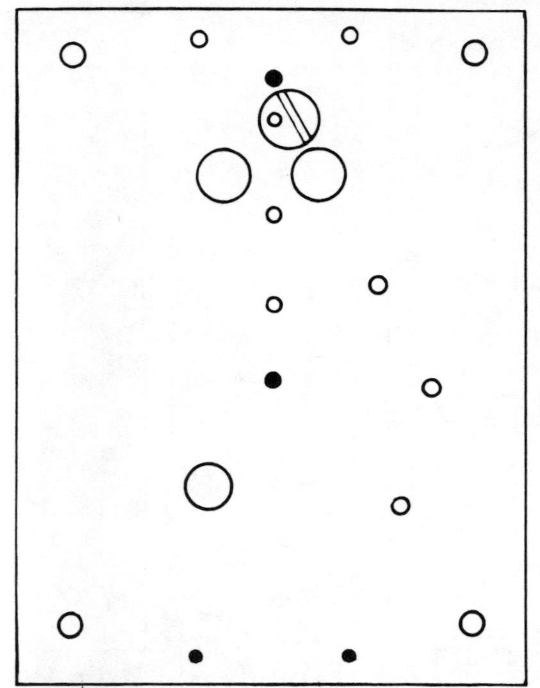

Plate 1667A 4-Ball Pendulum
USE .0036" (.091mm) HOROLOVAR
Unit 11B (18 x 38)
See Appendix 81

Georg Wurthner c 1952

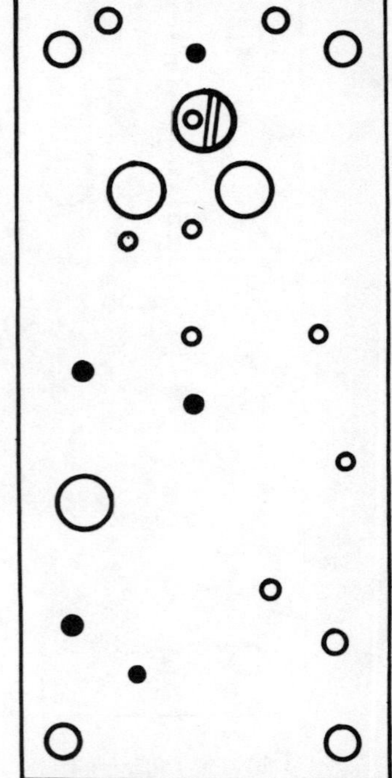

Plate 1669 4-Ball Pendulum
USE .004" (.102mm) HOROLOVAR
Unit 26 (19 x 36)
See Appendix 67

Georg Wurthner c 1952

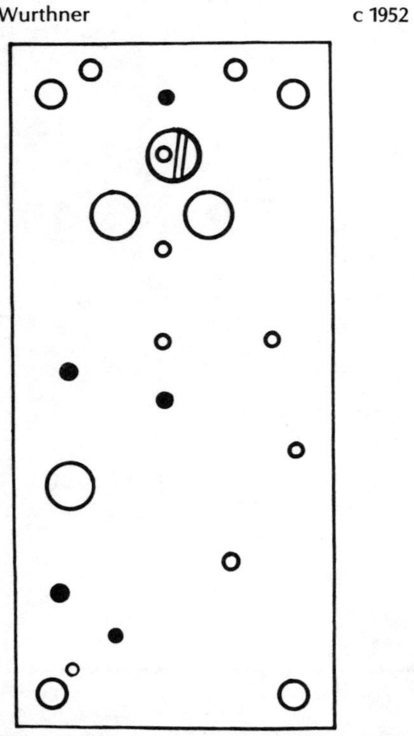

Plate 1670 4-Ball Pendulum
USE .004" (.102mm) HOROLOVAR
Unit 26 (19 x 36)
See Appendix 67

Edgar Henn c 1953

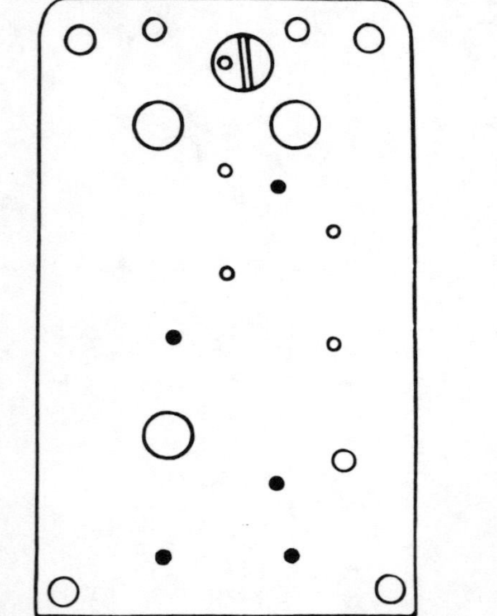

Plate 1671 4-Ball Pendulum Miniature Clock
USE .0023" (.058mm) HOROLOVAR
Unit 29C (19 x 32)

Uhrenfabrik M. Reiner c 1954

Plate 1672 4-Ball Pendulum Miniature Clock
USE .0028" (.071mm) HOROLOVAR
Unit 28C (16 x 36)
See Appendix 25, 55, 109

W. Petersen c 1954

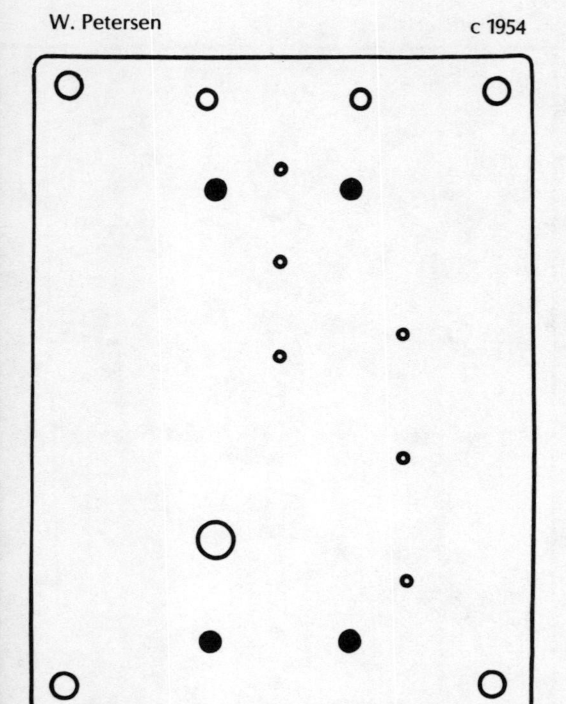

Plate 1673 4-Ball Pendulum
USE .0033″ (.084mm) HOROLOVAR
Units 18B, 18C (19 x 38)
See Appendix 59, 65

W. Petersen c 1954

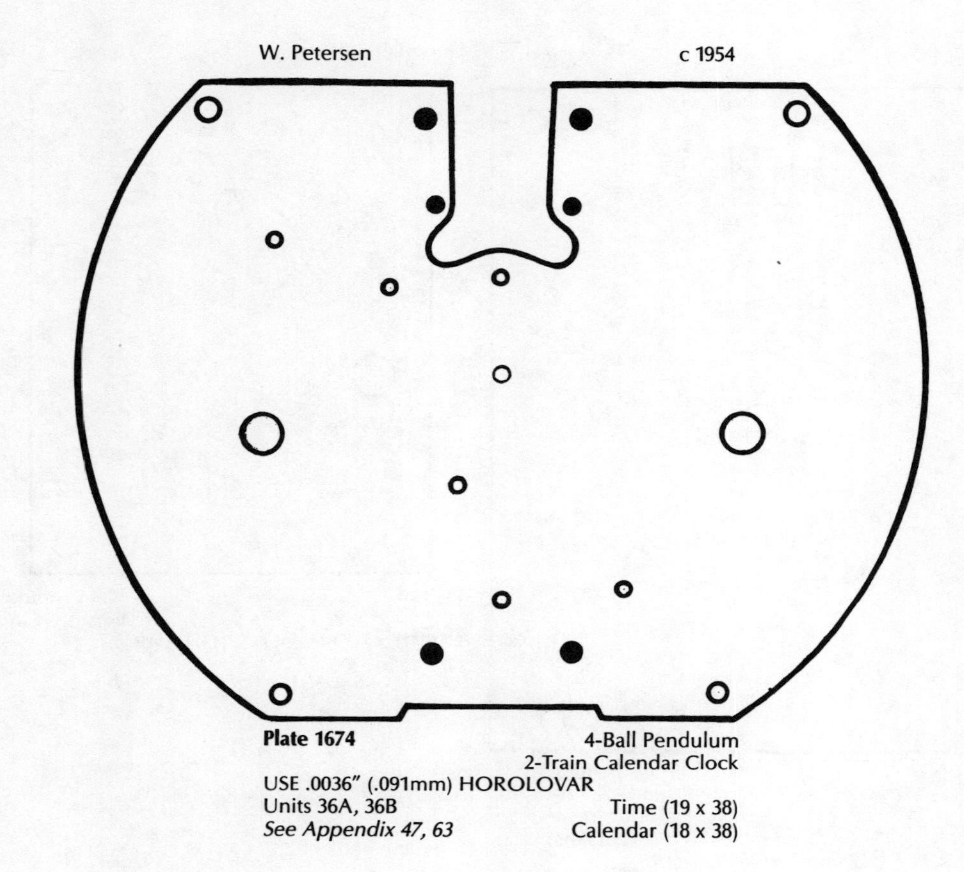

Plate 1674 4-Ball Pendulum
2-Train Calendar Clock
USE .0036″ (.091mm) HOROLOVAR
Units 36A, 36B Time (19 x 38)
See Appendix 47, 63 Calendar (18 x 38)

Uhrenfabrik Herr c 1951

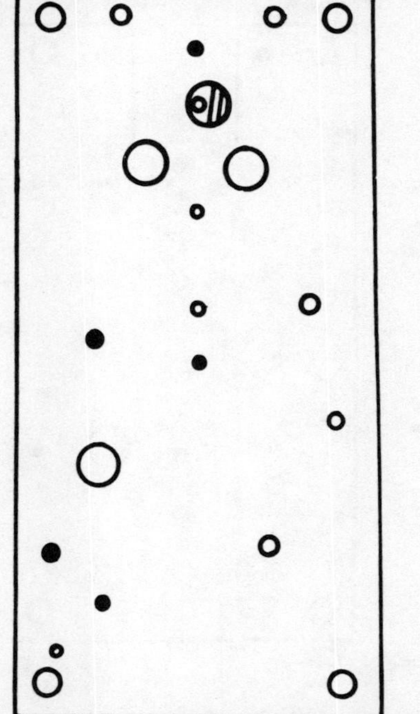

Plate 1675 4-Ball Pendulum
Dome Clock
USE .0035″ (.089mm) HOROLOVAR
Units 19, 20A, 21
Square Clock
USE .0033″ (.084mm) HOROLOVAR
Unit 20B (19 x 36)
See Appendix 109, 110

Uhrenfabrik Herr c 1953

Plate 1676 4-Ball Pendulum
USE .0038″ (.097mm) HOROLOVAR
Unit 27B (19 x 36)
See Appendix 106, 109, 110

Uhrenfabrik Neueck c 1956

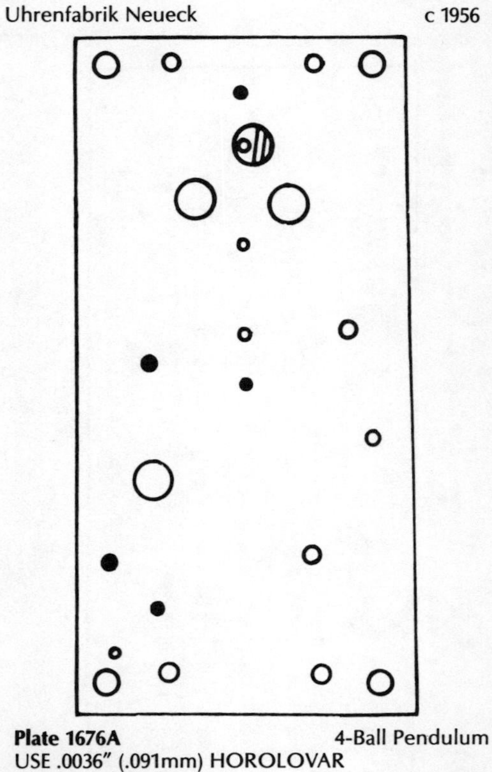

Plate 1676A 4-Ball Pendulum
USE .0036″ (.091mm) HOROLOVAR
Unit 27C (19 x 36)
See Appendix 43, 110

Badische Uhrenfabrik c 1910

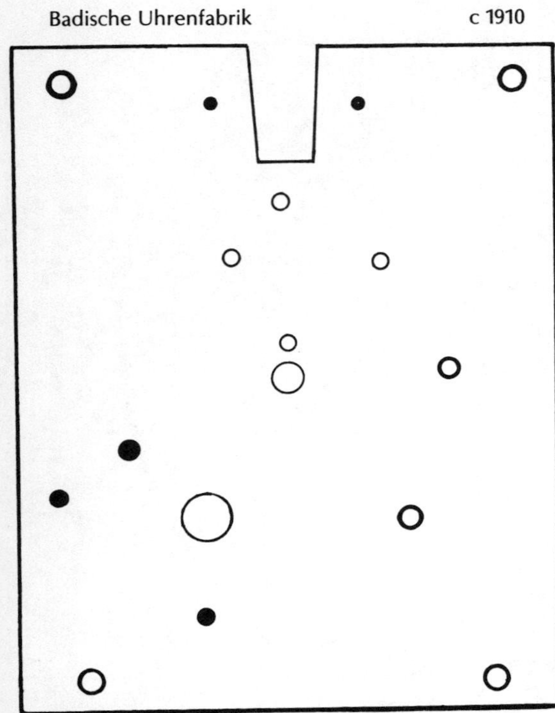

Plate 1677 4-Ball Pendulum
Pin Pallet Escapement
Solid Pallet Escapement
Lantern Pinions
USE .0035"* (.089mm*) HOROLOVAR
See Appendix 63 (20 x 38)

Uhrenfabrik Herr c 1950

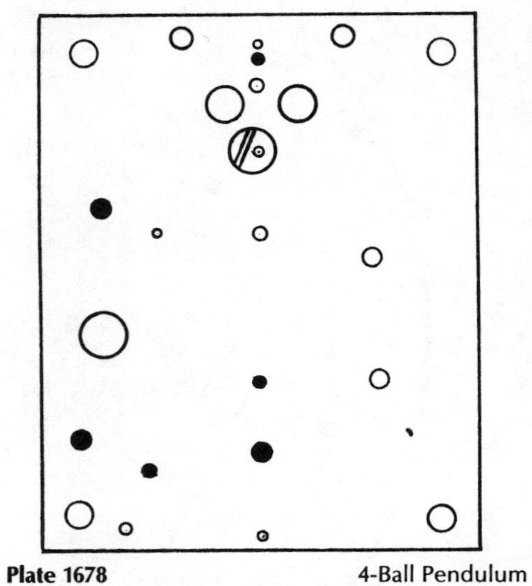

Plate 1678 4-Ball Pendulum
Miniature Clock
USE .0025" (.064mm) HOROLOVAR
Units 28A, 28B (16 x 36)
See Appendix 25, 72

Thomas Haller c 1882

Plate 1680 Special Pendulum
Underslung Lever Escapement
Front Wind
See Appendix 99

Edgar Henn c 1952

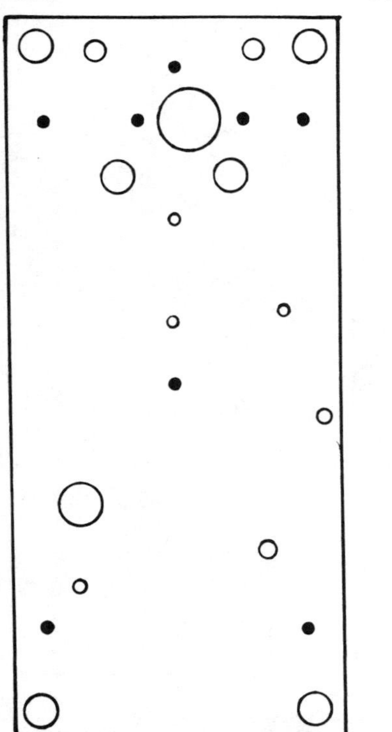

Plate 1681 4-Ball Pendulum
USE .0035" (.089mm) HOROLOVAR
Unit 31 (19 x 38)
See Appendix 39

Jahresuhrenfabrik c 1898

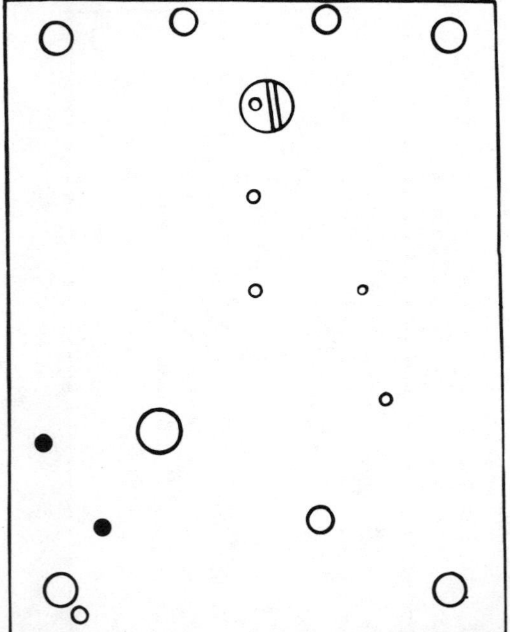

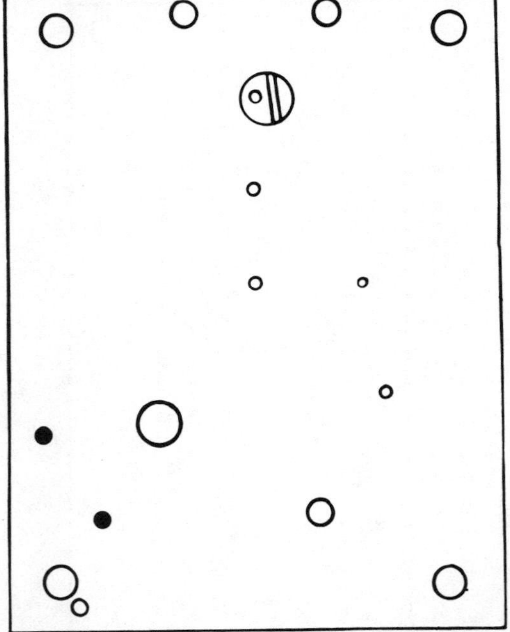

Plate 1681A Disc Pendulum
USE .004" (.102mm) HOROLOVAR
(19 x 36)

Konrad Mauch c 1954

Plate 1682 4-Ball Pendulum
Miniature Clock
USE .0032" (.081mm) HOROLOVAR
Unit 14A
USE .003" (.076mm) HOROLOVAR
Unit 14B Original (13 x 32)
See Appendix 52, 63, 64, 85 Revised (15 x 32)

Konrad Mauch c 1950

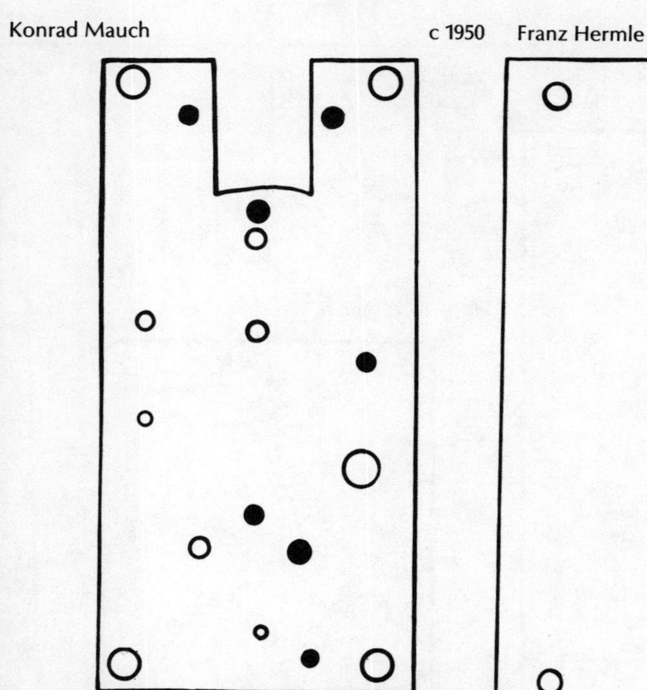

Plate 1683 4-Ball Pendulum
USE .0035" (.089mm) HOROLOVAR
Units 13A, 13B, 13C (20 x 38)
See Appendix 63, 64, 84

Franz Hermle c 1951

Plate 1684 4-Ball Pendulum
USE .0033" (.084mm) HOROLOVAR
Unit 25A (18 x 38)
See Appendix 71

Jauch & Haller c 1948

Rivet

Plate 1685 4-Ball Pendulum
USE .0038" (.097mm) HOROLOVAR
Unit 33 (19 x 38)
See Appendix 63

Manufacturer Not Known c 1953

Rivet

Rivet

Plate 1687 4-Ball Pendulum
Verge Escapement
60-Day Clock

USE .005"—* (.127mm—*) HOROLOVAR
See Appendix 48

Manufacturer Not Known c 1905

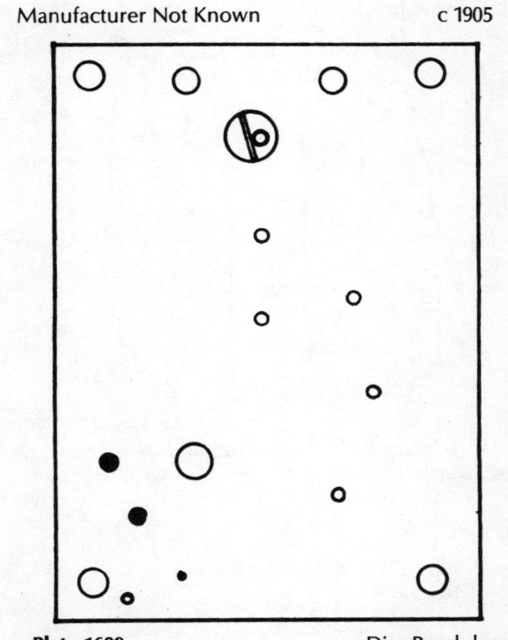

Plate 1689 Disc Pendulum
Miniature Clock

USE .0025" (.064mm) HOROLOVAR
See Appendix 73

Ishihara Clock Co. c 1952

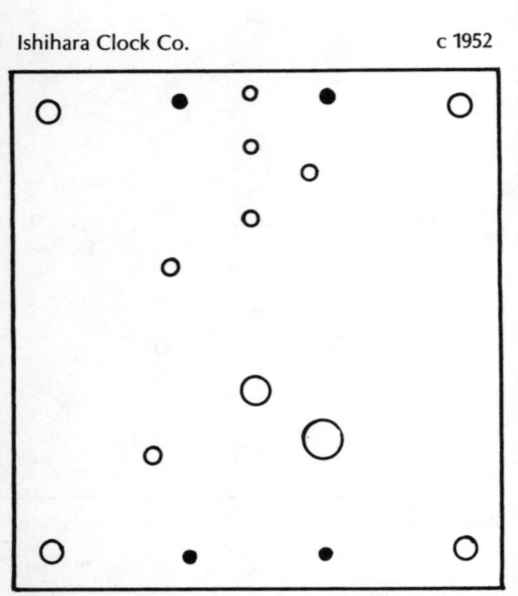

Plate 1691 3-Ball Pendulum
Pin Pallet Escapement
USE .0036" (.091mm) HOROLOVAR
Unit 40A
See Appendix 24, 102

Georg Wurthner c 1956

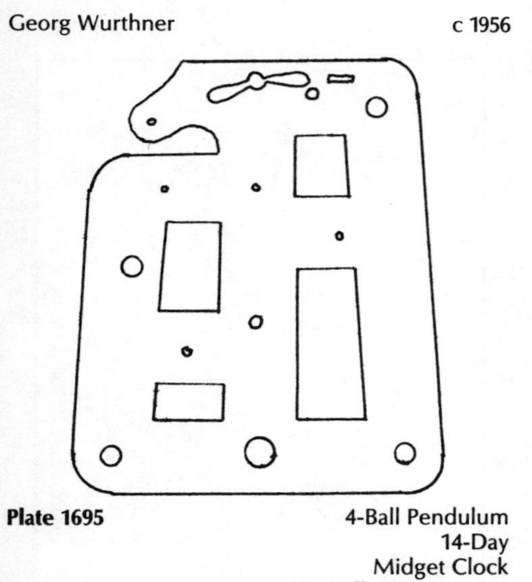

Plate 1695 4-Ball Pendulum
14-Day
Midget Clock
Pin Pallet Escapement
USE .002" (.051mm) HOROLOVAR
See Appendix 104

Jauch & Haller c 1948

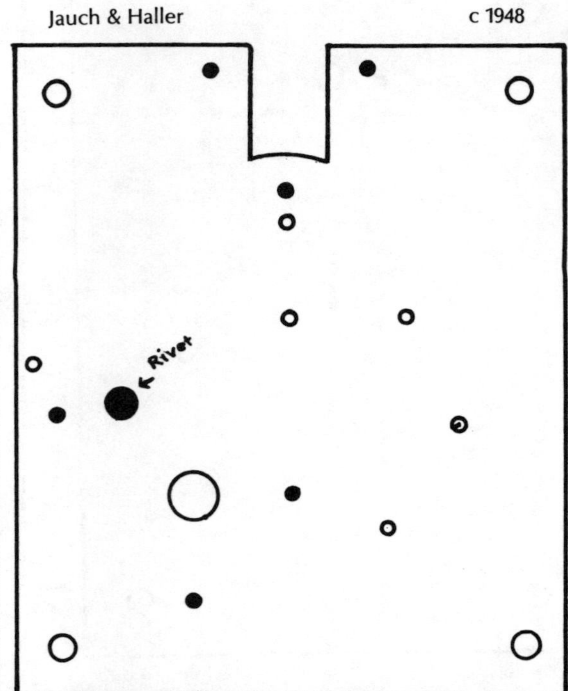

Plate 1693 4-Ball Pendulum
USE .0038" (.097mm) HOROLOVAR
Unit 33 (19 x 38)
See Appendix 63

Uhrenfabrik J. Kaiser c 1954

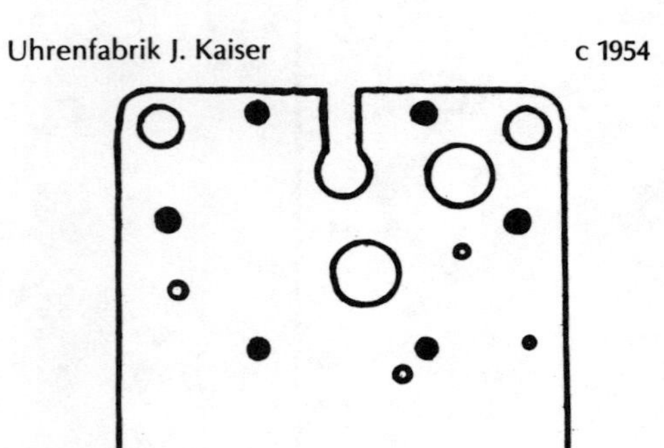

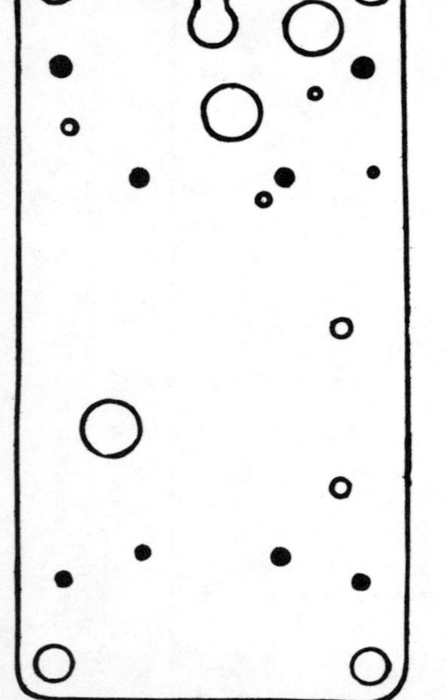

Plate 1696 Ball Pendulum
USE .003" (.076mm) HOROLOVAR
Unit 34 (19 x 38)
See Appendix 21, 34, 63, 68

Georg Wurthner c 1957

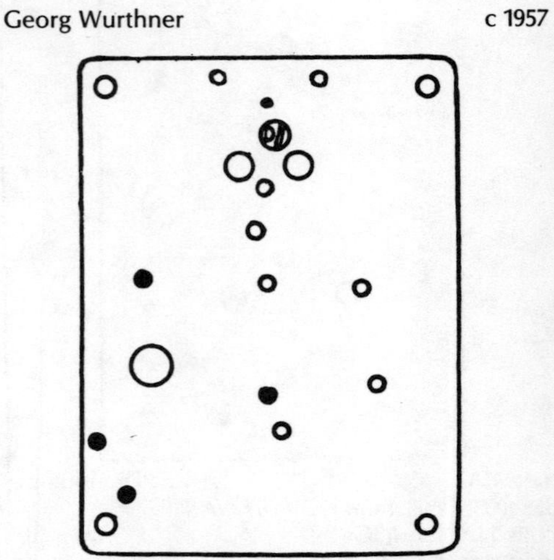

Plate 1694 4-Ball Pendulum
Midget Clock
USE .002" (.051mm) HOROLOVAR
Unit 37 (9.5 x 27)
See Appendix 27

Konrad Mauch c 1957

Plate 1697 3-Ball Pendulum
Midget Clock
Pin Pallet Escapement
USE .0022" (.056mm) HOROLOVAR
Units 38, 38A Original (12 x 25)
See Appendix 63, 64, 86 Revised (14 x 25)

Phillipp Haas c 1908

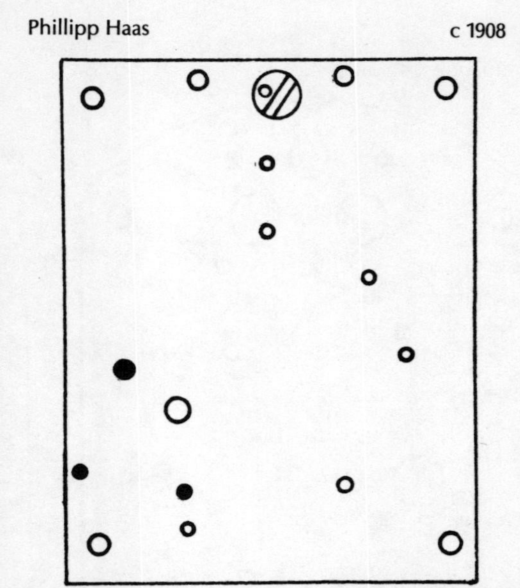

Plate 1699

Disc Pendulum
Miniature Clock

See Appendix 118

Sigfried Haller c 1970

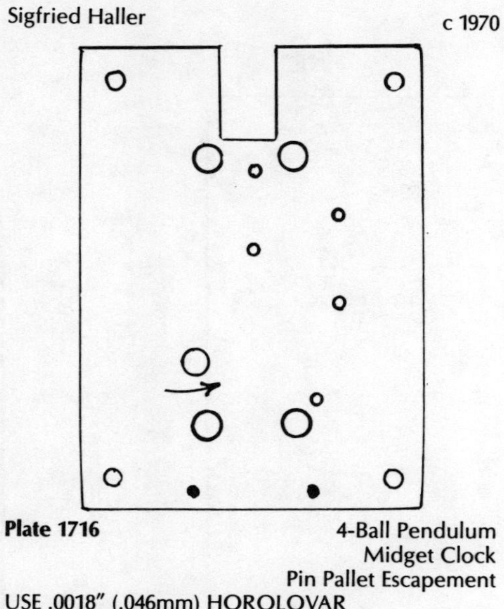

Plate 1716

4-Ball Pendulum
Midget Clock
Pin Pallet Escapement

USE .0018" (.046mm) HOROLOVAR
Units 44, 44A (12 x 24)
See Appendix 63, 64

Badische Uhrenfabrik c 1900

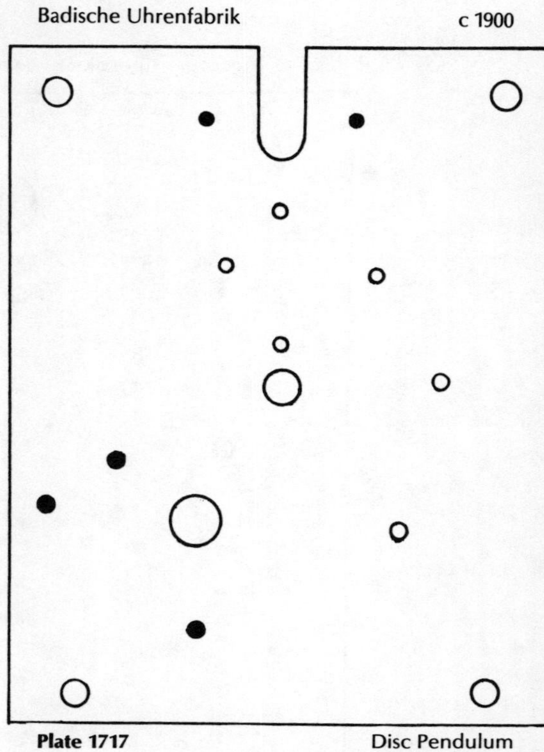

Plate 1717

Disc Pendulum
4-Ball Pendulum
Pin Pallet Escapement
Lantern Pinions

USE .0038"* (.097mm*) HOROLOVAR
See Appendix 63, 96 (20 x 38)

Manufacturer Not Known c 1905

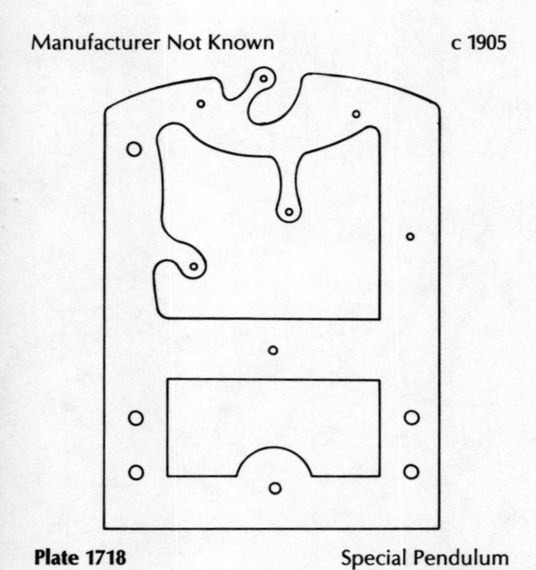

Plate 1718

Special Pendulum
30-Day
Miniature Clock

USE .003"* (.076mm*) HOROLOVAR
See Appendix 73, 103

Badische Uhrenfabrik c 1910

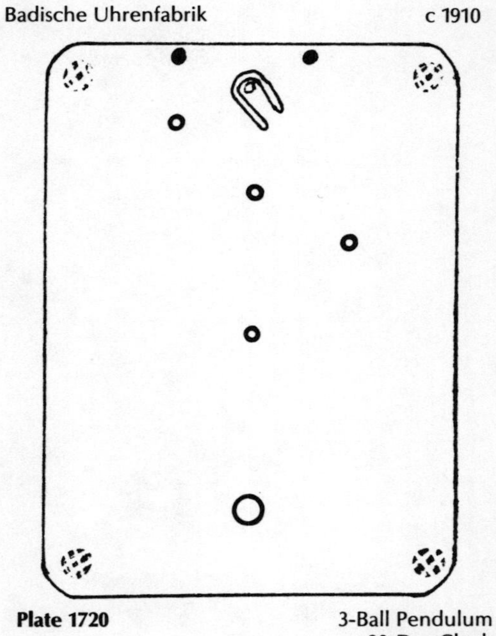

Plate 1720

3-Ball Pendulum
30-Day Clock
Pin Pallet Escapement
Lantern Pinions

USE .0032"* (.081mm*) HOROLOVAR
See Appendix 50

Manufacturer Not Known c 1910

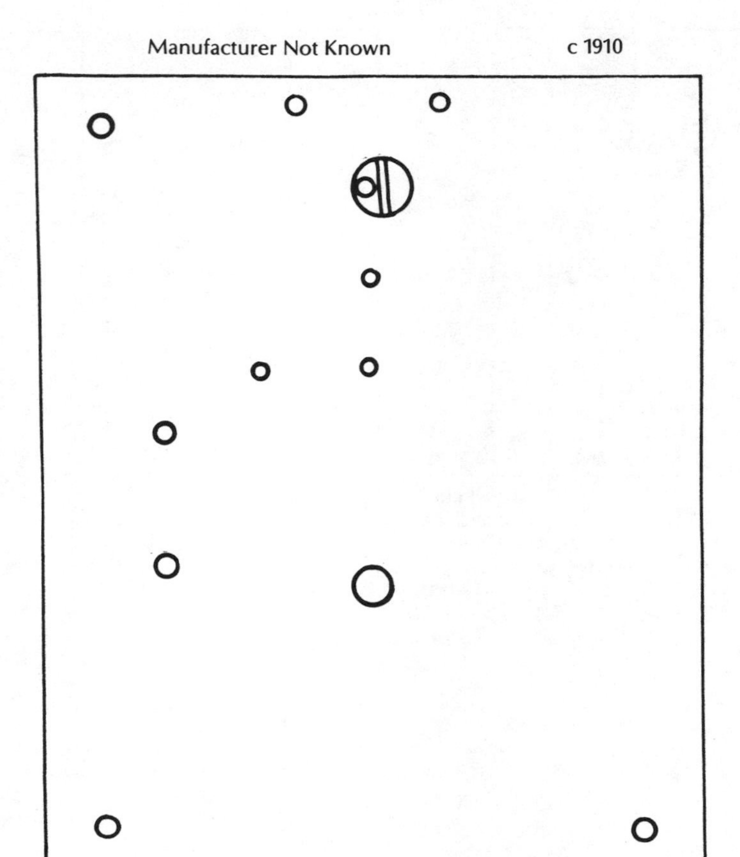

Plate 1722 4-Ball Pendulum Wall Clock

USE .0045" (.114mm) HOROLOVAR

See Appendix 73, 98

Uhrenfabrik J. Kaiser c 1954

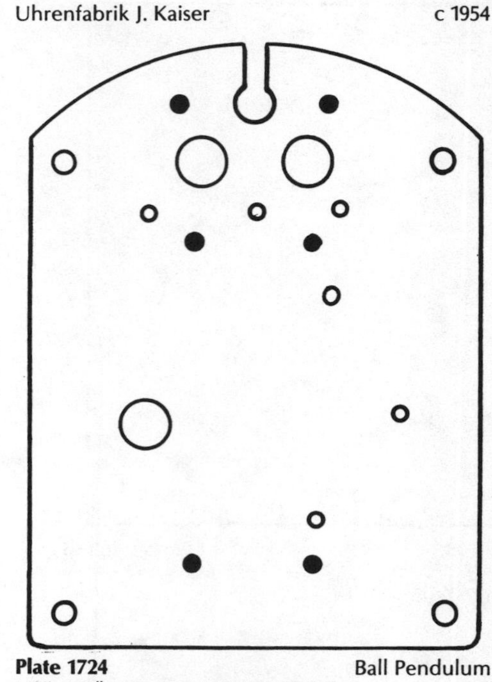

Plate 1724 Ball Pendulum

USE .003" (.076mm) HOROLOVAR

Unit 34 (19 x 38)

See Appendix 34, 63, 68

Georg Wurthner c 1957

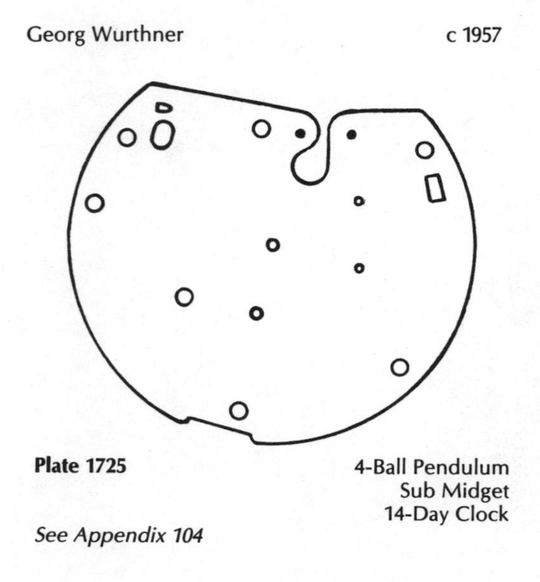

Plate 1725 4-Ball Pendulum Sub Midget 14-Day Clock

See Appendix 104

A. Willmann Co.? c 1879

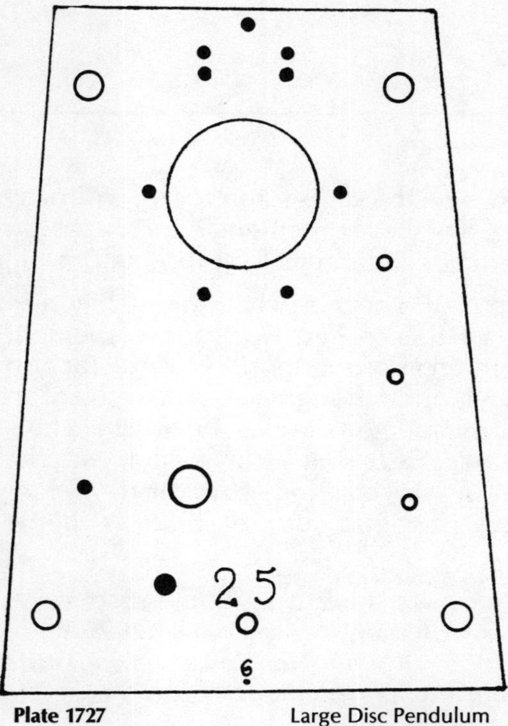

Plate 1727 Large Disc Pendulum
Verge Escapement
(19 x 36)

See Appendix 91, 116

Manufacturer Not Known c 1885

4574
R

(½ actual size)

Plate 1729 Disc Pendulum
Hour and Half Hour Gong Strike
USE .0045"— (.114mm—) HOROLOVAR

Plate 1730 Disc Pendulum
Hour and Half Hour Bell Strike
USE .005"— (.127mm—) HOROLOVAR

See Appendix 63, 73

Gustav Becker? c 1880

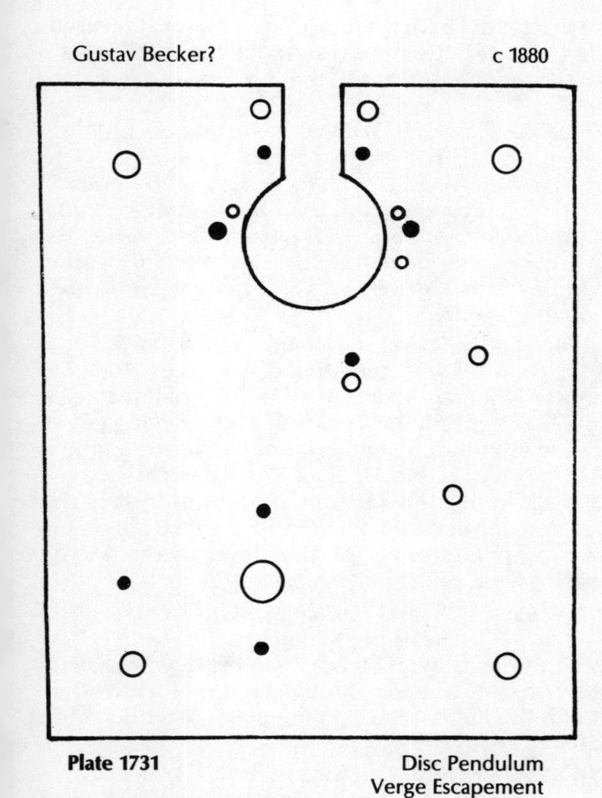

Plate 1731 Disc Pendulum
Verge Escapement

See Appendix 122, 128

A. Willmann Co.? c 1880

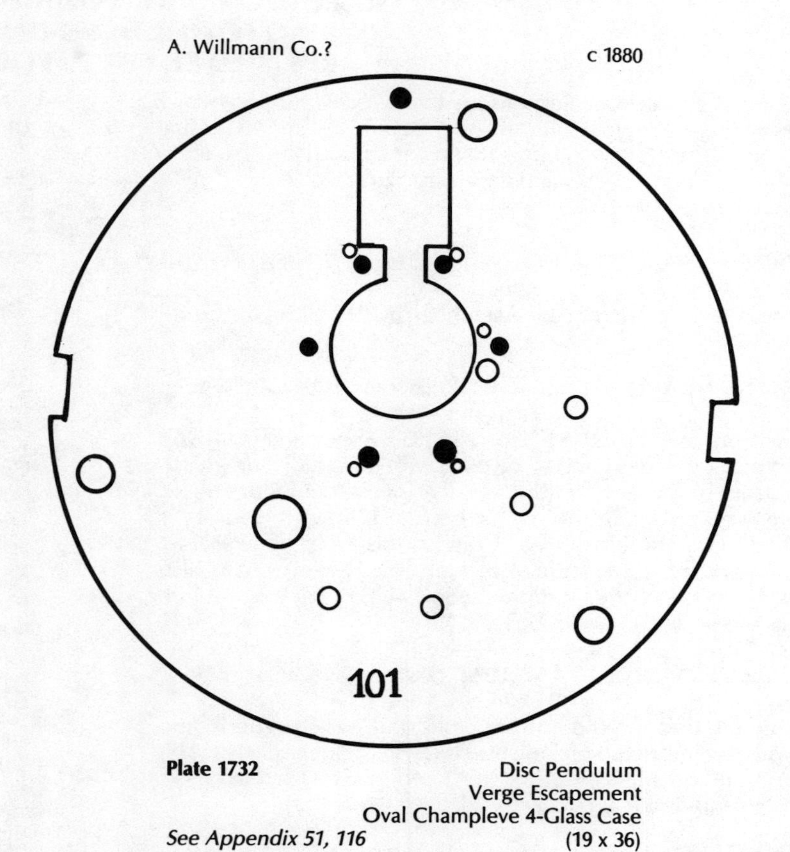

Plate 1732 Disc Pendulum
Verge Escapement
Oval Champleve 4-Glass Case
(19 x 36)

See Appendix 51, 116

SECTION 10 400-Day Clock Suspension Spring Unit Identification

This section contains accurate illustrations of the manufacturers' pendulum suspension spring Units for practically all clocks imported in the U.S., from 1949 until the time of publication of this Guide. **It does not include suspension Units from older clocks, because no original manufacturers' samples are available, and because it is almost impossible to tell whether or not Units in the available old clocks have been altered at one time or another.**

The important contribution which these illustrations offer to the repairer is that, since they have been reproduced in exact size, they make it possible, when replacing a suspension spring, to cut it to the correct length and also set the fork in the correct position on the spring. A quick examination of the illustrations will show that the distance between top block and fork varies considerably from clock to clock. A variation of more than one sixteenth of an inch in the correct position of the fork on the spring may mean the difference between a correct adjustment and one which will give all kinds of trouble.

The position of the fork on the suspension spring is directly related to the operation of the escapement. A fork, if set too low, will cause the anchor to flutter; if set too high, it will cause a jerky motion of the anchor which ultimately may cause the clock to stop. Either of these two poor escapement actions, caused only by an improper position of the fork on the suspension spring, is often thought to be the result of an incorrectly adjusted escapement. It will be worth while, therefore, to check the correct Unit in this section with the one in the clock you are repairing, particularly before coming to the conclusion that your problem is in the escapement.

For those who do not have time to assemble suspension Units, The Horolovar Company makes available 70 completely assembled Units, all with *Horolovar Temperature Compensating Suspension Springs.* (See Section 17)

SUSPENSION UNITS FOR MECHANICAL (SPRING POWERED) 400-DAY CLOCKS ARE LISTED BELOW IN BOLD FACE.

Note: Selection of the proper Suspension Units for Kundo Standard, Miniature, and Midget Clocks (Units **1** through **5F**) is difficult. See Section 18, Appendix No. 123 for detailed information concerning the characteristics of the various Kundo clocks and their associated Units.

1- KUNDO STANDARD 49 — USE .0032" (.081mm) HOROLOVAR
8 Beats per Minute
Fits all standard clocks from the first model, made in the 1920's, until about 1952.

2- KUNDO STANDARD 52 — USE .0032" (.081mm) HOROLOVAR
8 Beats per Minute
Used with only a few thousand clocks in 1953. A special stirrup-type bottom block was designed to be used with a new locking guard. This bottom block required the use of a special pendulum hook. There are no screws in the top or bottom blocks; the spring is "pinched" into the blocks. No Units of this type are available. Replacement must be made with Unit 3A. However, it is also necessary to replace the pendulum hook with a Type B pendulum hook as shown in Appendix No. 123.

3A- KUNDO STANDARD 53 — USE .0032" (.081mm) HOROLOVAR
8 Beats per Minute
Has fixed pin through the bottom block and was designed to be used with the improved locking guard (Type E, Appendix No. 123) Unit 3A is interchangeable with Unit 1 and may be used in clocks with other Kundo locking guards.

3B- KUNDO STANDARD 54 — USE .0032" (.081mm) HOROLOVAR
8 Beats per Minute
Used with the first model having a locking pendulum. It is exactly the same as Unit 3A except there is no fixed pin through the bottom block. Instead, a cotter pin holds the bottom block in the pendulum hook.

3C- KUNDO STANDARD 55 — USE .0032" (.081mm) HOROLOVAR
8 Beats per Minute
Used with the final locking pendulum design which entered production in 1954 and continued in use until the end of spring driven clock production. A loose brass pin was first used in the bottom block but this was soon replaced by steel pin with an upset center portion which was driven into the bottom block. (See Section 18, Appendix No. 2.)

4- KUNDO JUNIOR 52 — USE .0023" (.058mm) HOROLOVAR
10 Beats per Minute
Used with only a few thousand clocks in 1953. A special stirrup-type bottom block was designed to be used with a new locking guard. There are no screws in the top and bottom block; the spring is "pinched" into the blocks. No Units of this type are available. Replacement must be made with Unit 5A. However, it is also necessary to replace the pendulum hook with a Type B pendulum hook as shown in Appendix No. 123. Unit 4 was used only with movements having a pin pallet escapement.

5A- KUNDO JUNIOR 53 — USE .0023" (.058mm) HOROLOVAR
10 Beats per Minute
Has a fixed pin through bottom block. It was exactly the same as Unit 5A except there is no fixed pin through bottom block. Instead, a cotter pin holds the bottom block in the pendulum hook. (See Note under 5E.)

5B- KUNDO JUNIOR 54 — USE .0023" (.058mm) HOROLOVAR
10 Beats per Minute
Used with the first model having locking pendulum. It is exactly the

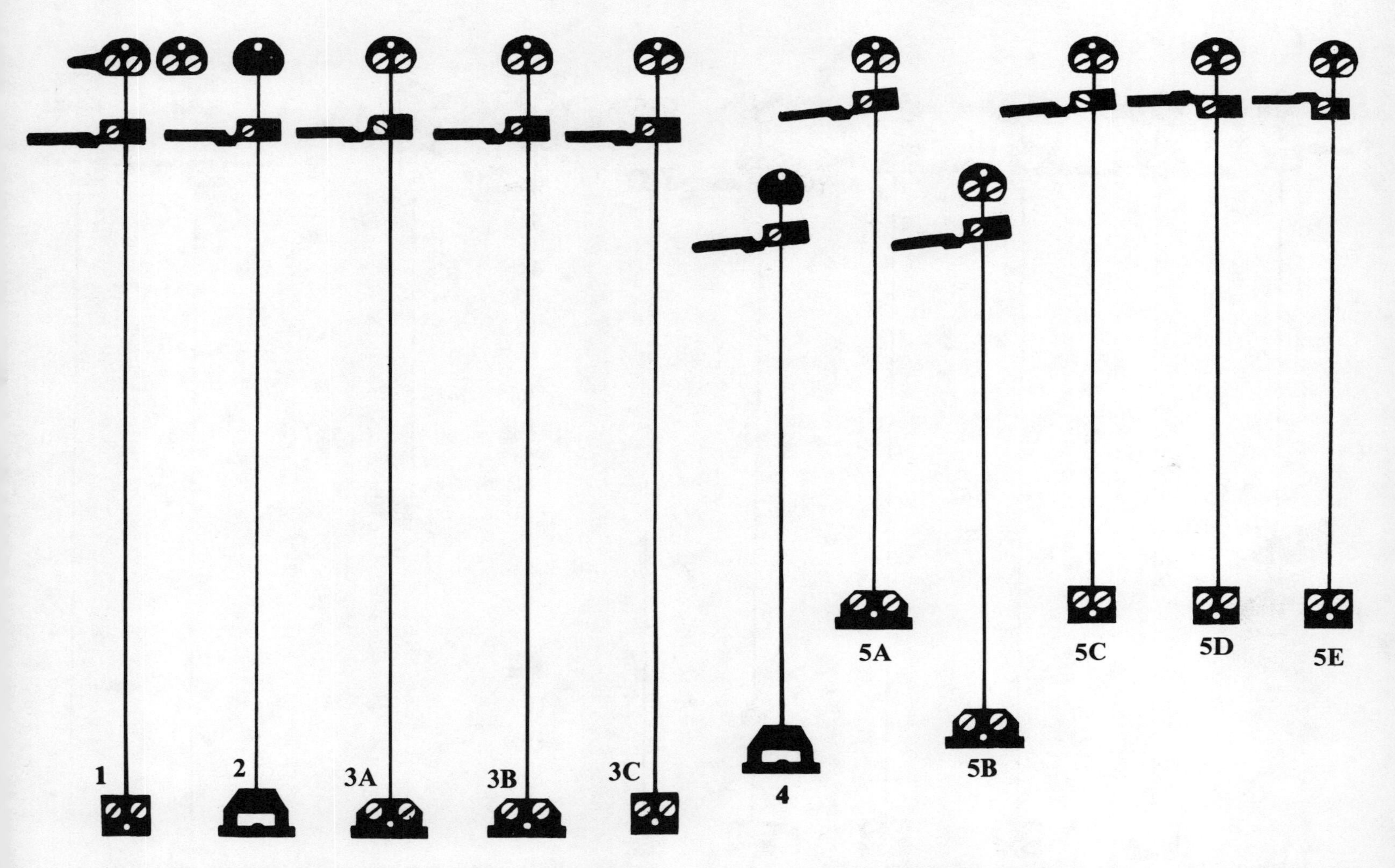

the same as Unit 5A except there is no fixed pin through bottom block. Instead, a cotter pin holds the bottom block in the hook, the latter having a larger threaded hole to fit the stronger pendulum center rod. Note that Unit 5B with pendulum locking mechanism is identified by pendulum locking lever which is ABOVE the movement platform. (See NOTE under Unit 5E.)

5C- KUNDO JUNIOR 55 — USE .0023" (.058mm) HOROLOVAR
10 Beats per Minute

The first of three Units used with the final locking pendulum design which entered production in 1955 and continued in use until the end of spring driven production. This Unit used a loose brass pin to retain it in the pendulum hook. **Unit 5E is recommended.**

5D- KUNDO JUNIOR 55 — USE .0023" (.058mm) HOROLOVAR
10 Beats per Minute

The second of three Units used with the final locking pendulum design. Note that the fork is in an inverted position. (The screw in some forks enters from the opposite side to the one illustrated.) Units 5C, 5D and 5E are interchangeable. **Unit 5E is recommended.**

5E- KUNDO JUNIOR 56 — USE .0023" (.058mm) HOROLOVAR
10 Beats per Minute

Third and last of the three Units used with the latest locking pendulum model. The only change is with the fork which is lighter in weight than forks of eariler Units. Kundo found that the lighter fork is less likely to bend the suspension spring between the top block and fork during severe transportation shocks. The most recent fork is designed with a closed end which not only prevents the anchor pin from slipping out, but also helps to keep the tines from becoming bent. Units 5E, 5C, and 5D are interchangeable. **Unit 5E is recommended.**

All movements using Unit 5E should have straight, upright anchor pins. If the pin is bent offset to the rear, the clock is a Kundo Midget and the correct Unit is 5F.

Note: If a clock with Units 5A, 5B, 5C, 5D or 5E does not produce good pendulum motion (320 to 450 dgrees), notice the position of the Anchor, Escape Wheel and Third Wheel. If they are arranged as shown in "A", below, they should be replaced with ones as shown in "B". Best pendulum motion is obtained when the anchor pin is as near as possible to the back plate.

Arrangements of *Kundo Miniature* Anchor, Escape Wheel and Third Wheel

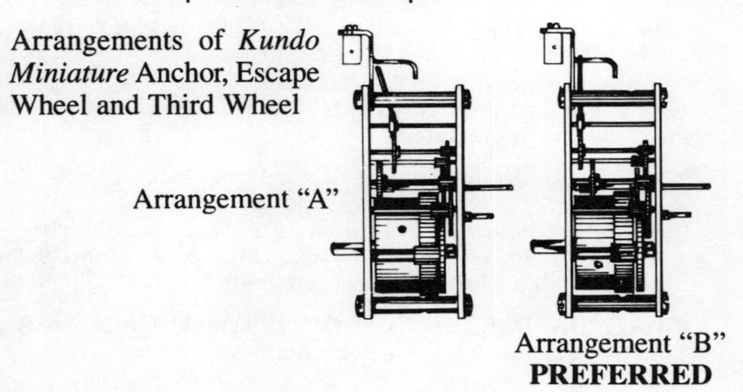

Arrangement "A"

Arrangement "B"
PREFERRED

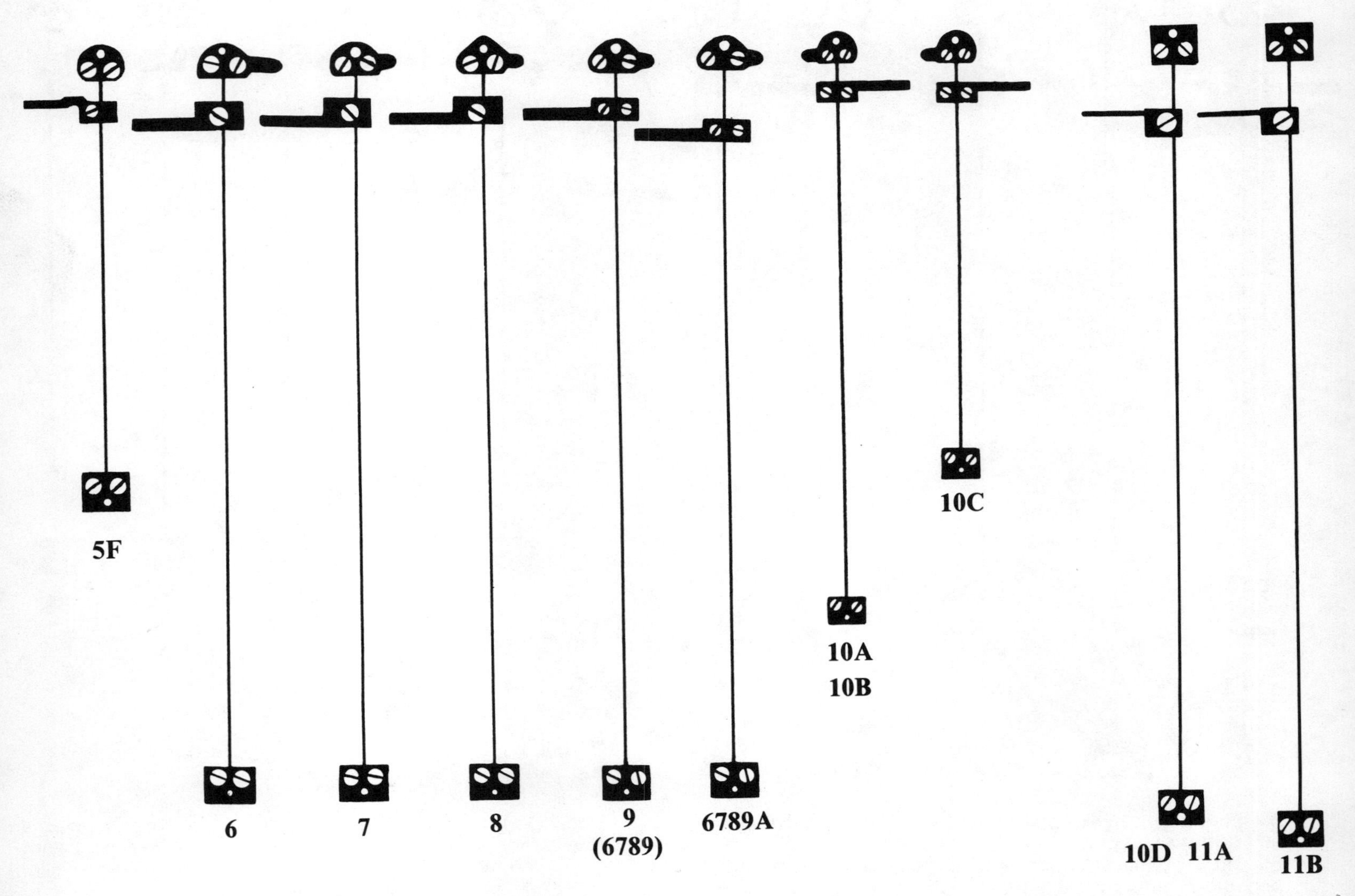

5F- KUNDO MIDGET 58 — USE .0022" (.056mm) HOROLOVAR
10 Beats per Minute
All movements using the 5F Unit have anchor pins with offset bends to the rear. The pin is so designed to make contact with the suspension fork slightly *behind* the back plate.

6- SCHATZ STANDARD 49 — USE .004" (.102mm) HOROLOVAR
8 Beats per Minute
This is the original Unit used with the 1949 model clock. Units 6, 7, 8 and 9 are interchangeable. **Unit 6789 is recommended.**

7- SCHATZ STANDARD 52 — USE .004" (.102mm) HOROLOVAR
8 Beats per Minute
Second unit used with the 1949 model clock. The only change is in the shape of the top block. Units 6, 7, 8 and 9 are interchangeable. **Unit 6789 is recommended.**

8- SCHATZ STANDARD 53 — USE .004" (.102mm) HOROLOVAR
8 Beats per Minute
Third unit used with the 1949 model clock. Again, the only change is in the shape of the top block. Units 6, 7, 8 and 9 are interchangeable. **Unit 6789 is recommended.**

9- SCHATZ STANDARD 54 — USE .004" (.102mm) HOROLOVAR
8 Beats per Minute
This Unit was first used with the model 1954 clock. A change was made in the design of the fork. Units 6, 7, 8 and 9 are interchangeable. **Unit 6789 is recommended.**

6789- SCHATZ STANDARD 49-54 — USE .004" (.102mm) HOROLOVAR
8 Beats per Minute
This is the number assigned to the Unit made by Horolovar for use with any Schatz Standard model.

6789A- JAHRESUHRENFABRIK — USE .004" (.102mm) HOROLOVAR
8 Beats per Minute
This Unit may be used with several Pre-World War I Jahresuhrenfabrik clocks having the saddle on top of the suspension bracket (See Section 15, No. 10).

10A- SCHATZ MINIATURE 53 — USE .0023" (.058mm) HOROLOVAR
10 Beats per Minute
This Unit is used with the glass- or plastic-domed clock; also with the "London Coach" and "Bermuda Carriage" models.

10B- SCHATZ 1000-DAY 54 — USE .0024" (.061mm) HOROLOVAR
10 Beats per Minute

10C- SCHATZ MIDGET 57 — USE .0022" (.056mm) HOROLOVAR
10 Beats per Minute

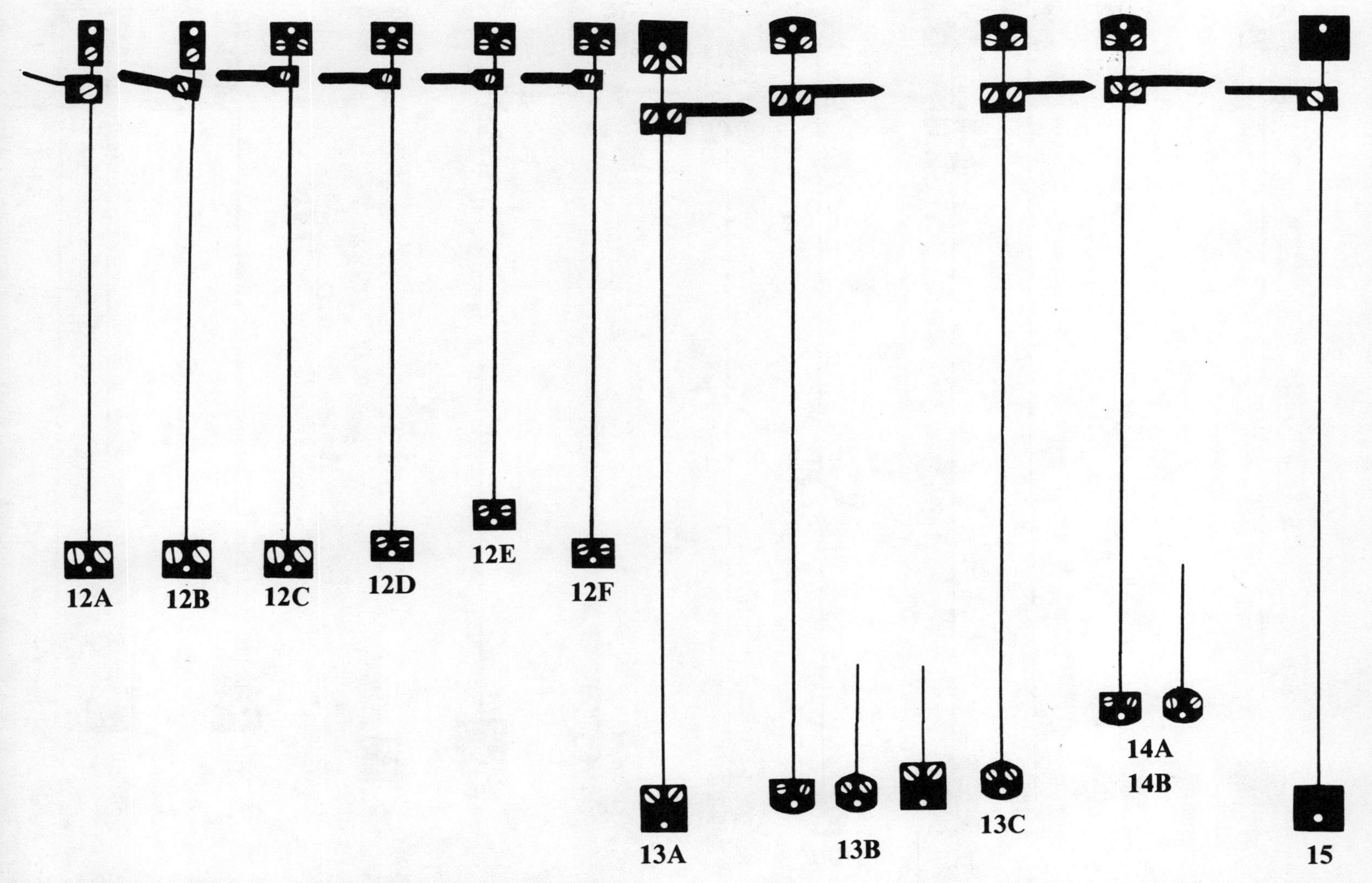

10D- SCHATZ STANDARD 81 — USE .0036" (.091mm) HOROLOVAR
8 Beats per Minute
This clock, made by Jahresuhrenfabrik, Aug. Schatz & Sohne, to celebrate the company's one hundred years of clockmaking, is an approximate reproduction of the Jahresuhrenfabrik 1885 model. (See Section 6, Clock 86).

11A- KERN STANDARD 50 — USE .0036" (.091mm) HOROLOVAR
8 Beats per Minute
This Unit should be used only with models having detachable pendulum.

11B- KERN STANDARD 54 — USE .0036" (.091mm) HOROLOVAR
8 Beats per Minute
This Unit should be used only with models equipped with locking pendulum mechanisms. It is slightly longer than Unit 11A.

12A- KERN MINIATURE 54 — USE .002" (.051mm) HOROLOVAR
6 Beats per Minute
Note that the fork tines are supposed to be bent upward as illustrated. Units 12A, 12B and 12C are interchangeable. **Unit 12C is recommended.**

12B- KERN MINIATURE 56 — USE .002" (.051mm) HOROLOVAR
6 Beats per Minute
The fork in this Unit is preferred to the one illustrated in 12A, because it is lighter in weight. Units 12A, 12B and 12C are interchangeable. **Unit 12C is recommended.**

12C- KERN MINIATURE 57 — USE .002" (.051mm) HOROLOVAR
6 Beats per Minute
The top block of this Unit is preferred to the one used in 12A and 12B, because it will hold the spring tighter. Units 12A, and 12B and 12C are interchangeable. **Unit 12C is recommended.**

12D- KERN MINIATURE 58 — USE .0019" (.048mm) HOROLOVAR
6 Beats per Minute
In addition to the lighter spring strength, this Unit is approximately 2 millimeters shorter than Units 12A, 12B and 12C.

12E- KERN MIDGET 58 — USE .0023" (.058mm) HOROLOVAR
12 Beats per Minute
This is the shortest of any of the Kern Units.

12F- KERN MINIATURE 61 — USE .0023" (.058mm) HOROLOVAR
8 Beats per Minute
This Unit is used with the only Kern movement that has the mainspring barrel on the right side when viewed from the back. With few exceptions, this movement is used only in a model with a brass and plastic-sided case.

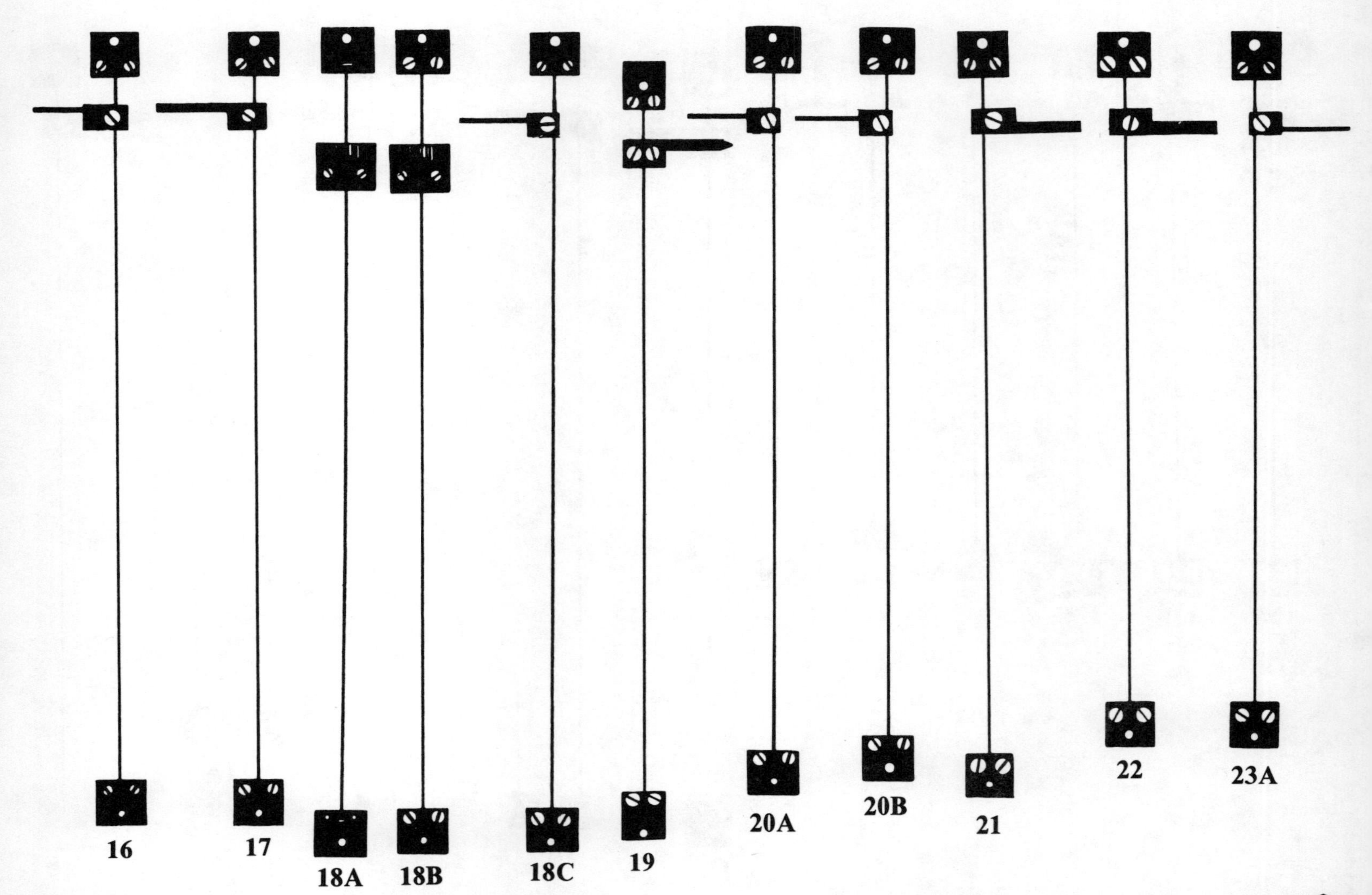

13A- KOMA STANDARD 50 — USE .0035″ (.089mm) HOROLOVAR
8 Beats per Minute
This is the original Unit used with the 1950 model. Subsequently, the shape of the top and bottom suspension blocks were changed as shown in Unit 13B. Units 13A and 13B are interchangeable. **Unit 13B is recommended.**

13B- KOMA STANDARD 52 — USE .0035″ (.089mm) HOROLOVAR
8 Beats per Minute
Second Unit used with the Koma Standard model. Units 13A and 13B are interchangeable. **Unit 13B is recommended.**

13C- KOMA STANDARD 56 — USE .0035″ (.089mm) HOROLOVAR
8 Beats per Minute
This Unit should be used only with models equipped with locking pendulum mechanisms. It is slightly shorter than Units 13A and 13B.

14A- KOMA MINIATURE 54 — USE .0032″ (.081mm) HOROLOVAR
8 Beats per Minute
This Unit is used with two different model clocks—one with a pendulum having a slightly longer axis than the other. Whereas the Units are interchangeable with both model clocks, the model having the longer pendulum axis also has slightly higher pillars which raise the movement.

14B- KOMA MINIATURE 65 — USE .003″ (.076mm) HOROLOVAR
8 Beats per Minute
The model requiring this Unit, with lighter spring than 14A, has pendulum weights shaped like this ....

15- LINK STANDARD 52 — USE .0037″ (.094mm) HOROLOVAR
8 Beats per Minute
First Unit used with the Link Standard clock, there are no screws in the top and bottom blocks. The suspension spring is "pinched" tight. No replacement units of this type are available. Replacement must be made with Unit 16 or 17. Units 15, 16 and 17 are interchangeable. **Unit 16 is recommended.**

16- LINK STANDARD 52 — USE .0037″ (.094mm) HOROLOVAR
8 Beats per Minute
The only difference between Units 16 and 17 is in the shape of the forks. Units 16 and 17 are interchangeable. **Unit 16 is recommended.**

17- LINK STANDARD 53 — USE .0037″ (.094mm) HOROLOVAR
8 Beats per Minute
The only difference between Units 17 and 16 is in the shape of the forks. Units 17 and 16 are interchangeable. **Unit 16 is recommended.**

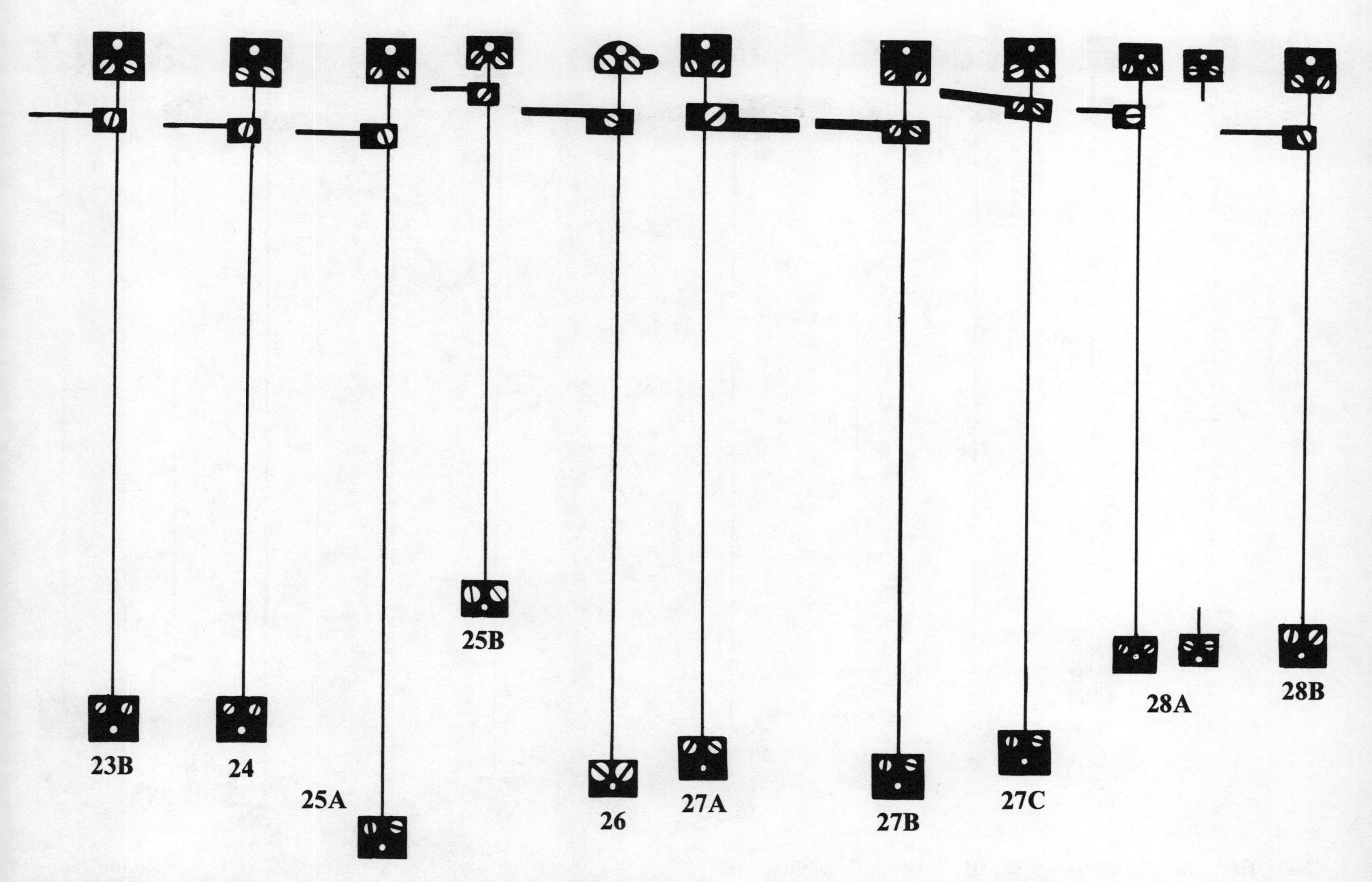

18A- PETERSEN STANDARD 53 — USE .0033″ (.084mm) HOROLOVAR
8 Beats per Minute
First Unit used with the Petersen Standard clock, there are no screws in the top and bottom blocks. The suspension spring is "pinched" tight. No replacement Units of this type are available. Replacement must be made with Units 18B or 18C. Units 18A, 18B or 18C are interchangeable. **Unit 18C is recommended.**

18B- PETERSEN STANDARD 54 — USE .0033″ (.084mm) HOROLOVAR
8 Beats per Minute
The only difference between Units 18B and 18C is in the shape of the fork. The design of the Petersen fork, in Units 18B and 18A, makes it necessary to set the suspension saddle position 90° from that normal to all other clocks. Since there is no functional reason for this different design, if a Petersen fork is not available, use Unit 18C. Units 18A, 18B and 18C are interchangeable. **Unit 18C is recommended.**

18C- PETERSEN STANDARD 53 — USE .0033″ (.084mm) HOROLOVAR
8 Beats per Minute
This Unit is functionally the same as Unit 18A and 18B, but it is provided with a conventional type fork. Units 18A, 18B and 18C are interchangeable. **Unit 18C is recommended.**

19- HERR STANDARD 53 — USE .0035″ (.089mm) HOROLOVAR
8 Beats per Minute
This Unit was used only with the 1953 model. The movement is Standard size, but with narrow plates. From the shape of the blocks and fork, it would appear as though Herr temporarily used those manufactured by Koma. (Note similarity with Unit 13A). Units 19, 20A and 21 are interchangeable. **Unit 20A is recommended.**

20A- HERR STANDARD 54 — USE .0035″ (089mm) HOROLOVAR
8 Beats per Minute
The only difference between Unit 20A and 19 is in the shape of the blocks and fork. It differs from Unit 21 only in the shape and position of the fork on the suspension spring. Units 20A, 19 and 21 are interchangeable. **Unit 20A is recommended.**

20B- HERR STANDARD 54 — USE .0033″ (.084mm) HOROLOVAR
8 Beats per Minute
This Unit is used with the Herr "Atlantis"—a Standard, narrow plate movement in a four-glass case, with mirror back, and a handle on top.

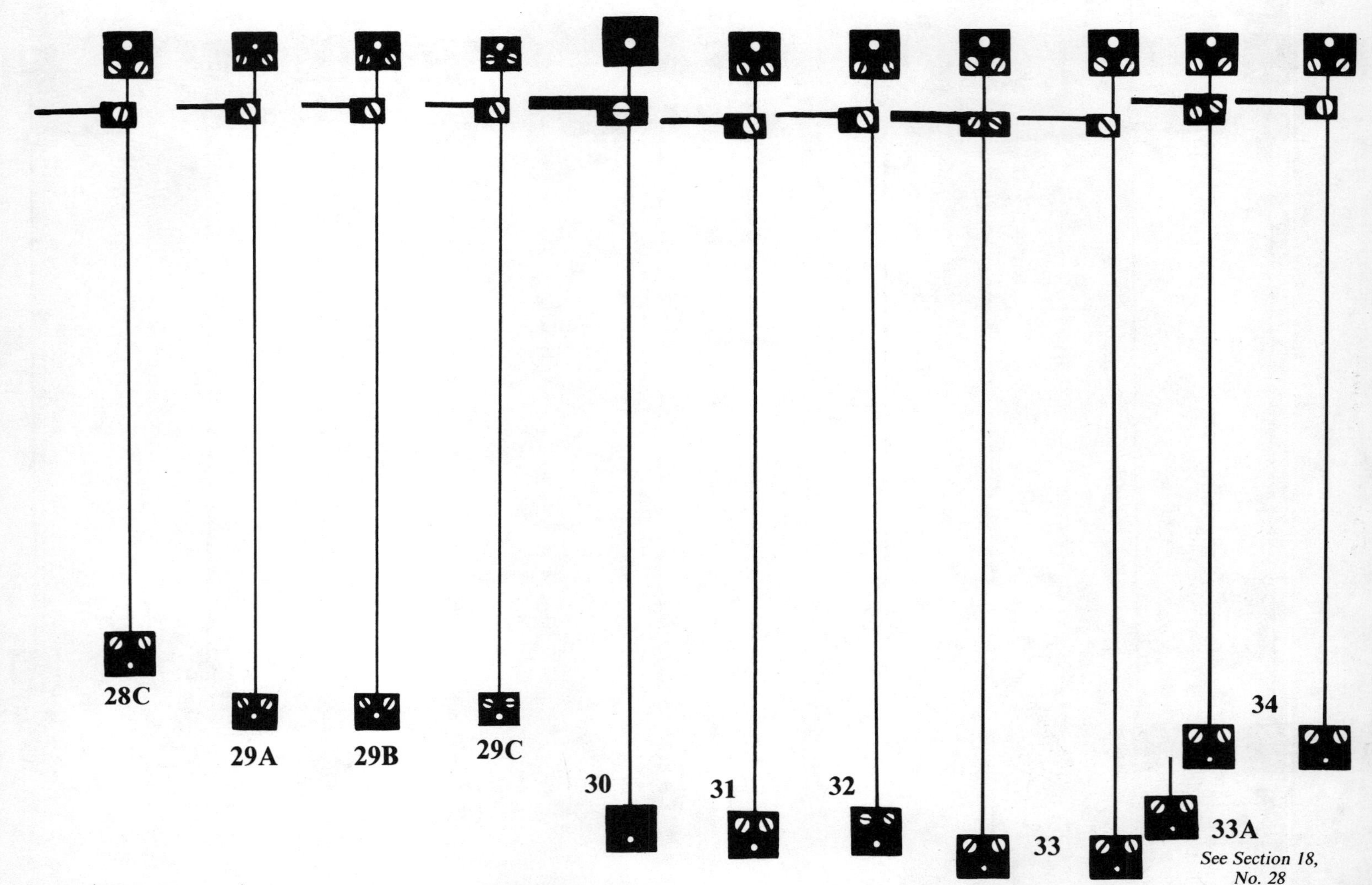

21- HERR STANDARD 53 — USE .0035″ (.089mm) HOROLOVAR
8 Beats per Minute
This Unit differs from Unit 20A only in the shape and position of the fork on the suspension spring. Units 21, 19 and 20A are interchangeable. **Unit 20A is recommended.**

22- REINER STANDARD 53 — USE .003″ (.076mm) HOROLOVAR
8 Beats per Minute
See details about the clocks with miniature movement using this Unit under Unit 23A. Units 22, 23A and 24 are interchangeable. **Unit 23A is recommended.**

23A- REINER STANDARD 53 — USE .003″ (.076mm) HOROLOVAR
8 Beats per Minute
This Unit is used with the Standard size clock with miniature movement. Most of these clocks are under the Standard 5½″ x 10½″ glass dome, but another model, known as the "Venita," has a glass sided "hood," with a curved top, which fits over a flat base. This miniature movement, which is also made by Herr, is their first model, identified as such by the fact that the anchor arbor pivot hole, in the back plate, is *below* the top suspension guard holding screw. (See Section 9, Plate 1091.) Units 23A, 22 and 24 are interchangeable, the only difference being in the shape of the fork or in its position on the suspension spring. The alternate positions of the fork do not affect the functioning of the escapement. **Unit 23A is recommended.** For the miniature clock (under a 4⅝″ x 8″ glass dome) with this movement (See Section 9. Plate 1572), use Units 28A or 28B.

23B- REINER STANDARD 56 — USE .0032″ (.081mm) HOROLOVAR
8 Beats per Minute
This Unit should be used with the improved miniature movement (larger escape wheel and anchor) in the clocks described above for Unit 23A. This movement (which is also made by Herr) is identified by the fact that the anchor pivot hole, in the back *above* the top suspension guard holding screw. (See Section 9, Plate 1092.) When this change in the movement was made, the weight of the pendulum, for some unknown reason was increased. Thus it is necessary to use the slightly heavier suspension spring. For the miniature clock (under a 4⅝″ x 8″ glass dome) with this movement (See Section 9, Plate 1093), use Unit 28C.

24- REINER STANDARD 53 — USE .003″ (.076mm) HOROLOVAR
8 Beats per Minute
See details about the clocks using this Unit under Unit 23A. Units 24, 22 and 23A are interchangeable. **Unit 23A is recommended**

25A- HERMLE STANDARD 53 — USE .0033″ (.084mm) HOROLOVAR
8 Beats per Minute

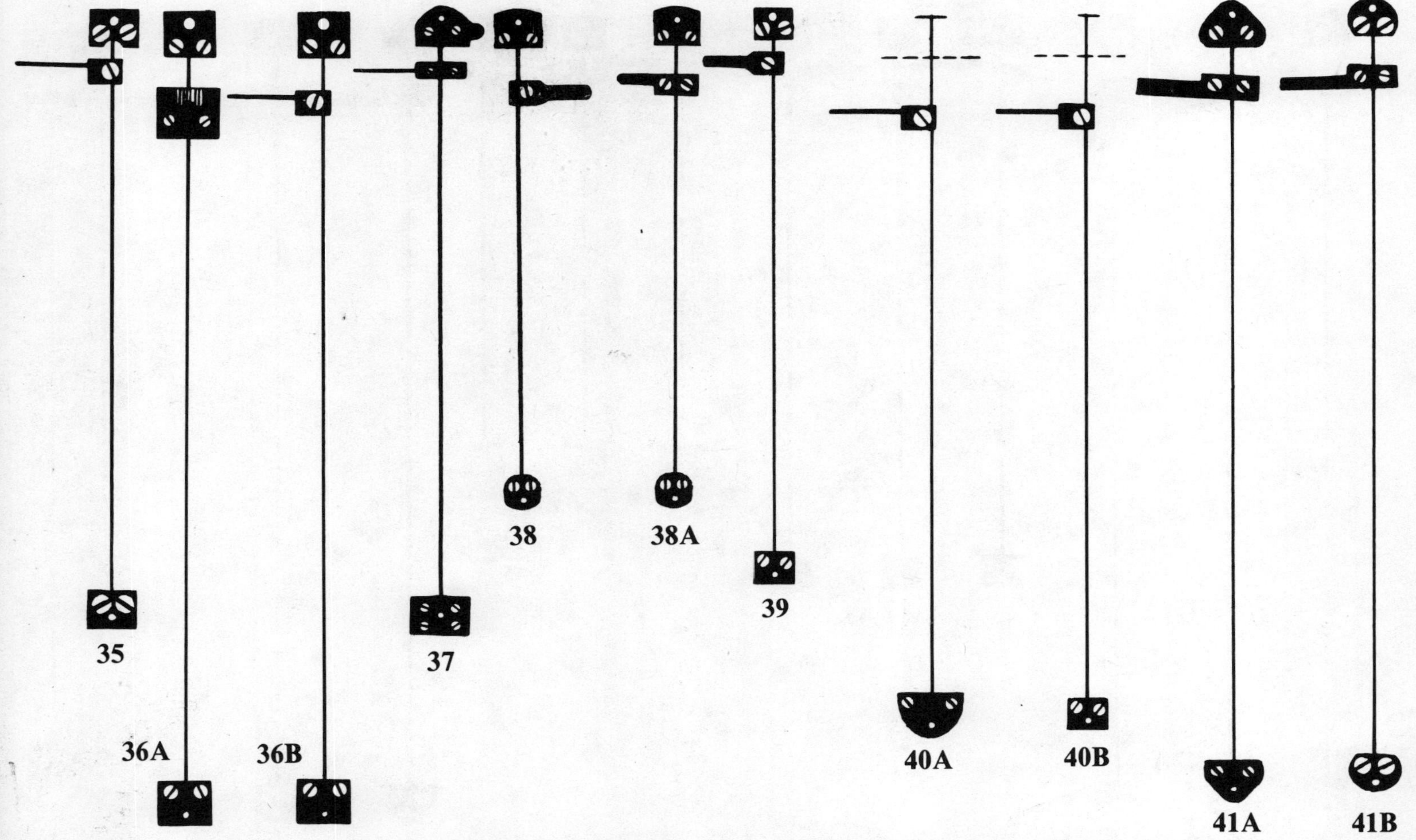

25B- HERMLE MINIATURE 54 — USE .022" (.056mm) HOROLOVAR
8 Beats per Minute
It is necessary for the top suspension block to be cut out as indicated in order for the spring to be of the correct length above the fork, and for the fork to contact the anchor pin at the proper point. If a Hermle block is not available, a square block can be filed to this shape.

26- WURTHNER STANDARD 53 — USE .004" (.102mm) HOROLOVAR
8 Beats per Minute

27A- HERR STANDARD 53 — USE 004" (.102mm) HOROLOVAR
8 Beats per Minute
This Unit is used with all Herr Standard clocks having wide plate movements.

27B- HERR STANDARD 55 — USE .0038" (.097mm) HOROLOVAR
8 Beats per Minute
This Unit is from one of several clocks with pendulum regulating nuts threaded in the opposite direction from most others. To make the pendulum go faster, you must turn the nut *clockwise.*

27C- NEUECK 57 — USE .0036" (.091mm) HOROLOVAR
8 Beats per Minute
The trade mark WILMAC or NEUECK appears on the back plates of most of these clocks.

28A- HERR MINIATURES 53 — USE .0025" (.064mm) HOROLOVAR
8 Beats per Minute
This Unit differs from Unit 28B only in the size of its top and bottom blocks. Units 28A and 28B are interchangeable.

28B- HERR MINIATURE 54 — USE .0025" (.064mm) HOROLOVAR
8 Beats per Minute
This Unit differs from Unit 28A only in the size of its top and bottom blocks. Units 28A and 28B are used in the first model miniature Herr clock, identified as such by the fact that the anchor pivot hole in the back plate is *below* the top suspension guard holding screw. (See Section 9, Plate 1243.) If the pivot hole is *above* the screw, use Unit 28C.

28C- HERR MINIATURE 56 — USE .0028" (.071mm) HOROLOVAR
8 Beats per Minute
This Unit is used with the latest model miniature Herr clock. It is distinguished from the first model by the fact that the anchor arbor pivot hole in the back plate is *above* the top suspension guard holding screw. (See Section 9, Plate 1244.) If the pivot hole is *below* the screw, use Unit 28A or 28B. This Unit is from one of several recently manufactured clocks with pendulum regulating nuts threaded in the opposite direction from most others. To make the pendulum go faster, you must turn the nut *clockwise.* This unit also fits the model in the six-sided case.

29A- HENN MINIATURE 52 — USE .0023" (.058mm) HOROLOVAR
8 Beat per Minute
The only difference between Unit 29A and 29B is in the smaller diameter of the hole in the top block. The top block holding

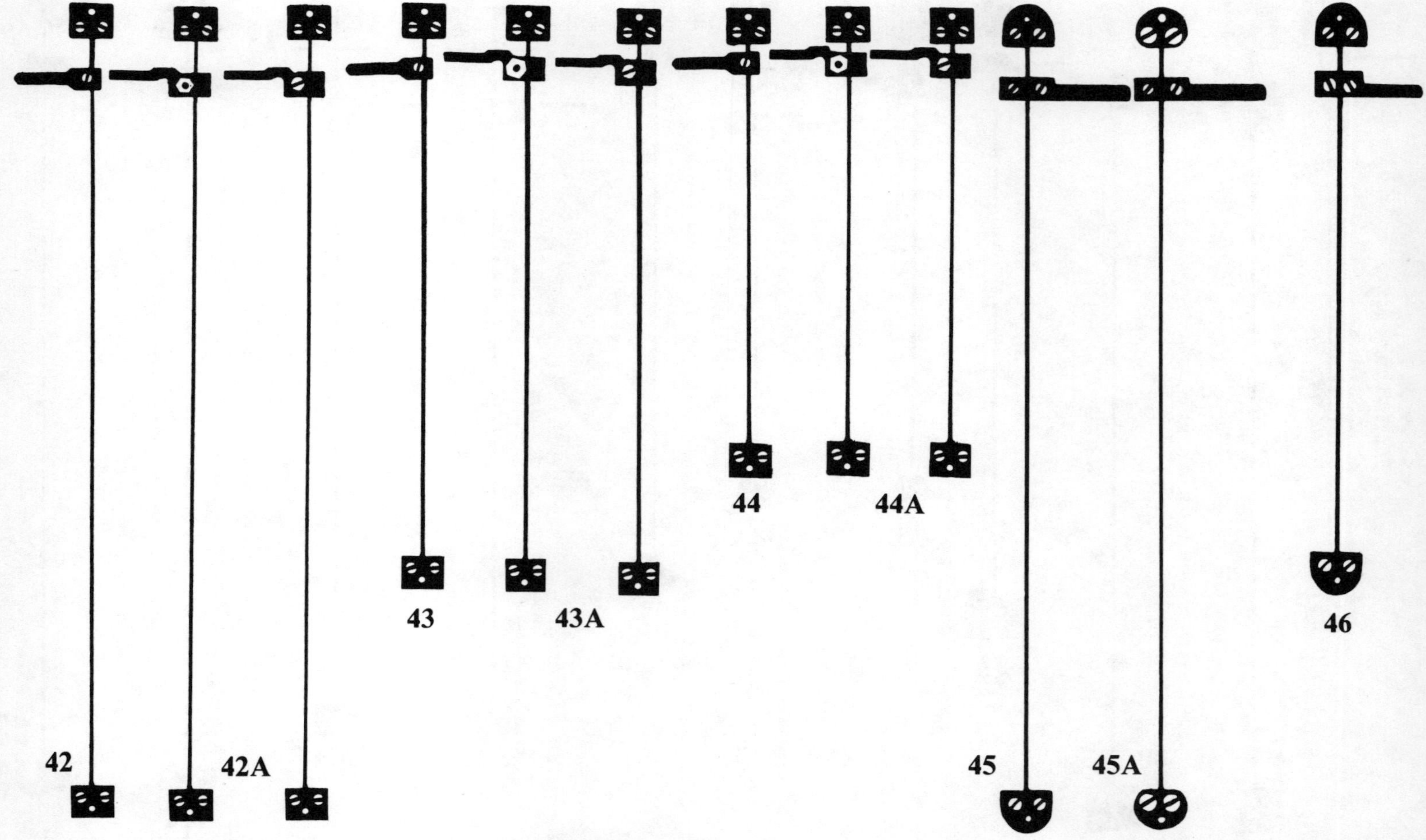

screw for Unit 29A is 1.4mm in diameter. Units 29B and 29C are functionally interchangeable with 29A.

29B- HENN MINIATURE 53 — USE .0023" (.058mm) HOROLOVAR
8 Beats per Minute
The only difference between Units 29B and 29A is in the larger diameter of the hole in the top block. The top block holding screw for 29B is 1.9mm in diameter. Some of the pendulums for this clock were made with regulating nuts threaded in the opposite direction from most others. To make these pendulums go faster, you must turn the nut *clockwise.* Units 29B and 29C are interchangeable.

29C- HENN MINIATURE 52-53 — USE .0023" (.058mm) HOROLOVAR
8 Beats per Minute
The Henn top and bottom blocks were made of a special size and are no longer available. The blocks in this Unit are slightly smaller than the originals, but Unit 29C is functionally interchangeable with Units 29A and 29B.

30- HENN STANDARD 51 — USE .0035" (.089mm) HOROLOVAR
8 Beats per Minute
First Unit used with the wide plate movement. There are no screws in the top and bottom blocks. The suspension spring is "pinched" tight. No replacement Units of this type are available. Replacement must be made with Units 31. Units 30 and 31 are interchangeable.

31- HENN STANDARD 52 — USE .0035" (.089mm) HOROLOVAR
8 Beats per Minute
Second Unit used with the wide plate movement. Units 31 and 30 are interchangeable. **Unit 31 is recommended.**

32- HENN STANDARD 54 — USE .0037" (.094mm) HOROLOVAR
8 Beats per Minute
Used with the narrow plate movement. The clock is equipped with locking pendulum.

33- JAUCH & HALLER 54 — USE .0038" (.097mm) HOROLOVAR

33A- JAUCH & HALLER 56 — USE .003" (.076mm) HOROLOVAR
8 Beats per Minute
This Unit is approximately ¼" shorter than Unit 33 and has a different style of fork than Units 33 and 33B.

33B- JAUCH & HALLER 56 — USE .0032" (.081mm) HOROLOVAR
8 Beats per Minute
This Unit is approximately ¼" shorter than Unit 33 and has a fork which is in a slightly higher position than Units 33 and 33A.

34- KAISER STANDARD 54 — USE .003" (.076mm) HOROLOVAR
8 Beats per Minute
This Unit is used with the model having the globe (world) pendulum.

35- LINK MINIATURE 56 — USE .0021" (.053mm) HOROLOVAR
6 Beats per Minute

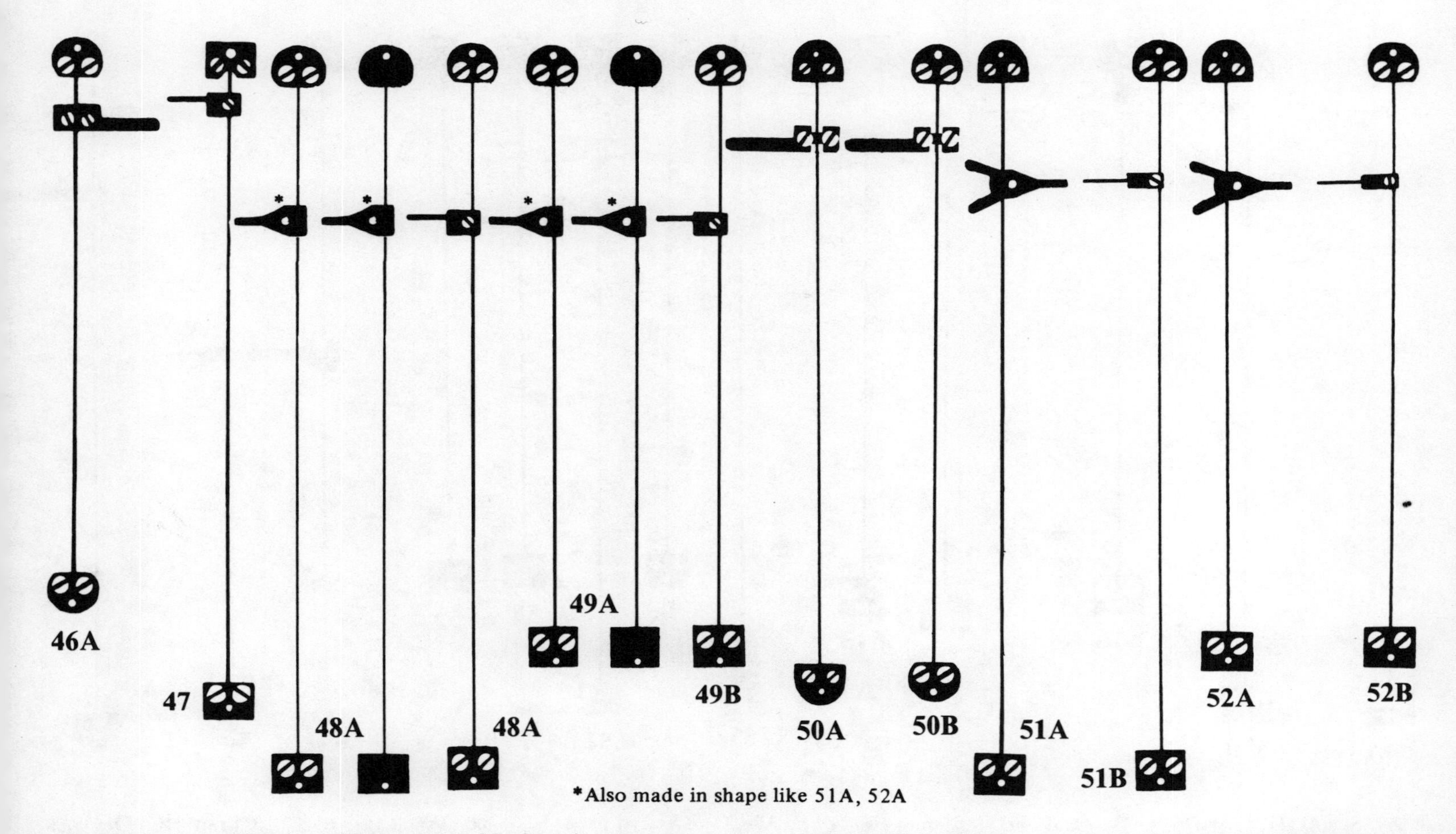

*Also made in shape like 51A, 52A

It is necessary for the top suspension block to be cut out as indicated in order for the spring to be of the correct length above the fork, and for the fork to contact the anchor pin at the proper point. If a Link block is not available, a square block can be filed to this shape.

36A- PETERSEN STANDARD 55 — USE .0036″ (.091mm) HOROLOVAR
8 Beats per Minute

The only difference between Units 36A and 36B is in the shape of the fork. The design of the fork in this calendar clock makes it necessary to set the suspension saddle position 90° from that normal to all other clocks. Since there is no functional reason for this different design, if a Petersen fork is not available, use Unit 36B. Units 36A and 36B are interchangeable. **Unit 36B is recommended.**

36B- PETERSEN STANDARD 55 — USE .0036″ (.091mm) HOROLOVAR
8 Beats per Minute

This Unit is functionally the same as Unit 36A, but it is provided with a conventional type fork. Units 36B and 36A are interchangeable. **Unit 36 B is recommended.**

37- WURTHNER MIDGET 57 — USE .002″ (.051mm) HOROLOVAR
10 Beats per Minute

Only a very small number of these clocks were produced.

38- KOMA MIDGET 57 — USE .0022″ (.056mm) HOROLOVAR
15 Beats per Minute

38A- KOMA MIDGET 73 — USE .0022″ (.056mm) HOROLOVAR
15 Beats per Minute

This Unit is functionally the same as Unit 38, but is equipped with a better designed fork.

39- SIEGFRIED HALLER 57 — USE .0019″ (.048mm) HOROLOVAR
6 Beats per Minute

Only a very small number of these clocks were produced.

40A- CRESENT STANDARD 55 — USE .0036″ (.091mm) HOROLOVAR
? Beats per Minute

This clock was manufactured in Japan by Ishikara Clock Co., Ltd. The suspension saddle is designed to hold the top of the suspension spring like a vise, so there is no top suspension block for this Unit. The dotted line indicates the location of the bottom of the saddle. Units 40A and 40B are interchangeable. **Unit 40B is recommended.**

40B- CRESENT STANDARD 55 — USE .0036″ (.091mm) HOROLOVAR
? Beats per Minute

Parts for this Unit are more available than they are for Unit 40A. Units 40A and 40B are interchangeable. **Unit 40B is recommended.**

41A- MASTER STANDARD 55 — USE .0035″ (.089mm) HOROLOVAR
8 Beats per Minute

This clock was manufactured in Japan by Nisshin Clock Industrial Co., Ltd. The plates of this movement have square corners. The plates of a similar movement that uses Unit 45 have rounded corners. (See Section 9, Plate 1461E.) Units 41A and 41B are interchangeable. **Unit 41B is recommended.**

41B- MASTER STANDARD 55 — USE .0035″ (.089mm) HOROLOVAR
8 Beats per Minute

Parts for this Unit are more available than they are for Unit 41A. Units 41A and 41B are interchangeable. **Unit 41B is recommended.**

42- S. HALLER STANDARD 70 — USE .003″ (.076mm) HOROLOVAR
8 Beats per Minute

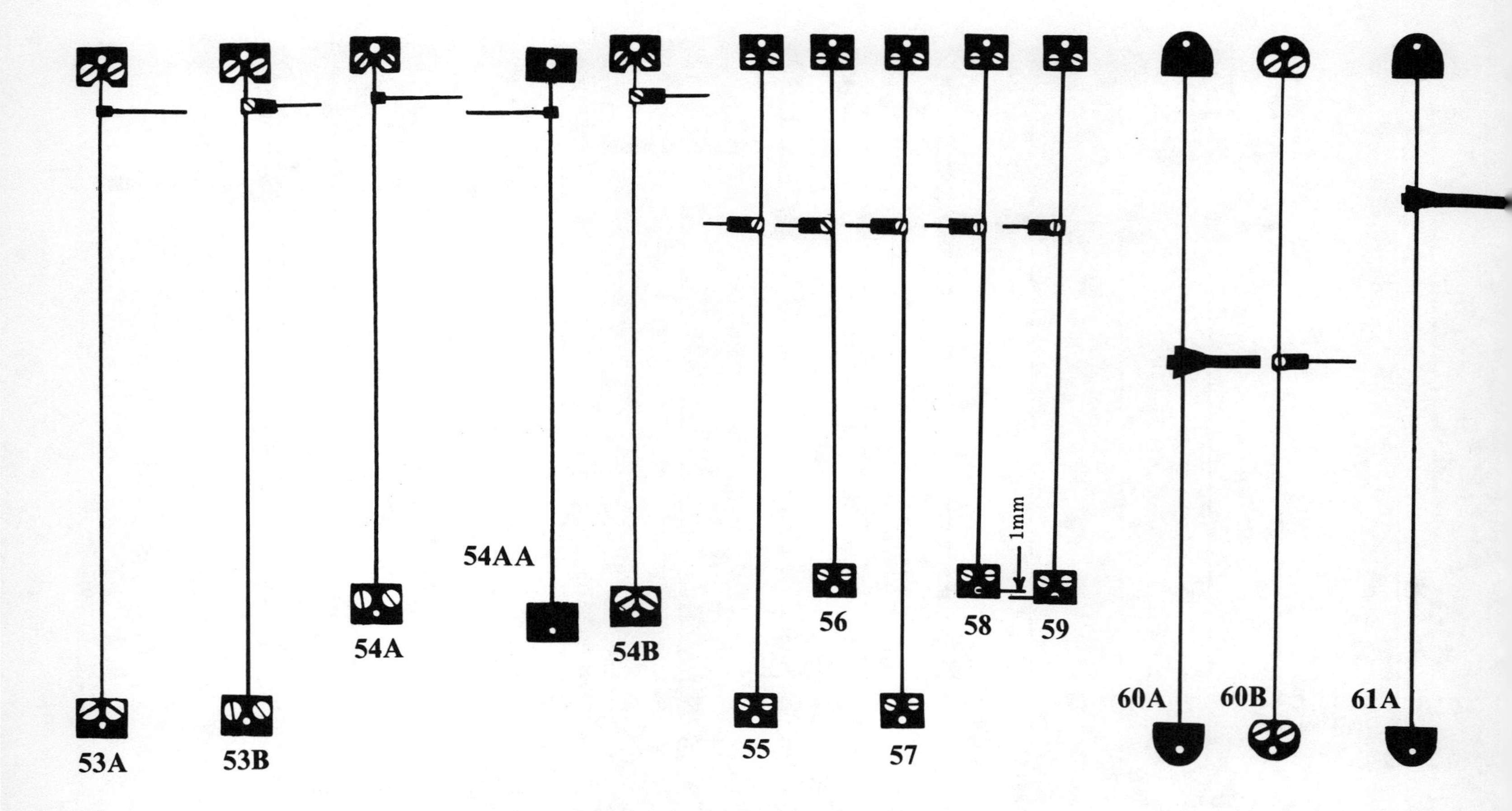

42A- S. HALLER STANDARD 73 — USE .003″ (.076mm) HOROLOVAR
8 Beats per Minute
This Unit is functionally the same as Unit 42, but is equipped with a better designed fork.

43- S. HALLER MINIATURE 70 — USE .0022″ (.056mm) HOROLOVAR
10 Beats per Minute

43A- S. HALLER MINIATURE 73 — USE .0022″ (.056mm) HOROLOVAR
10 Beats per Minute
This Unit is functionally the same as Unit 43, but is equipped with a better designed fork.

44- S. HALLER MIDGET 70 — USE .0018″ (.046mm) HOROLOVAR
10 Beats per Minute

44A- S. HALLER MIDGET 73 — USE .0018″ (.046mm) HOROLOVAR
10 Beats per Minute
This Unit is functionally the same as Unit 44, but is equipped with a better designed fork.

45- MASTER STANDARD 70 — USE .0034″ (.086mm) HOROLOVAR
8 Beats per Minute
The plates of this movement have rounded corners. The plates of a similar movement that uses Unit 41A and 41B have square corners. (See Sction 9, Plate 1461C.) Units 45 and 45A are interchangeable. **Units 45A is recommended.**

45A- MASTER STANDARD 70 — USE .0034″ (.086mm) HOROLOVAR
8 Beats per Minute
Parts for this Unit are more available than they are for Unit 45. Units 45 and 45A are interchangeable. **Unit 45A is recommended.**

46- NEW MASTER 100-DAY — USE .0028″ (.071mm) HOROLOVAR
12 Beats per Minute

46A- NEW MASTER 100-DAY — USE .0028″ (.071mm) HOROLOVAR
12 Beats per Minute
Parts for this Unit are more available than they are for Unit 46. Units 46 and 46A are interchangeable. **Unit 46A is recommended.**

47- HERMLE MINIATURE 71 — USE .0022″ (.056mm) HOROLOVAR
KOMA 74 — USE .0022″ (.056mm) HOROLOVAR

48A- KUNDO STANDARD 73 —
This Unit has non-detachable fork. **Replace with Unit 48B.**

48B- KUNDO STANDARD 73 — USE .0032″ (.081mm) HOROLOVAR
This Unit is interchangeable with Unit 48A.

49A- KUNDO MINIATURE 73 —
This Unit has a non-detachable fork. **Replace with Unit 49B.**

49B- KUNDO MINIATURE 73 — USE .0023″ (.081mm) HOROLOVAR
This Unit is interchangeable with Unit 49A.

50A- MASTER STANDARD 75 —
This Unit is inerchangeable with Unit 50B. **Unit 50B is recommended.**

50B- MASTER STANDARD — 75 USE .0028″ (.071mm) HOROLOVAR
? Beats per Minute
This Unit is interchangeable with Unit 50A.

51A- KUNDO STANDARD 77 —
This Unit is equipped with a non-detachable fork. **Replace with Unit 51B.**

51B- KUNDO STANDARD 77 — USE .0022″ (.056mm) HOROLOVAR
This Unit is interchangeable with Unit 51A.

52A- KUNDO MINIATURE 77 —
This Unit has non-detachable fork. **Replace with Unit 52B.**

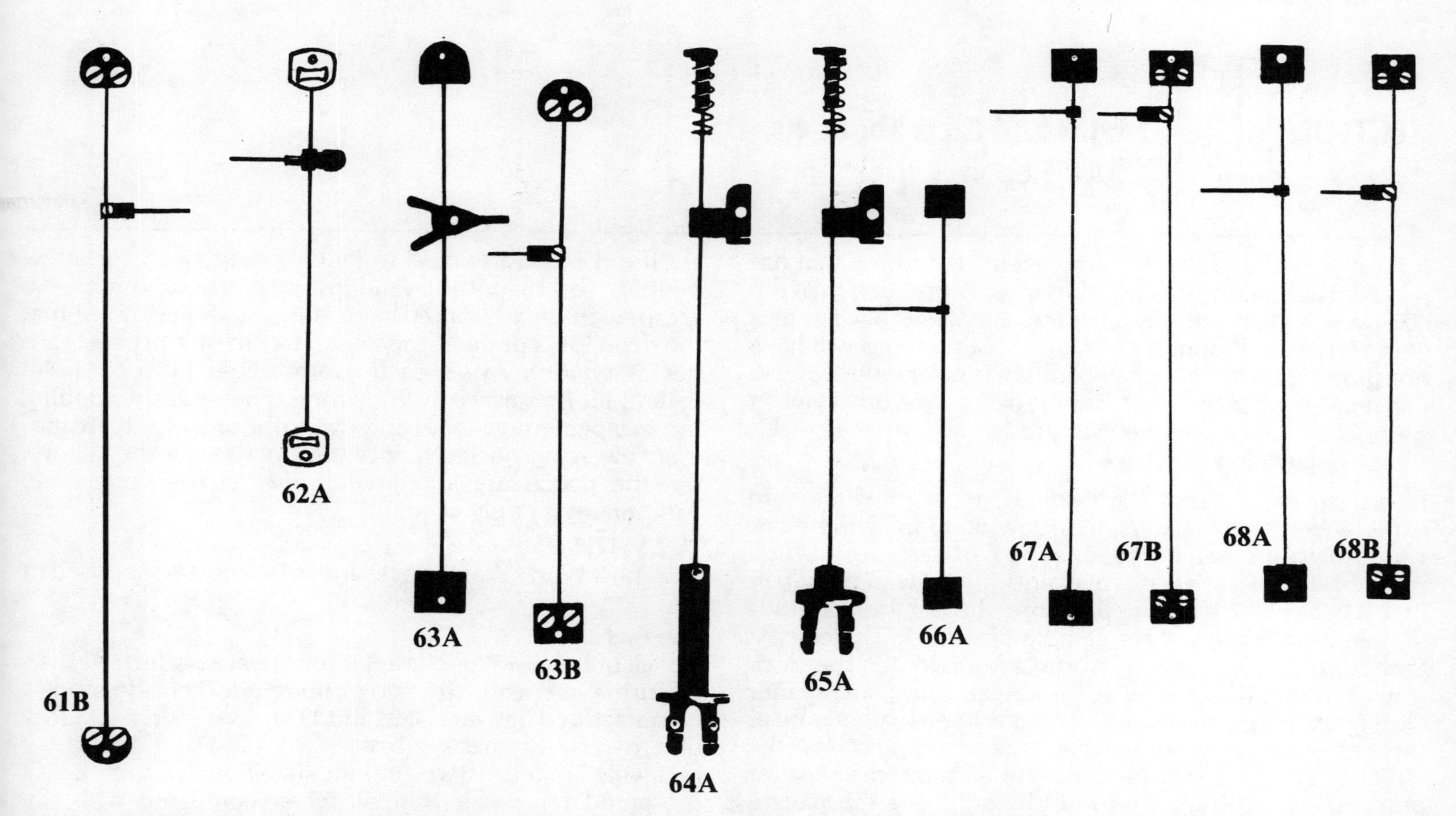

52B- KUNDO MINIATURE 77 — USE .0022" (.056mm) HOROLOVAR
This Unit is interchangeable with Unit 52A.

53A- HERMLE 75 —
This Unit has non-detachable fork. **Replace with Unit 53B.**

53B- HERMLE 75 — USE .003" (.076mm) HOROLOVAR
Interchangeable with Unit 53A.

54A- HERMLE 77 —
This Unit has non-detachable fork. **Replace with Unit 54B.**

54AA- HERMLE 78 —
This unit has non-detachable parts. **Replace with Unit 54B.**

54B- HERMLE 77 — USE .0028" (.071mm) HOROLOVAR
Interchangeable with Unit 54A and 54AA.

55- KERN STANDARD 75 — USE .0022" (.056mm) HOROLOVAR

56- KERN MINIATURE 75 — USE .0019" (.048mm) HOROLOVAR

57- KERN STANDARD 77 — USE .0022" (.056mm) HOROLOVAR

58- KERN MINIATURE 77 — USE .0019" (.048mm) HOROLOVAR
Same as Unit 59, but 1mm **shorter.**

59- KERN MINIATURE 78 — USE .0019" (.048mm) HOROLOVAR
Same as Unit 58, but 1mm **longer.**

60A- MASTER 75 —
This Unit has non-detachable parts. **Replace with Unit 60B.**

60B- MASTER 75 — USE .0023" (.058mm) HOROLOVAR
Interchangeable with Unit 60A.

61A- MASTER 77 —
This Unit has non-detachable parts. **Replace with Unit 61B.**

61B- MASTER 77 — USE .0018" (.046mm) HOROLOVAR
Interchangeable with Unit 61A.

62A- KOMA MINIATURE 78 —
This Unit has non-detachable blocks. **No HOROLOVAR Unit is available.**

63A- KUNDO MINIATURE 76 —
This Unit has non-detachable parts. **Replace with Unit 63B.**

63B- KUNDO MINIATURE 76 — USE .0023" (.081mm) HOROLOVAR
Interchangeable with Unit 63A.

64A- KUNDO STANDARD 79 —
This Unit has maximum rotation guard. **No HOROLOVAR Unit is available.**

65A- KUNDO MINIATURE 79 —
This Unit has maximum rotation guard. **No HOROLOVAR Unit is available.**

66A- KOMA MINIATURE 80 —
This Unit has non-detachable parts. **No HOROLOVAR Unit is available.**

67A- HERMLE 80 —
This Unit has non-detachable parts. **Replace with 67B.**

67B- HERMLE 80 — USE .0028" (.071mm) HOROLOVAR
Interchangeable with Unit 67A.

68A- HERMLE 83 —
This Unit has non-detachable parts. **Replace with 68B.**

68B- HERMLE 83 — USE .0081" (.048mm) HOROLOVAR
Interchangeable with Unit 68A.

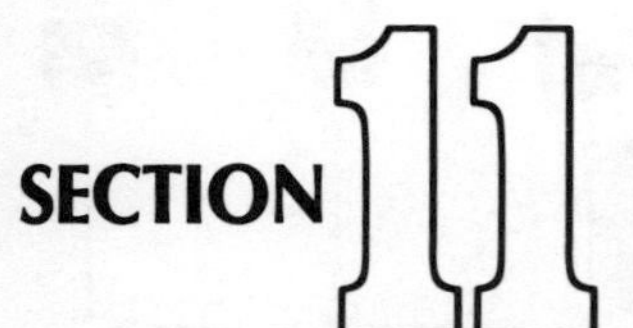

Myths and Facts about 400-Day Clock Repair

It is unlikely that there has ever been a timepiece that has received more unjustifiable criticism over the years than the 400-Day Clock. Practically all of the complaints have originated with a small number of clock repairers who will have nothing to do with the repair of this particular type of clock. For reasons of their own, these repairers are unwilling to spend the few minutes necessary to learn about the clock's simple peculiarities.

In all fairness to these individuals, most of whom are experienced *watchmakers,* their low opinion of the clock can usually be traced to one bad experience, which probably follows this series of events: Faced with the repair of a 400-Day Clock for the first time, the repairer carefully dismantles, cleans, and reassembles the clock. The clock is then started only to find that, after a few minutes, the clock stops running. Following watchmaking practice, our repairer closely observes the escapement and immediately notices the rather deep locks of the escape wheel teeth on the pallets. Remembering watchmaking training, the repairer proceeds to "correct" the fault by adjusting the pallets, moving the eccentric nut, filing down the escape wheel teeth, grinding the faces of the pallets, or various combinations of these actions. In short, everything is done to throw the escapement (which was probably originally in good order) completely out of adjustment. When the clock does not run after all of this misguided effort, our repairer concludes that the clock is no good and swears that a 400-Day Clock will never cross his or her workbench again. Perhaps to get out of trouble with the customer, our repairer takes the clock to an experienced 400-Day Clock repairer with the explanation that someone else has "worked on the clock." Finally, our repairer joins the anti-400-Day Clock group whose aim is to promulgate false accusations about the clock to whomever will listen.

If this sounds like fiction, be assured that it is not. Ask any repair person who won't repair 400-Day Clocks as to the reasons *why.* Chance are the stated reason will be that "the clock is no good," and one or more of the following myths will be expounded. Be assured that all are myths!

THE MYTH
"The factory doesn't set the escapements right."

The Fact
The escapements are set as well in 400-Day Clocks as they are in any other clock with a Graham escapement. The problem which faces our friend above, however, is not uncommon. Yet if our repairer would study the escapement for a few minutes (or read the well-illustrated discussion of the escapement in Section 7), he or she would quickly see that the deep locks of the escape wheel teeth on the pallets are necessary. Otherwise, each tooth would fall on to the pallet's lifting surface at the wrong time, quickly stopping the escapement action. Moreover, not only is the escapement always inspected before the clock leaves the factory, but the pallets are also precision-set at the same time. Adjustment is rarely needed.

THE MYTH
"The post World War II clocks are not as good as they used to be."

The Fact
Not all of the pre World War I models were perfect! And, to be sure, there were also two or three poorly made models manufactured between 1954 and 1958. Fortunately, production of any of these sub-standard 400-Day Clocks was relatively limited. Therefore, the likelihood of their being presented for repair is small when compared with the literally millions of good 400-Day Clocks that have been made since 1900. In general, the later clocks are equally as good as those made 30 to 50 years ago. In addition, later clocks have several advantages over the old ones. They have more accessible spare parts, their pendulums are permanently attached to the clock (which reduces suspension spring damage or breakage) and they are mounted on bases (or in cases) having a leveling device. All things considered, 400-Day Clocks made after World War II are much *better* than they used to be.

THE MYTH
"The Movement plates are thinner than they used to be."

The Fact
The plate thickness of 400-Day Clocks has never been standardized. There is no evidence to support the claim, however, that plates made in the later years are thinner than they used to be. The three largest manufacturers who produced standard size 400-Day Clocks prior to World War I and who also produce them after World War II have used the same thickness of plates continuously: Schatz-.079", Kundo-.068", and Kienzle (now Kern)-.060". The difference of .019" between the thickest plate and the thinnest has absolutely no connection with the quality of the movement.

THE MYTH
"You have to clean all the burrs on the wheels."

The Fact

You can, if you're fussy, but you don't *have* to! In most clocks, the wheels are clean. Occasionally, burrs will be found which are the result of the multiple wheelcutting operation. True, this condition is offensive to the fine mechanic, but the clock is not a precision made timepiece. Therefore, there is no need to waste time removing the burrs if they are not interfering with the mesh of the pinion leaves.

THE MYTH

"You have to bush every hole in the plates."

The Fact

This is ridiculous, although all too often holes are bushed by uninformed repairers. It is normal for the pivot holes for the anchor, the escape wheel, and the centerwheel to be somewhat larger than one would expect to see, for instance, in a finely made 8-day movement. The experience of The Horolovar Company may be of interest on this point. After repairing several thousand 400-Day Clocks, both old and new, we have *never* found it necessary to bush a hole in a plate! Wear on 400-Day Clock pivot holes is negligible because the movement action is so slow.

THE MYTH

"The arbors have too much end-shake."

The Fact

There does appear to be slightly more end-shake with 400-Day Clock arbors than with arbors in fine movements. However, as long as the spacing of the wheels on the arbors is sufficient to prevent interference anywhere in the train, there's absolutely no harm caused by the end-shake. Consequently, there is no justification for this complaint.

THE MYTH

"They're terrible timekeepers."

The Fact

Just because 400-Day Clocks are said to need winding only once a year, many people think that they will keep good time for a year. Yet, one must reset the hands of a French 8-day clock occasionally; so must this be done with the 400-Day Clock. If the clock is expected to keep good time, it should be given its final regulation in the place it will occupy permanently. If the clock is equipped with a *Horolovar Temperature Compensating Suspension Spring,* and is regulated properly (details on how to regulate are in Section 8), it should keep time within a minute a week or better for at least two months. For best timekeeping, the clock should be wound at least every two months. Considering the fact that the clock is not precision made, it is far from being a "terrible timekeeper."

THE MYTH

"It takes too much time to find (or to 'stone down') a spring to fit."

The Fact

Although this is an infrequently made complaint, it could be made only by someone who is not aware of the existence of *The Horolovar 400-Day Clock Repair Guide.* With the Repair Guide, such a waste of time is eliminated. It takes only a few seconds to find, in Section 9, the illustration of the back plate of the clock being repaired, and directly under the illustration the exact strength *Horolovar Temperature Compensating Suspension Spring* needed for this particular clock. It is easier to learn immediately what strength spring is required than to waste time trying several springs by trial and error, or trying to "stone down" a heavier strength spring to fit. This is particularly true in view of the fact that there are 24 different spring strengths, assuring that there is a proper spring for each type of clock.

THE MYTH

"We lost all of our profit because we had to make good on the clocks that the customers wrecked after we repaired them."

The Fact

You do not have to be responsible for the fact that there are a lot of clumsy clock owners whose fingers are all thumbs. When you turn the clock over to the customer, take two minutes to expalin (1) how to attach and detach the pendulum (if it is not permanently attached); (2) how to start the clock; and (3) how to regulate it. Then, *call particular attention* to the delicate suspension spring and say (also write it on the paid receipt), "This clock has our customary guarantee *except for abuse of the suspension spring."* A form, "400-Day Clock Repair Warranty with Operation Instructions," is described in Section 8. Copies are available from The Horolovar Company.

THE MYTH

"There are so many adjustments to make that I will never get them all right."

The Fact

All of the adjustments are properly made at the factory and unless you are certain that someone else has "fooled" with the clock, there is no need to make any changes. If adjustments are required, carefully follow the instructions in Section 7.

SECTION 12 Names and Trade Marks on 400-Day Clock Dials

Of the millions of 400-Day Clocks that have been made during the past one hundred years, relatively few have markings on their dials. An exception is the clock's country of origin which, after October 1913, was a requirement of U.S. Customs. The country (Germany, West Germany, Japan, etc.), sometimes preceded by "Made in," is usually positioned below the numeral 6 or VI in the smallest type allowed by Customs regulations.

The use of names or trade names of manufacturers, importers and others on dials was indiscriminate, possibly depending on the number of dials involved, the additional cost and the time needed to mark the dials. When the dial carries a manufacturer's or importer's name, or the clock is of a quality to be sold by a jeweler such as Tiffany & Co., and the jeweler's name and perhaps location are shown, the collector has valuable added information about his acquisition.

The following list has been complied from numerous sources and probably contains omissions and/or inaccuracies, due largely to the length of time that has passed since the clocks were being marketed and the scarcity of surviving records. The listing shows names and trade name and their owners, the manufacturers of the clock carrying the name and whether it was made before World War I or after that time.

Name or Trade Name on Dial	Owned by	Manufacturer of Clock	Before/After World War I
4 jewels	Manufacturer	Edgar Henn Co.	After
400 days clock	Manufacturer	R. Schneckenburger	Before
Adoria	Importer?	Jahresuhrenfabrik	Before
Adorna	Importer?	Konrad Mauch	After
American Specialty Co.	American Specialty Co. U.S. Importer	Kieninger & Obergfell	Before?
Anniversary	Bowler & Burdick U.S. Importer	Jahresuhrenfabrik	Before
Annual	Importer?	Kieninger & Obergfell	Before?
B	Manufacturer	Badische Uhrenfabrik	Before
Bentima	English Importer	Kern & Sohne	After
Birks	Canadian Importer	Kieninger & Obergfell	After
Black Forest	Importer?	Kieninger & Obergfell S. Haller Georg Wurthner	After
BuK	Importer?	Kieninger & Obergfell	After
Bucherer	Importer?	Konrad Mauch	After
Bulova	Bulova Watch Co. U.S. Importer	Kieninger & Obergfell	After
C.D. Peacock	C.D. Peacock U.S. Importer	Kieninger & Obergfell?	Before
Clebar	Importer?	Georg Wurthner	After
Crescent	Manufacturer	Ishisara Clock Co. (Japan)	After
Cresta	Importer?	Sigfried Haller	After
Crown	Manufacturer	Unknown (Japan)	After
D.R.P. 2437 R.L. Patent 2182 U.S. Patent 269052	Manufacturer	Jahresuhrenfabrik	Before
DeBruce	Walter Kocher Co. U.S. Importer	Kieninger & Obergfell	After
Dial	Importer?	Uhrenfabrik Herr	After
Diana	Importer?	Konrad Mauch	After
Elo	Importer?	Sigfried Haller	After
Elvia	Importer?	Kieninger & Obergfell	After
Eve	Importer?	Walter Petersen	After
Florn	Florn Clock Co. U.S. Importer	Kieninger & Obergfell	After
Forestville	Forestville Clock Co. U.S. Importer	Kieninger & Obergfell	After
Fortuna	Manufacturer?	Jahresuhrenfabrik	Before
GB	Manufacturer	Gustav Becker	Before
Gloria	Importer?	Kienzle Clock Factories	Before
Grivolas/Paris	Manufacturer	C. Grivolas (France)	Before
Gruttert/Bremen	German Exporter	Kieninger & Obergfell	Before
Gufa	Manufacturer	Gutenbacher Uhrenfabrik, C.H. Schatz KG	After
H. Dzialoszynski/ Czestochowa	Polish Retailer?	Gustav Becker	Before
Haller	Manufacturer	Jauch & Haller Sigfried Haller	After
Hamilton	Hamilton Watch Co. U.S. Importer	Sigfried Haller	After
Harder/ Ransen bei Steinau a/O	Inventor	A. Willman & Co.?	Before
Harder/ Ransen bei Steinau a/O/ D.R. Patent No. 2437	Inventor	Gustav Becker?	Before
Heco	Henry Coehler Co. U.S. Importer	Kieninger & Obergfell Kern & Sohne	After
Heirloom	Heirloom Clock Co. U.S. Importer	Uhrenfabrik Herr Uhrenfabrik M. Reiner	After
Hermle	Manufacturer	Franz Hermle	After
Herr	Manufacturer	Uhrenfabrik Herr	After
Hervo	Importer?	Georg Wurthner	After
Hudson	J.L. Hudson Co. U.S. Retailer	Kieninger & Obergfell	After

Name or Trade Name on Dial	Owned by	Manufacturer of Clock	Before/After World War I
Jauch Haller	Manufacturer	Jauch & Haller	After
John Elkan	English Retailer	Kienzle Clock Factories	Before
John Wanamaker	John Wanamaker Co. U.S. Retailer	Uhrenfabrik M. Reiner Uhrenfabrik Herr Georg Wurthner	After
Kaiser	Manufacturer	J. Kaiser & Co.	After
Kembro	Importer?	Uhrenfabrik Herr	After
Kern	Manufacturer	Kern & Sohne	After
Kienzle	Manufacturer	Badische Uhrenfabrik	Before
Koma	Manufacturer	Konrad Mauch	After
Kuehl Clock Co.	Kuehl Clock Co. U.S. Importer	Gustav Becker and Others	Before
Kundo	Manufacturer	Kieninger & Obergfell	After
Lepold Company	Lepold Co. U.S. Importer	Kienzle Clock Factories	Before
Linco	Manufacturer	J. Link & Co.	After
London	Retailer?	J. Link & Co.	After
Louis Philipp	Importer?	Kieninger & Obergfell	Before
Ludwig Simon/ Berlin W.	Ludwig Simon Berlin Retailer	Jahresuhrenfabrik	Before
Master	Manufacturer	Nisshindo Watch Co. (Japan)	After
Montrose	P.R. Myers & Co. U.S. Importer	Kieninger & Obergfell	After
Omax	Emo Gottleib U.S. Importer	Uhrenfabrik Herr	After
Orris	Importer?	Uhrenfabrik Herr	After
Oskar Dold	German Retailer?	J. Link & Co.	After
Overocean	Overocean Clock Co. U.S. Importer	Konrad Mauch	After
Pendule 400 Jours	Manufacturer	C. Grivolas (France)	Before
Peneron	Importer?	Sigfried Haller	After
Perfecta	Perfecta Watch and Clock Co. U.S. Importer	Konrad Mauch Kieninger & Obergfell Kern & Sohne	After
Phinney-Walker	Semca Clock Co. U.S. Importer	Sigfried Haller	After
Precisa	Importer?	Kieninger & Obergfell	After
Rensie	Rensie Watch Co. U.S. Importer	Kieninger & Obergfell	After
S. Fisher Ltd.	London Retailer	Kienzle Clock Factories	Before
S. Kind & Sons/ Phila.	S. Kind & Sons U.S. Retailer	Jahresuhrenfabrik	Before
SM	Sokol Montag & Co. U.S. Importer	Kieninger & Obergfell Jahresuhrenfabrik	Before
Schatz	Manufacturer	Aug. Schatz & Sohne	After
Selva	German Exporter	Kern & Sohne	After
Seth Thomas	General Time, Inc. U.S. Importer	Kieninger & Obergfell	After
Singer	Importer?	Uhrenfabrik Herr	After
Style King	Cuckoo Clock Mfg. Co. U.S. Importer	Several Manufacturers	After
Thomas Haller/ Schwenningen/ Wurttenberg/ D.R.P. No. 21340	Manufacturer	Thomas Haller	Before
Treasureland	Importer?	Kieninger & Obergfell	After
Umpire	Importer?	Phillipp Haas	Before
Urania	Manufacturer?	Jahresuhrenfabrik? Phillipp Haas?	Before
Vandor	Importer?	J. Link & Co.	After
Violeta	Importer?	Uhrenfabrik Herr	After
Welby	Welby Co. U.S. Importer	Franz Hermle Kieninger & Obergfell	After
Wright Kay & Co.	U.S. Retailer	Jahresuhrenfabrik	Before
Yearlong	Manufacturer	Jahresuhrenfabrik	Before

SECTION 13 About 400-Day Clock Pendulums

The manufacturers did not realize at first that the ideal pendulum, from a technical standpoint, should be designed with the maximum of its weight concentrated in its periphery. Just as a swinging pendulum must have weight concentrated in its bob, so must the periphery of a torsion pendulum have greater weight in order to maintain the required overswing necessary for a safe unlocking of the escapement.

The early manufacturers were obviously not aware of this reqirement, because many pendulums were made with much of their weight close to their axes. For instance, some Jahresuhrenfabrik disc pendulums of the late 1880's, turned from 2½" to 3⅜" solid brass stock, were hollowed out on a lathe to make the thickness of the outer wall only ¼"—and above the center was a heavy 6-pillar gallery. The very first pendulums without a gallery, actually offered better operation.

Over the years, there were more changes in pendulum shapes and sizes than on any other part of the clock. The desire, of course, was to have an ornamental pendulum that was also as technically correct as possible, but increasing the technical quality required avoidance of decorative appendages. The result has been a compromise in which decoration has won over efficiency and technical correctness in most cases.

The relatively poor timekeeping offered by the first 400-Day Clocks was recognized very soon after production of the clocks began. The problem in timekeeping was thought to be the result of expansion and contraction of the suspension spring due to changes in temperature. As a result, several designs of pendulums were developed and patented which were intended to compensate for changes in the length of the suspension spring (See Section 4). Of these various designs, the "twin loop" pendulum design, patented by Andreas Huber (Pendulum 10 on the following page) became by far the most popular.

Many manufacturers allowed the retail seller to "mix and match" pendulums to the customer's taste (special temperature compensating pendulums were available at extra cost). However, the temperature compensating pendulums were largely ineffective and their extra cost cancelled any slight technical advantage. As a result, none of these special pendulums remained in production very long. In the meantime, the 4-ball pendulum came into being and gained fast acceptance for being not only more decorative than the disc pendulum, but also for having the technical advantage of concentrating most of its weight away from the center.

One of many disc pendulums made prior to World War I by Kienzle Clock Factories contained six or eight thin iron washers, similar in shape to the rubber rings used on Mason jars, that could be added or removed to supplement regulation. Typically, when these Kienzle pendulums are found today, several of these iron washers are missing.

The disc pendulum with gallery went out of production in 1912. After 1949, the 4-ball (sometimes 3-ball) pendulum, was used by almost all manufacturers. The 400-Day Clock pendulums illustrated in this section are representative of the types used over one hundred years of manufacture. The sizes are not to scale.

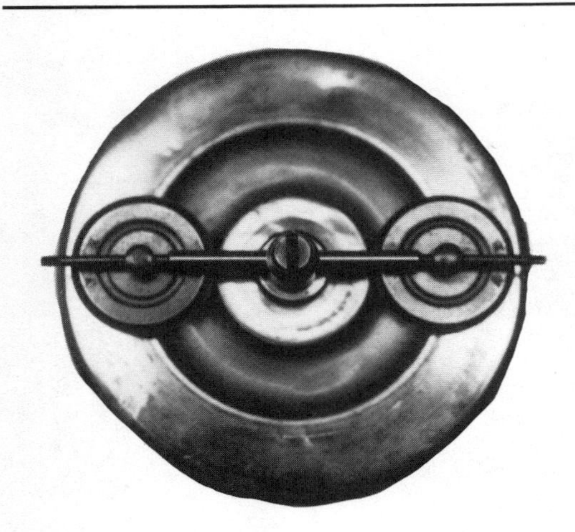

1 A. Willmann Co.? c.1879
Inventor's Model
Plate 1727

2 Jahresuhrenfabrik c. 1881
First Production
Plate 1475

3 Jahresuhrenfabrik c. 1885
First Gallery
Plate 1475

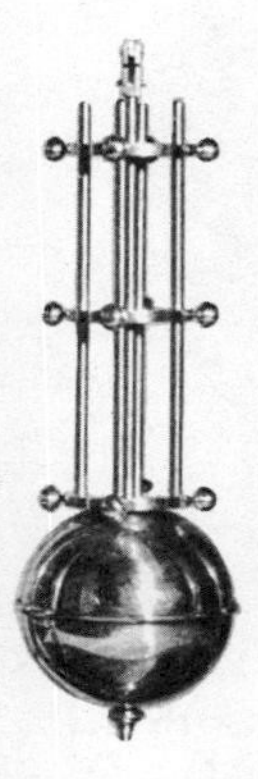

4 Jahresuhrenfabrik C. 1885
Year Striking Clock
Plate 1473A

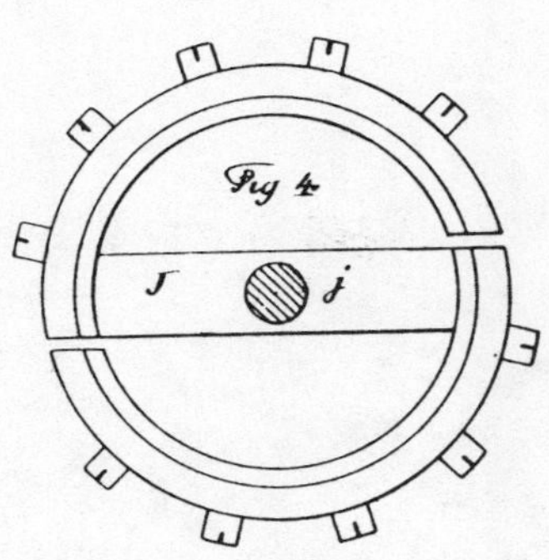

5 J.J. Meister Patent c. 1892
First Temperature
Compensating
Plate 1635

6 Phil. Hauck c. 1894
Temperature Compensating
Plate 1415

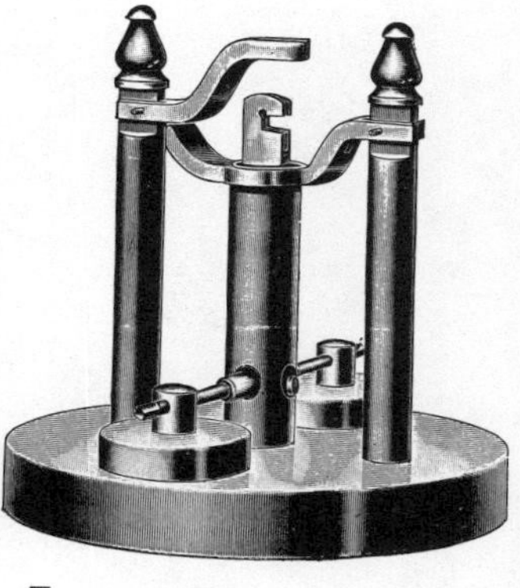

7 Manufacturer Unknown
c. 1904
Temperature Compensating
Plate 1471

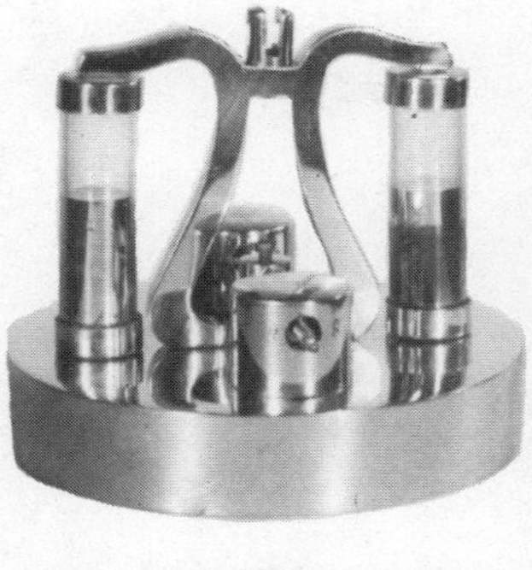

8 Gustav Becker c. 1904
Temperature Compensating?
Plate 1199

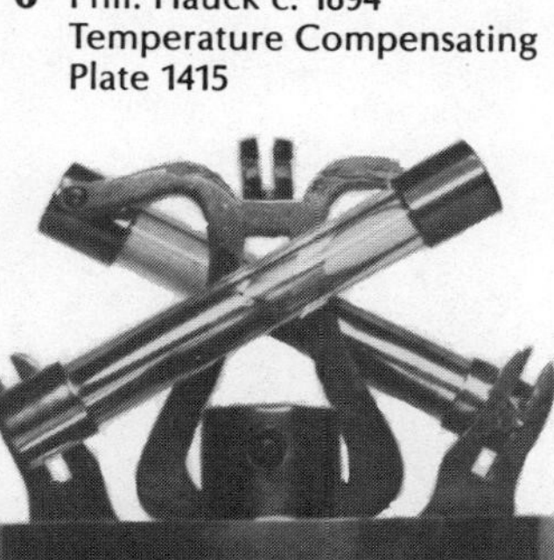

9 Gustav Becker c. 1905
Temperature Compensating
Plate 1201

10 Andreas Huber Patent
c. 1902
Temperature Compensating
Plate 1471

11 Manufacturer Unknown
c. 1900
Constant Force Escapement
Plate 1470E

12 Kienzle Clock Factories
c. 1910
30-Day Clock
Plate 1175

13 Kienzle Clock Factories
c. 1905
Has Guide Slots for Reg.
Weights
Plate 1423

14 Manufacturer Unknown
c. 1905
30-Day Clock
Plate 1718

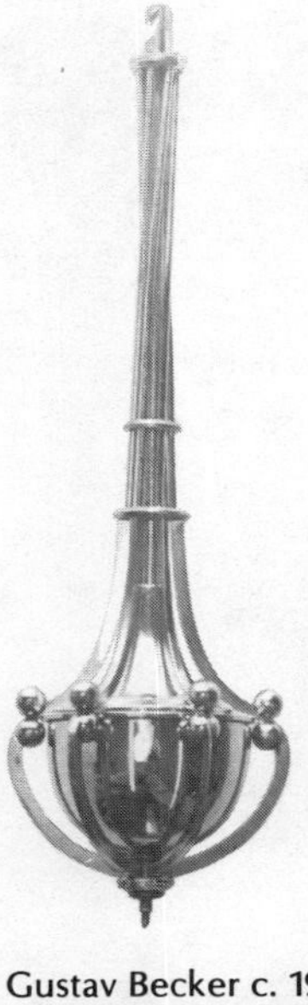

15 Gustav Becker c. 1903
Wood Case Wall Clock
Plate 1193

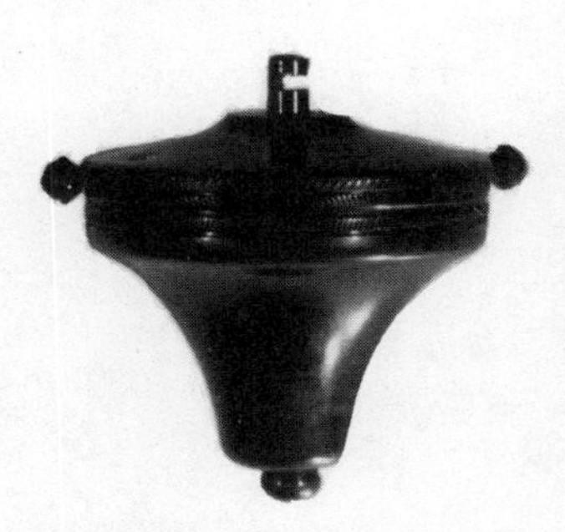

16 Phillipp Haas c. 1910
Internal Regulating Weights
Plate 1419

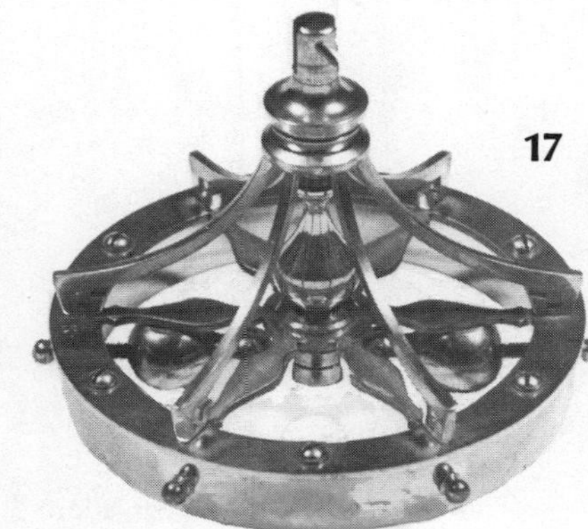

17 Gustav Becker c. 1907
Skeleton Clock
Plate 1189

18 Kienzle Clock Factories
c. 1906
2½" Diameter
Plate 1183

19 Phillipp Haas c. 1905
Narrow Gallery
Plate 1419

20 Manufacturer Unknown
c. 1902
Used with Several Clocks
Plate 1613

21 Jahresuhrenfabrik c. 1904
Most Popularly Used
Disc Pendulum
Plate 1471

22 Jahresuhrenfabrik c. 1905
Bowler & Burdick Clock
Plate 1009

23 Gustav Becker c. 1902
Most Popularly Used
Disc Pendulum
Plate 1195

24 Manufacturer Unknown
Bowler & Burdick Clock
Plate 1009A

25 Badische Uhrenfabrik
c. 1898
Not Well Designed
Plate 1663

26 Phillipp Haas
c. 1905
Relatively Heavy Balls
Plate 1607

27 Jahresuhrenfabrik c. 1906
Temperature Compensating?
Plate 1613

28 Badische Uhrenfabrik
c. 1902
Not Well Designed
Plate 1019

29 Gustav Becker c. 1908
Plate 1207A

30 Gustav Becker c. 1908
Variation of 29
Plate 1207A

31 Junghans c. 1910
"Governor" regulation
Plate 1171

32 Badische Uhrenfabrik
c. 1900
Plate 1717

33 Kienzle Clock Factories
c. 1908
Plate 1163

34 Gustav Becker c. 1908
The only Becker 4-Ball Pendulum
Plate 1199

35 Kieninger & Obergfell c. 1930
Plate 1491

36 Jahresuhrenfabrik c. 1910
Plate 1257

37 Kieninger & Obergfell c. 1911
Plate 1317

38 Kienzle Clock Factories c. 1910
Plate 1431

39 Jahresuhrenfabrik c. 1905
Plate 1603

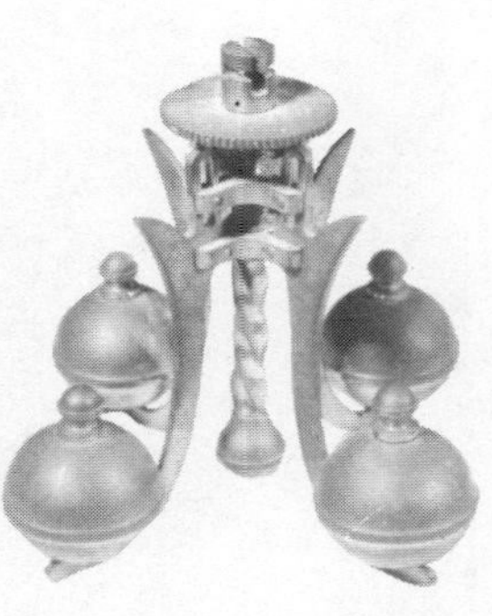

40 Kern Standard c. 1949
Plate 1343

41 Kieninger & Obergfell c. 1930
Plate 1534

42 Georg Wurthner Standard c. 1953
Plate 1591

43 Crescent Standard c. 1952
Plate 1691

44 Henn Standard c. 1951
Plate 1059
Hermie Standard c. 1954
Plate 1126

45 Koma Standard c. 1950
Plate 1394

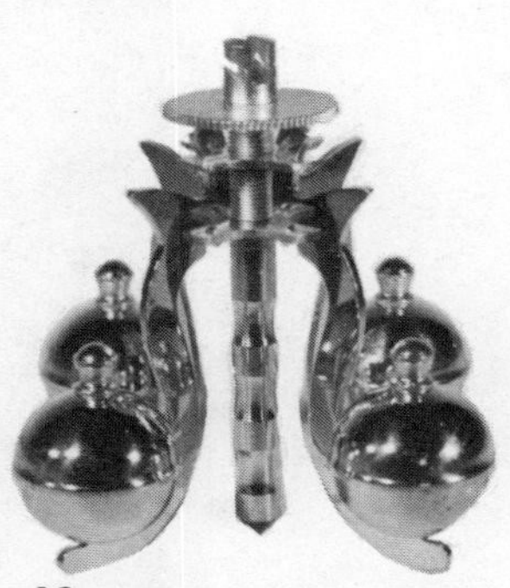

46 Kern Standard c. 1954
Plate 1405

47 Kundo Standard c. 1952
Plate 1371

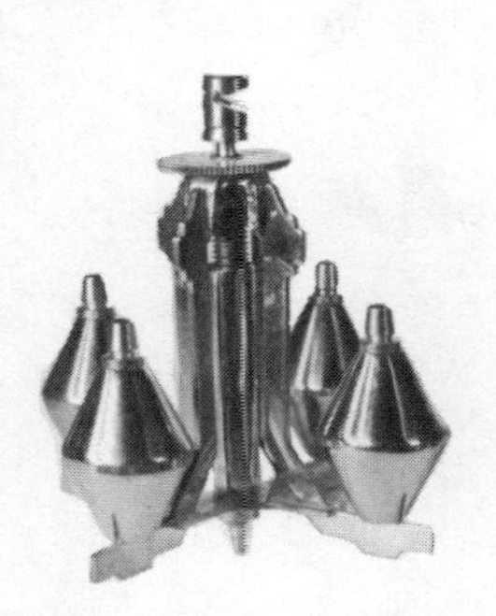

48 Koma Standard c. 1956
Plate 1395B

49 Kundo Miniature c. 1953
Plate 1379B

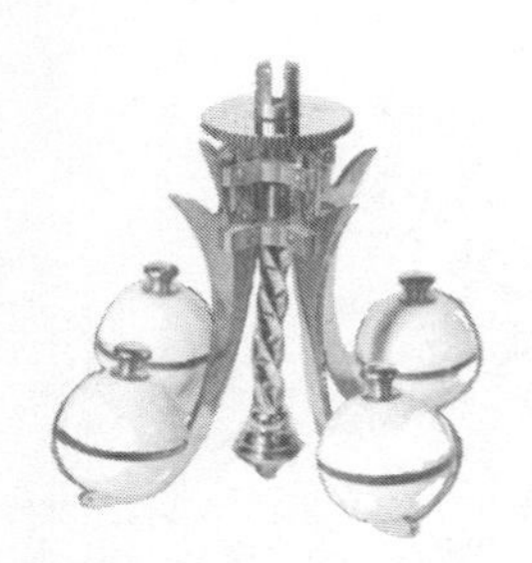

50 Hermie Standard c. 1954
Plate 1126

51 Fr. Vossler c. 1910
Plate 1208A

52 Fr. Vossler c. 1910
Plate 1145

53 Kern Midget c. 1958
Plate 1350

54 Nisshindo Miniature c. 1968
Plate 1461G

55 Schatz Miniature c. 1953
Plate 1299

56 Nisshindo 100-Day c. 1957
Plate 1469E

57 Kern Miniature c. 1954
Plate 1406

58 Schatz Miniature c. 1954
Plate 1013

59 Henn Miniature c. 1953
Plate 1088

60 Haller Midget c. 1972
Plate 1522

61 Henn Miniature c. 1952
Plate 1087

62 Hermie Miniature c. 1954
Plate 1104

63 Kern Midget c. 1958
Plate 1406C

64 Kundo Miniature c. 1954
Plate 1231

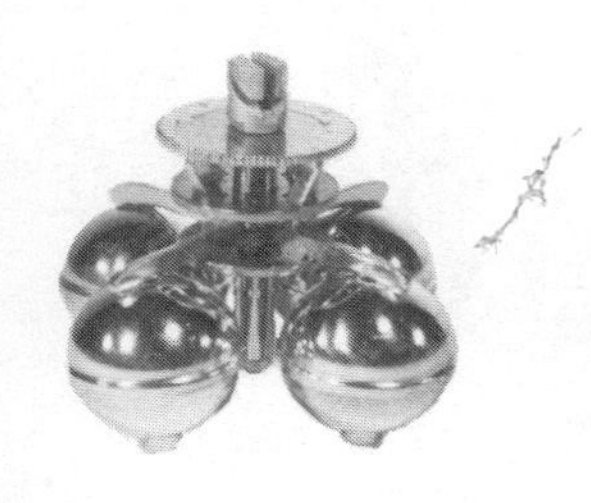

65 Haller Miniature c. 1970
Plate 1521

66 Koma Miniature c. 1965
Plate 1398

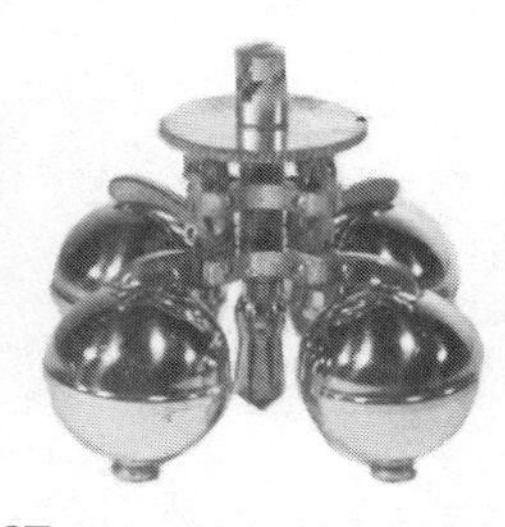

67 Kern Miniature c. 1954
Plate 1406

68 Schatz Miniature c. 1954
Plate 1013A

69 Kundo Midget c. 1965
Plate 1407H

70 Kundo Midget c. 1976
Plate 1382A

71 Kundo Miniature c. 1952
Plate 1379

72 Kundo Miniature c. 1953
Plate 1379

73 Kundo Miniature c. 1955
Plate 1379B

74 Herr Miniature c. 1950
Plate 1678

75 Kundo Miniature c. 1976
Plate 1381

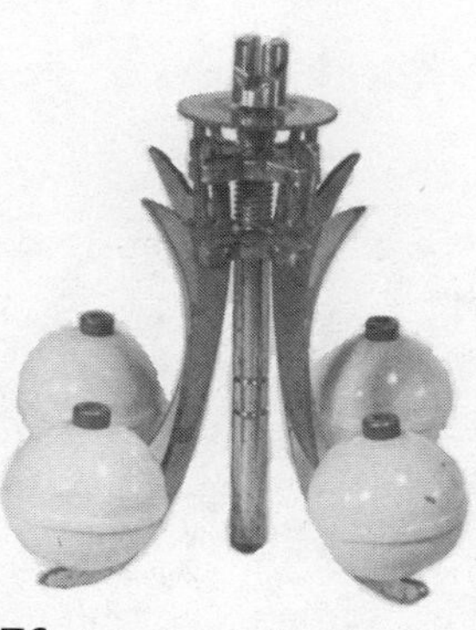

76 Jauch & Haller Standard
c. 1952
Plate 1308A

77 Haller Standard c. 1960
Plate 1520

78 Koma Miniature c. 1954
Plate 1398

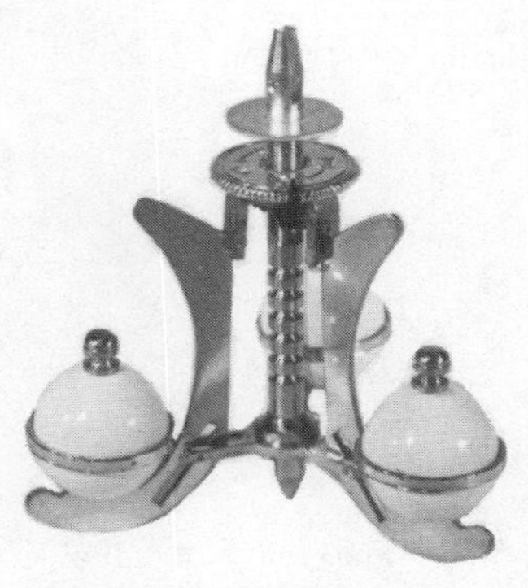

79 Nisshindo 100-Day c. 1972
Plate 1469E

80 Koma Standard c. 1952
Plate 1395B

81 Petersen Standard c. 1952
Plate 1583

82 Link Standard c. 1952
Plate 1077

83 Kaiser Standard c. 1954
Plate 1309

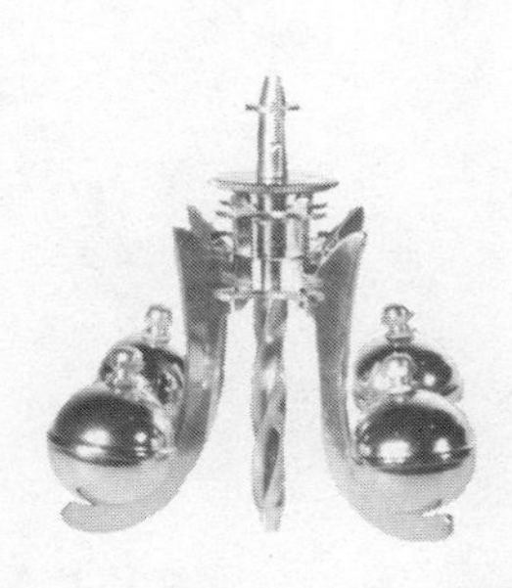

84 Nisshindo Standard c. 1955
Plate 1461C

85 Riner Standard c. 1951
Plate 1574AA

86 Herr Standard c. 1951
Plate 1212

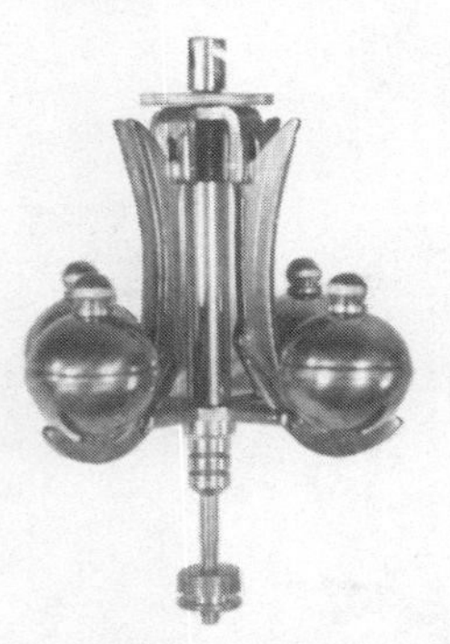

87 Wurthner Standard c. 1952
Plate 1591

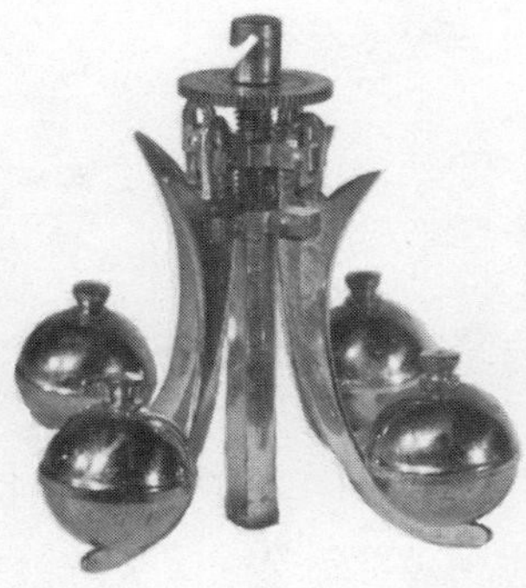

88 Henn Standard c. 1952
Plate 1246

89 Reiner Standard c. 1953
Plate 1325

90 Schatz 1000-Day c. 1954
Plate 1010

91 Kern Standard c. 1970
Plate 1345

92 Koma Standard c. 1970
Plate 1395A

93 Wurthner Standard c. 1952
Plate 1591

94 Schatz Standard c. 1949
Plate 1271

95 Schatz Standard c. 1950
Plate 1014

96 Schatz Standard c. 1950
Plate 1014A

97 Kundo Standard c. 1947
Plate 1359

98 Kundo Standard c. 1952
Plate 1407B

99 Kundo Standard c. 1954
Plate 1407B

100 Kundo Standard c. 1955
Plate 1407B

101 Herr Standard c. 1951
Has Four Steady Pins
Plate 1453

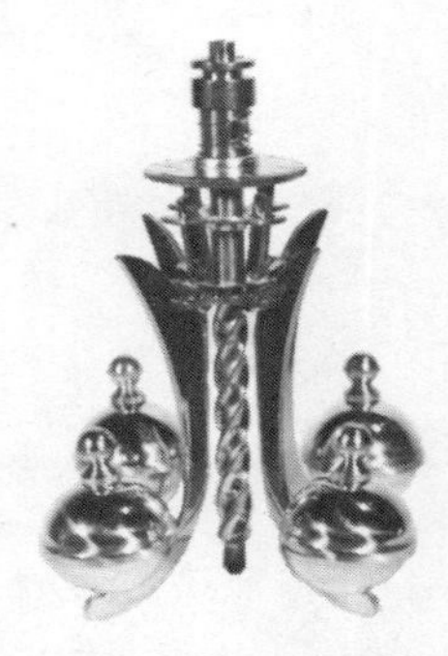

102 Kundo Standard c. 1976
Plate 1380

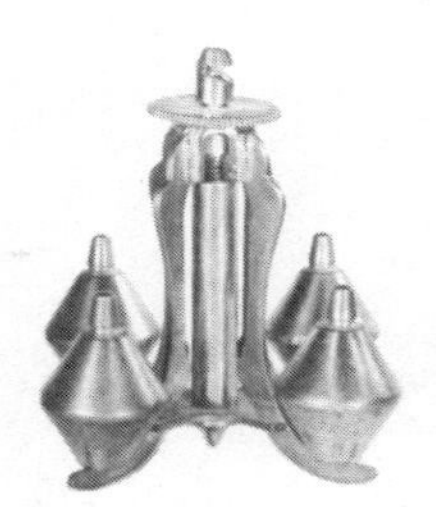

103 Koma Standard c. 1951
Plate 1395A

104 Wurthner Midget c. 1957
Plate 1725

105 Koma Midget c. 1959
Plate 1393C

106 Kern Miniature c. 1962
Plate 1340E

107 Koma Midget c. 1957
Plate 1393B

SECTION 14 400-Day Clock Ratchet Wheels

The ratchet wheels illustrated in this Section are all from 400-Day Clocks manufactured since 1949. Ratchet wheels from older clocks are functionally the same, but can usually be identified because of the smoother surface of the brass, finer teeth, or sometimes because the underside of the square has been hollowed out slightly or there is a decorative groove on the face of the wheel.

The majority of ratchet wheels are brass, but beginning in the mid-1970's both Kieninger & Obergfell and Uhrenfabrik S. Haller changed the material to steel. Steel wheels are die-stamped instead of gang-cut on a milling machine and have the advantage that their teeth hold their shape better. Also at the same time, Kundo and Haller saved one operation in making their barrel arbors by milling the ratchet wheel flats on only two sides instead of four. Thus the holes in the ratchet wheels are shaped to conform to the configuration of the arbor as can be seen by the illustrations. Usually these ratchet wheels will interchange with those with square holes from earlier productions or they can be made to fit by filing the holes slightly.

The diameter of the wheel and its tooth count, even the shape of the teeth, are not critical. The shapes are more a matter of manufacturers' taste than they are of theoretical design. The clock must cause a positive lock when it engages the ratchet teeth during the winding backlash. However, the size of the hole is important. It should fit on the arbor with minimum play. If the hole is slightly small, it can be enlarged by filing to make a snug fit. If the hole is too large, the ratchet wheel should be replaced, as the hole may become larger with use and eventually lose its grip.

It is not at all unusual to find the tips of brass teeth bruised in well-used clocks, but unless the teeth are broken, they may still function well. The ratchet wheel is an inexpensive part, so if the clock is being overhauled, the job will have a more professional appearance if the wheel is replaced. If the teeth are badly worn, the wheel *must* be replaced in order to prevent severe damage to the teeth in both barrel and first wheel if a tooth should break and the mainspring unwind suddenly while under tension.

The identification by manufacturer, model, diameter, hole size and tooth count on this table makes it possible to identify a 400-Day Clock ratchet wheel that carries no marking. Wheels are illustrated from either back or front randomly, but their position in the movement is obvious.

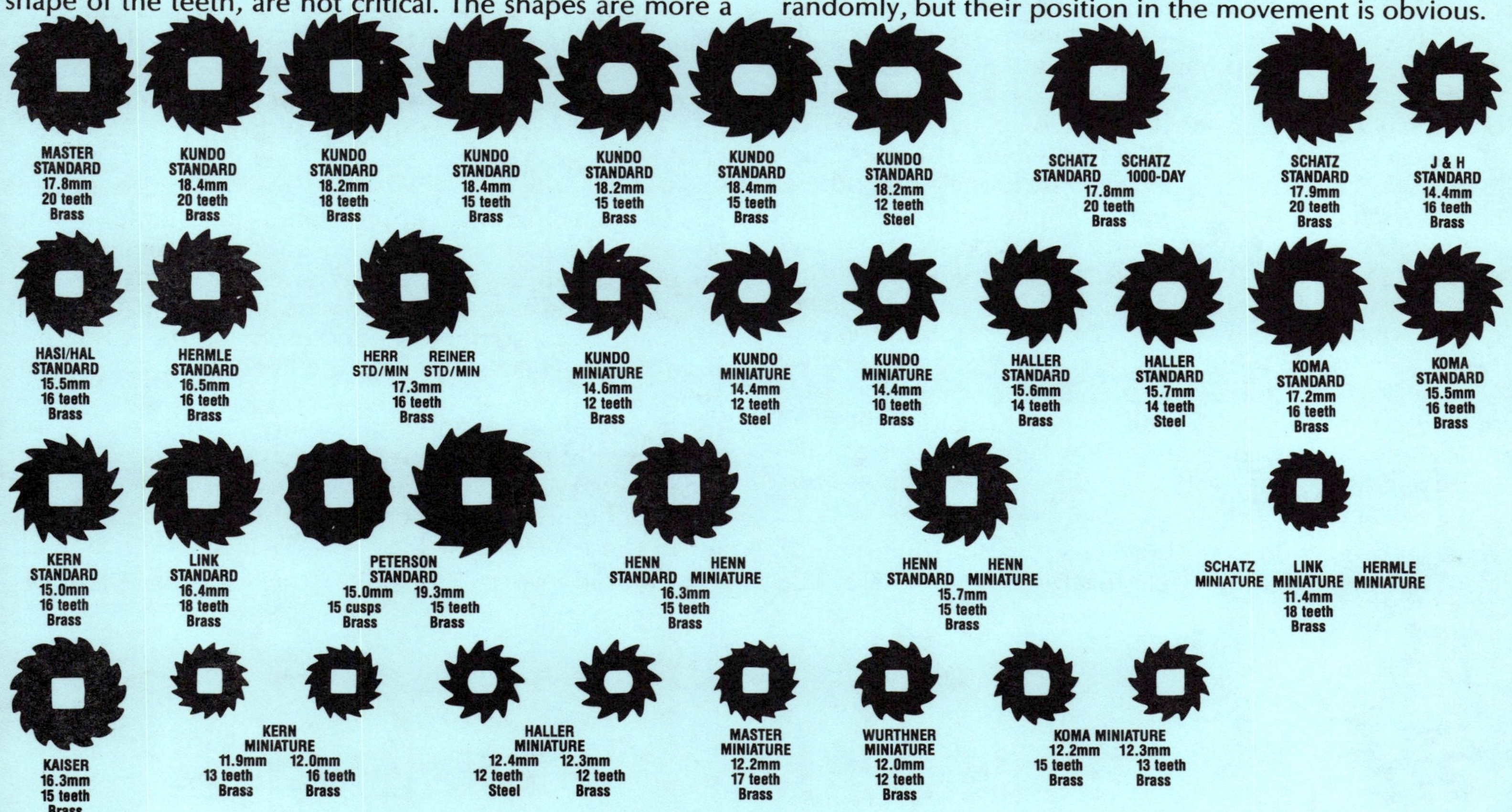

SECTION  400-Day Clock Suspension Brackets and Saddles

The top of the suspension spring in a 400-Day Clock is held in position behind and above the movement back plate. Manufacturers have made it possible to attach the top of the suspension spring in a variety of ways, as indicated by the many bracket and saddle illustrations shown below. When the first models were made there was no "saddle," so the adjustment for beat could be made only by twisting the spring with tweezers just below the top block. It is interesting that some manufacturers were slow to make their brackets with saddles, as indicated in the illustrations of Becker's No. 4 and Koma's No. 21, but since neither of these is to be seen frequently, they must have been replaced quickly with Nos. 5 and 22, which do have saddles.

Most of the bracket and saddle units illustrated differ in design, but are functionally the same. The saddle held the top of the suspension spring, usually in a block, either above or below the bracket platform. If above, the saddle is friction tight or held in position by a set screw through the side. If below, the saddle is usually held to the platform by a screw through the top.

Three bracket and saddle units used in pre-World War I clocks were designed so the pendulum will hang straight even when the clock is not perfectly level. One claim made for these designs is that the suspension spring is less likely to become bent when the clock is picked up to be moved. No. 14 is designed with the saddle in gimbals. There are also gimbaled brackets on clocks made by Kienzle and others. No. 16, used only by Kienzle, has double action. A flat bar-shaped saddle, attached to the top of the bracket, projecting outward about 17mm, can be adjusted a few degrees left and right. The top suspension block is in three pieces, the lower two of which clamp the top of the suspension spring with two screws, while the top piece rests on a small steel ball lying in a cavity in the bar. Because the top of the block rests on the steel ball, the pendulum will hang straight even though the base is not exactly level. However, this complicated design provided very little adjustment of beat and was actually less effective than other simpler versions.

Another pre-World War I bracket and saddle unit seen on clocks of more than one manufacturer is No. 15. The top suspension block, shaped somewhat like the letter E, adjusts the suspension if the clock is not in a level position, somewhat like the bracket with steel ball previously described. The top of the E is equipped with a set screw, the point of which fits into a depression in the saddle, which is adjustable in the bracket. The center of the E fits into a groove in the end of the saddle and turns with it. The lower part of the E has two sections and holds the top of the suspension spring with two screws. Detailed drawings of this saddle and the saddle with gimbals are in Section 18, No. 118.

An interesting bracket, No. 8, was used by Gustav Becker in many of their Standard models over a period of several years. Loosely known as Becker's "overhead suspension," the bracket requires a specially designed top suspension block that hangs on a steel post and has a forked end that straddles a vertical rod that is adjustable from the top. The adjustment makes it possible to change the position of the top suspension block for beat setting. Functionally, the design works very well, but it presents two problems: (1) If the suspension spring breaks, there is nothing to keep the top suspension block on the post, so it often becomes lost. Replacements were probably available before the war, but today the complicated piece has to be made by hand. (2) The unit has 20 individual pieces, some requiring special bending and machining, so it is expensive to reproduce. Detailed drawings of this bracket and the top suspension block are in Section 18, No. 11.

Several manufacturers included a cross piece in their brackets which contained the anchor pivot hole, thereby replacing the eccentric nut in the back plate. While putting the pivot hole in the bracket may have made a small saving in manufacturing cost, it now adds repair time to a movement overhaul, because the locks and drops of the escapement have to be checked before the bracket is again firmly attached.

Most of the bracket and saddle units in clocks made since 1949 have distinctive features that identify the manufacturer, but prior to World War I there were several brackets that will be found on the movements of more than one manufacturer. These include Nos. 1, 11, 14, 15 and possibly others. The illustrations are 25% to 30% smaller than the originals.

1 Badische Std.

2 Badische Std.

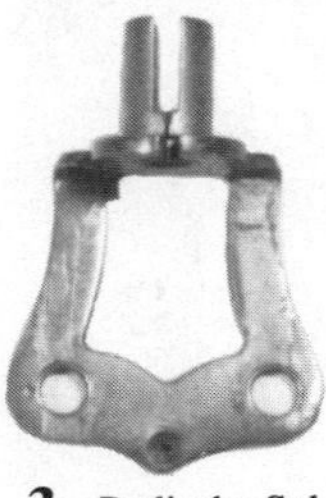

3 Badische Std.

4 Becker Std.

5 Becker Std.

6 Becker Std.

7 Becker Std.

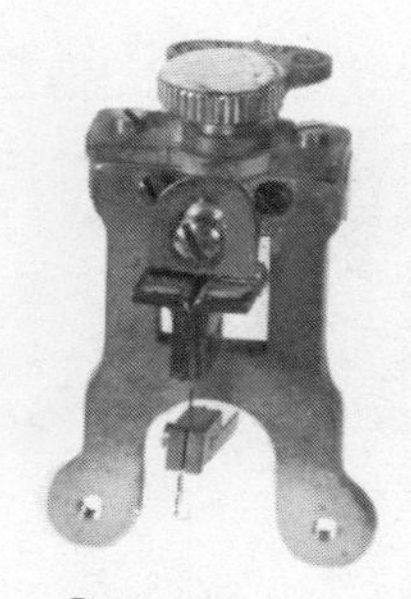

8 Becker Std.

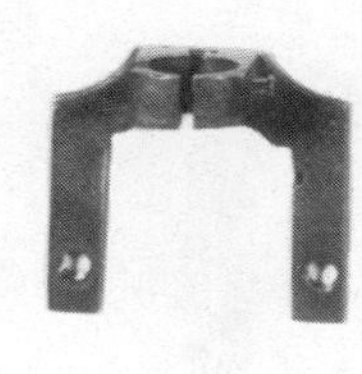

9 Jahresuhrenfabrik Std.

10 Jahresuhrenfabrik Std.

11 Jahresuhrenfabrik Std.

12 Jahresuhrenfabrik Min.

13 Fr. Vossler Min.

14 Kienzle Std.

15 Kienzle Std.

16 Kienzle Std.

17 Haas Std.

18 Unknown Std.

19 Unknown Min.

20 Kaiser Std.

21 Koma Std.

22 Koma Std.

23 Koma Min.

24 Jauch/Haller Std.

25 Schatz Std.

26 Schatz Min./1000-Day

27 Kundo Std.

28 Kundo Std.

29 Kundo Std.

30 Kundo Min.

31 Kundo Midget

32 Henn Std.

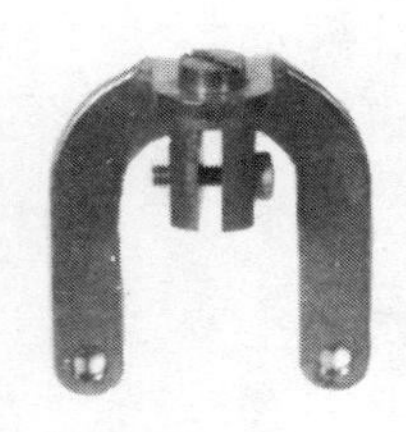

33 Henn Std.

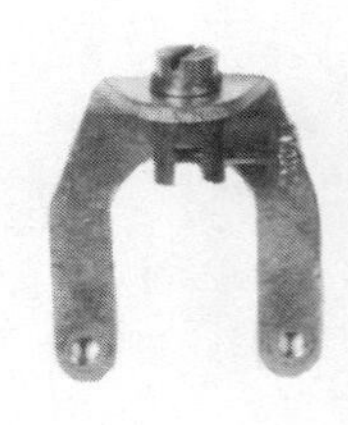

34 Henn Min.

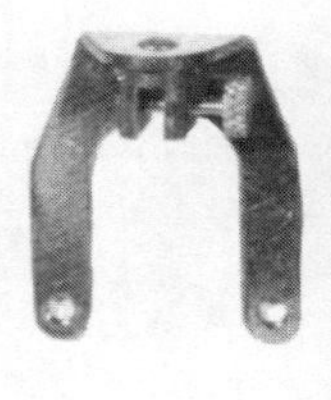

35 Henn Min.

36 Henn Min.

37 Hermle Std.

38 Hermle Min.

39 Kern Std.

40 Kern Min.

41 Kern Min.

42 Nisshindo Std.

43 Nisshindo 100-Day

44 Nisshindo Min.

45 Herr Std.

46 Link Std.

47 Haller Std.

48 Haller Min./Midget

49 Petersen Std.

50 Würthner Std.

51 Würthner Std.

52 Würthner Midget

53 Würthner Midget

SECTION 16 About 400-Day Clock Spare Parts and Repairs

Spare Parts

The Horolovar Company is the largest supplier in the world of key-wind 400-Day Clock parts. We not only stock parts for clocks in current production, but we also have a large stock of parts for clocks out-of-production. Most of the latter have been obtained by cannibalizing old movements. Certain parts no longer available, for which there is still a demand, we have remade to order.

In 1953, peak year for 400-Day Clock manufacture, there were 13 German clock factories each making a variety of models. Most of these clocks reached the U.S., either through direct importation or as gifts purchased by servicemen at Post Exchanges throughout the world. With each passing year the maintenance of a spare parts inventory is an increasing problem. See "WANTED" below.

How to Order

Order parts through your local watch material dealer or direct from The Horolovar Company. Always identify the clock by the back plate number in Section 9 of this Repair Guide and describe, carefully, the part(s) needed. The listing of movement parts shown in Section 18 can help in describing parts which are needed. Using this system, it should not be necessary to submit a sample.

Samples may be needed for clocks of pre-World War I manufacture whose movement back plates are marked only with a serial number or are not marked with the name of a manufacturer or importer.

Repairs

The Horolovar Company discontinued repair service in December, 1964. For several years prior to that date, we had solicited only the repair of "problem" clocks from the trade. Since the answer to each problem was usually to be found somewhere in the pages of this Guide, or could easily have been located by checking through the answer to Question 21 in Section 8, it was felt that this service was not being used for the purpose intended.

If you are faced with a specific repair problem which you feel is not covered somewhere in this Guide, or if you have other questions about the clock, we will gladly try to help. Write: Question Department, The Horolovar Company, P.O. Box 264, St. Clair Shores, Michigan 48080. If the question is about repair, give as many details about the symptoms, and the steps you have taken, as you possibly can. And...*be sure to enclose a self-addressed, stamped return envelope.*

WANTED

OUT-OF-PRODUCTION MOVEMENTS AND PENDULUMS

We are interested in purchasing movements and pendulums only, from certain out-of-production clocks, *whether they are in working order or not.* If you have any for sale, *please do not send them.* Instead, write us, giving the back plate numbers of the movements, as shown in Section 9, and state the approximate condition of both movement and pendulum. Address your letter to: **Purchasing Department, The Horolovar Company, P.O. Box 264, St. Clair Shores, MI 48080.** We will tell you how much the parts are worth to us.

SECTION 

About Horolovar 400-Day Clock Products

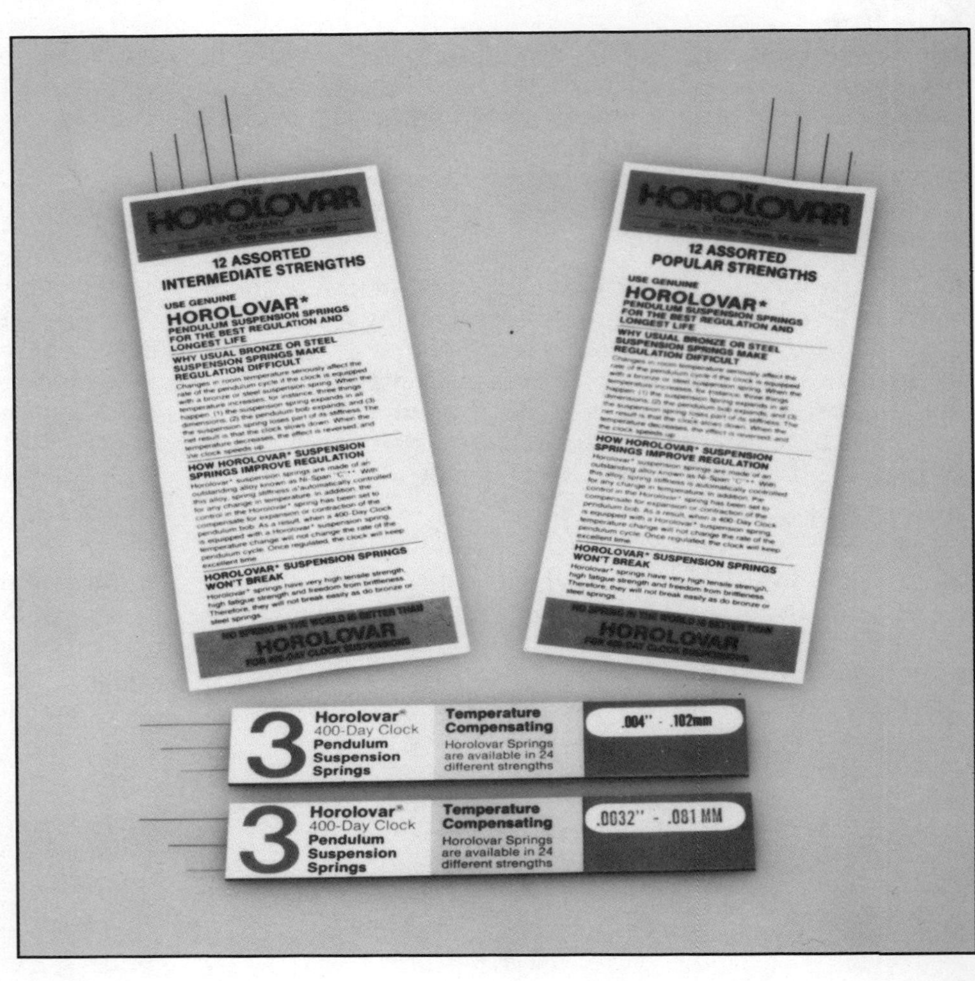

HOROLOVAR* TEMPERATURE COMPENSATING 400-DAY CLOCK SUSPENSION SPRINGS

SINGLE STRENGTH

3 Springs per Package

.0018"	-	.046mm
.0019"	-	.048mm
.0020"	-	.051mm
.0021"	-	.053mm
.0022"	-	.056mm
.0023"	-	.058mm
.0024"	-	.061mm
.0025"	-	.064mm
.0028"	-	.071mm
.0030"	-	.076mm
.0031"	-	.079mm
.0032"	-	.081mm
.0033"	-	.084mm
.0034"	-	.086mm
.0035"	-	.089mm
.0036"	-	.091mm
.0037"	-	.094mm
.0038"	-	.097mm
.0040"	-	.102mm
.0045"	-	.114mm
.0050"	-	.127mm
.0055"	-	.140mm
.0060"	-	.152mm

POPULAR ASSORTED STRENGTHS

12 Springs per Package

.0020"	-	.051mm
.0023"	-	.058mm
.0023"	-	.058mm
.0025"	-	.064mm
.0030"	-	.076mm
.0032"	-	.081mm
.0032"	-	.081mm
.0035"	-	.089mm
.0035"	-	.089mm
.0037"	-	.094mm
.0040"	-	.102mm
.0040"	-	.102mm

ASSORTED INTERMEDIATE STRENGTHS

12 Springs per Package

.0018"	-	.046mm
.0019"	-	.048mm
.0021"	-	.053mm
.0022"	-	.056mm
.0024"	-	.061mm
.0028"	-	.071mm
.0031"	-	.079mm
.0033"	-	.084mm
.0036"	-	.091mm
.0038"	-	.097mm
.0045"	-	.114mm
.0050"	-	.127mm

*Trade Mark Registered, U.S. Patent Office

WHY BRONZE OR STEEL SPRINGS MAKE REGULATION DIFFICULT

Changes in room temperature seriously affect the rate of the pendulum cycle if the clock is equipped with a bronze or steel suspension spring. When the temperature increases, for instance, three things happen: (1) the suspension spring expands in all dimensions; (2) the pendulum bob expands; and (3) the suspension spring loses part of its stiffness. The net result is that the clock slows down. When the temperature decreases, the effect is reversed, and the clock speeds up.

HOW HOROLOVAR* SUSPENSION SPRINGS IMPROVE REGULATION

Horolovar* suspension springs are made of an outstanding alloy known as Ni-Span "C"**. With this alloy, spring stiffness is automatically controlled for any change in temperature. In addition, the control in the Horolovar* spring has been set to compensate for expansion or contraction of the pendulum bob. As a result, when a 400-Day Clock is equipped with a Horolovar* suspension spring, temperature change will not change the rate of the pendulum cycle. Once regulated, the clock will keep excellent time.

HOROLOVAR* SUSPENSION SPRINGS WON'T BREAK

Horolovar* springs have very high tensile strength, high fatigue strength and freedom from brittleness. Therefore, they will not break easily as do bronze or steel springs.

LABORATORY TESTS

TEST	HOROLOVAR*	STEEL	BRONZE
Time Error Per Day For 5° F. Temperature Change	2 seconds	35 seconds	60 seconds
Tensile Strength Index	100	133	73
Fatigue Strength Index	100	133	50
Brittleness Resistance (Ductility) Index	100	24	67
Corrosion Resistance	Good	Rusts	Good
Magnetic Property	Slight	Strong	None

NO SPRING IN THE WORLD IS BETTER THAN HOROLOVAR* FOR 400-DAY CLOCK PENDULUM SUSPENSION!

*HOROLOVAR is a Registered Trade Mark owned by The Horolovar Co., St. Clair Shores, MI 48080
**NI SPAN "C" is a Registered Trade Mark of The International Nickel Co., Inc.

HOROLOVAR* LONG ROUND TAPERED PINS—BRASS AND STEEL

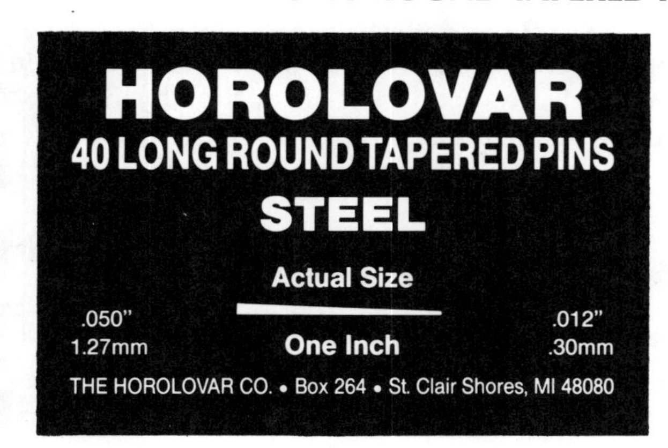

Save Time
No need to hunt for—or waste time filing—a steel or brass pin to fit. These long, round, taperered pins come in three sizes, fit practically any size hole in any clock; hand holding dial lug, pillar, intermediate wheel post, suspension blocks, etc.

Look Better
New pins look neater, hold better. Bent or chewed pins make your best work look sloppy.

Economical, too
The pins are all long, can often be cut in two pieces, each piece being used separately.

ENVELOPE COLOR	METAL	QUANTITY	LENGTH	SMALL DIAMETER	LARGE DIAMETER	ACTUAL SIZE
GREEN	STEEL	40	1.00" (25.4mm)	.012" (.30mm)	.050" (1.27mm)	
RED	BRASS	50		.030" (.76mm)	.065" (1.65mm)	
BLACK	STEEL	40				
BLUE				.045" (1.44mm)	.075" (1.91mm)	

*Trade Mark Registered, U.S. Patent Office

HOROLOVAR* 400-DAY CLOCK MAINSPRINGS

AVAILABLE FOR EVERY MODEL

	Width of Spring		Inside Diameter of Barrel	Length	Thickness
mm	**12**	**x**	**24**	584	.38
in.	1/2	x	1	23	.015
mm	**12**	**x**	**25**	697	.33
in.	1/2	x	1	27	.013
mm	**12**	**x**	**32**	1232	.30
in.	1/2	x	1-1/4	48	.012
mm	**13**	**x**	**30**	972	.36
in.	17/32	x	1-3/16	38	.014
mm	**13**	**x**	**32**	714	.55
in.	17/32	x	1-1/4	28	.022
mm	**14**	**x**	**25**	697	.33
in.	9/16	x	1	27	.013
mm	**14**	**x**	**28**	720	.41
in.	9/16	x	1-1/8	28	.016
mm	**14**	**x**	**30**	815	.36
in.	9/16	x	1-3/16	32	.014
mm	**14**	**x**	**32**	1041	.46
in.	9/16	x	1-1/4	41	.018
mm	**15**	**x**	**32**	760	.50
in.	19/32	x	1-1/4	30	.020
mm	**16**	**x**	**36**	1039	.46
in.	5/8	x	1-7/16	41	.018
mm	**18**	**x**	**38**	1118	.46
in.	11/16	x	1-1/2	44	.018
mm	**19**	**x**	**32**	951	.41
in.	3/4	x	1-1/4	38	.016
mm	**19**	**x**	**36**	1143	.43
in.	3/4	x	1-7/16	45	.017
mm	**19**	**x**	**38**	1346	.41
in.	3/4	x	1-1/2	53	.016
mm	**20**	**x**	**38**	1346	.41
in.	25/32	x	1-1/2	53	.016
mm	**21**	**x**	**32**	2540	.38
in.	13/16	x	1-15/16	100	.015

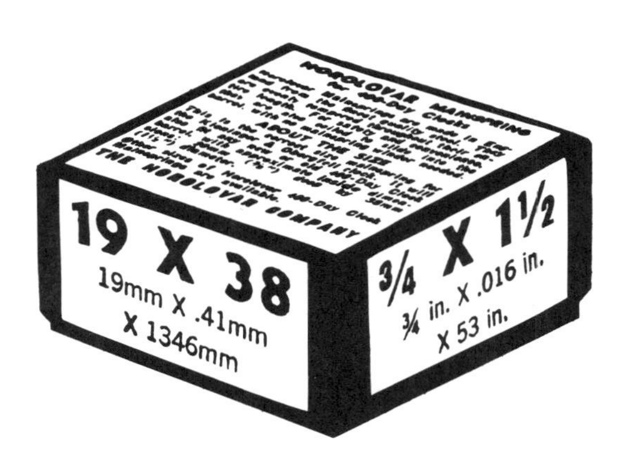

HERE'S WHY YOU WILL PREFER HOROLOVAR 400-DAY CLOCK MAINSPRINGS

SAFE
Horolovar 400-Day Clock mainsprings are of the finest grade steel, tempered evenly throughout their entire length.

NO MAINSPRING WINDER NEEDED
Each mainspring is coiled to a diameter small enough to slide into the barrel easily. No need to risk distortion by inserting it in the barrel by hand. Saves time.

CORRECTLY SIZED
The box label clearly shows the size of each mainspring. All dimensions are given in both millimeters and inches.

WON'T RUST
Coated with a harmless preservative, wrapped in a moisture-proof bag, and individually packaged in a convenient box, each mainspring is protected three ways against rust.

Note: When ordering, it is necessary to specify only the width of the spring and the inside diameter of the barrel, i.e. 12 x 25, 19 x 38, etc. See Section 18, Appendix No. 75 for the sizes of mainsprings used by 400-Day Clock manufacturers since 1949.

HOROLOVAR* 400-DAY CLOCK BEAT-SETTING TOOL (Patent No. 196385)

With this tool, the final and most important operation in 400-Day Clock repair—putting the pendulum in beat**—becomes simple, quick, accurate, and permanent! Moreover, it can be used on any model 400-Day Clock suspension saddle, old or new.

This precision tool expressly designed to make the day-by-day life of the clock repairer easier and more profitable—is the result of 10 years of study and experimentation with many different shapes and sizes of suspension saddles.

HOW TO USE IT

IT'S SIMPLE! Just follow the easy steps explained in the instructions included with every tool.

IT'S QUICK! The tool saves the time usually wasted by frequent over-adjustments and under-adjustments with poor fitting pliers. It remains firmly attached to the saddle during beat-setting operation, thereby making it easier to watch escapement action and pendulum rotation.

IT'S ACCURATE! Its long arm gives you greater control, makes micrometer adjustments possible. Adjustments are smooth—not jerky.

IT'S PERMANENT! It makes your final adjustment permanent, because you have tightened the saddle before the beat is set.

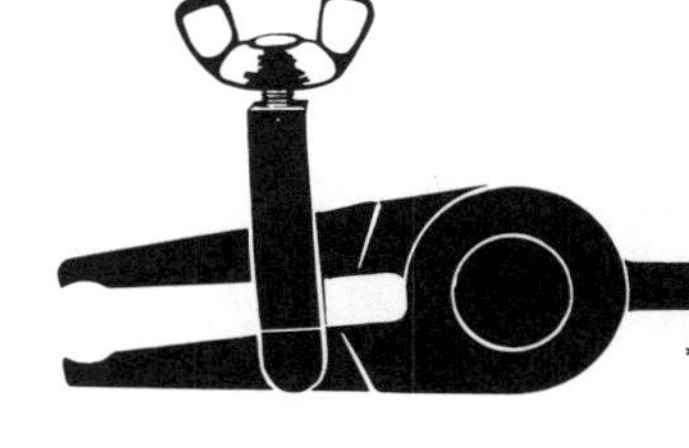

**Full details on how to put the pendulum in beat appear in Section 8, Question 6.

HOROLOVAR* ULTRA SENSITIVE BEAT AMPLIFIERS

The task of setting a 400-Day Clock in beat is made much easier by the use of a Horolovar Beat Amplifier. In operation, the Horolovar Beat Amplifier greatly amplifies the sound that the clock makes as the escape wheel teeth strike the anchor pallets. This distinct sound makes it easy to properly adjust the beat of the clock using the procedure described in Section 8, Question 6. The Horolovar Beat Amplifier includes a battery, speaker, and an ear phone along with two extemely sensitive pick-ups:

An Acoustic Pick-Up used for setting mechanical clocks
An Inductive Pick-Up used for working on electronic clocks.

This versatile tool is not limited to use with 400-Day Clocks as it can be used on all mechanical clocks and watches. Figure 1 shows the Horovolar Beat Amplifier and Beat Setting Tool being used to set the beat on a 400-Day Clock.

Figure 1. Setting the beat using a Horolovar Beat Setting Tool and a Horolovar Ultra Sensitive Beat Amplifier.

*Trade Mark Registered, U.S. Patent Office.

HOROLOVAR* 400-DAY CLOCK SUSPENSION SPRING UNITS

These Suspension Units are equipped with a Horolovar* Temperature compensating Suspension Spring, and are assembled in individual jigs. The position of the blocks and fork, and the *strength*** of the spring, are guaranteed to be accurate in every respect for the clock and model specified. Repairers whose time is at a premium will find these complete Horolovar* Suspension Units a great convenience. **No complete Suspension Units are available for clocks made prior to 1949.**

Unit No.	Make and Model of Clock
1	Kundo Standard 49
3A	Kundo Standard 53
3B	Kundo Standard 54
3C	Kundo Standard 55
5A	Kundo Miniature 53
5B	Kundo Miniature 54
5E	Kundo Miniature 56
5F	Kundo Midget 58
6789	Schatz Standard 49
10A	Schatz Miniature 53
10B	Schatz 1000-Day 54
10C	Schatz Midget 57
10D	Schatz Standard 81
11A	Kern Standard 50
11B	Kern Standard 54
12C	Kern Miniature 57
12D	Kern Miniature 58
12E	Kern Midget 58
12F	Kern Miniature 61
13B	Koma Standard 52
13C	Koma Standard 56
14A	Koma Miniature 54
14B	Koma Miniature 65
16	Link Standard 52
18C	Petersen Standard 54
20A	Herr Standard 54
20B	Herr Standard 54
23A	Reiner Standard 53
23B	Reiner Standard 56
25A	Hermle Standard 53
25B	Hermle Miniature 54
26	Wurthner Standard 53
27A	Herr Standard 53
27B	Herr Standard 55
27C	Neuck Standard 57
28A	Herr Miniature 53
28B	Herr Miniature 54
28C	Herr Miniature 56
29A	Henn Miniature 52
29B	Henn Miniature 53
31	Henn Standard 52
32	Henn Standard 54
33	Jauch & Haller 54
34	Kaiser Standard 54
35	Link Miniature 56
36B	Petersen Standard 55
38A	Koma Midget 57
40B	Crescent Standard 55 (Japan)
41B	Master Standard 55 (Japan)
42A	Haller Standard 70
43A	Haller Miniature 70
44A	Haller Midget 70
45A	Master Standard 70 (Japan)
46A	New Master 70 (Japan)
47	Hermle Miniature 71
48B	Kundo Standard 73
49B	Kundo Miniature 73
50B	Master Standard 72 (Japan)
51B	Kundo Standard 76 Quartz
52B	Kundo Miniature 76 Quartz
53B	Hermle/Koma 76 Electronic
54B	Hermle/Koma 77 Quartz
55	Kern Standard 75 Electronic
56	Kern Miniature 75 Electronic
57	Kern Standard 77 Quartz
58	Kern Miniature 77 Quartz I
59	Kern Miniature 78 Quartz II
60B	Master 75 Electronic
61B	Master 77 Quartz
63B	Kundo Miniature 76 Electronic
67B	Hermle 80 Quartz
68B	Hermle 83 Quartz

Unit Numbers in Bold Face Indicate Battery Clocks

See Section 10 of this Guide for illustrations and descriptions of these Suspension Units.

Unit No. 3C

*Trade Mark Registered, U.S. Patent Office

**The width and thickness of the Horolovar* Suspension Spring may not be the same as for the spring which it replaces, but its torsion characteristics are nevertheless correct.

SECTION 18 APPENDIX

This Appendix contains added helpful repair information (and historical facts) about specific 400-Day Clocks not documented in other Sections of this *Guide*. Most of the information relates directly to clocks whose movement back plates are illustrated in Section 9.

The section can be used in two ways. The references can be read as attention is called to them by the "See Appendix" numbers under back plates. Or, the Section will be found to be interesting and educational reading, as it gives useful additional information about many clocks.

The parenthesis following the Appendix number shows a typical movement back plate number in Section 9, or the Section to which the information refers.

1. (1007) About this clock, which was imported by Bowler & Burdick, Cleveland, Ohio, see Section 1.

2. (Section 10, Unit 3C) With the first production of the Kundo locking pendulums, the bottom suspension block holding pin was designed to be loose so that it could easily be removed and replaced, but when the clocks were shipped, sometimes the holding collar would jog loose and the pin would fall out, releasing the pendulum to do considerable damage. To prevent this from happening, the factory replaced the brass pin with a steel pin having an upset center. The pin was then forced into the block under pressure making it very difficult to remove. A simple way to remove the pin is to tap it out with a hammer and nail-set punch. Before replacing the pin, file off the upset portion so that it can be removed again and replaced with ease.

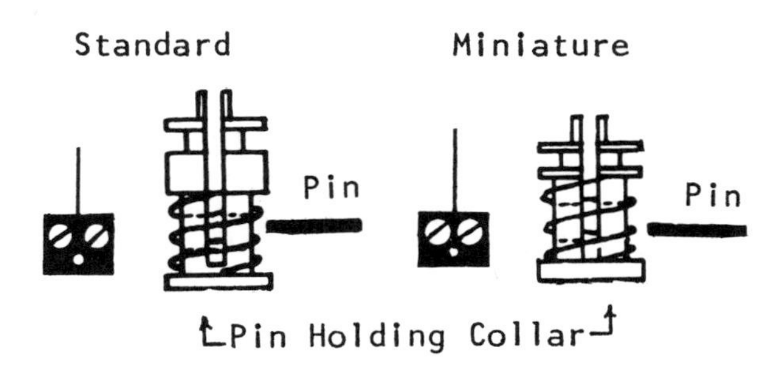

3. (1193) This movement is from a Gustav Becker wood-cased wall clock. Its unusual shaped pendulum is illustrated in Section 13, No. 15.

4. (1010) About how to remove the little cannon pinion from the Schatz Miniature and 1000-Day centerwheel arbors. If you are only cleaning the clock, it is not necessary to remove the pinion. Just be sure to flush the cleaning fluid thoroughly around the pivot hole. If it is necessary to remove the pinion, there are two methods: (1) Pry it off, but with great care! Use two small screwdrivers of similar widths and straight blades, or two offset screwdrivers of the same size. While protecting the plate with thin leather or other soft material, place two blades under the pinion so that pressure can be applied evenly on opposite sides of the pinion. A slight twisting or prying motion should loosen the pinion. (2) Pound it off, but with great care! Support the under side of the front plate as close to the centerwheel as possible, with material that will not bruise or scratch. Replace the hand nut, then tap the end of the centerwheel gently and straight down with a fiber or plastic hammer until the pinion becomes loose.

   To replace the cannon pinion, reassemble the wheel train and plates, oil the front centerwheel pivot (if you haven't already oiled all pivots), and set the back plate on some soft material over an anvil with a hole positioned under the centerwheel pivot. The pivot extends through the plate and will become bent if this precaution is not taken. The centerwheel is now supported by the back plate and the pinion can be driven on with any suitable hollow punch. If the pinion is being replaced, the factory hole may be slightly small. If so, broach the hole very lightly, because you can easily overdo it and destroy the friction fit. See Appendix 66.

5. (1013) About Schatz Coach Clocks. There were two case models. London Coach, with straight edge roof, and Bermuda Carriage, with curved edge roof. Both were made with movements having back plate Nos. 1010A and 1013, but with variations in dials and hands. Those

London Coach   Bermuda Carriage

made in the 1950's had glass front and side panels, but later models were equipped with plastic.

6. (1214) Unlike other Standard Schatz clocks, this model was especially made for Henry Coehler Co., New York importer. Its distinguishing features are (1) two leveling screws in the base instead of three, and (2) pendulum locking by a lever under the base.

7. (1325, 1326) This is the clock that John Wanamaker sold for $13.75 in 1953 and, following the Wanamaker inventory auction, was sold for $6.99 by S. Klein-on-the-Square. (See Section 1; also Appendix 114 and 115.)

8. (1385) Some factories sold movements made by other factories.

9. (1017) Some factories sold movements made by other factories. This movement marked Baduf (Badische Uhrenfabrik) was made by Petersen (Plate 1583).

10. (1663) This is an early Badische Uhrenfabrik movement since it has no eccentric nut.

11. (1149) This Gustav Becker clock *may* be equipped with an overhead suspension bracket with special attachment for adjusting the pendulum beat. This illustration shows the overhead bracket (A), the adjustable attachment (B), the special top suspension block (C) which (B) interlocks and the suspension guard (D) with bottom suspension lock (E).

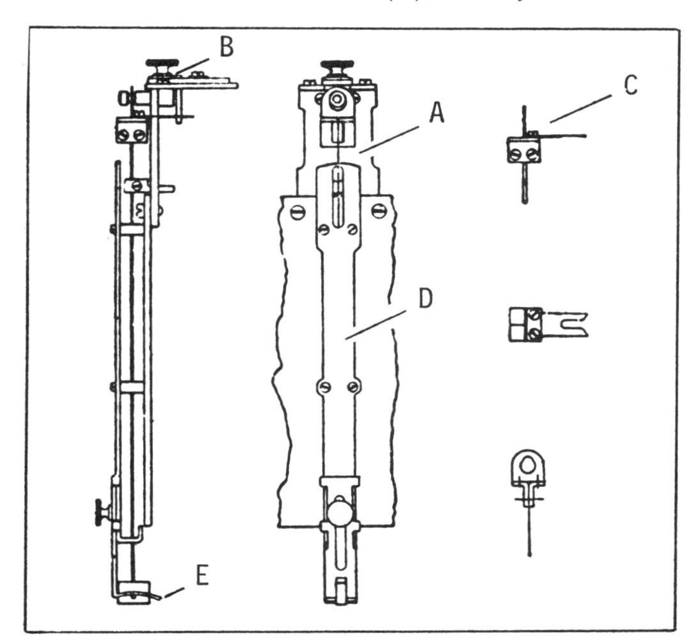

12. (1016A) On some back plates of this movement, these patent numbers will be found:

DRGM	DRGM
484408	502714

These patent papers have not been located. However, they *may* have been for a regulating device in the bottom suspension block that makes it possible to shorten or lengthen the suspension spring. (See Appendix 96.) The pendulum used with the device cannot be regulated. (See Section 13, Pendulum 32.)

13. (1043) These German and U.S. patents:

    D.R.P. No. 144688
    U.S.P. No. 751686

    refer to the "twin loop" temperature compensating pendulum. (See Section 4, Patent 18; Section 13, Pendulum 10.) The fact that the patent numbers appear on the back plate does not necessarily mean that this pendulum was with the clock when it was originally sold.

14. (1049) This German patent:

    D.R.P. 144687

    (also D.R.F.G. 208297), refer to a temperature compensating pendulum. (See Section 4, Patent 17; Section 13, Pendulum 7.) The fact that the patent number appears on the back plate does not necessarily mean that this pendulum was with the clock when it was originally sold.

15. (1316) Only a few hundred of these miniature clocks reached the market before Link went out of business in 1957.

16. (1382) The regulating nut on this pendulum for this midget Kundo clock must be turned *clockwise* to make it go faster which is just the opposite of what most other regulating nuts are designed to do.

17. (1014, 1287) These are the only wide plate *Standard* Schatz clocks with two jewels—one in each of the anchor pivot holes. All of the Schatz *Miniature, Midget,* narrow plate *Standard* and *1000-Day* models are equipped with two jewels.

18. (1287) The little numerals 1 55, 9 54, 53, etc. indicate the month or year of manufacture.

19. (1010) The letters $^{R}{}_{R}$ refer to Remington Rand. Their Electric Shaver Division (with outlets all over the U.S.) was the exclusive factory representative for Schatz from about 1950-1958. (Some back plates have 49R in the trademark.)

20. (1309A) The anchor in this Kaiser movement is in the normal position. (See 1309.)

21. (1309) The anchor in this Kaiser movement is in an off-center position and is especially counterbalanced. The reason for this unusual design has never been made clear. The same movement with anchor in normal position may be seen in 1309A.

22. (1312) Some factories sold movements made by other factories. This movement marked J. Kaiser was made by W. Petersen (1583).

23. (1475) This back plate is from one of the clocks in the first successful Jahresuhrenfabrik production. Serial numbers 913R to 2991R have been identified for clocks having disc pendulums with no gallery, c. 1882—c.1884. Those with serial numbers 4722R to 5558R have been identified for later clocks with pendulums having a 3-pillar gallery, c. 1885.

24. (1111) This is an early production Link movement without an eccentric nut.

25. (1406A) The eccentric nut in this movement holds the escape wheel pivot rather than the anchor pivot. To increase or decrease the "drops" of the escape wheel teeth off of the pallets, one must do the opposite of that recommended in the "Trouble Shooting Chart" in Section 7.

26. (1145) The anchor of this Vossler clock is attached to the crutch. Its tall, pillared case was probably made by a special case manufacturer, for it is exactly the same case as used for a similar clock made by Jahresuhrenfabrik. (See 1257.)

27. (1157) Very few of these midget 400-Day Clocks were made by Georg Wurthner before they went out of business in 1958.

28. (From Section 10, Unit 33A)

One may find Jauch & Haller clocks that use Units shorter than Unit 33, as per these illustrations. For Unit 33A, USE .003"* (.076mm*) HOROLOVAR. For Unit 33B (with fork in slightly higher position), USE .0032" (.081mm) HOROLOVAR.

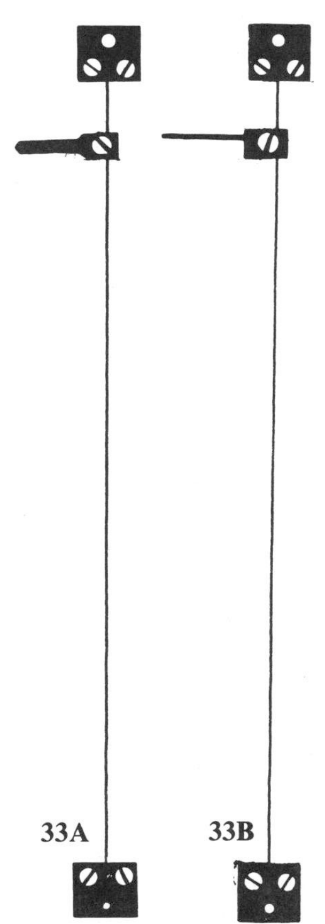

29. (1171) The "governor" type pendulum on this Junghans clock (See Section 13, Pendulum 31) was promoted in trade paper advertising as being temperature compensating, but it is obviously not.

30. (1175) This midget clock by Kienzle is the smallest 30-day key-wind, torsion pendulum clock ever made.

31. (1181) This year striking (hour and half hour) clock is illustrated in the Jahresuhrenfabrik catalog for 1905. Very few of them have survived. (See Section 3, Clock 8.)

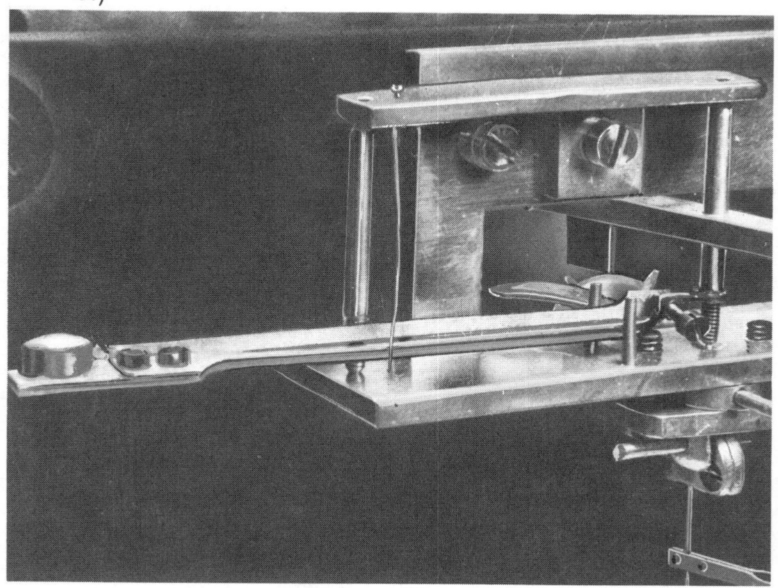

32. (1504) The back plate of this Schneckenburger movement with detent escapement is distinctive from the movement with duplex escapement (1303) because of the long oblong hole near the top. For clocks with this movement, see Section 3, Clocks 2, 3, 5 and 7.

33. (1303) The back plate of this Schneckenburger movement with duplex escapement is distinctive from the movement with detent escapement (1504) because of the large cut-out at the top. For a clock with this movement, see Section 3, Clock 6.

34. (1309) The J. Kaiser Universe clock may go down in history as the best 400-Day Clock ever made. It was imported into the United States in late 1953 - early 1954 with considerable promotion through several importers. However, a U.S. watch importer soon called Kaiser's attention to the fact that it was the owner of the registered trademark "Universe" and Kaiser's use of the word had to be withdrawn. The clock was introduced at a time when importers were pressuring 400-Day Clock manufacturers to lower their factory prices, so it was not the time to introduce a higher priced, quality made clock. It is doubtful that more than a few thousand of the clocks were produced before it was discontinued.

The design of both movement and pendulum was unique and its departure from the Harder patent, which almost all other manufacturers followed, is indicated in many ways. The two major differences, the ones that give repairmen the most trouble, are with the motion train assembly including hand tension, and the pendulum.

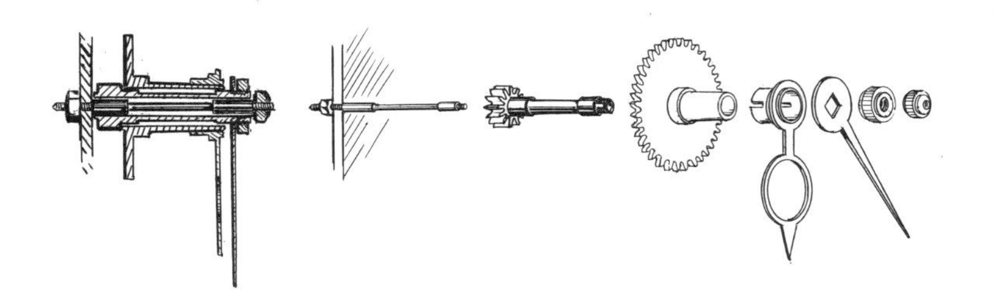

The cannon pinion, hour wheel and hands are mounted on a long lug bolted to the front plate. When assembled, the train should be just slightly loose.

Tension is provided by the intermediate wheel unit. The intermediate wheel shaft is an extension through the front plate of the third wheel arbor. The sequence of parts is lock washer, dished tension washer, intermediate wheel, spacer washer, lock washer. When assembled, the hand tension should be just sufficient to allow the minute hand to be turned with the little finger.

For the pendulum assembly, see Appendix 68.

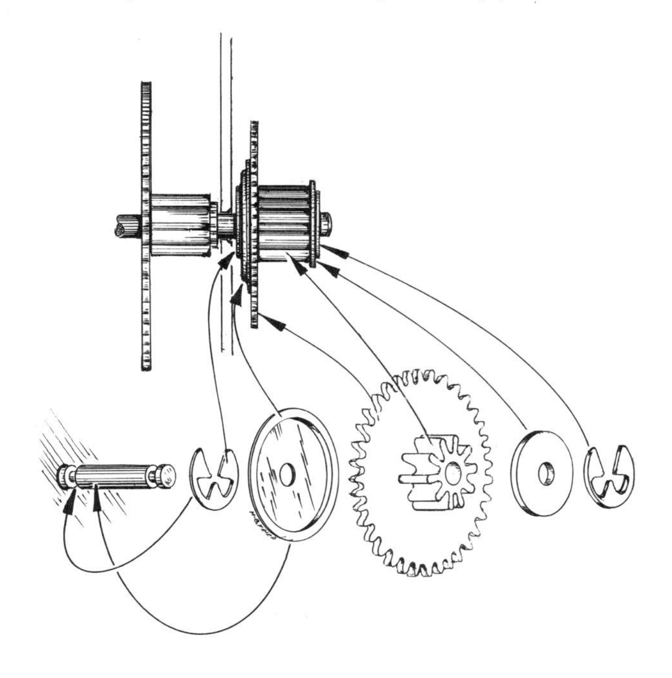

35. (1371AA) This Kundo wide plate represents the first transitition between wide plate and narrow plate movements. There is no provision for ratchet wheel or click on the outside of the plate. The click spring screw hole is tapped, but is not used. The ratchet wheel, click and click spring are now between the plates.

36. (1407D) This Kundo wide plate is one of the few that remained in the factory before the complete transition from wide plate to narrow plate movements. The ratchet wheel, click and click spring are between the plates. (See 1377.)

37. (1471A) This is the only complete 400-Day Clock made in France. (See Section 4, Patent 24; Section 6, Clock 62.)

38. (1314) The original Link Standard movement was made with 4 jewels, but only a few were produced. The importers ordered their clocks made with no jewels (See 1077) in order to avoid having to pay extra duty.

39. (1038) The Henn Standard clock is the only 400-Day Clock ever made with a replaceable eccentric nut.

Models with pendulum locking appear as per illustration. When Lever B is in the locked position behind plunger-button A, the pendulum center rod is in a square hole in raised cup with regulating nut F firmly against pendulum locking bracket E. When Lever B is moved over to Button C, the cup is lowered and the pendulum becomes free to be turned.

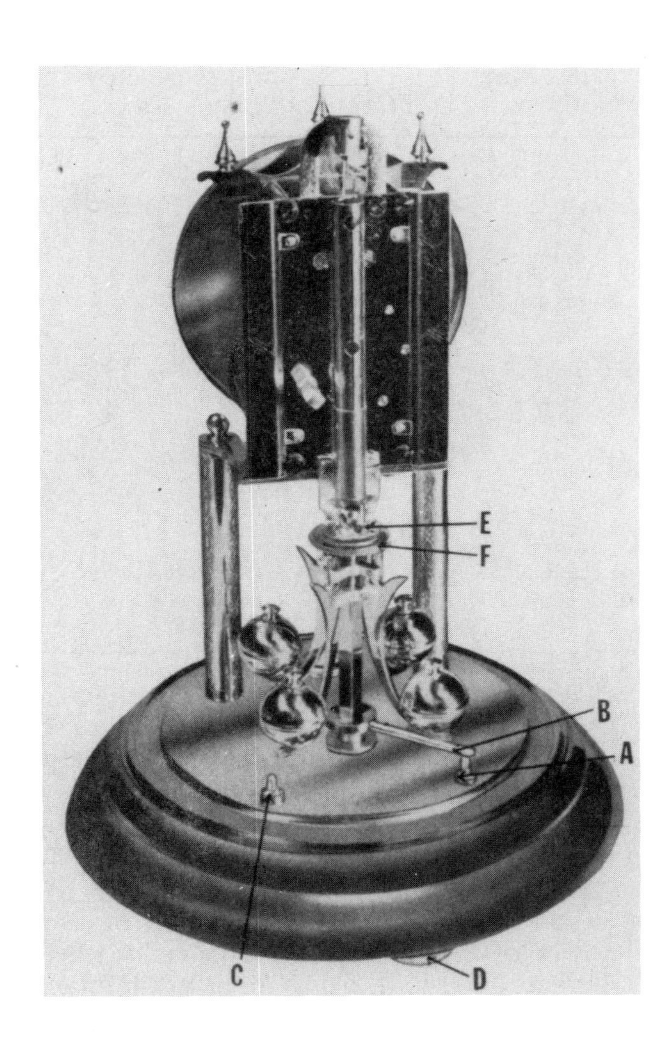

40. (1013B) This was the last new model 400-Day Clock movement produced by Schatz before they gave up making key wind 400-Day Clocks.

41. (1602) The Kienzle calendar clock (Section 6, Clock 21) has this cadrature:

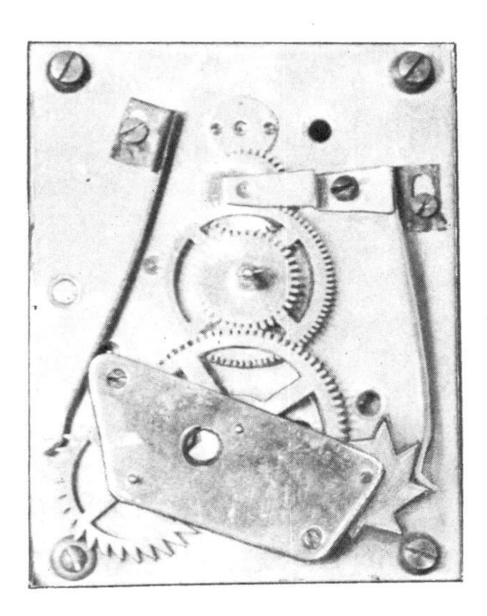

42. (1531) It is believed that the firm of Haller & Dorsch may have been associated with Jahresuhrenfabrik about 1900. It's conjecture, but the Carp trade mark may have evolved from a bad translation of the word Dorsch which actually means cod in German.

43. (1572B) Uhrenfabrik Herr, Gutenbach and (1574) Uhrenfabrik M. Reiner, Furtwangen, complemented each other in output. The principlas were, in fact, related. A third factory, (1469C) Uhrenfabrik Neueck, supplied parts to both firms. When Herr and Reiner went bankrupt in 1954 and 1955, Uhrenfabrik Neueck continued to make the Standard model for several months, chiefly for two customers, The Around-the-World Shoppers Club and Wilmac Co., exporter (1582D).

44. (1469C) This Neueck movement was made for several months following the bankruptcy Uhrenfabrik Herr.

45. (1252) This plate is believed to be a misstamping of the XHM trade mark for a German exporter. See 1594A.

46. (1633) This clock, with cylinder escapement, is one of the early attempts to make a successful 400-Day Clock. It is believed to have been made by A. Willmann Co., Freiburg, Silesia, about 1879.

47. (1674) This two-train calendar 400-Day Clock made by W. Petersen Uhrenfabrik, Schwenningen, is unquestionably the most complicated 400-Day Clock ever made. Only a very limited number of clocks reached the market and it was soon discontinued, because there was insufficient power in the time train to trip the calendar every 24 hours. The clock has become a collector's item as one that never ran.

48. (1687) This clock is shown complete in Section 6, Clock 92. Other than its verge escapement, this clock has features that suggest it was made in the 1950's, but none of its characteristics indicates who might have made it and in what country. Even the dome size is not common for a 400-Day Clock: 5¾" x 12½". What looks to be a bell at the top is actually just a finial.

49. (1007) According to history (See Section 1), Bowler & Burdick, Cleveland, Ohio, owners of the trade mark "Anniversary," made an exclusive contract with Jahresuhrenfabrik for purchase of their movement and pendulum units (1009). However, Bowler & Burkick also purchased movements from Phillipp Haas (1007), Kienzle Clock Factories (1008) and from an unknown manufacturer (1009A, 1009AA) that might possibly have been French.

50. (1720) This is an inexpensively made 30-day clock.

51. (1732) This clock, with verge escapement, is one of the early attempts to make a successful 400-Day Clock. In addition to this round plate movement, this escapement is also found in movements with both rectangular (1731) and trapezoid (1727) plates. They are believed to have been made by A. Willmann Co., Freiburg, Silesia, about 1879-80.

The wheel and pinion counts for movements with both round and trapezoid plates are:

WHEEL and PINION Count -- RP (Round Plate), TP (Trapezoid Plate)

RP	Barrel	TP	RP	First Wheel	TP	RP	Second Wheel	TP
86	teeth	86	66	teeth	66	64	teeth	64
43.5	diam.	43.5	30.9	diam.	30.9	27.6	diam.	27.6
23.6	width	23.9		Pinion			Pinion	
6.5	gear hole	6.2	12	leaves	12	12	leaves	12
6.3	cover hole	7.0	6.2	diam.	6.2	6.0	diam.	6.0
RP	**Third Wheel**	**TP**	**RP**	**Fourth Wheel**	**TP**	**RP**	**Crown Wheel**	**TP**
60	teeth	60	96	teeth	64	20	teeth	15
23.3	diam.	24.3	22.3	diam.	19.9	17.6	diam.	17.2
	Pinion			Pinion			Pinion	
10	leaves	10	8	leaves	8	?	leaves	?
4.7	diam.	4.7	3.5	diam.	3.7	?	diam.	2.9
RP	**Cannon Pinion**	**TP**	**RP**	**Intermediate Wheel**	**TP**	**RP**	**Hour Wheel**	**TP**
36	teeth	40	36	teeth	40	96	teeth	72
15.0	diam.	17.3	26.1	diam.	17.4	26.1	diam.	31.0
				Pinion				
			8	leaves	6			
			3.7	diam.	3.4			

52. (1006) In the 1960's, Konrad Mauch changed the design of their miniature movement to increase the power of the mainspring. The mainspring size was increased in width from 13mm to 15mm with corresponding increase in the size of the barrel. Otherwise, the layout of the movement was not changed.

53. (1393B) In the 1960's, Konrad Mauch changed the design of their midget movement to increase the power of the mainspring. The mainspring size was increased in width from 12mm to 14mm with corresponding increase in the size of the barrel. Otherwise, the layout of the movement was not changed.

54. (1091) This is the first Reiner model which is identified by the fact that the position of the anchor pivot hole is *below* the top guard holding screw. The escape wheel has a small diameter, ranging from 13.3mm to 13.6mm depending upon the production. See 1092 for an example of the second Reiner model which is identified by the fact that the position of the anchor pivot hole is *above* the top guard holding screw. The escape wheel has a larger diameter, 16.8mm, and the anchor is correspondingly larger.

55. (1092) This is the second Reiner model which is identified by the fact that the position of the anchor pivot hole is *above* the top guard holding screw. The escape wheel has a large diameter, 16.8mm. See 1091 for an example of the first Reiner model which is identified by the fact that the anchor pivot hole is *below* the top guard holding screw. The escape wheel has a smaller diameter, ranging from 13.3mm to 13.6mm depending upon production.

56. (1071) This is an hour striker. The extra hole in the upper right part of the plate is the pivot hole for the hammer arbor. Unfortunately, the mainspring in this clock does not have sufficient power to cause the hammer to be lifted every hour for more than a short period of time. Eventually, the loss of power stops the clock. Although the striking feature received promotional attention at the time, the clock was not long on the market.

57. (1461C) This movement uses a slightly shorter and slightly heavier suspension spring than similar movement 1416E. The movement can be identified by the fact that the plates have *square* corners.

58. (1461E) This movement uses a slightly longer and slightly weaker suspension spring than similar movement 1416C. The movement can be identified by the fact that the plates have *round* corners.

59. (1312) This movement is designed with the anchor near the front plate. The "eccentric nut" is the little triangular plate attached to the front plate. The position of the pivot hole can be adjusted by its holding screws.

60. (1097) This 4-jewel movement was made specifically for a special promotion. However, the extra duty required because of the jewels was not thought by the importers to be worth it, so very few of these clocks were produced.

61. (1591A) The name "Semester" was probably a trade mark used by Phillipp Haas. This miniature movement used with a Standard sized base and pendulum did not make a particularly attractive clock as is illustrated in Section 6, Clock 13. However, when this movement was used with a Miniature Sized pendulum, the proportions were more pleasing. Use a 0.0034 inch (0.086mm) Horolovar suspension spring with the Miniature sized pendulum.

62. (1208) This plate and 1014B were stamped with the same dies. Gutenbacher Uhrenfabrik was affiliated with Jahresuhren-Fabrik Aug. Schatz & Sohne. The first was known as Gufa, the latter as J.U.F. At the time (1952), exclusive U.S. distribution of the Schatz clocks was through the Electric Shaver Division of Remington Rand. But by special agreement, Aug. Schatz could sell clocks under the Gufa umbrella to other U.S. customers at a lower price. One such sale was to Henry Coehler Co. (1214). Changes in the leveling screws and pendulum locking mechanism made the clock sufficiently different from those being made by Jahresuhren-Fabrik for Remington Rand.

63. (1005) The "eccentric nut" in this movement is in the bracket. The position of the anchor pivot hole should be adjusted before the holding screws are tightened.

64. (1213A) This movement is designed with its anchor nearer to the front plate than the back. The "eccentric nut" is in the front plate. The anchor pivot holes should be adjusted with the "eccentric nut" or with the bracket before the holding screws are tightened.

65. (1472) The W. Petersen factory in Schwenningen never had time to refine its production of 400-Day Clocks before it went out of business in 1954. They made both wide plate and narrow plate movements, but one will find many variations in the parts used. For instance, ratchet wheels with cusps instead of teeth as shown in Section 14; suspension fork shaped like the one in Unit 36A, Section 10; intermediate wheel on arbor pivoted through both plates; knurled nuts on all four back pillar screws and on the suspension saddle holding screw; suspension bracket holding screws too long; anchor "eccentric nut" in front plate; bushing-like oil sinks in back plate pivot holes; escape wheel not crossed out. Some Petersen movements were sold by, and carried the name of J. Kaiser (1312).

66. (1013) For reasons not clear, the Schatz miniature, midget and 1000-Day movements were made with motion trains having two sets of dimensions. There were no changes in the wheel tooth count or wheel diameters.

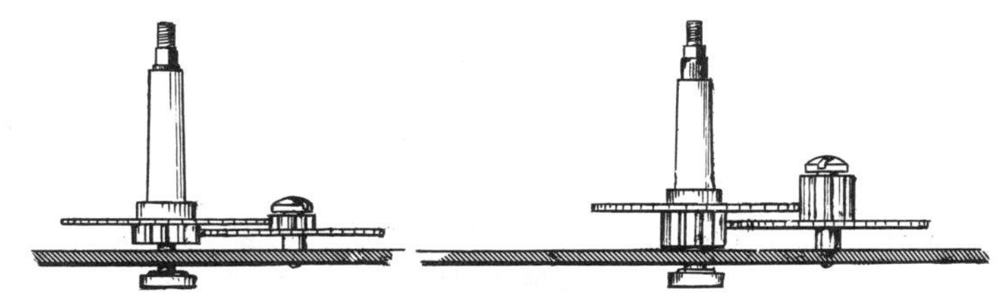

The 2.4mm Cannon Pinion can be substituted for the 1.5mm pinion, but the overall length of the hour wheel will have to be filed shorter so that its end will be inside of the minute hand square on the center-wheel arbor; otherwise, the minute hand will bind on the end of the hour wheel pipe.

The intermediate wheels cannot be interchanged, because the lengths of the holding screws are different.

The longer hour wheel can be shortened, if necessary, but not visa versa.

67. (1591) Some clocks made by Georg Wurthner use a piano wire click spring. Anchored around the head of the pillar screw, coiled once around the click screw, it is held against the click with a reverse bend.

68. (1310) The globe pendulum used with the J. Kaiser clock has many parts. This drawing shows how they are assembled:

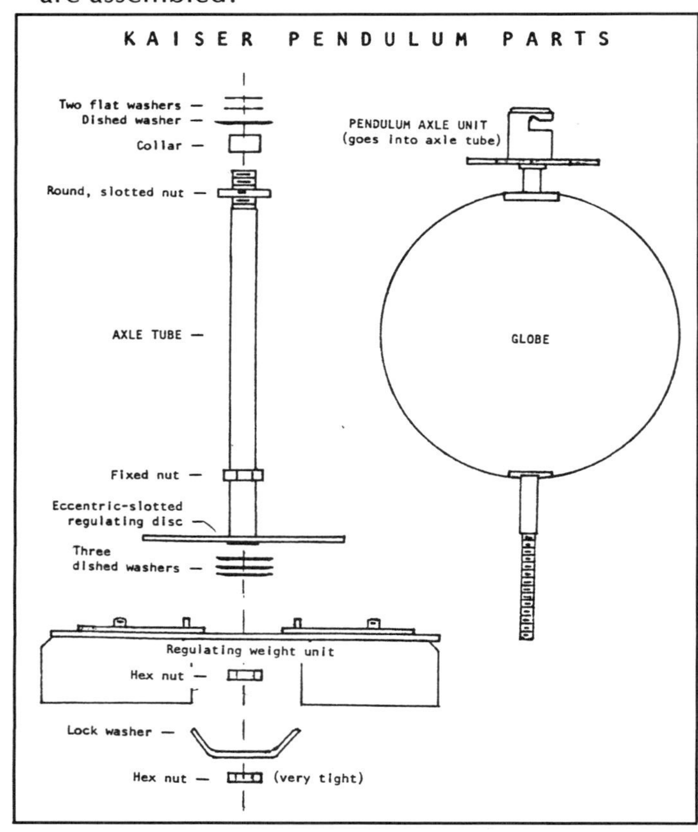

69. (1033) This is a Becker skeleton pendulum that goes with the skeleton clock. Practically every surface is finished and highly polished.

70. (1206) These Gustav Becker serial numbers are within the approximate year of manufacture:

Serial number	Year	Serial number	Year
1,500,000	1900	2,150,000	1907
1,590,000	1901	2,250,000	1908
1,680,000	1902	2,350,000	1909
1,770,000	1903	2,450,000	1910
1,860,000	1904	2,550,000	1911
1,950,000	1905	2,650,000	1912
2,050,000	1906		

Note: Some Becker back plates do not have serial numbers. Others may have incomplete numbers with either one or two of the first or last digits missing, possibly because of defective stamping.

The pair of click spring holes will be found in at least six different positions in the Becker plates and bear no relation to the date of manufacture.

71. (1103) The "eccentric nut" in this movement is in the bridge attached to the back plate. The position of the anchor pivot hole should be adjusted with the bracket before tightening the screws.

72. (1678) This Herr miniature clock could have been experimental. Very few of them were made.

73. (1461) This movement could have been made by one of the manufacturers who were known to have made 400-Day Clocks, but who did not stamp any identifying marks on their back plates. Among others, they include:

Fortuna	Thomas Haller	Ideal	Wurth
Friberger	Phil. Hauck	Lenzkirch	

74. (1032) Gustav Becker was born 2 May 1819 in Ols, Silesia in eastern Germany. His first shop was in Freiburg. He died 14 September 1885. Following his death, his two sons, Albert and Paul, continued the business. In 1890, a branch was opened in Braunan, Bohemia. About 1898, the firm Freiburg Vereinigte Uhrenfabrik A.G. was formed. In 1925, the business was purchased by Gebr. Junghans, Schramberg, Black Forest.

75. (From Section 17) Sizes of mainsprings used by all 400-Day Clock manufacturers since 1949:

SIZEmm*	MANUFACTURER	MODEL
12 x 24	S. Haller	Miniature and Midget
12 x 25	Kern	Miniature
	Koma	Miniature (Original)
12 x 32	Nisshindo	New Master (100-Day)
13 x 30	Hermle	Miniature
	Schatz	Miniature and Midget
13 x 32	Koma	Miniature (Original)
14 x 25	Kern	Midget
	Koma	Midget (Revised)
14 x 28	Link	Miniature
14 x 30	Kundo	Miniature and Midget
14 x 32	Nisshindo	Miniature
15 x 32	Koma	Miniature (Revised)
16 x 36	Herr	Miniature
	Reiner	Miniature

SIZEmm*	MANUFACTURER	MODEL
18 x 38	S. Haller	Standard
	Hermle	Standard
	Jauch & Haller	Standard
	Kern	Standard
19 x 32	Henn	Miniature
19 x 36	Herr	Standard
	Schatz	Standard
	Würthner	Standard
19 x 38	Henn	Standard
	Kaiser	Standard
	Kundo	Standard
	Link	Standard
	Petersen	Standard
20 x 38	Koma	Standard
21 x 50	Schatz	1000-Day

*(Width of spring)
x (Inside diam. of barrel)

76. Kundo Standard Movement Parts

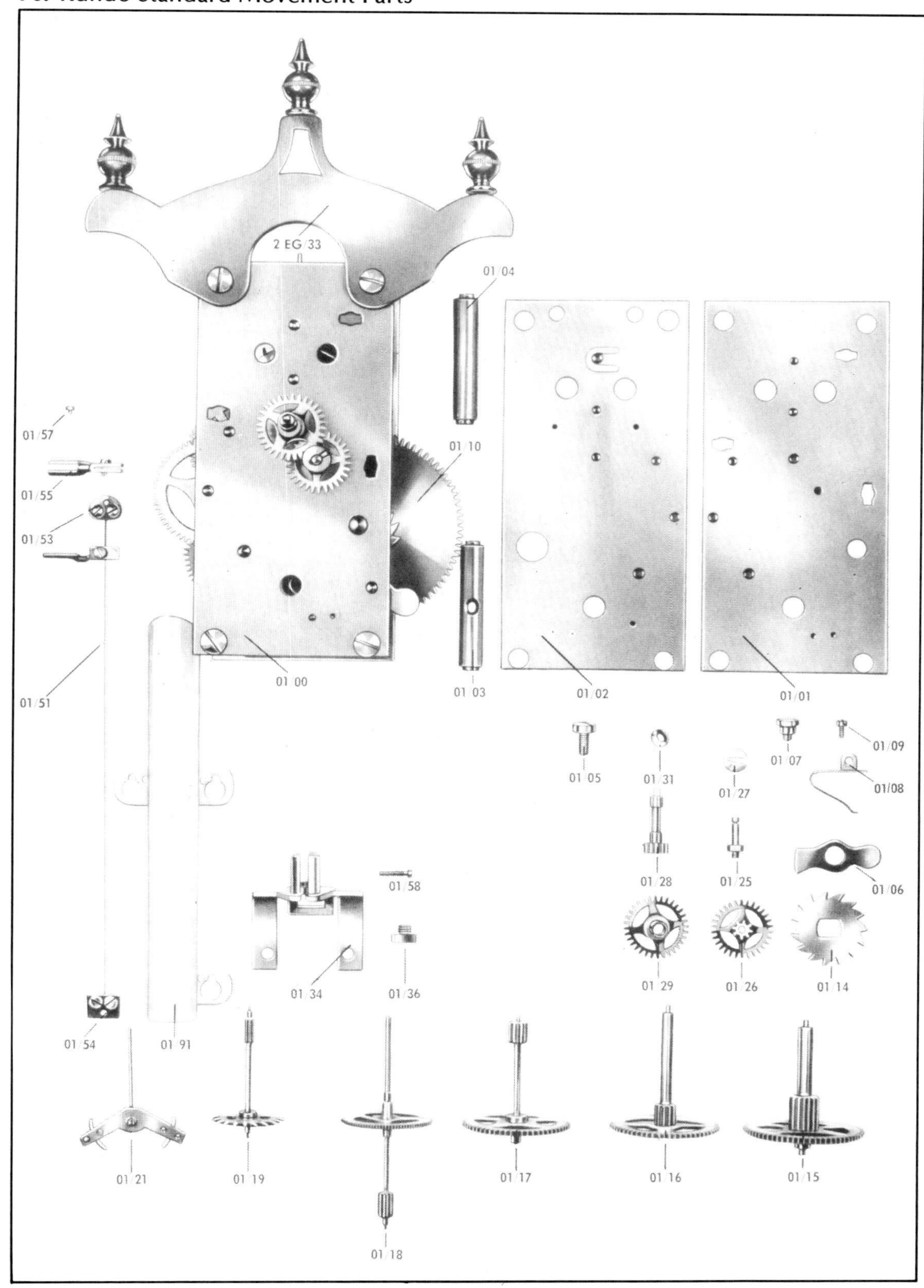

The illustrations show a movement of about 1976 production. With minor variations, the individual parts are the same for earlier narrow plate movements as well as the original wide plate movements. Major variations in design are in the dial lugs, suspension bracket and suspension guard.

01-00 – Narrow Plate Movement
01-01 – Front Plate
01-02 – Back Plate
01-03 – Lower Pillar
01-04 – Upper Pillar
01-05 – Pillar Screw
01-06 – Click
01-07 – Click Screw
01-08 – Click Spring
01-09 – Click Spring Screw
01-10 – Mainspring Barrel
01-14 – Ratchet Wheel
01-15 – First Wheel
01-16 – Second Wheel
01-17 – Third Wheel
01-18 – Center Wheel (Fourth Wheel)
01-19 – Escape Wheel
01-21 – Anchor
01-21E – Entrance Pallet
01-21X – Exit Pallet
01-21Y – Pallet Holding Strap
01-21Z – Strap Screw
01-25 – Intermediate Wheel Post
01-26 – Intermediate Wheel (Minute Wheel)
01-27 – Intermediate Wheel Lock Washer
01-28 – Cannon Pinion
01-29 – Hour Wheel
01-31 – Hand Tension Washer
01-34 – Suspension Bracket
01-36 – Bracket Holding Screw
01-51 – Suspension Unit
01-51S – Suspension Spring
01-53 – Top Suspension Block
01-53S – Block Screw
01-54 – Bottom Suspension Block
01-54S – Block Screw
01-55 – Suspension Fork
01-55S – Fork Screw
01-57 – Suspension Guard Holding Screw
01-58 – Suspension Unit Holding Screw
01-91 – Suspension Spring Guard
2EG33 – Narrow Plate Arch

77. Kundo Miniature/Midget Movement Parts

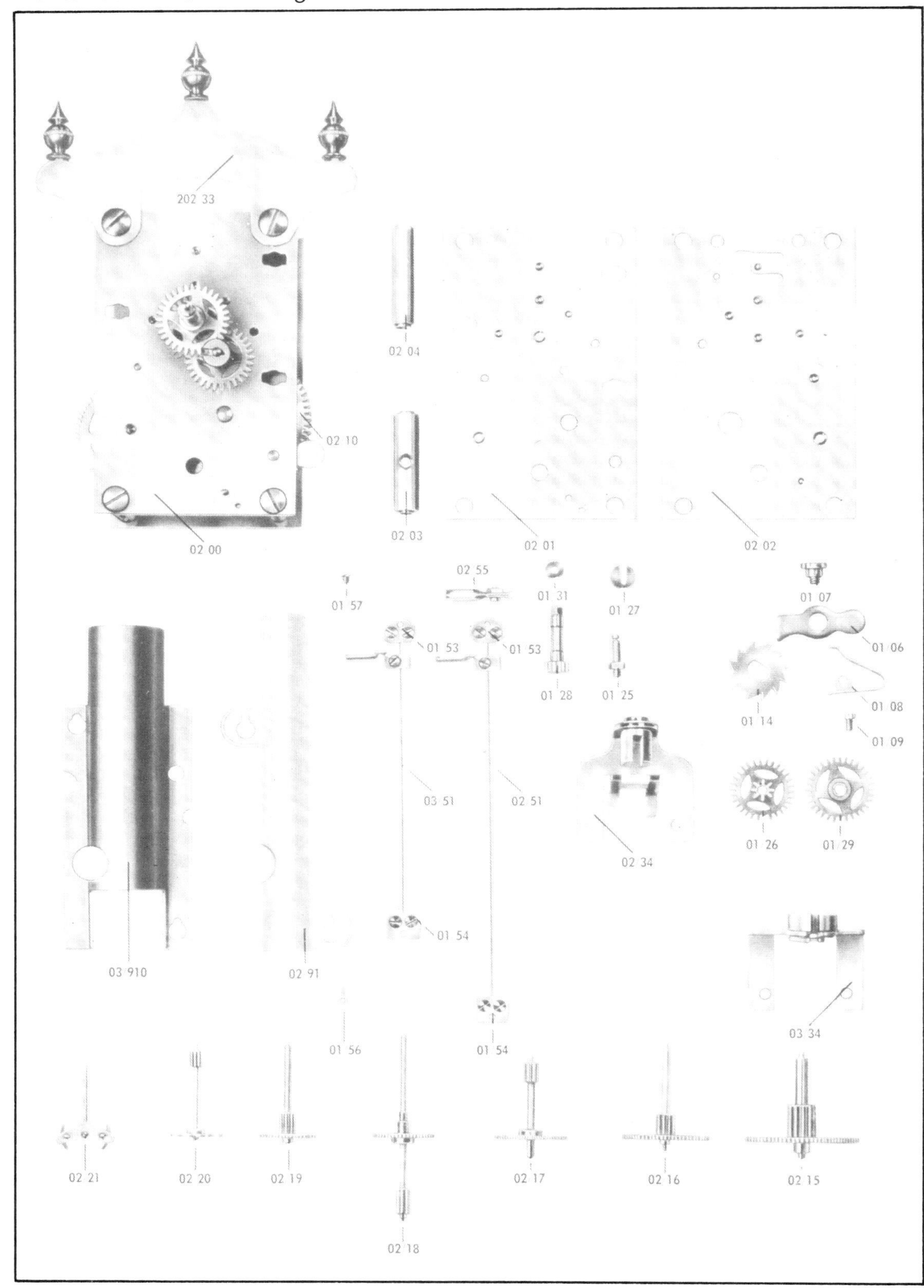

The illustrations show a movement of about 1976 production. With minor variations, the individual parts are the same as for earlier miniature movements. Major variations in design are in the dial lugs, suspension bracket and suspension guard.

01/06 –Click
01/07 –Click Screw
01/08 –Click Spring
01/09 –Click Spring Screw
01/14 –Ratchet Wheel
01/25 –Intermediate Wheel Post
01/26 –Intermediate Wheel (Minute Wheel)
01/27 –Intermediate Wheel Lock Washer
01/28 –Cannon Pinion
01/29 –Hour Wheel
01/31 –Hand Tension Washer
01/53 –Top Suspension Block
01/53S–Block Screw
01/54 –Bottom Suspension Block
01/54S–Block Screw
01/56 –Guard Holding Screw
01/57 –Block Screw
02/00 –Miniature Movement
02/01 –Front Plate
02/02 –Back Plate
02/03 –Lower Pillar
02/04 –Upper Pillar
02/10 –Mainspring Barrel
02/16 –Second Wheel
02/17 –Third Wheel
02/17 –Third Wheel
02/18 –Center Wheel (Fourth Wheel)
02/19 –Fifth Wheel
02/20 –Escape Wheel
02/21 –Anchor
02/21E–Entrance Pallet
02/21X–Exit Pallet
02/21Y–Pallet Holding Strap
02/21Z–Strap Screw
02/34 –Miniature Suspension Bracket
02/51 –Suspension Unit
02/51S–Suspension Spring
02/55 –Suspension Fork
02/55S–Fork Screw
02/91 –Suspension Spring Guard
03/34 –Midget Suspension Bracket
03/910–Midget Suspension Guard
202/33–Miniature Arch

78. Schatz Standard Movement Parts

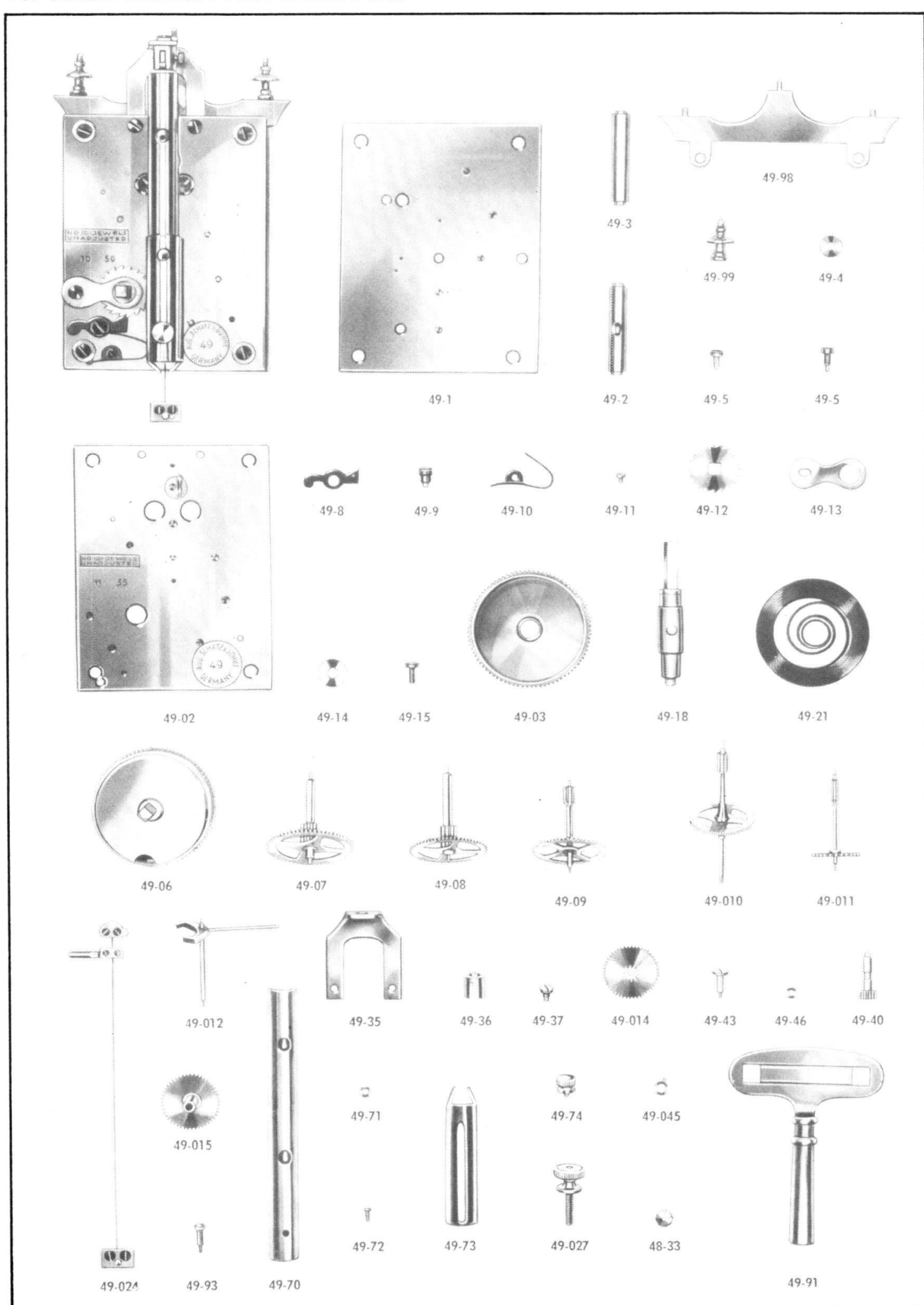

The illustrations show a movement of about 1950 production. With minor variations, the individual parts are the same as for later narrow plate movements. Major variations in design are in the pendulum locking mechanism.

49/ 1 –Front Plate
49/02 –Back Plate
49/ 2 –Lower Pillar
49/03 –Barrel
49/ 3 –Upper Pillar
49/ 4 –Pillar Washer
49/5A –Pillar Screw
49/5B –Pillar Screw
49/06 –Barrel Complete
49/07 –First Wheel
49/08 –Second Wheel
49/ 8 –Click
49/09 –Third Wheel
49/ 9 –Click Screw
49/010–Center Wheel (Fourth Wheel)
49/ 10–Click Spring
49/011–Escape Wheel
49/ 11–Click Spring Screw
49/012–Anchor
49/ 12–Ratchet Wheel
49/ 13–Ratchet Wheel Bridge
49/014–Intermediate Wheel (Minute Wheel)
49/ 14–Ratchet Bridge Washer
49/015–Hour Wheel
49/ 15–Ratchet Bridge Holding Screw
49/ 18–Mainspring Arbor
49/ 21–Mainspring
49/024–Suspension Spring Unit
49/024S–Suspension Spring
49/024T–Top Suspension Block
49/024TS–Block Screw
49/024F–Suspension Fork
49/024FS–Fork Screw
49/024B–Bottom Suspension Block
49/024BS–Block Screw
49/027–Movement Holding Screw
49/ 33–Hand Nut
49/ 35–Suspension Bracket
49/ 36–Suspension Saddle
49/ 37–Saddle Holding Screw
49/ 40–Cannon Pinion
49/ 43–Intermediate Wheel Holding Screw
49/045–Dial Lug Collar with Screw
49/ 46–Hand Tension Washer
49/ 70–Suspension Spring Guard
49/ 71–Guard Holding Washer
49/ 72–Guard Holding Screw
49/ 73–Bottom Suspension Block Guard
49/ 74–Guard Locking Thumb Screw
49/ 91–Key
49/ 93–Suspension Unit Holding Screw
49/ 98–Arch
49/ 99–Arch Finial

79. Schatz Miniature/Midget Parts

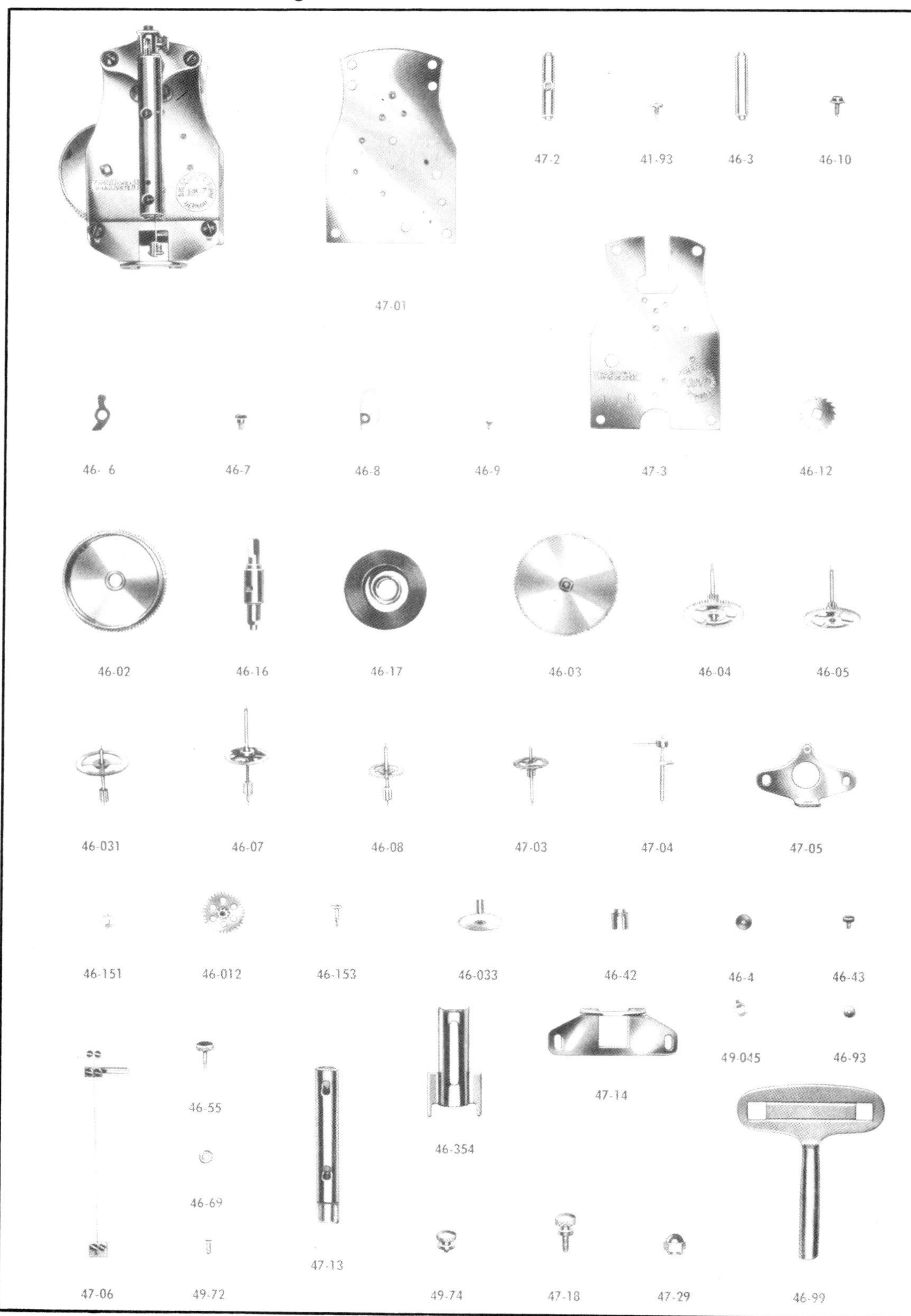

The illustrations show movements of about 1957 production. With minor variations, the individual parts are the same as for movements made prior to or after this production. Major variations in design are in a miniature model with front and back plates incorporating curved legs.

41/ 93–Pillar Screw
46/ 02–Barrel
46/ 03–Barrel Cover
46/ 3–Upper Pillar
46/ 04–First Wheel
46/ 4–Saddle Washer
46/05 –Second Wheel
46/ 6 –Click
46/07 –Center Wheel (Fourth Wheel)
46/ 7–Click Screw
46/ 08–Fifth Wheel
46/ 8–Click Spring
46/ 9–Click Spring Screw
46/ 10–Pillar Screw
46/012–Intermediate Wheel (Minute Wheel)
46/ 12–Ratchet Wheel
46/ 16–Mainspring Arbor
46/ 17–Mainspring
46/031–Third Wheel
46/033–Hour Wheel
46/ 42–Saddle
46/ 43–Saddle Holding Screw
46/ 55–Suspension Unit Holding Screw
46/ 69–Guard Holding Washer
46/ 93–Hand Nut
46/ 99–Key
46/151–Cannon Pinion
46/153–Intermediate Wheel Holding Screw
46/354–Pendulum Locking Guard
47/01 –Front Plate
47/ 2 –Lower Pillar
47/03 –Escape Wheel
47/ 3 –Back Plate
47/04 –Anchor
47/05 –Suspension Bracket
47/06 –Suspension Unit
47/06S–Suspension Spring
47/06T–Top Suspension Block
47/06TS–Block Screw
47/06F –Suspension Fork
47/06FS–Fork Screw
47/06B –Bottom Suspension Block
47/06BS–Block Screw
47/13 –Suspension Guard
47/14 –Pendulum Locking Bracket
47/18 –Movement Holding Screw
47/29 –Crown Nut for Case
49/045–Dial Lug Collar with Screw
49/72 –Guard Holding Screw
49/74 –Guard Locking Thumb Screw

80. Schatz 1000-Day Movement Parts

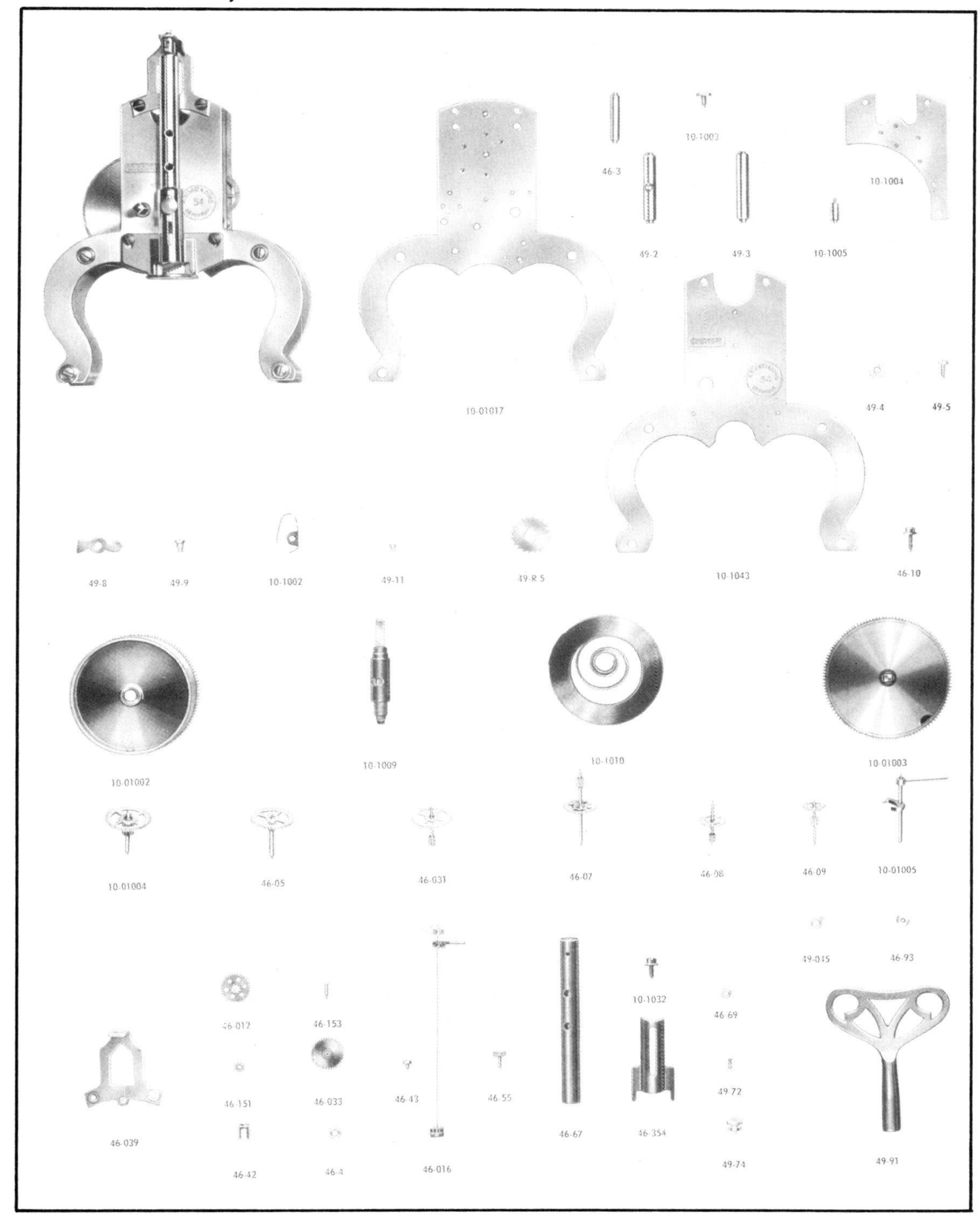

The illustrations show a movement of about 1955 production. The same movement was also made with angled instead of curved supports. (See Appendix 113.)

10/01002–Barrel
10/01003–Barrel Cover
10/01004–First Wheel
10/01005–Anchor
10/01017–Front Plate
10/ 1002–Click Spring
10/ 1003–Pillar Screw
10/ 1004–Intermediate Plate
10/ 1005–Intermediate Plate Pillar
10/ 1009–Mainspring Arbor
10/ 1010–Mainspring
10/ 1032–Pillar Screw
10/ 1043–Back Plate
46/ 3–Intermediate Plate Pillar
46/ 4–Saddle Washer
46/ 05–Second Wheel
46/ 07–Center Wheel (Fourth Wheel)
46/ 08–Fifth Wheel
46/ 9–Escape Wheel
46/ 10–Pillar Screw
46/ 012–Intermediate Wheel (Minute Wheel)
46/ 016–Suspension Unit
46/ 016S–Suspension Spring
46/ 016T–Top Suspension Block
46/ 016TS–Top/Bottom Block Screw
46/ 016F–Suspension Fork
46/ 016FS–Fork Screw
46/ 016B–Bottom Suspension Block
46/ 031–Third Wheel
46/ 033–Hour Wheel
46/ 039–Suspension Bracket
46/ 42–Saddle
46/ 43–Saddle Holding Screw
46/ 55–Suspension Unit Holding Screw
46/ 67–Suspension Guard
46/ 69–Suspension Guard Holding Washer
46/ 93–Hand Nut
46/ 151–Cannon Pinion
46/ 153–Intermediate Wheel Holding Screw
46/ 354–Pendulum Locking Guard
49/ 2–Lower Pillar
49/ 3–Upper Pillar
49/ 4–Pillar Washer
49/ 5–Pillar Screw
49/ R5–Ratchet Wheel
49/ 8–Click
49/ 9–Click Screw
49/ 11–Click Spring Screw
49/ 045–Dial Lug Collar with Screw
49/ 72–Suspension Guard Holding Screw
49/ 74–Guard Locking Thumb Screw
49/ 91–Key

81. Kern Standard Movement Parts

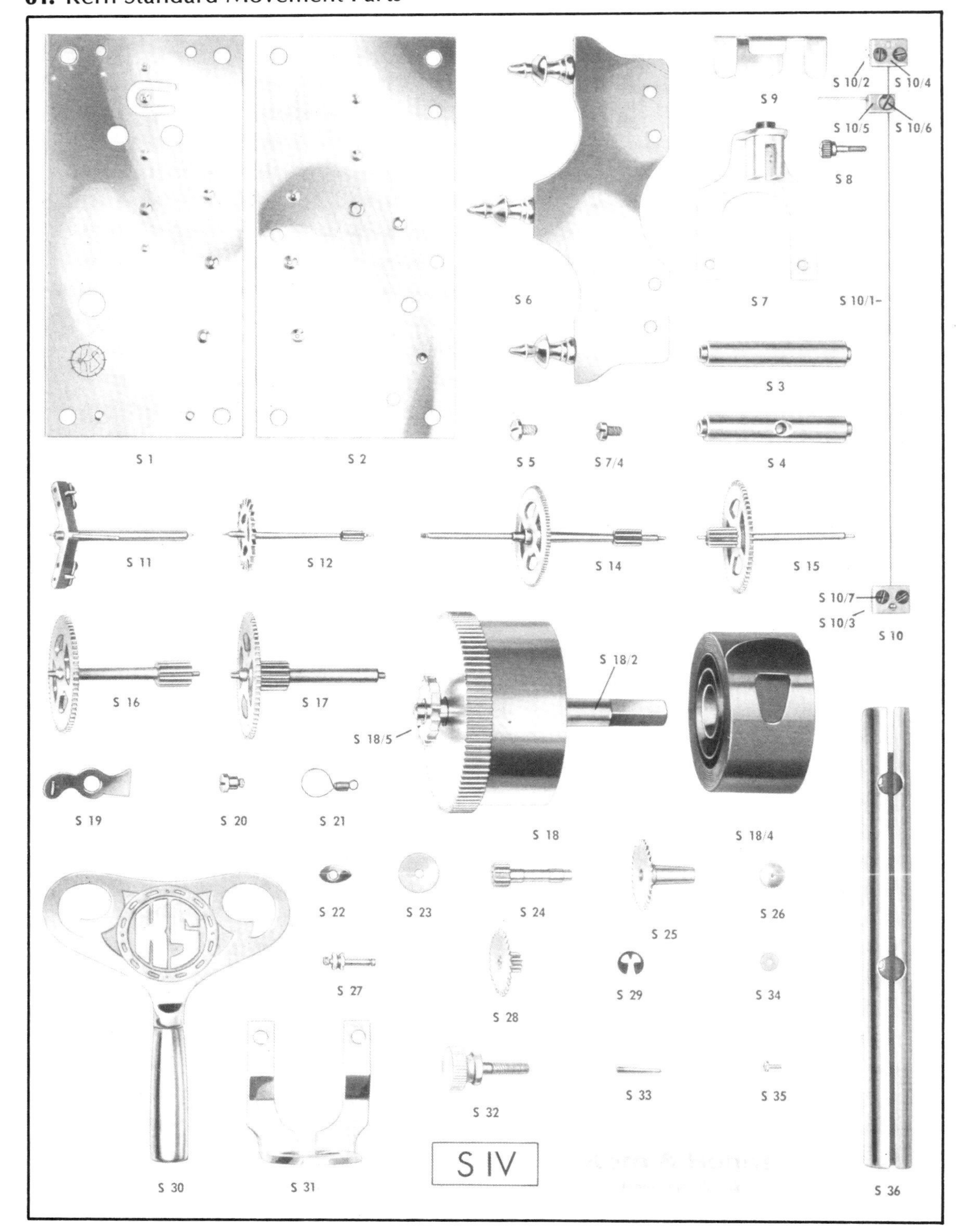

The illustrations show a movement of about 1960 production. With minor variations, the individual parts are the same as for the original wide plate movements, but parts S 9 and S 31 were not included.

S- 1–Back Plate
S- 2–Front Plate
S- 3–Upper Pillar
S- 4–Lower Pillar
S- 5–Pillar Screw
S- 6–Arch
S- 7–Suspension Bracket
S- 7A–Saddle
S- 7B–Saddle Washer
S- 7C–Saddle Holding Screw
S- 8–Suspension Unit Holding Screw
S- 9–Escapement Guard
S- 10–Suspension Unit
S- 10/1–Suspension Spring
S- 10/2–Top Suspension Block Lower Half
S- 10/3–Bottom Block Lower Half with Pin
S- 10/4–Top/Bottom Block Upper Half
S- 10/5–Suspension Fork
S- 10/6–Fork Screw
S- 10/7–Block Screw
S- 11–Anchor
S- 11E–Entrance Pallet
S- 11X–Exit Pallet
S- 11Y–Pallet Holding Strap
S- 11Z–Strap Screw
S- 12–Escape Wheel
S- 14–Center Wheel (Fourth Wheel)
S- 15–Third Wheel
S- 16–Escape Wheel
S- 17–First Wheel
S- 18–Mainspring Barrel
S- 18/1–Barrel Cover
S- 18/2–Mainspring Arbor
S- 18/4–Mainspring
S- 18/5–Ratchet Wheel
S- 19–Click
S- 20–Click Screw
S- 21–Click Spring
S- 22–Hand Tension Washer
S- 23–Intermediate Wheel Holding Washer
S- 24–Cannon Pinion
S- 25–Hour Wheel
S- 26–Hand Nut
S- 27–Intermediate Wheel Post
S- 28–Intermediate Wheel (Minute Wheel)
S- 29–Intermediate Wheel Lock Washer
S- 30–Key
S- 31–Pendulum Locking Bracket
S- 32–Movement Holding Screw
S- 33–Dial Lug Pin
S- 34–Guard Holding Washer
S- 35–Guard Holding Screw
S- 36–Suspension Guard

82. Kern Miniature Movement Parts

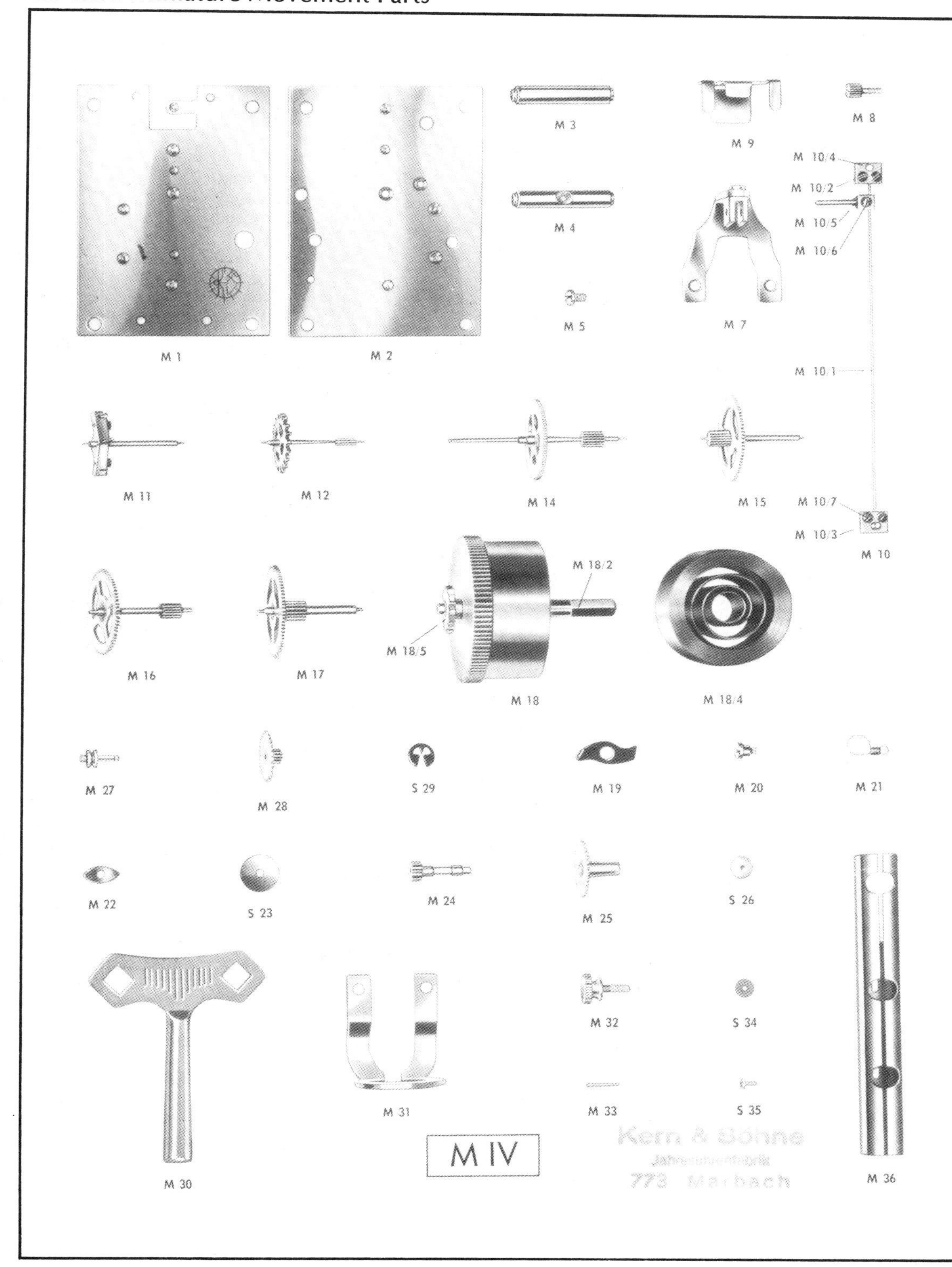

The illustrations show a movement of about 1958 production. With minor variations, the individual parts are the same as in movements made prior to and after this production.

M- 1–Back Plate
M- 2–Front Plate
M- 3–Lower Pillar
M- 4–Upper Pillar
M- 5–Pillar Screw
M- 7–Suspension Bracket
M-7A–Saddle
M-7B–Saddle Washer
M-7C–Saddle Holding Screw
M- 8–Suspension Unit Holding Screw
M- 9–Escapement Guard
M-10/1–Suspension Spring
M-10/2–Top Suspension Block Lower Half
M-10/3–Bottom Block Lower Half with Pin
M-10/4–Top/Bottom Block Upper Half
M-10/5–Suspension Fork
M-10/6–Fork Screw
M-10/7–Block Screw
M-11–Anchor
M-11E–Entrance Pallet
M-11X–Exit Pallet
M-11Y–Pallet Holding Strap
M-11Z–Strap Screw
M-12–Escape Wheel
M-14–Center Wheel (Fourth Wheel)
M-15–Third Wheel
M-16–Second Wheel
M-17–First Wheel
M-18–Mainspring Barrel
M-18/1–Barrel Cover
M-18/2–Mainspring Arbor
M-18/4–Mainspring
M-18/5–Ratchet Wheel
M-19–Click
M-20–Click Screw
M-21–Click Spring
M-22–Hand Tension Washer
S-23–Intermediate Wheel Holding Washer
M-24–Cannon Pinion
M-25–Hour Wheel
S-26–Hand Nut
M-27–Intermediate Wheel Holding Pin
M-28–Intermediate Wheel
M-30–Key
M-31–Pendulum Locking Bracket
M-32–Movement Holding Screw
M-33–Dial Lug Pin
S-29–Intermediate Wheel Locking Clip
S-34–Guard Holding Washer
S-35–Guard Holding Screw

83. Kern Midget Movement Parts

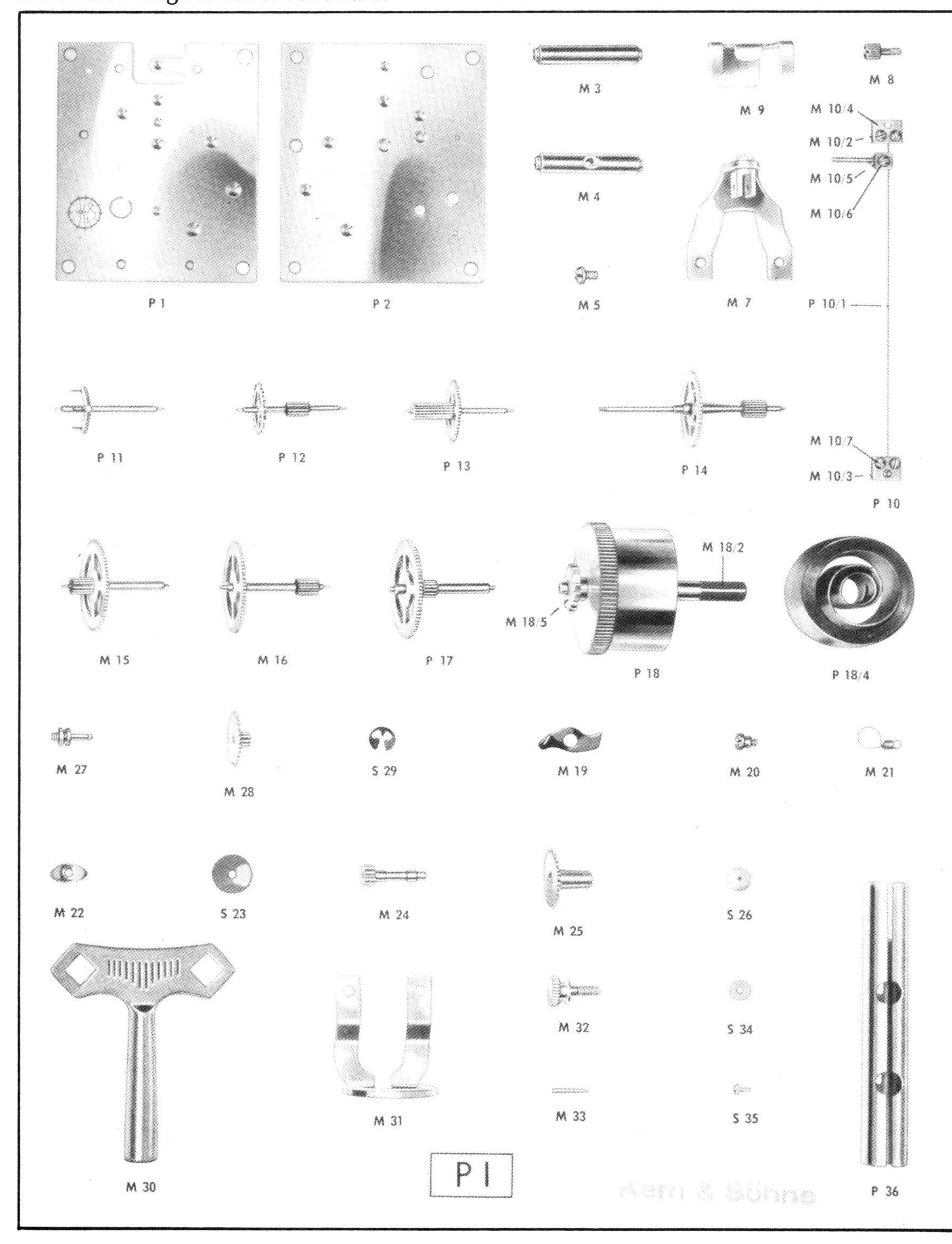

The illustrations show a movement of about 1958 production. With minor variations, the individual parts are the same as for movements made since this production.

P- 1–Back Plate
P- 2–Front Plate
M- 3–Lower Pillar
M- 4–Upper Pillar
M- 5–Pillar Screw
M- 7–Suspension Bracket
M-7A–Saddle
M-7B–Saddle Washer
M-7C–Saddle Holding Screw
M- 8–Suspension Unit Holding Screw
M- 9–Escapement Guard
P-10/1–Suspension Spring
M-10/2–Top Suspension Block Lower Half
M-10/3–Bottom Block Lower Half with Pin
M-10/4–Top/Bottom Block Upper Half
M-10/5–Suspension Fork
M-10/6–Fork Screw
M-10/7–Block Screw
P-11–Anchor
M-11E–Entrance Pallet
M-11X–Exit Pallet
M-11Y–Pallet Holding Strap
M-11Z–Strap Screw
P-12–Escape Wheel
P-13–Fifth Wheel
P-14–Center Wheel (Fourth Wheel)
M-16–Second Wheel
P-17–First Wheel
P-18–Mainspring Barrel
P-18/1–Barrel Cover
M-18/2–Mainspring Arbor
P-18/4–Mainspring
M-18/5–Ratchet Wheel
M-19–Click
M-20–Click Screw
M-21–Click Spring
M-22–Hand Tension Washer
S-23–Intermediate Wheel Holding Washer
M-24–Cannon Pinion
M-25–Hour Wheel
S-26–Hand Nut
M-27–Intermediate Wheel Holding Post
M-28–Intermediate Wheel (Minute Wheel)
S-29–Intermediate Wheel Lock Washer
M-30–Key
M-31–Pendulum Locking Bracket
M-32–Movement Holding Screw
M-33–Dial Lug Washer

84. Koma Standard Movement Parts

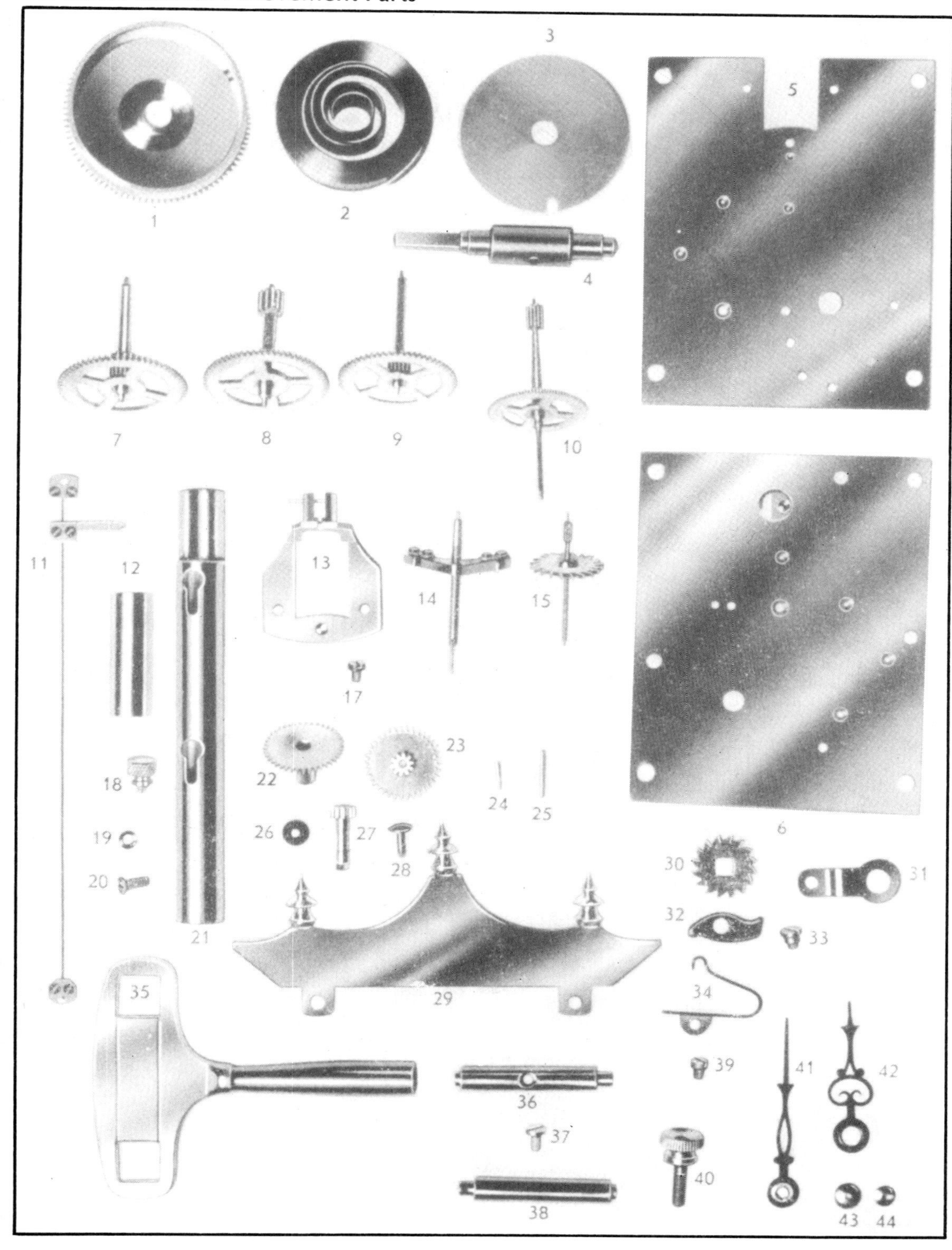

The illustrations show a movement of about 1950 production. With minor variations, the individual parts are the same as for later narrow plate movements. Major variations in design are with the pendulum locking mechanism.

1 –Barrel
2 –Mainspring
3 –Barrel Cover
4 –Mainspring Arbor
5 –Back Plate
6 –Front Plate
7 –First Wheel
8 –Second Wheel
9 –Third Wheel
10–Center Wheel (Fourth Wheel)
11–Suspension Unit
11S–Suspension Spring
11T–Top Suspension Block
11TS–Top/Bottom Block Screw
11F–Suspension Fork
11FS–Fork Screw
11B–Bottom Block
12–Suspension Guard Extension
13–Suspension Bracket with Saddle
14–Anchor
14E–Entrance Pallet
14X–Exit Pallet
14Y–Pallet Holding Strap
14Z–Strap Screw
15–Escape Wheel
17–Bracket Holding Screw
18–Suspension Guard Locking Thumb Screw
19–Suspension Guard Holding Washer
20–Suspension Guard Holding Screw
21–Suspension Guard
22–Hour Wheel
23–Intermediate Wheel (Minute Wheel)
24–Pillar Pin
25–Suspension Unit Holding Pin
26–Hour Wheel Spacing Washer
27–Cannon Pinion
28–Intermediate Wheel Holding Screw
29–Arch
30–Ratchet Wheel
31–Ratchet Wheel Bridge
32–Click
33–Click Screw
34–Click Spring
35–Key
36–Lower Pillar
37–Pillar Screw
38–Upper Pillar
39–Click Spring Screw
40–Movement Holding Screw
41–Minute Hand
42–Hour Hand
43–Hand Washer
44–Hand Nut

85. Koma Miniature Movement Parts

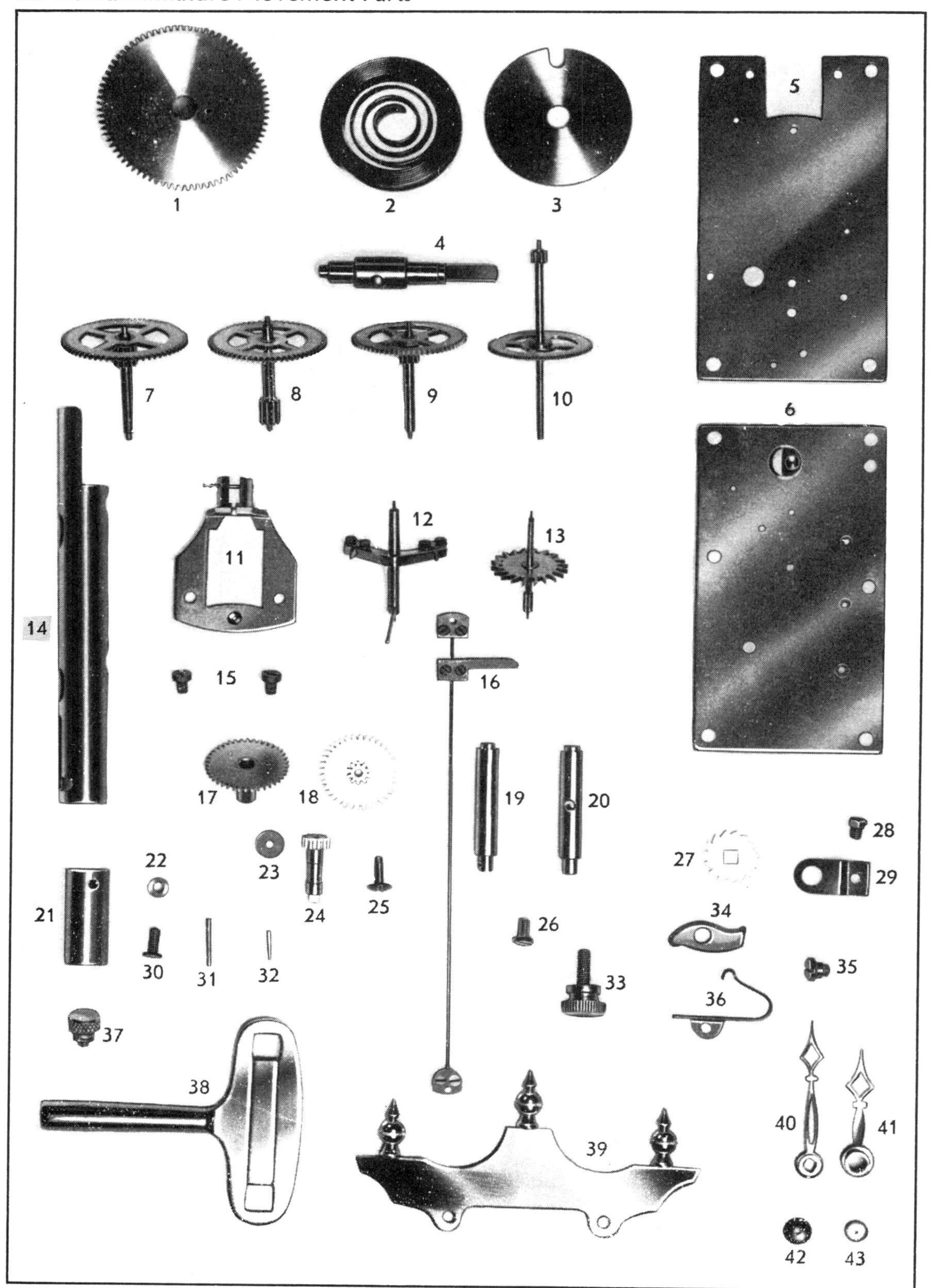

The illustrations show a movement of about 1960 production. With minor variations, the individual parts are the same as for other Koma miniature movements made prior to and after this production.

About 1954, the 13 × 32 mainspring in the miniature movement was increased in size in order to increase the power to the escapement. The revised 15 × 32 mainspring not only necessitated an increase in the width of the barrel, but also in the size of the arbor, as well as changes in the design of the second wheel, third wheel, centerwheel and escape wheel units. The shape of these revised units differs from those pictured above, but otherwise all parts are the same.

1 – Mainspring Barrel
2 – Mainspring
3 – Barrel Cover
4 – Mainspring Arbor
5 – Back Plate
6 – Front Plate
7 – First Wheel
8 – Second Wheel
9 – Third Wheel
10 – Center Wheel (Fourth Wheel)
11 – Suspension Bracket with Saddle
12 – Anchor
12E – Entrance Pallet
12X – Exit Pallet
12Y – Pallet Holding Strap
12Z – Strap Screw
13 – Escape Wheel
14 – Suspension Guard
15 – Bracket Holding Screw
16 – Suspension Unit
16S – Suspension Spring
16T – Top Suspension Block
16TS – Top/Bottom Block Screw
16F – Suspension Fork
16FS – Fork Screw
16B – Bottom Block
17 – Hour Wheel
18 – Intermediate Wheel (Minute Wheel)
19 – Upper Pillar
20 – Lower Pillar
21 – Suspension Guard Extension
22 – Suspension Guard Holding Washer
23 – Hour Wheel Spacing Washer
24 – Cannon Pinion
25 – Intermediate Wheel Holding Screw
26 – Pillar Screw
27 – Ratchet Wheel
28 – Ratchet Wheel Bridge Screw
29 – Ratchet Wheel Bridge
30 – Suspension Guard Holding Screw
31 – Suspension Unit Holding Pin
32 – Pillar Pin
33 – Movement Holding Screw
34 – Click
35 – Click Screw
36 – Click Spring
37 – Suspension Guard Locking Thumb Screw
38 – Key
39 – Arch
40 – Minute Hand
41 – Hour Hand
42 – Hand Washer
43 – Hand Nut

86. Koma Midget Movement Parts

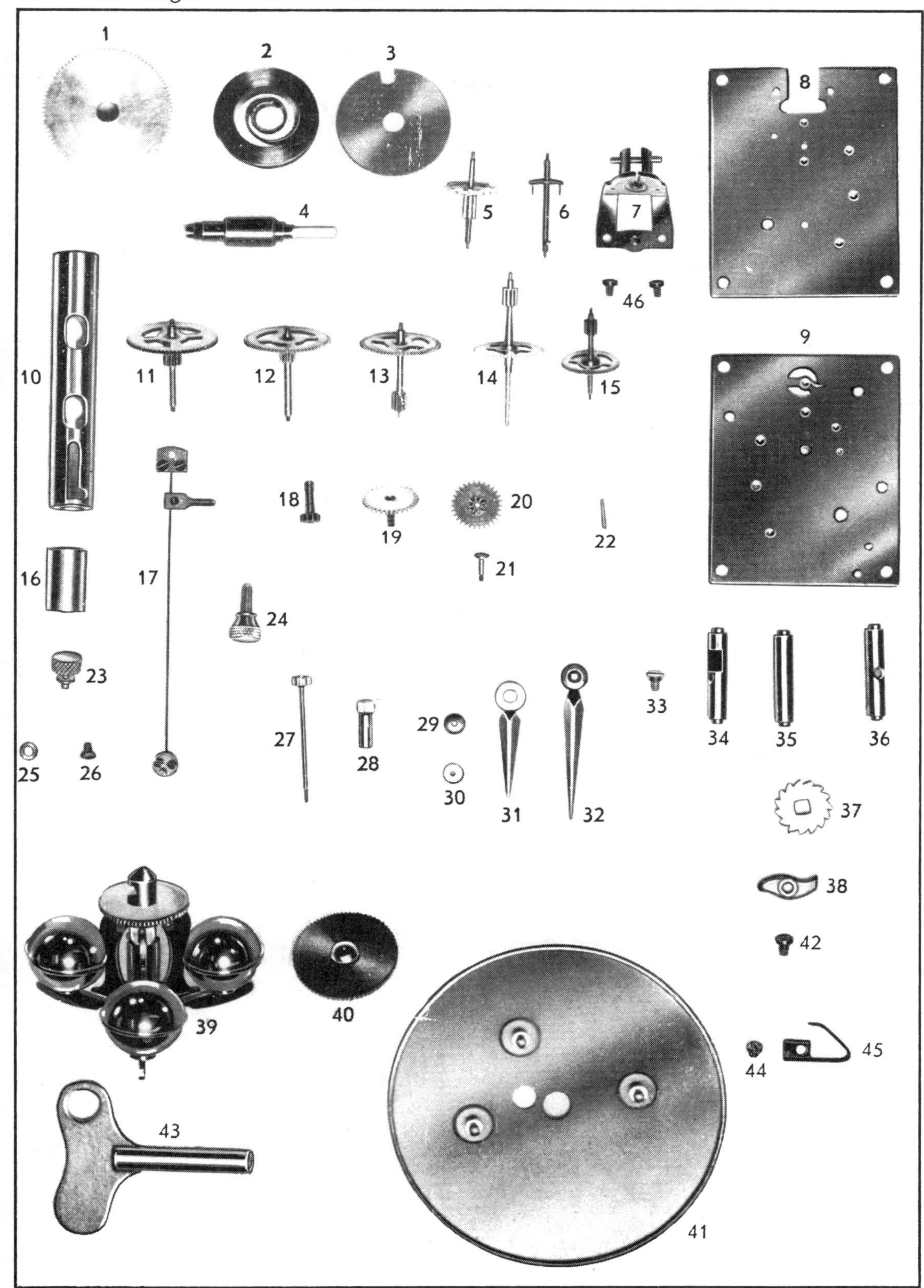

The illustrations show a movement of about 1974 production. With minor variations, the individual parts are essentially the same as for other Koma midget movements made after this production.

About 1968, the 12 × 25 mainspring in the midget movement was increased in size in order to increase the power to the escapement. The revised 14 × 25 mainspring necessitated an increase in the width of the barrel. Because of the limited amount of space available, the plates of the revised movement were increased in separation by 1mm which necessitated an increase from 20.0mm to 21.0mm in the shoulder-to-shoulder length of every arbor in the wheel train from barrel through anchor.

1 – Mainspring Barrel
2 – Mainspring
3 – Barrel Cover
4 – Mainspring Arbor
5 – Escape Wheel
6 – Anchor
7 – Suspension Bracket with Saddle
8 – Back Plate
9 – Front Plate
10 – Suspension Guard
11 – First Wheel
12 – Second Wheel
13 – Third Wheel
14 – Center Wheel (Fourth Wheel)
15 – Fifth Wheel
16 – Suspension Spring Guard Extension
17 – Suspension Unit
17S – Suspension Spring
17T – Top Suspension Block
17TS – Top/Bottom Block Screw
17F – Suspension Fork
17FS – Fork Screw
17B – Bottom Block
18 – Cannon Pinion
19 – Hour Wheel
20 – Intermediate Wheel (Minute Wheel)
21 – Intermediate Wheel Holding Screw
22 – Dial Lug Pin
23 – Suspension Guard Locking Thumb Screw
24 – Movement Holding Screw
25 – Suspension Guard Holding Washer
26 – Suspension Guard Holding Screw
27 – Hand Turning Arbor with Gear
28 – Hand Turning Knob
29 – Hand Washer
30 – Hand Nut
31 – Hour Hand
32 – Minute Hand
33 – Pillar Screw
34 – Lower Pillar (barrel side)
35 – Upper Pillar
36 – Lower Pillar
37 – Ratchet Wheel
38 – Click
39 – Pendulum
40 – Leveling Disc
41 – Dial Plate
42 – Click Screw
43 – Key
44 – Click Spring Screw
45 – Click Spring

87. (1199) For Becker Standard models under glass domes, there may be one of these decorative brass pieces screwed to the front plate under the dial:

88. (1009A) This back plate is equipped with two banking pins.

89. (1607) On some Phillipp Haas clocks, there may be a small guard protecting the top suspension block. Also, there may be a small, specially formed brass piece secured with a screw to the top left of the back plate into which the bottom suspension block can be held friction tight during shipping.

90. (1008A) This back plate is equipped with *adjustable* banking pins. Also, on some models of this clock, the arch, instead of being attached to the top of the front plate with screws, just rests upright behind the dial as shown by the tracing below.

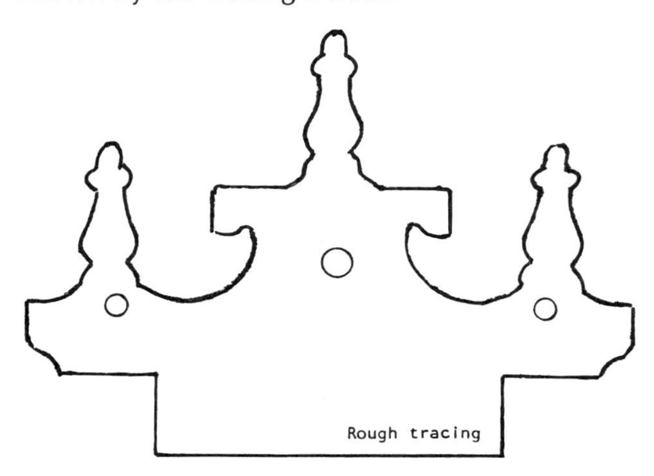

91. (1727) This very early 400-Day Clock with its verge escapement is in the collections of the Hopkins Observatory, Williams College, Williamstown, Massachusetts. Individual movement parts are shown below while the complete clock is shown in Section 6, Clock 1. The pendulum is shown in Section 13, Number 1 and the movement assembled is shown in Section 18, Appendix 116.

92. (1470E) This back plate is from an interesting constant force torsion pendulum clock by an unknown maker. Only two examples are known to exist, one in a Leipzig museum. The complete clock is illustrated in Section 6, No. 91; the unusual pendulum in Section 13, No. 11. Details of the clock were first documented in 1953 on pages 141-143 of *The Horolovar Collection,* photographs and text (freely translated) through courtesy of Felix Schmid of Leipzig and Dr. P.H. Witthoft, publisher of *Die Uhr.*

The "escape wheel" of the clock is actually a three-tooth rack which, by force of gravity, escapes through a pair of cylindrical, dead beat pallets. When the third tooth escapes, the rack is reset through the action of a duplex escape wheel. Thus the force to the pallets is constant and determined by the wieght of the rack. The impulse is transmitted to the suspension spring through a linkage unusual in form, but identical in action to that of an ordinary 400-Day Clock. The pendulum makes three full oscillations per minute, one for each of the rack teeth. Thus when the train is released to reset the rack, the minute hand indicates jump minutes.

The clock represents an ingenious and well-constructed solution to the wrong problem. It was the temperature compensating suspension spring, not equal power to the escapement, that eventually led to accurate timekeeping of torsion pendulum clocks.

93. (1145) This Franz Vosseler movement uses a crutch attached to the anchor which is very similar to the crutch in a typical American clock. In this design, the fork points towards the rear of the clock rather than towards the front of the clock as is usual. The suspension unit for this clock is made up as per drawing below. The top suspension block fits friction tight into a small, simple bracket riveted to the back plate. (See Section 15, Bracket 13.) The suspension spring (Horolovar .003"—.076mm) is held into the top block and fork with tapered pins. The shape of the bottom block is not important.

94. (1257) The suspension unit for this clock is made up as per drawing below. The clock that was tested could be regulated with a .0027"—.069mm Horolovar spring.

95. (1087) The pendulum most frequently used with this clock is illustrated in Section 13, No. 61. However, in some four-glass case models, pendulum No. 59 was used. To regulate this pendulum, it is necessary for the bottom of the square axis to be inserted into the locking cup and the disc rotated to bring the weights in or out. (This design was quite unsatisfactory and many pendulums have become broken.)

96. (1663, 1717) Early pendulums used with these clocks (See Section 13, No. 32) were not designed for regulating. Regulation was made possible with a special bottom suspension block which made it possible to shorten or lengthen the suspension spring.

97. (1207B) This is an unusual and rare Becker movement with no eccentric nut.

98. (1722) This unusually large movement is in a wood case wall clock and has no identification. The distance between the plates is also large for a 400-Day Clock—1.40".

99. (1680) This back plate is from a very unusual, front wind, torsional pendulum, wood case wall clock with a rare, lever escapement. (See Section 6, Clock 8.) The dial is marked

Thomas Haller
Schwarzwald
Wurttemberg
D.R.P. 21340

Patent number D.R.P. 21340 has not been found.

100. (1376) The letters H, R or W were stamped on some Kundo back plates as a factory identification that the clocks were purchased by importers Hallcraft, Rensie or Welby.

101. (1380) The "D" shaped holes in this plate shows that plastic "snap in" suspension guards were used on this clock.

102. (1691) This was an inexpensively made clock by Ishikara Clock Co., Tokyo, Japan. The movement plates are only 13mm apart. The approx. 14mm mainspring is in a housing attached to the front plate behind the dial. The 3-ball pendulum is very similar to the one used by Nisshindo, Section 13, No. 43. The name "Crescent" appears on the dial and also on the manufacturer's printed label with the words "400 Days." The Ishikara company was apparently very small and short lived.

103. (1718) This is a rare 30-Day clock, poorly made and hard to keep running. The arm that holds the movement is so thin that it bends with the weight of the movement, thereby making it difficult to keep the pendulum from touching the base. It takes a 4½" x 11" dome. (See Section 6, Clock 64; Section 13, Pendulum 14.)

104. (1157A, 1695) These Wurthner midget 14-Day clocks were being produced in very small quantities just before Wurthner stopped making torsion pendulum clocks in 1957.

105. (1591) Three unusual features *may* be found with some Wurthner Standard clocks:

(1) About 1952, John Wanamaker, New York and Philadelphia, sold a thousand or more clocks made especially for them by Wurthner. These clocks had a moon-phase dial similar to the type often used in grandfather clocks. The moon dial is turned by a 20-tooth intermediate gear driven by a second 20-tooth gear attached friction tight to an extension of the third wheel arbor through the front plate. The pendulum locking device on this clock is designed so that when the clock is picked up a flat spring pushes the guide cup against the bottom of the pendulum which in turn forces the regulating disc up against an adjustable horse shoe-shaped bracket below the movement platform. When the clock is put back down on a flat surface, the reverse takes place and the pendulum is free to rotate.

(2) Another Wurthner locking device has pliers action controlled by a cam attached to a knurled thumb disc under the base. In the locked position, the "jaws" grip a brass collar attached to the extended axis of the pendulum (See Section 13, Pendulum 87). In the unlocked position, the cam opens the pliers and the pendulum is free to rotate.

(3) See Section 13, Pendulum 42. A half-round thumb screw is attached to the hook on some Wurthner pendulums that makes it possible to lock the bottom suspension block in the hook. Thus the block remains in the hook when the pendulum is locked and the suspension spring becomes relaxed.

106. (1676) This Uhrenfabrik Herr clock with suspension Unit 27B may have a pendulum with regulating nut threaded in the opposite direction from most others. To make the pendulum go faster, you must turn the regulating nut *clockwise.*

107. (1003) Uhrenfabrik Wintermantel and Jahresuhrenfabrik apparently had some working agreement, for parts in both movements are interchangeable.

108. (1521AA) There are often extra, unused holes in Sigfried Haller back plates.

109. (1323, 1325) Both Standard and Miniature movements made by Herr and Reiner have centerwheel arbors with two different lengths to accommodate models with dials of different distances from the front plate. The factories made each of the four models with slight differences in the motion trains, although the wheel diameters and tooth counts are the same. Special note: *Hand tension is obtained from a pinched cannon pinion in the Standard model with long centerwheel arbor. All other models obtain tension from a bent hand washer behind the hand nut.* (See answer to Section 8, Question 13.)

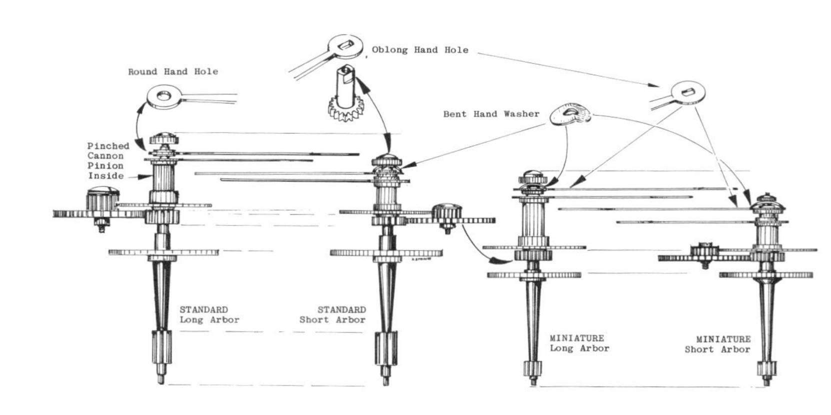

CENTERWHEEL	CANNON PINION Overall Length	Pinion Thickness	HOUR WHEEL Overall Length	INTERMEDIATE WHEEL Thickness
Standard				
Long Arbor	16.6*	2.8	11.8	A - 6.8
Short Arbor	12.5**	1.9	8.0	B - 4.6
Miniature				
Long Arbor	16.8**	2.0	12.3	B - 4.6
Short Arbor	12.5**	1.9	8.0	C - 3.3

*Has round end for minute hand bushing with round hole
**Has O end for minute hand bushing with oblong hole

110. (1012) Uhrenfabrik Herr Standard and Uhrenfabrik Neueck movement parts:

1 Front Plate
2 Back Plate
3 Arch
4 Bracket and Saddle Unit
5 Barrel
6 Barrel Cover
7 Barrel Arbor
8 First Wheel
9 Second Wheel
10 Third Wheel
11 Center Wheel (Fourth Wheel)
12 Escape Wheel
13 Hour Wheel
14 Intermediate Wheel
15 Intermediate Wheel Holding Screw
16 Click
17 Click Screw
18 Ratchet Wheel Bridge
19 Ratchet Wheel
20 Click Spring
21 Cannon Pinion
22 Movement Holding Screw
23 Guard Holding Screw
24 Hand Washer
25 Hand Nut
26 Upper Pillar
27 Lower Pillar
28 Mainspring
29 Minute Hand
30 Hour Hand
31 Suspension Unit Holding Screw
32 Suspension Unit
32S Suspension Spring
32T Top Suspension Block
32F Suspension Fork
32B Bottom Suspension Block
33 Anchor

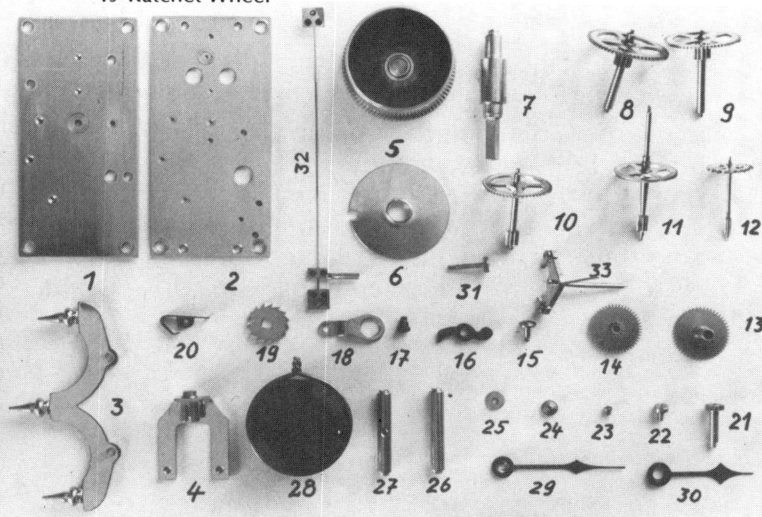

111. (1379) Kundo Miniature movements were first made in 1952 with pin pallet escapements. A few months later, the factory changed production to Graham escapements. Small numbers of pin pallet movements continued to be produced until 1955. The pin pallet movement can be converted to Graham by exchanging *three* parts: Anchor, escape wheel and third wheel, the latter necessary because the old third wheel will interfere with the Graham escape wheel.

The original positions of the pin pallet anchor and escape wheel were approximately the same as shown in Section 10, under Unit 5E, Arrangement "A". When converted to Graham escapement the movement should look like Arrangement "B".

112. (From Section 1) This portrait is of Johann Anton Harder, inventor of the 400-Day Clock. (See Section 4, Patent 5.) He was born 16 September 1811 in St. Petersburg, Russia and died 23 June 1888 in Ransen bei Steinau/Oder, Germany. The photograph is courtesy

of Gerd-Harald Ludendorff of Munich, Germany, great-great-grandson of Harder.

113. (1010, 1010A) Only two models of the Schatz 1000-Day Clock were made. With the first production, the dome and case sides were of glass. With later production, they were of plastic. Movements were made with supports curved (as illustrated) and straight.

114. (1325, 1326) John Wanamaker's advertisement for the Uhrenfabrik Reiner clock—25 August 1953. (See Section 1.)

115. (1325, 1326) S. Klein-on-the-Square's advertisement for the John Wanamaker clock following the inventory auction—7 February 1955. (See Section 1.)

116. (1727) Rear view of an early Willmann? experimental 400-Day Clock with verge escapement. (See Section 6, Clock 1.)

117. (1063) This back plate identifies a movement manufactured by Uhrenfabrik Herr, but sold by Edgar F. Henn. It has an applied, metallic E.F.H. trade mark on the back plate. One can only assume that, at that time, Henn had more orders for clocks than they could fill, so they purchased an unknown number of movements from Herr, identifying them for U.S. Customs with the applied Henn trade mark. The number of these clocks imported was undoubtedly very small.

118. (From Section 15) Detailed drawings of some of the particularly complicated suspensions.

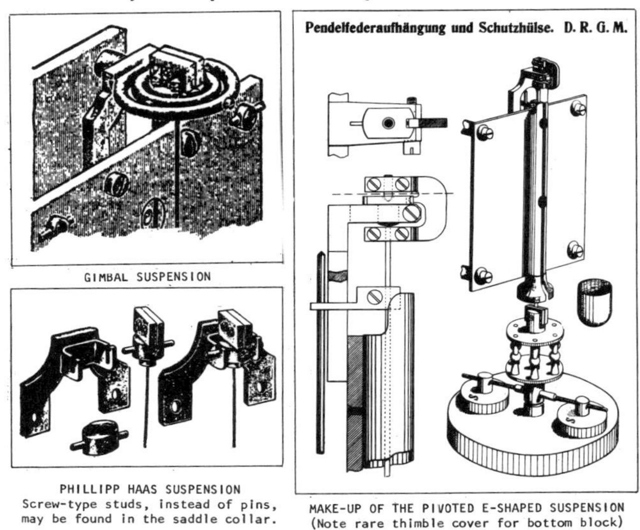

GIMBAL SUSPENSION

PHILLIPP HAAS SUSPENSION
Screw-type studs, instead of pins, may be found in the saddle collar.

MAKE-UP OF THE PIVOTED E-SHAPED SUSPENSION
(Note rare thimble cover for bottom block)

119. (1077) The click spring in the Standard Link clock is made of a small length of piano wire, both of its ends bent at right angles, its center coiled around a holding screw on the outside of the back plate. One end fits into a small hole and anchors the wire. The other end goes through the plate via a larger hole and acts on the click, which is loosely riveted to the inside of the back plate.

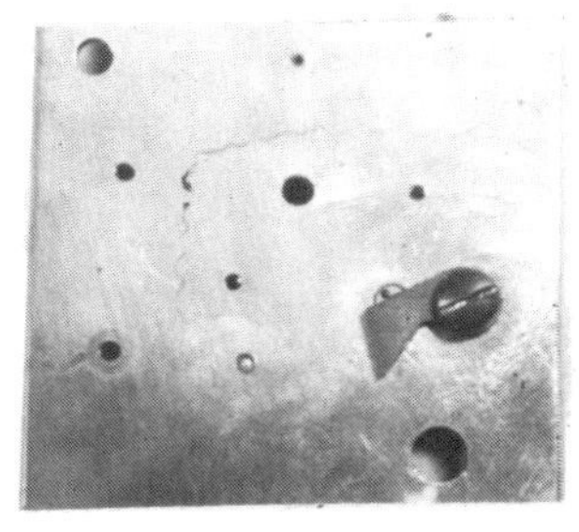

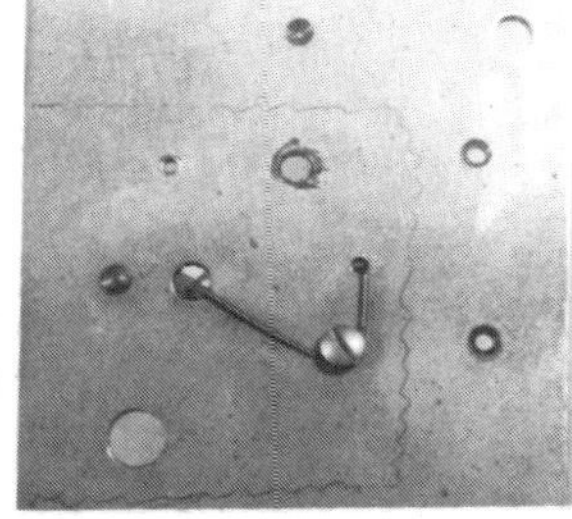

120. (1471) During the period before the temperature compensating "twin loop" pendulum patent was issued in 1902, the movement plates were marked "Patent Pending." After the patent was issued, the plates carried the patent numbers. (See Section 9, Plate 1055.) Also see Section 18, Appendix No. 13.

121. (1469E) This inexpensively made 100-Day movement by Nisshindo, Japan, has pin pallet escapement, mainspring housing riveted to the back plate and a self-contained alarm clock type winding key. The pendulum is shown in Section 13, Pendulum 79.

122. (1731) This clock with verge escapement is shown complete in Section 6, Clock 2. The back of the movement is shown in Section 18, Appendix 116.

Photographs of this clock are courtesy of Gerd-Harald Ludendorff of Munich, Germany, great-great-grandson of Anton Harder, inventor of the 400-Day Clock. (See Section 4, Patent 5.)

123. (Section 10) As the production of 400-Day Clocks increased after World War II, breakage of glass domes and other damage caused by loose pendulums became a serious problem to the 400-Day Clock manufacturers. In order to solve this problem, Kieninger & Obergfell (Kundo) began to experiment with different systems designed to better retain the pendulum during shipping. These retaining systems included the suspension unit locking guard, the suspension unit lower block, and the pendulum hook. The following illustrations show the various systems as they related to the Kundo suspension Units and their approximate years of manufacture.

Unit 1 - KUNDO STANDARD 49

This Unit fits all Standard clocks from the first model made in the late 1920's until about 1952. The bottom locking guard (A) was replaced with an improved design (B) in models made after World War I.

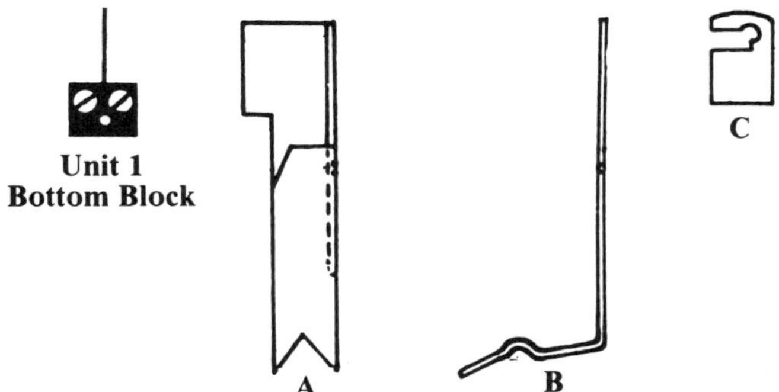

Unit 2 - KUNDO STANDARD 52

This Unit was used with only a few thousand clocks in 1953. A special stirrup-type bottom block was designed to be used with an improved locking guard (D). A special pendulum hook (E), which passed through the bottom block, was used. There were no screws in either the top or bottom blocks and the suspension spring was "pinched" in place.

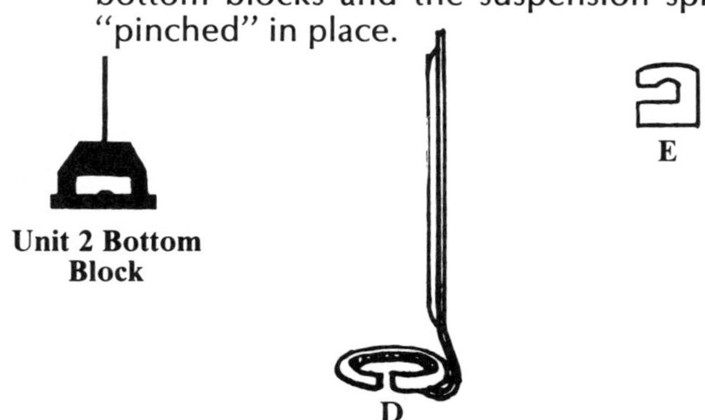

Unit 3A - KUNDO STANDARD 53

This 1953 Unit has a fixed pin through the bottom block and was designed to be used with the improved locking guard (D). Some clocks were manufactured with a stronger pendulum center rod. Pendulum hook (BB) is used with this larger diameter pendulum center rod. The filled circles under the pendulum hooks show the diameter of the two pendulum center rod threads.

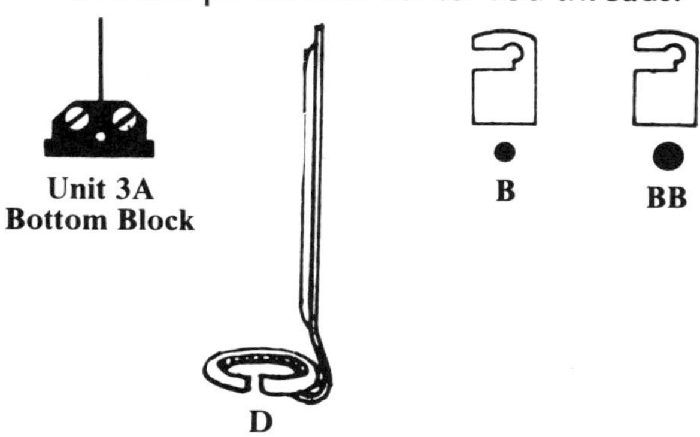

Unit 3B - KUNDO STANDARD 54

This 1954 Unit was used with the first clock to have a locking mechanism built into the movement platform (E). A bottom block shaped the same as Unit 3A was used except there was no fixed pin in this unit. Instead, a brass cotter pin (F) was used to hold the bottom block into the pendulum hook (G or GG). Here again, clocks were manufactured with two different diameters of pendulum center rods. The earliest platform with the locking mechanism (E) can be identified as having the locking lever ABOVE the platform.

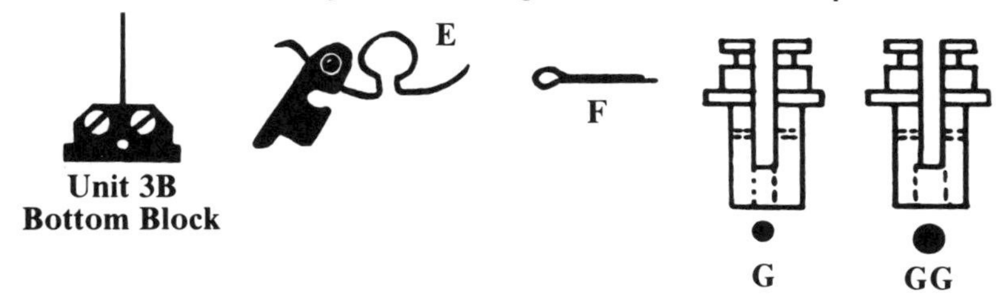

Unit 3C - KUNDO STANDARD 55

This Unit, introduced in late 1954, was the final design which was used until Kundo stopped making key wound 400-Day Clocks. The pendulum locking lever is attached BELOW the movement platform (H). A loose brass pin (M), 10mm long, holds the bottom block in the pendulum hook (J). The brass rod is held in place by the collar (K) which, with pressure from the coil spring (L), locks the ends of the pin so that it will not fall out. In order to detach the pendulum, the collar must be raised until the end of the pin is exposed and can be withdrawn. In 1958, pendulum hook (J) was changed slightly in design but the pin, collar, and coil spring continued to perform the same functions. (See Section 18, Appendix No. 2.)

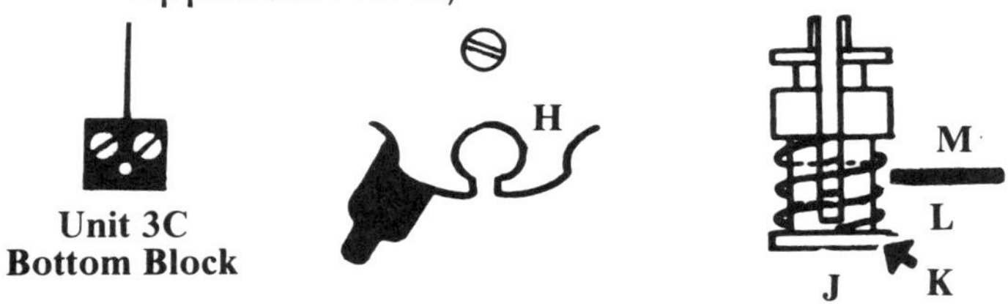

Unit 3C
Bottom Block

Unit 4 - KUNDO JUNIOR 52

This Unit was used in only a few thousand clocks in 1952. A special stirrup-type bottom block was designed to be used with the improved locking guard (D). This special block required the use of a special pendulum hook (E) which passed through the bottom block. There are no screws in the top or bottom blocks and the suspension spring was "pinched" into the blocks. Unit 4 was used only with movements having pin pallet escapements. (See Section 18, Appendix No. 111.)

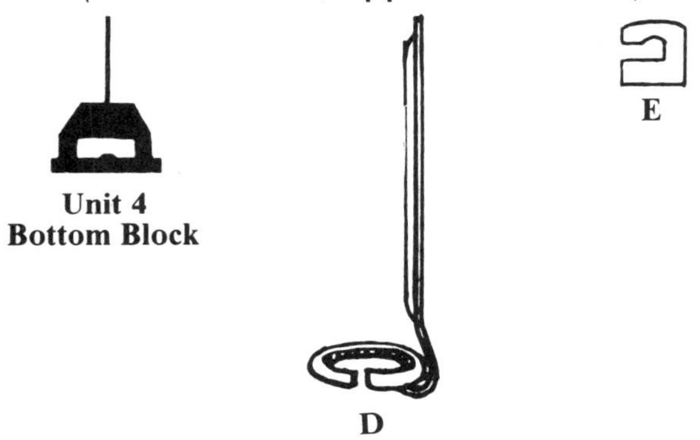

Unit 4
Bottom Block

Unit 5A - KUNDO JUNIOR 53

This Unit was used in both the 1953 model with pin pallet escapement (See Section 18, Appendix No. 111) and the 1954 model with the Graham escapement.

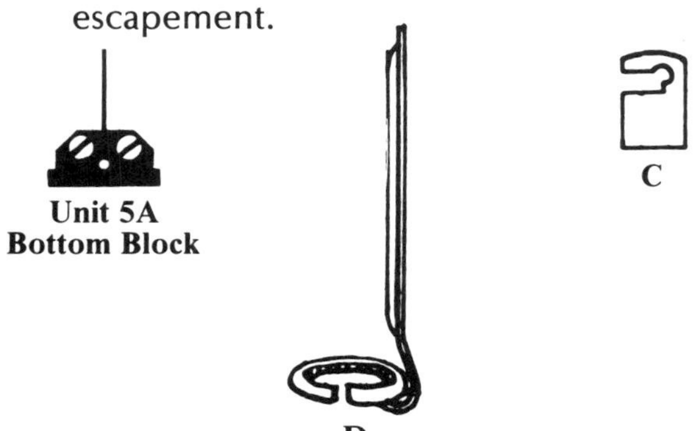

Unit 5A
Bottom Block

Unit 5B - KUNDO JUNIOR 54

This Unit, which was introduced in 1954, was the first model to have a locking pendulum. A brass cotter pin (F) holds the bottom block in the pendulum hook (P and PP). Pendulum hooks (P) and (PP) differ in the size of the thread which attaches to the pendulum center rod. The filled circles under the pendulum hooks are equal to the diameter of the pendulum center rod thread. This model can be identified by the locking lever which is attached ABOVE the movement platform (N).

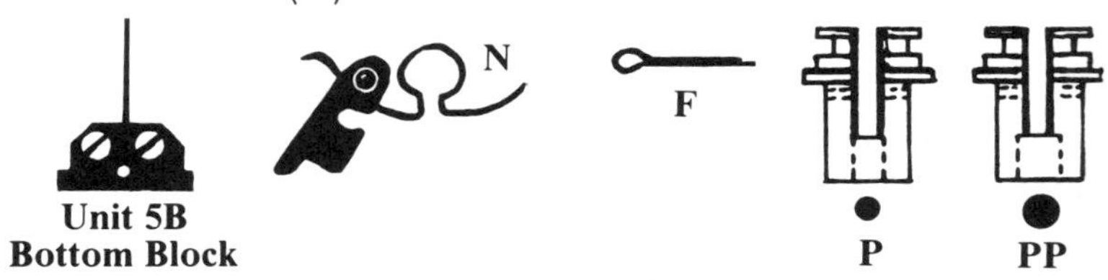

Unit 5B
Bottom Block

Units 5C, 5D, and 5E - KUNDO JUNIOR 55

These Units were first introduced in 1955. They used a loose brass pin (M) which was held in place in the pendulum hook (R) by the collar (K) and coil spring (L). Clocks using this system can be identified by the locking lever which is attached BELOW the movement platform (Q).

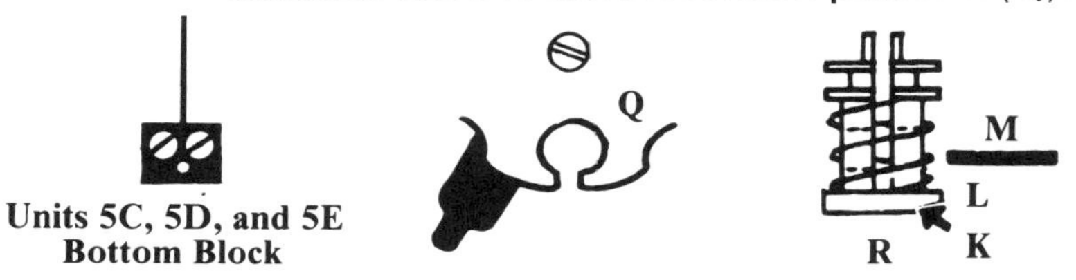

Units 5C, 5D, and 5E
Bottom Block

All movements using Unit 5E should have straight, upright anchor pins. If the pin is bent offset to the rear, the clock is a Kundo Midget and the correct Unit is 5F.

Unit 5F - KUNDO MIDGET 58

All movements using the 5F Unit have anchor pins with offset bends to the rear. The pin is so designed to make contact with the suspension fork slightly behind the back plate.

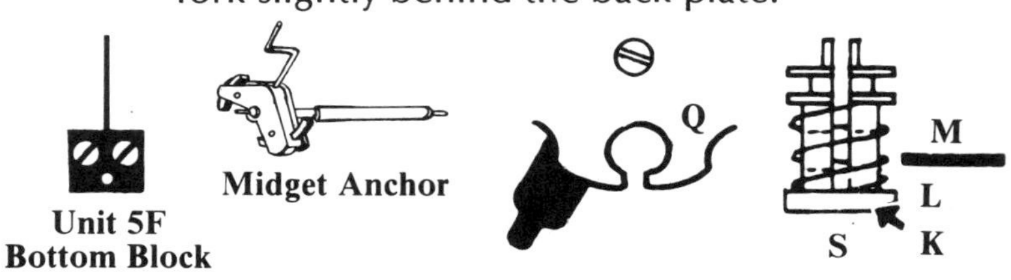

Unit 5F
Bottom Block

124. (1325, 1326) This shows how Reiner was able to make a "Standard" size clock at such a low factory cost. Note the miniature movement with fins added to make it appear larger.

125. (Section 3, Clock 7) The name "Sylvester" on the dial of this clock may have something to do with the fact that, in southern Germany, "Sylvester" means New Year's Eve. Perhaps this name was viewed as a means of avoiding the Bowler and Burdick patent on the name "Anniversary".

126. (1431) The disc pendulums used with early Kienzle movements had several large diameter, thin steel washers in the inside of the disc. Regulation of the clock was accomplished by adding or removing one or more of the washers. Since the washers were not attached to either the clock or the pendulum, they are frequently missing. As a consequence, the required strength of the replacement suspension spring may vary from clock to clock.

127. (1043) These movements may be found on clocks marked "Urania" on the dial.

128. (1731) Rear view of an early Gustav Becker experimental 400-Day Clock with a verge escapement. (See Section 6, Clock 2.)

129. (Section 9) Many early 400-Day Clocks are found with one or two digit numbers stamped on parts of the movement such as the insides of the plates and barrels, wheel spokes, sides of anchors, etc. These numbers are believed to identify groups of workers within the factory who formed teams. These teams then bid for the right to manufacture and assemble a batch of clocks using the designs, materials, manufacturing equipment, and other facilities belonging to the factory. Efficient teams obviously won the right to assemble more clocks and, thus, to earn more money.

130. (1241) It is generally believed that Herr used to work for Schatz before leaving in the early 1950's and establishing his own 400-Day Clock factory in Gutenbach. When he left Schatz, Herr took the blueprints of the Schatz standard sized, wide plate movement. It should be noted that the retention of these blueprints was probably more of a convenience than an outright theft since all patents covering this movement had expired by this time. Whatever the circumstances, the story is somewhat confirmed by the fact that practically every part in the Herr wide plate movement is interchangeable with the Schatz wide plate movement.